Mobile Development for SAP®

SAP PRESS

SAP PRESS is a joint initiative of SAP and Galileo Press. The know-how offered by SAP specialists combined with the expertise of the Galileo Press publishing house offers the reader expert books in the field. SAP PRESS features first-hand information and expert advice, and provides useful skills for professional decision-making.

SAP PRESS offers a variety of books on technical and business-related topics for the SAP user. For further information, please visit our website: *www.sap-press.com*.

Sanjeet Mall, Tzanko Stefanov, Stanley Stadelman
Mobilizing Your Enterprise with SAP
2012, 409 pp., hardcover
ISBN 978-1-59229-419-0

James Wood, Shaan Parvaze
Web Dynpro ABAP: The Comprehensive Guide
2013, 784 pp., hardcover
ISBN 978-1-59229-416-9

Thomas Jung, Rich Heilman
Next Generation ABAP Development (Second Edition)
2011, 735 pp., hardcover
ISBN 978-1-59229-352-0

Dr. Berg and Penny Silvia
SAP HANA: An Introduction (Second Edition)
2013, 525 pp., hardcover
ISBN 978-1-59229-865-5

Dave Haseman, Ross Hightower

Mobile Development for SAP®

Galileo Press

Bonn • Boston

Galileo Press is named after the Italian physicist, mathematician, and philosopher Galileo Galilei (1564–1642). He is known as one of the founders of modern science and an advocate of our contemporary, heliocentric worldview. His words *Eppur si muove* (And yet it moves) have become legendary. The Galileo Press logo depicts Jupiter orbited by the four Galilean moons, which were discovered by Galileo in 1610.

Editor Kelly Grace Harris
Copyeditor Miranda Martin
Cover Design Graham Geary
Photo Credit iStockphoto.com/11592827/parfynov
Layout Design Vera Brauner
Production Graham Geary
Typesetting Publishers' Design and Production Services, Inc.
Printed and bound in the United States of America, on paper from sustainable sources

ISBN 978-1-59229-448-0

1st edition 2013

Library of Congress Cataloging-in-Publication Data
Haseman, William D.
Mobile development for SAP / William D. Haseman and Ross Hightower. — 1st edition.
pages cm
ISBN-13: 978-1-59229-448-0
ISBN-13: 978-1-59229-647-7
ISBN-13: 978-1-59229-648-4
ISBN-10: 1-59229-448-0
1. Mobile computing. 2. Business—Computer programs. 3. SAP AG.
I. Hightower, Ross. II. Title.
QA76.59.H374 2013
004—dc23
2013006519

Contents at a Glance

PART I Getting Started

1 Introduction to Mobile App Development 23
2 Introduction to Sybase Unwired Platform 49
3 Installing the Sybase Unwired Platform Runtime 63
4 Installing and Connecting the Sybase Mobile SDK 91
5 Installing the Hybrid Web Container Application 117

PART II Building Hybrid Apps

6 Working with the Hybrid Web Container 143
7 Building Mobile Business Objects 167
8 Creating Hybrid Apps 225
9 Examples of Hybrid App Development 269

PART III Customizing Hybrid Apps

10 Customizing Hybrid App User Interfaces 315
11 Customizing Hybrid App Functionality 347

PART IV Developing Native Apps

12 Introduction to Developing Object API Native Apps 385
13 Developing Android Apps 421
14 Developing iOS Apps 473

PART V Advanced Topics

15 Using the Mobile Analytics Kit 515
16 SAP Support for Open Platforms 553
17 CRM Lite Case Study 565

Dear Reader,

As an editor of technology books, I have a theory: programmers have a special skill for writing. In my experience, some of SAP PRESS' most talented authors are also programmers, and they always display the same strengths: their objectives are well considered, their writing is clear, and their explanations successfully navigate the fine line between comprehensiveness and concision. I suspect that, as it turns out, crafting a book is not all that different from crafting code.

For evidence in support of my theory, I'm happy to give you the writing of Dave Haseman and Ross Hightower. Editing this book was a true pleasure—and, dear reader, I trust you'll find that reading it is also one. Even more importantly, you'll walk away with a brand new skill set: developing mobile applications for SAP. As programmers and as educators with years of SAP experience, I'm confident you'll find that Dave and Ross wrote exactly the book you needed.

Of course, we at SAP PRESS are always interested in your opinion. What do you think about *Mobile Development for SAP*? How could it be improved? As your comments and suggestions are the most useful tools to help us make our books the best they can be, we encourage you to visit our website at *www.sap-press.com* and share your feedback.

Thank you for purchasing a book from SAP PRESS!

Kelly Grace Harris
Editor, SAP PRESS

Galileo Press
Boston, MA

kelly.harris@galileo-press.com
www.sap-press.com

Contents

Introduction 17
Acknowledgments 19

PART I Getting Started

1 Introduction to Mobile App Development 23

1.1 The Mobile Market and SAP 23
1.1.1 Growth of Smart Devices 23
1.1.2 SAP and Mobility 26
1.2 Mobile Devices and Data Sources 27
1.2.1 Device Platforms 27
1.2.2 Backend Data Sources 30
1.3 Types of Mobile Apps 32
1.3.1 Game Apps 32
1.3.2 Utility Apps 32
1.3.3 Real-Time Apps 33
1.4 Types of Mobile App Development 34
1.4.1 Native App Development 34
1.4.2 Web App Development 35
1.4.3 Hybrid App Development 36
1.5 Introduction to SAP Mobile Platform 36
1.5.1 Architecture 37
1.5.2 Mobile App Development for SAP Mobile Platform 39
1.6 Summary 47

2 Introduction to Sybase Unwired Platform 49

2.1 Sybase Mobile SDK 50
2.1.1 Unwired Workspace 51
2.1.2 Mobile SDK Library: Object API 53
2.1.3 Mobile SDK Library: OData 54
2.2 Unwired Platform Runtime 55
2.2.1 DMZ 56
2.2.2 Unwired Server 58

2.2.3 SAP Afaria Server 61
2.2.4 Sybase Control Center (SCC) 62
2.3 Summary 62

3 Installing the Sybase Unwired Platform Runtime 63

3.1 Choosing the Unwired Platform Landscape 63
3.1.1 Single-Server Installation 63
3.1.2 Simple Load-Balancing Cluster 64
3.1.3 Standard Microsoft Failover Cluster 65
3.1.4 Microsoft Failover Cluster with Shared Hosts 66
3.2 Choosing Licensing Options 66
3.2.1 Types of Environments 67
3.2.2 Product Editions 67
3.2.3 License Types 68
3.2.4 Mapping Environment to Product Editions and License Types 69
3.2.5 Examples: Single-Developer Development and Team Development 70
3.3 Typical Installation Steps 72
3.3.1 Hardware and Software Requirements 72
3.3.2 Generating License Keys 73
3.3.3 Downloading Installation Files 76
3.3.4 Installation of the Unwired Platform Runtime 76
3.4 Sybase Control Center 82
3.4.1 Creating a New Administrative User 84
3.4.2 Creating a New Domain 87
3.5 Summary 89

4 Installing and Connecting the Sybase Mobile SDK 91

4.1 Getting Started 91
4.1.1 Hardware and Software Requirements 92
4.1.2 Backward Compatibility 92
4.2 Steps for Installation and Connection 93
4.2.1 Installing the Unwired Workspace 93
4.2.2 Connecting the Unwired Workspace 95

4.3 Installing and Connecting a Cloud-Based Sybase Unwired Platform 107
4.3.1 Creating the SUP Server Instance 108
4.3.2 Installing the Unwired Workspace 109
4.3.3 Managing the Server 110
4.3.4 Configuring Connections in the Local Unwired Workspace 114
4.4 Summary 116

5 Installing the Hybrid Web Container Application 117

5.1 Android Devices 117
5.1.1 Installing Android Emulators 118
5.1.2 Installing and Configuring the Container Application on the Emulator 127
5.1.3 Installing the Container Application on the Device 133
5.2 iOS Devices 134
5.2.1 Installing the Simulator 134
5.2.2 Installing and Configuring the Container Application on the Simulator 136
5.2.3 Installing the Container Application on the Device 139
5.3 Summary 139

PART II Building Hybrid Apps

6 Working with the Hybrid Web Container 143

6.1 Mobile Architectures 143
6.1.1 Native Apps 143
6.1.2 Mobile Web Apps 147
6.1.3 Hybrid Apps 150
6.2 Hybrid Web Container Architecture 153
6.3 Hybrid App Design for the Hybrid Web Container 156
6.4 Hybrid App Patterns for the Hybrid Web Container 160
6.4.1 Online Lookup 160
6.4.2 Server Notification 162
6.4.3 Cached Data 164
6.5 Summary 165

7 Building Mobile Business Objects 167

7.1 Introduction to Mobile Business Objects 168
7.1.1 Defining MBOs 168
7.1.2 MBO Development Concepts 170
7.2 MBO Development Tools 172
7.2.1 Eclipse Menu 172
7.2.2 Workspace Navigator View 174
7.2.3 Enterprise Explorer View 175
7.2.4 Mobile Application Diagram Editor 176
7.2.5 Properties View 176
7.2.6 Palette 177
7.3 Creating MBOs 178
7.3.1 Database MBO 179
7.3.2 Web Service MBO 187
7.3.3 SAP MBO 192
7.4 Controlling Data Flow and Performance 202
7.4.1 Load Arguments 203
7.4.2 Synchronization Parameters 206
7.4.3 Personalization Keys 208
7.4.4 Cache and Cache Groups 210
7.4.5 Join within an MBO 211
7.5 Creating Database Operations 213
7.6 Creating Object Queries 216
7.6.1 FindAll 217
7.6.2 FindByPrimaryKey 218
7.6.3 Designer Developed Queries 218
7.7 Creating Relationships between MBOs 219
7.8 Summary 224

8 Creating Hybrid Apps 225

8.1 Designing a Hybrid App: Process Overview 225
8.1.1 Creating Hybrid Apps 226
8.1.2 Designing Screen Flow 228
8.1.3 Designing Screens 230
8.1.4 Generating Mobile Packages 233
8.1.5 Testing Apps 235

8.2 Designing Screen Flow 236
8.2.1 Flow Design Page Properties 236
8.2.2 Starting Points 239
8.2.3 Drag and Drop 241
8.2.4 Palette Options 246
8.3 Designing Screens 247
8.3.1 Defining General Properties 248
8.3.2 Adding Controls 250
8.3.3 Defining Values for Control Properties 252
8.3.4 Adding Menu Items and Custom Actions 259
8.3.5 Defining Values for Menu Item and Custom Action Properties 260
8.4 Generating a Hybrid App 265
8.5 Summary 268

9 Examples of Hybrid App Development 269

9.1 Employee Database Example 269
9.1.1 MBO 269
9.1.2 Flow Design 270
9.1.3 Screen Design 271
9.1.4 Review App 275
9.2 Customer/Sales Order Database Example 276
9.2.1 MBOs 276
9.2.2 Flow Design 278
9.2.3 Screen Design 280
9.2.4 Review App 285
9.3 General Ledger Account Balances from SAP Example 291
9.3.1 MBO 291
9.3.2 Flow Design 296
9.3.3 Screen Design 296
9.3.4 Review App 300
9.4 Sales Order Notification Database Example 301
9.4.1 MBO 301
9.4.2 Create Hybrid App and Define Server Notification 304
9.4.3 Flow Design 308
9.4.4 Test App 309
9.5 Summary 311

PART III Customizing Hybrid Apps

10 Customizing Hybrid App User Interfaces 315

10.1 Generating the Project Code 316
10.1.1 Project Files 319
10.1.2 HTML Files 320
10.1.3 JavaScript Files 322
10.1.4 CSS Files 324
10.2 Adding a jQuery Button Group 325
10.2.1 Modifying the Hybrid App 325
10.2.2 Adding the HTML 326
10.2.3 Hiding the Menu Items 328
10.3 Creating a Custom List 329
10.3.1 Modifying the Hybrid App 329
10.3.2 Adding Custom JavaScript 330
10.4 Displaying a Map Using a jQuery Plug-in 335
10.4.1 Copying the jQuery Map Plug-in 335
10.4.2 Modifying the Hybrid App 336
10.4.3 Adding Custom JavaScript 337
10.5 Implementing Conditional Navigation 341
10.5.1 Modifying the Hybrid App 342
10.5.2 Adding Custom JavaScript 343
10.6 Summary 345

11 Customizing Hybrid App Functionality 347

11.1 Working with Data 347
11.1.1 Understanding the WorkflowMessage Structure 350
11.1.2 Accessing Data using the JavaScript API 352
11.2 Using Custom Actions 360
11.2.1 Two Custom Actions 361
11.2.2 One Custom Action 366
11.3 Using the SUPStorage API 369
11.3.1 Saving the Chosen Postage Type as a Default Value 370
11.3.2 Saving the Postage Types on the Device 373
11.3.3 Using Dynamic Credentials 379
11.4 Summary 381

PART IV Developing Native Apps

12 Introduction to Developing Object API Native Apps 385

12.1 Synchronization 387
12.2 Object API App Functionalities 389
12.2.1 Registering an App on the Unwired Server 389
12.2.2 Accessing MBO Data 393
12.2.3 Manipulating Data 395
12.3 APIs Included in the Object API 397
12.3.1 Application API 397
12.3.2 Afaria APIs 402
12.3.3 Connection API 402
12.3.4 Authentication API 405
12.3.5 Personalization API 405
12.3.6 Synchronization API 406
12.3.7 Message-Based Synchronization API 407
12.3.8 Log Record APIs 408
12.3.9 Change Log APIs 408
12.3.10 Security APIs 408
12.3.11 Callback and Listener APIs 408
12.3.12 Query APIs 415
12.3.13 Persistence API 416
12.4 Summary 419

13 Developing Android Apps 421

13.1 Getting Started 421
13.1.1 Components of an Android App 422
13.1.2 Preparing the Development Environment 423
13.1.3 Introducing the Example App 425
13.2 Creating the MBOs 427
13.3 Generating the Object API Classes 428
13.4 Creating the Project 431
13.5 Importing Libraries 434
13.6 Implementing the App 435
13.6.1 Defining the App Structure in AndroidManifest.xml 435
13.6.2 Creating the Layout Files 437
13.6.3 MainActivity 448

13.6.4 CustomerListActivity 455
13.6.5 CustomerListAdapter Activity 458
13.6.6 CustomerDetailsActivity 460
13.6.7 CustomerSearchActivity 465
13.7 Running the App 469
13.8 Summary 471

14 Developing iOS Apps 473

14.1 Getting Started 474
14.1.1 iOS App Components 474
14.1.2 Preparing the Development Environment 476
14.1.3 Introducing the Example App 476
14.2 Generating the Object API Classes 478
14.3 Creating the Project 481
14.4 Importing Libraries 483
14.5 Importing Generated Code 485
14.6 Implementing the App 486
14.6.1 CustomerAppAppDelegate 486
14.6.2 ConnectionController 489
14.6.3 CustomerAppViewController 493
14.6.4 ItemListViewController 498
14.6.5 DetailViewController 504
14.7 Running the App 510
14.8 Summary 511

PART V Advanced Topics

15 Using the Mobile Analytics Kit 515

15.1 Introducing the Example App 516
15.2 Creating the Hybrid App 518
15.3 Copying MAKit Files 521
15.3.1 Copying CSS Files 523
15.3.2 Copying Images Associated with CSS 524
15.3.3 jQuery Style Sheets 524
15.3.4 General Images 525
15.3.5 Copying JavaScript Files 525
15.4 Creating a Simple Column Chart 527

15.4.1 Modifying the Current App 527
15.4.2 Developing Code 529
15.4.3 Testing the App 535
15.5 Creating a Zooming Chart 538
15.5.1 MBOs 539
15.5.2 Flow Design Changes 539
15.5.3 Developing Code 541
15.5.4 Testing the App 548
15.6 Summary 550

16 SAP Support for Open Platforms 553

16.1 Introduction to Third-Party Frameworks 553
16.2 SAP Partnerships 556
16.2.1 Adobe PhoneGap 556
16.2.2 Sencha Touch 557
16.2.3 Appcelerator Titanium 561
16.3 REST Services Apps 562
16.4 Summary 563

17 CRM Lite Case Study 565

17.1 The Scenario 565
17.2 Defining MBOs 566
17.2.1 Queries and Server Queries 567
17.2.2 Customer MBO 570
17.2.3 Products and Categories MBOs 571
17.2.4 Employee MBO 573
17.2.5 Sales Orders and Sales Order Items MBOs 574
17.2.6 Purchase Orders MBO 575
17.3 Developing the Hybrid App 576
17.3.1 Designing the Standard App 576
17.3.2 Customizing the App 584
17.4 Summary 608

The Authors 609
Index 611

Introduction

This book is written for developers and educators of developers, and aims to give them a complete picture of building mobile apps using SAP Mobile Platform. The coverage includes everything from installing the Sybase Unwired Platform to complete explanations of building and customizing hybrid apps, as well as developing both iOS and Android native apps. After learning the steps, you'll be able to solidify your understanding with a case study that details the building of a live CRM mobile app.

Part I is an introduction to SAP Mobile Platform, with a detailed discussion of the installation of the required components for designing, deploying, and running mobile apps. In **Chapter 1**, we talk about the mobile market, device platforms, and backend data sources that are important for business mobile apps. We also introduce SAP Mobile Platform—a leader in the mobile application development platform arena. In **Chapter 2**, we introduce Sybase Unwired Platform, which provides for the development, deployment, and execution of mobile apps. The discussion includes an overview of both the Sybase Mobile SDK and the Unwired Platform Runtime. In **Chapter 3**, we discuss the various landscapes and licensing options supported by the Unwired Platform Runtime and present the detailed steps required for a typical installation. In **Chapter 4**, we discuss the installation of the Unwired Workspace using the Sybase Mobile SDK, the development platform for designing and deploying hybrid apps. Included in this discussion is the configuration of the connections to the backend data systems. In **Chapter 5**, we describe the installation of Android and iOS emulators, as well as the installation and configuration of the Hybrid Web Container on iOS and Android platforms.

Part II provides details about developing hybrid apps. In **Chapter 6**, we go into detail about the concept of the Hybrid Web Container, which allows a single application to run on all four major mobile device platforms. In **Chapter 7**, we introduce the concept of the Mobile Business Object (MBO), which forms the business logic for defining data from backend systems. In **Chapter 8**, we walk you through the entire process of building a hybrid app, with a special focus on the user interface. In **Chapter 9**, we bring the pieces together and present examples of several complete hybrid apps.

Part III describes how to extend the hybrid apps developed in the previous part by adding additional user interface or app functionality features. In **Chapter 10**, we discuss the addition of custom JavaScript to create customized user interfaces, such as button groups and Google Maps. In **Chapter 11**, we discuss customizations that deal with how the app functions, such as manipulating messages sent to and from the Unwired Server.

Part IV describes how SAP Mobile Platform supports the development of native apps. In **Chapter 12**, we introduce how native apps that take advantage of SAP Mobile Platform can be developed, specifically utilizing MBOs, the Unwired Platform Runtime, and Object API libraries. In **Chapter 13**, we demonstrate the usage of the Android Object API to explain design concepts and develop an example native Android app. In **Chapter 14**, we demonstrate the use of the Object API for iOS to explain design concepts and develop an example of a native iOS app using Xcode, an IDE that runs only on the Apple platform.

Part V presents a collection of advanced concepts related to SAP Mobile Platform. In **Chapter 15**, we discuss the Mobile Analytics Kit (MAKit) for HTML5. MAKit allows you to graphically visualize and manipulate data in your mobile app. In **Chapter 16**, we discuss how SAP has made SAP Mobile Platform an open platform and is encouraging third-party platforms to access SAP data via the Unwired Server. Finally, **Chapter 17** brings together many of the concepts presented in the book to present a case study entitled "CRM Lite." CRM Lite is a complete hybrid app that can be used by a service rep in the field.

Software Versions

As a relatively new product, SAP Mobile Platform is undergoing rapid improvements that are released either as service packs or new versions. As an example, recent changes that have taken place include "Workflows" being renamed to "Hybrid Apps" and "Mobile Workflow Forms Editor" being renamed to "Hybrid App Designer." While we focus on the latest version, there is a chance that new versions will require changes to what is presented in this book. All of the screen shots in the book and the hybrid apps presented in Part I and Part II were tested on version 2.2, service pack 02 (the ramp-up version). The code presented in Part III, Part IV, and Part V was tested on version 2.2 service pack 01 (the ramp-up version).

To assist the reader, the current versions and all future versions of all of the examples in the book will be added to the book's website at *www.sap-press.com* as new releases become publicly available.

Acknowledgments

I would like to dedicate this book to my wife, Barb. While authors often talk about the many evenings and weekends they spent writing their books, it should be obvious from whom those hours were taken. I would also like to dedicate this book to Mackenzie, Mackaelyn, Jacob, Emily, Chace, and Dylan—our six grandchildren, who are what our lives are all about.

I wish to, first of all, thank Ross, my co-author, with whom I spent many hours discussing and debating issues in person and over the phone. Trying to write a book about an emerging moving target is a challenge, and we both learned a great deal in this process. During the development, I received feedback on a related curriculum we developed for the SAP University Alliance from a number of faculty members, and I would like to thank all of them. In addition, I would also like to specifically thank the following individuals for their efforts related to the book: Gail Schemberger, Jerad Lumina, and Alex McLeod, Jr. And finally, I wish to thank Kelly, our editor, for her many contributions.

Dr. Dave Haseman

Writing this book occurred during a tumultuous time in my life, and, as a result, I could not have done it without the support of my family. I would especially like to thank my wife, Debby. She keeps me sane even during the crazy times. I would also like to thank my parents, Ross and Catherine, who are always there when I need them.

I would not have attempted this without my co-author, Dave, who has a keen eye for what is possible. Many other people contributed to this book by testing apps and providing feedback. They are too numerous to thank individually, but their contributions were indispensable. Finally, I would like to thank three people in particular for making this possible: Jerad Lumina for tireless technical support; Gail Schemberger, who has a keen eye for the English language; and Alex McLeod, Jr., who started us off on this path.

Dr. Ross Hightower

PART I
Getting Started

Before diving into mobile development, you should have a basic understanding of the mobile market and of the technical concepts involved in mobility. In this chapter, we'll introduce you to both.

1 Introduction to Mobile App Development

This chapter covers the basics of what you need to know before starting on your journey of developing mobile apps for SAP. We'll talk about the mobile market, device platforms, and backend data sources that mobile apps rely on; the different types of mobile apps that are out there; and the different ways you can develop mobile apps. In the last section of the chapter, we introduce you to SAP Mobile Platform and one of its main components, Sybase Unwired Platform, which is the focus of this book.

1.1 The Mobile Market and SAP

It's no secret that the world is going mobile. In this section, we examine the increasing popularity of smart devices, as well as SAP's place in a mobile world.

1.1.1 Growth of Smart Devices

It was not long ago that the IT world was driven by the desktop PC in the office and laptop computers in the field. Then, along came the next generation—the smart phone and tablet. In 2011, 563 million smart phones and tablets were sold worldwide, compared to 353 million PCs and laptops. By 2016, International Data Corporation (IDC) forecasts that 1.36 billion smart phones and tablets will be sold worldwide, compared to 518 million PCs and laptops. As seen in Figure 1.1, the growth of the PC and laptop is essentially flat, whereas growth of tablets and smart phones will continue into the future. Included in the IDC study was this statement by Bob O'Donnell, vice president, Clients and Displays at IDC:

> *Whether it's consumers looking for a phone that can tap into several robust "application" ecosystems, businesses looking at deploying tablet devices into their environments, or educational institutions working to update their school's computer labs, smart, connected, computer-capable devices are playing an increasingly important role in nearly every individual's life.*[1]

Given this dramatic growth, all business organizations need to develop a strategy for moving into this world of handheld smart devices and the apps driving them.

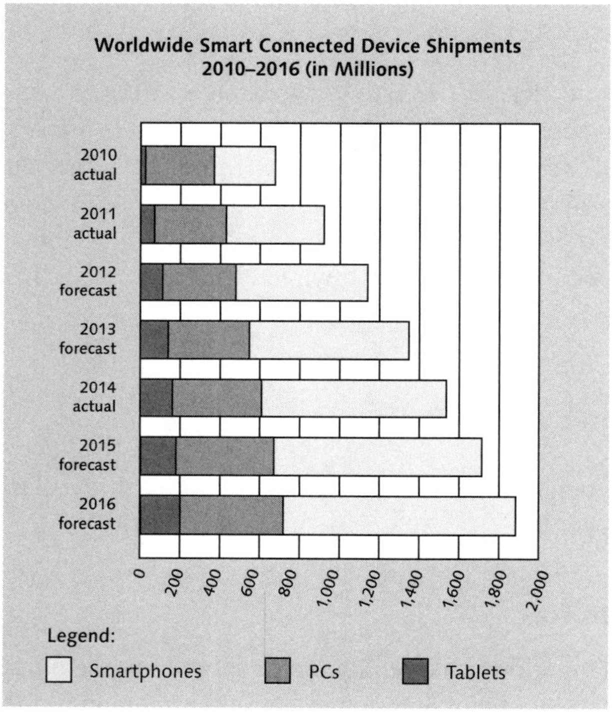

Figure 1.1 Growth of Smart Devices Broken Down by Category (adapted from IDC)

In order to better understand the differences between PCs and smart devices, it is important to compare their differences in technology:

- **Software**
  PCs run large application software packages, which are designed to perform many different functions. Most PCs operate using the Microsoft operating

1 *http://www.idc.com/getdoc.jsp?containerId=prUS23398412*

system, which serves as the standard. Hardware vendors build PCs around this standard. Smart phones and tablets, on the other hand, run *apps*, which are small, self-contained, single-function applications. Smart phones and tablets also run a number of different operating systems with limited or no standards between them. Most device manufacturers have their own proprietary software. It is interesting to note, however, that nearly all apps are developed on PC-type machines.

- **Computer power**
  Computer power available on both PCs and smart phones and tablets has grown so rapidly that, in general, this is no longer a differentiator.
- **Connectivity**
  When the PC was introduced, it came with a lot of "hard-wired technology" to allow connectivity to corporate servers and the outside world. Today's handheld devices utilize wireless technology, such as WiFi and 4G, and provide these devices with use-anywhere ability.
- **Screen**
  In the office, we have gotten used to viewing large screens (in some cases, multiple screens) primarily because we want to view multiple applications at the same time. As we move to the mobile app, screen size is a premium. In designing screens, we need to keep in mind that we are limited to one app at a time and only information that is really needed, as opposed to many applications and all the information that is available. Some of the technology solutions that are helping to bridge this gap in screen capability are the introduction of very high-quality screens and larger screen sizes that come with tablet devices such as the iPad.
- **Keyboard**
  The keyboard still remains the most efficient method for entering data. While Bluetooth and infrared can connect external keyboards to smart devices, most of these "portable" keyboards offer limited functionality. The voice recognition recently being developed by Apple and Google are promising. Most apps focus on limited data entry, such as texting short strings using messaging, using Twitter, or even sending email.

The function of the PC for the future will be to continue its current legacy role, namely, providing a platform for data entry, designing and deploying software for PCs, and developing apps that will run on smart devices. Mobile information workers will focus their demands on information from their smart phones and tablets.

1.1.2 SAP and Mobility

According to Gartner:

> *Mobile application development has become a top concern for enterprises and their CIOs. According to the 2012 Gartner CIO Survey, 61% of respondents plan to enhance their mobility capability during the next three years, and 48% believe they will become leaders in their industries by fully adopting innovative mobility solutions. This Magic Quadrant assesses the major vendors of application development platforms that enable enterprise developers to design, code, integrate, test, publish and manage business-to-consumer (B2C), B2B and business-to-enterprise (B2E) mobile applications.*[2]

The 2012 Gartner Magic Quadrant for Mobile Application Development Platforms, as shown in Figure 1.2, positions SAP as a leader.

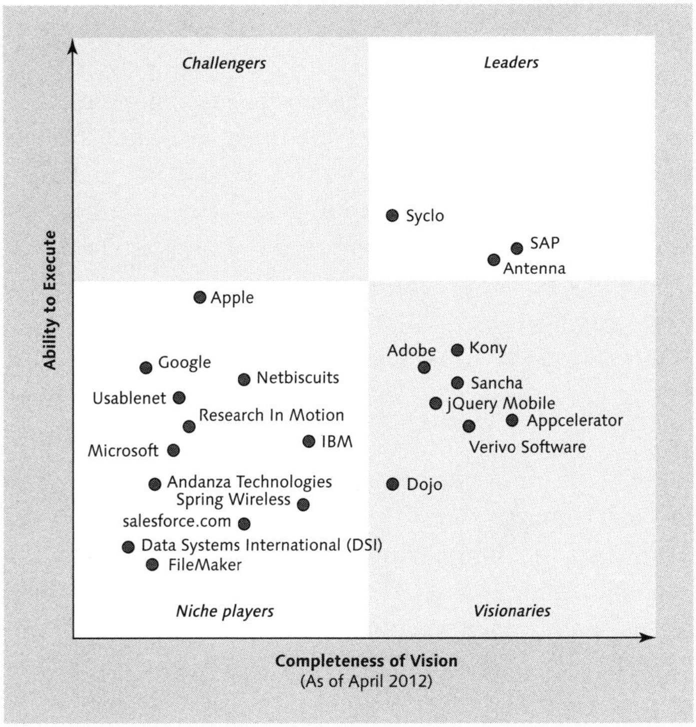

Figure 1.2 SAP as a Leader in the Magic Quadrant for Mobile App Development Platforms (© Gartner)

2 *http://www.gartner.com/technology/reprints.do?id=1-1AAJYX9&ct=120427&st=sb*

Recent Updates

Since this analysis was reported, SAP has acquired Syclo (another leader) and entered into partnership agreements with Adobe, Sencha, and Appcelerator (all viewed as visionaries).

Gartner also states that "SAP has one of the largest mobile development efforts, both in terms of devoted internal resources and partner management. Gartner observed partners actively integrating with SUP during 2011, and we anticipate that continuing in the long term." Finally, Gartner reports that SAP "has the most flexible application development environment, offering plug-ins for Eclipse and Visual Studio, as well as a proprietary studio...Sybase continues to have the broadest device support among all the multichannel vendors, and the strongest MDM offering, Afaria."[3]

One of the reasons SAP has gained this recognition was its early bet on mobility. SAP has one of the largest mobile development efforts, in terms of internal investment and the development of external partnerships. Much of this has been driven by the acquisitions of Sybase and Syclo, along with their various products. This places mobility alongside SAP's other major developments: SAP HANA and the cloud.

1.2 Mobile Devices and Data Sources

Every mobile business app needs two things: a mobile device to operate on and a data source to pull information from. In this section, we introduce these two core pieces of mobility.

1.2.1 Device Platforms

When we use the term *smart device* in this book, we are referring to smart phones and smart tablets (or pads). A number of the leading smart devices are shown in Figure 1.3.

3 *http://www.gartner.com/technology/reprints.do?id=1-1AAJYX9&ct=120427&st=sb*

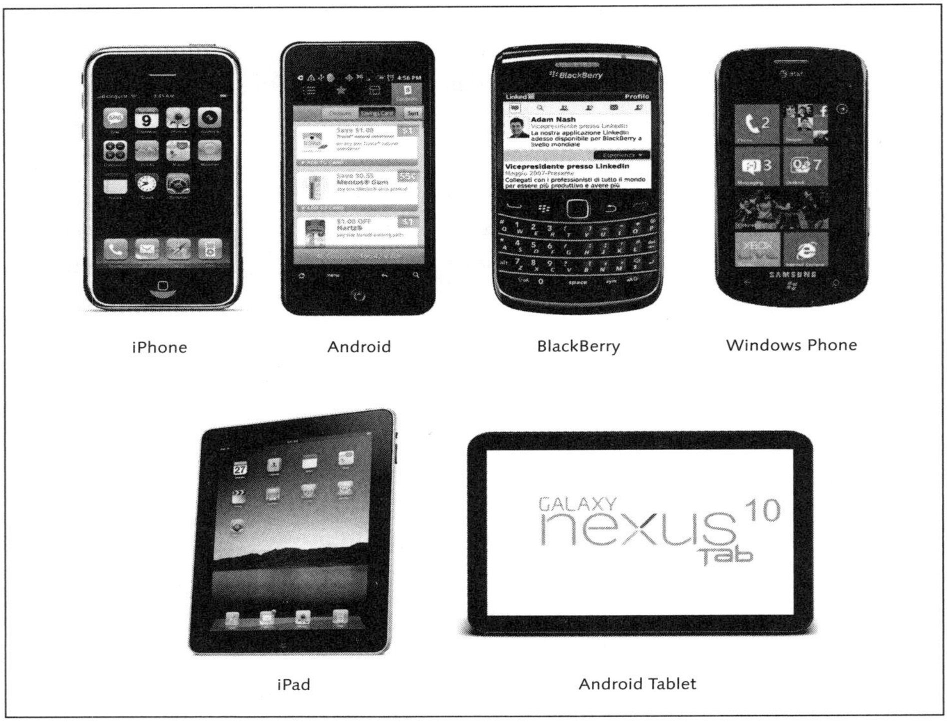

Figure 1.3 Leading Smart Phones and Tablets

Smart devices are often classified by the device platforms directly related to their operating systems. At this stage, several operating systems share a significant portion of the market, and each seems to have loyal followers. As the market matures, we will likely see this number of operating systems decrease. These leading platforms are shown in Figure 1.4 and are described as follows:

- **Android**
  The leading mobile operating system (OS) today is Android, which is Linux-based and currently being developed by Google. A number of different manufacturers produce Android devices. In order to differentiate themselves, some of these manufacturers have chosen to modify or enhance the operating system and software functionality of their devices. This can lead to software incompatibility with respect to apps. Android devices tend to be the least expensive smart devices. As of March 2013, there were 510,000 apps available from Google Play, and an estimated 25 billion downloads had taken place.

- **iOS**
  The iOS operating system is proprietary, has been developed by Apple, and is only available on Apple-produced devices, namely the iPhone, iPad, and iPod. The iOS devices are often viewed as high quality and well designed but tend to be the most expensive. The early 2013 numbers indicate that there are 750,000 apps in the Apple App Store and that, to date, more than 40 billion apps have been downloaded. While iOS has lost market share to Android in the smartphone market, it is still dominant in the tablet world because of the wide adoption of the iPad.

- **BlackBerry**
  BlackBerry OS is a proprietary operating system that runs only on BlackBerry devices. BlackBerry devices dominated the business world when email first became popular on phones. As the smart phone technology entered the market, BlackBerry was slow to react. In early 2013, the BlackBerry App World had 100,000 apps available and an estimated 3 billion downloads.

- **Microsoft**
  The Windows Phone operating system runs on a number of different manufacturers' platforms. The Microsoft mobile OS is closely linked to the OS that Microsoft delivers on the PC. While Microsoft has had a difficult time in the smart-device arena, a lot of articles have been written recently indicating that the new Windows Phone 8 may turn things around. In early 2013, about 40,000 apps were available for download from Microsoft's store.

When full-featured (i.e., not "smart") phones dominated the corporate world, it was very common for a company to declare that all phones must be of a specific type (for example, BlackBerry phones). Since it was common for the corporation to purchase, own, and pay the monthly bill for those devices, many companies also purchased long-term plans for these devices to save money. These policies led to the reluctance of corporations to try new devices as they came on the market.

As smart phones became popular, it became common for employees to purchase them for personal use and still carry their corporate, full-featured devices, as well. Employees began to complain about carrying two devices, and companies began allowing personal mobile devices in the work place (with and without company reimbursement) to access the company network. The latest trend is that many companies are getting out of the smart-device business altogether and asking their employees to provide their own devices. *Bring your own device* (BYOD) to work is what this phenomenon became known as.

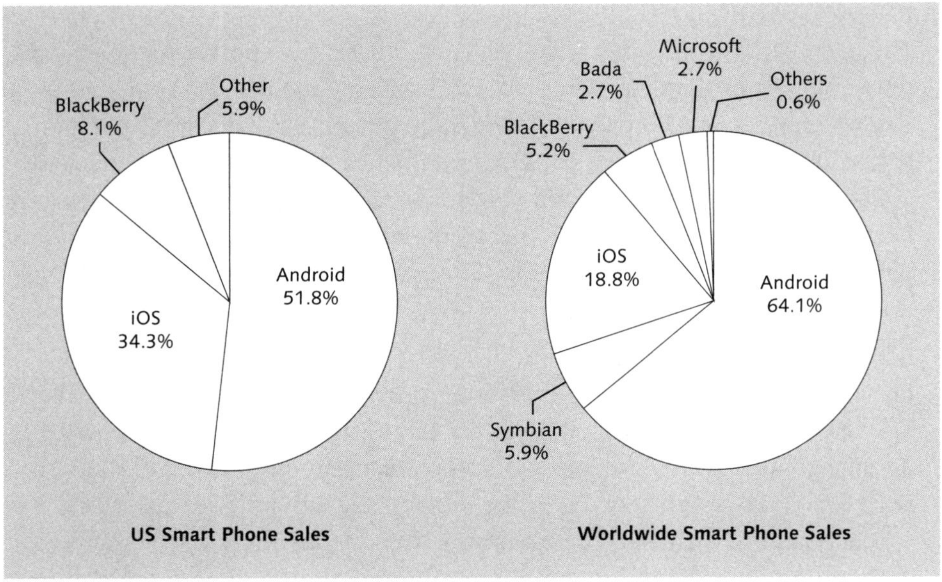

Figure 1.4 Worldwide and US Sales of Smart Phones by Platform (adapted from Nielsen and Gartner)

BYOD brings with it a significant number of issues that companies need to address. One concern is security, due to the need to access corporate data from those devices. Another concern is how employees are using their personal devices while on company time. Most relevant to this book, however, is a loss of company control of the platforms they support since company employees, not to mention its customers, are utilizing their platforms of choice. Organizations will, in fact, likely need to support all platforms. Any mobile solution developed by an organization, therefore, will have to address multiple device platforms.

1.2.2 Backend Data Sources

A unique characteristic of mobile business apps is their ability to interface with enterprise or backend data sources. When we talk about interface, we need to include the ability to perform the following operations:

- Access data for display
- Create new data
- Update existing data
- Delete unneeded data

SAPinsider has also reported on the data sources that were used for developing the top mobilized business apps. The results of this study, as shown in Figure 1.5, indicate that the primary data sources for the group studied were SAP systems. The most common types of data sources that need to be available when developing apps include the following:

- **SAP systems**
  Access is provided via Remote Function Calls (RFCs), Business APIs (BAPIs), or Business Object Repository (BOR).
- **Database systems**
  Access is provided to relational databases using JAVA Database Connectivity (JDBC).

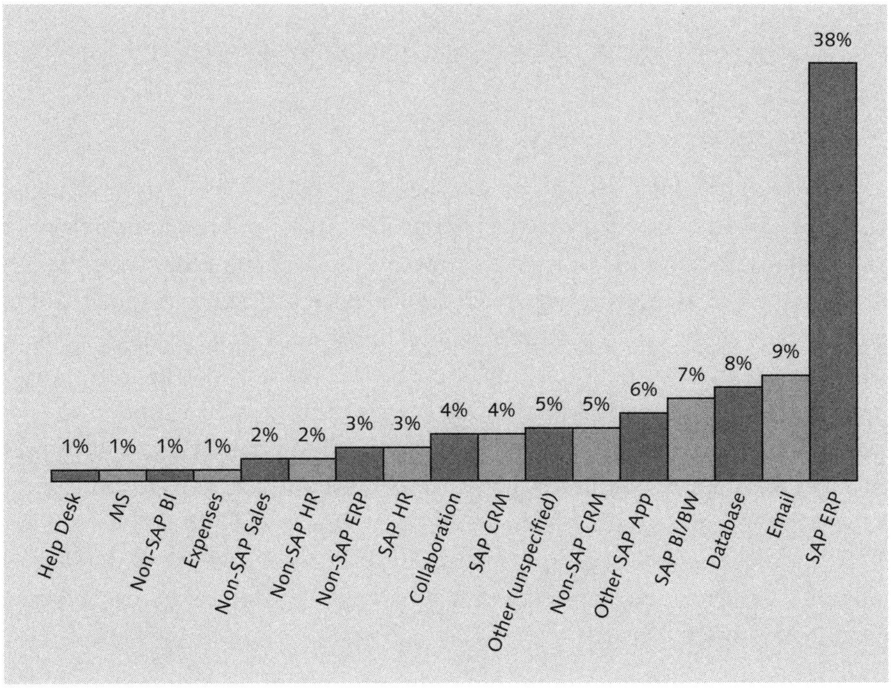

Figure 1.5 Connections with Backend Business Applications (adapted from *SAPinsider*)

- **Web services**
  Web service access is provided by Simple Object Access Protocol (SOAP) using WSDL.
- **REST**
  An alternative web service that utilizes HTTP verbs such as `GET`, `POST`, and `PUT` for data transfer.

Other issues that are important in the design of business apps include data security, data integrity, concurrent updates, and performance issues, such as caching.

1.3 Types of Mobile Apps

When grouping mobile apps according to the functionality they provide for the end user, we identify three major categories: game apps, utility apps, and real-time apps.

1.3.1 Game Apps

The largest group of apps is games, or so-called time killers. The history of many of these apps can be traced back to game consoles. Smart phones are often used for game playing because they provide a way to just pass the time, a distraction from other issues, an opportunity to relax, and a sense of accomplishment. Some of the important characteristics of successful gaming apps that might be applied to business apps to improve them include ease of use (no instruction manual required); incorporation of graphics and gestures; utilization of simple controls; and uncluttered, focused screens. One important differentiation between games and business apps is the number of copies likely to be in use. Successful games, such as version "X" of Angry Birds, have sold between 50 million and 75 million copies. At a dollar or two each, a lot of revenue is generated for research and development. On the other hand, the cost of developing business apps needs to be spread over many fewer users.

1.3.2 Utility Apps

The next largest group of apps (by functionality) fit into the category of utilities. These apps provide many of the features that were common in the PC world:

access to texting or email, photos, songs, movies, text documents, spreadsheets, and printing. Many of these utility apps are provided as built-in apps that come with the mobile device. These utility functions allow a smart phone to replace a number of devices from the past, including the camera, PDA, music player (iPod), calendar, watch, pager, voice recorder, and even the old "black book." Many of these apps have their own associated data files whose static content needs to be stored on the device. This content is often created by the device or downloaded from a store, saved on the device, and viewed or played a number of times on the device. Most of these apps do not deal with backend data sources.

1.3.3 Real-Time Apps

The final group of apps provides access to current and changing information, requiring access to backend data sources. For example, such apps include those commonly used during air travel. When a flight is canceled, the experienced traveler immediately turns to their smart device, not only to locate an alternative flight, but also to go ahead and book that flight. A group of apps provides information such as the quickest route between two points (given current travel conditions), where to go to lunch, the best place to buy that special gift, or what the weather will be tomorrow when that flight actually arrives. The common characteristic of most of these apps is that while they are interacting with backend database systems, most of the data is not confidential. Thus, security does not play a major role.

When it comes to apps that would be developed in the SAP environment, most of them would fall into this real-time category; for example, an app might be used by a salesperson in the field to check on the status of a customer's order or even modify that order. As such, this book focuses on real-time apps. One difference, however, is that we look at apps whose unique characteristic is the need to access backend databases that often contain private corporate data; here, issues such as security and privacy are very important. Many of these apps will be developed using data destined for corporate or partner use. The top mobile apps being developed that were listed in a recent survey by *SAPinsider* include examples of these apps, as shown in Figure 1.6.

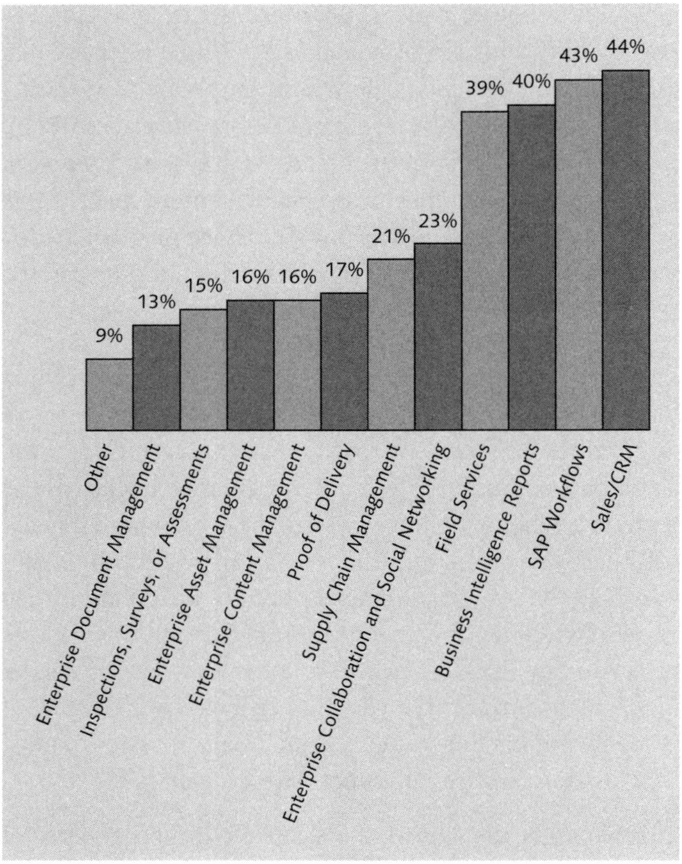

Figure 1.6 Top Mobilized Business Apps (adapted from *SAPinsider*)

1.4 Types of Mobile App Development

There are three main ways to develop mobile apps: native development, web development, and hybrid development (which is a combination of the first two approaches).

1.4.1 Native App Development

The majority of the apps currently available in the various app stores are developed using what is called *native app development*. All of the major mobile operating systems, as shown in Table 1.1, utilize their own conceptual model and a unique programming language and development tools. Development of a native app for a specific operating system requires knowledge of its own Integrated Development Environment (IDE) and associated tools, and as well as its own programming language. As a rule, native apps require the most time to develop, deliver the quickest app, and create the best-looking apps that take advantage of the unique characteristics of that specific platform. The skill set required for designers and developers is significant. While many advantages exist for native apps, the major disadvantage is that native apps run on only one operating system. If you want an app that works on, for example, iOS, Android, BlackBerry, and Windows Phone, you must develop and maintain four different native apps. Furthermore, since most developers are skilled on only one platform, four major development teams must be supported for the four versions of that app.

Mobile OS	IDE	Language
Android	ADT plug-in for Eclipse	Java
iOS	Xcode, runs on Mac	Objective C
BlackBerry	BlackBerry JDE	Java
Windows Phone	Visual Studio	C, C++

Table 1.1 Mobile Operating Systems, IDEs, and Languages

Note
We discuss native app development for SAP in Part IV of this book.

1.4.2 Web App Development

An alternative to native app development is *web app development*. Mobile web apps are developed using techniques commonly in use today for developing other web apps. With this approach, a web server is used to serve up pages that are then viewed on the smart device utilizing a web browser. The major difference between these sites and existing websites is that the screens are designed specifically to fit the smart device screen, as opposed to a PC monitor. Nearly all of today's smart devices have built-in or downloaded browsers that support HTML5 and new versions of JavaScript. With this approach, a single website can be accessed using a browser on each device platform. The major disadvantage is poor user experience. Nearly every user can tell the difference between a website (which requires each page to be downloaded separately when requested) and a mobile app (which doesn't), because their performance and appearance are quite different.

> **Note**
>
> Web app development for SAP is supported through SAP partnerships, and is not the focus of this book.

1.4.3 Hybrid App Development

There is also a development approach that combines native development and web development; this is referred to as *hybrid app development*. In this type of development a *container* app (which is itself a native app) is developed for each major operating system. The end user then downloads this container onto whichever device platform is being used. Each container maps that unique operating system's features to a common set of features and provides a file system for storing hybrid apps, a database for storing data, a messaging system, and a browser for displaying the hybrid app. From the inside of the container, all device platforms look the same. With this approach, it is possible to develop a single hybrid app that can be downloaded to any container and that runs on that device. The concept is simple: namely, develop a single hybrid app, and the same app will run on any device platform: iOS, Android, BlackBerry, and Windows Phone.

> **Note**
>
> Hybrid app development for SAP is the main focus of this book, and is discussed in Part II and Part III.

1.5 Introduction to SAP Mobile Platform

SAP's response to mobility has been the design and development of a single platform that can integrate SAP's many different platforms: *SAP Mobile Platform*. In this section, we briefly introduce you to SAP Mobile Platform and then focus specifically on Sybase Unwired Platform (SUP), which is part of SAP Mobile Platform and the main platform for mobile app development.

1.5.1 Architecture

SAP Mobile Platform is made up of several different components, as shown in Figure 1.7 and listed as follows:

- **Sybase Unwired Platform**
  Used for developing employee-focused business to employee (B2E) apps used in various business areas and industries.
- **Sybase 365**
  Used for business to consumer (B2C) apps in lines of business such as banking and marketing.
- **Syclo Agentry**
  A smaller mobile platform best suited for highly-specialized apps in field service and mobile asset management.
- **SAP Afaria**
  A mobile-device management software designed to aid in buying, deploying, securing, and maintaining smart devices.
- **SAP NetWeaver Gateway**
  Facilitates provisioning and consumption of business logic and content from SAP backend systems for mobile and web apps using the OData standard.
- **Third-party development tools**
  Makes sure the standard being developed is totally compatible with cross-platform tools like PhoneGap, Sencha, Appcelerator, and others.

Ultimately, SAP Mobile Platform will integrate all of these elements; currently, however, it is still a work in progress. While the long-term roadmap of SAP Mobile Platform is likely to change over time, we discuss the best information available to date.

The first products already integrated into SAP Mobile Platform are SUP and Afaria. The next step, which is now underway, is the integration of third-party products and OData. This will then be followed by porting the Syclo Agentry to SUP, and finally, the plan is to provide a cloud version, as well. Customers will then have a single, unified platform for developing, administering, provisioning, and deploying both B2C and B2E apps in a BYOD environment. Customers will also have the option to run those apps either on-premise or in the cloud.

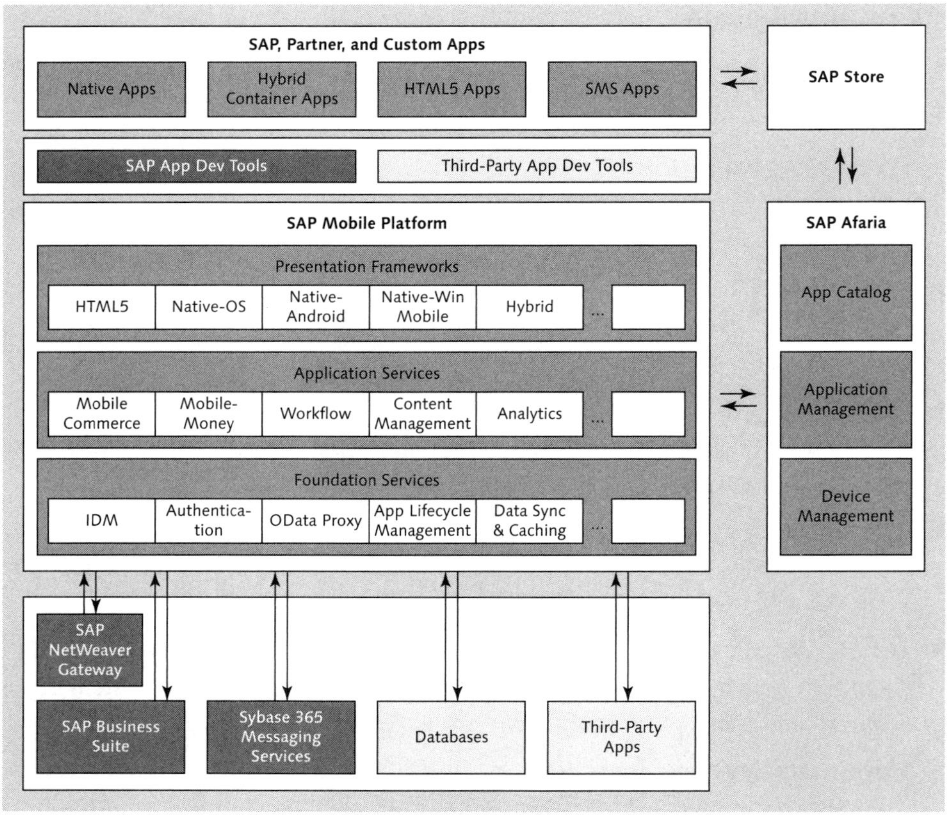

Figure 1.7 SAP Mobile Platform

The foundation layer of SAP Mobile Platform focuses on the data layer. Of course, central to this data is the SAP Business Suite. This includes the traditional access via BAPIs and RFCs, as well as the OData standard being proposed by the SAP NetWeaver Gateway technology. Another major source of data is from the many corporate standalone databases. This layer also includes the data associated with the Sybase 365 product and third-party apps.

The middle layer (sometimes referred to as *middleware*) of SAP Mobile Platform consists of foundation services, app services, and presentation frameworks. The foundation services provide the ability to communicate with the data layer. This includes issues such as security and performance. The app services are the various product lines. The presentation frameworks are the various techniques available for developing apps.

The tools layer includes both SAP and third-party app development tools. The primary SAP tool is SUP, and the third-party tools include those used for native apps, as well as integration of those provided by partners.

The top layer presents the various types of apps currently being supported by SAP Mobile Platform. The plan is for SAP and its partners to develop some of the apps, and also for the customer to develop their own apps. The role of the SAP store is for app developers to reach potential customers and generate revenue.

The final component, as discussed previously, is SAP Afaria.

1.5.2 Mobile App Development for SAP Mobile Platform

The specific focus of this book is mobile development using the Sybase Unwired Platform (SUP), the first complete component of SAP Mobile Platform. As you can see in Figure 1.8, SUP provides the developer with the following four major functions:

- **Connect**
  This function represents the connection to corporate data sources such as databases, web services, and software applications such as SAP. The platform provides for the creation of Mobile Business Objects (MBOs), which are used to logically define the attributes and operations required from each of the heterogeneous data sources. These MBOs can then be treated as homogeneous data structures. Support for the different access requirements, support for concurrent updates, performance features such as caching, and solutions to data security issues are all provided by the platform.
- **Create**
  This function is used to graphically create both MBOs and hybrid apps. The tool is an eclipse plug-in called Unwired Workspace, which is a component of the Sybase Mobile Software Development Kit (Sybase Mobile SDK). This tool runs on the developer's Windows-based PC and saves all the development components locally on the developer's machine. The first editor is used to create MBOs and includes support for creating relationships between MBOs and developing object queries to be processed on the MBOs. When complete, the editor tool provides the ability to deploy the MBOs to the common Unwired Server. The second editor is used to develop hybrid apps. This process involves screen design and the definition of the flow between screens. Many of the screens utilize data from the MBOs created previously. If it is determined that a native app is required (as opposed to a hybrid app), the MBOs can be downloaded to the native

development tool at the starting point. The native app will utilize the Unwired Server and inherit all of its benefits.

- **Consume**
  The primary apps developed by SUP are hybrid apps. These apps have been deployed to the Unwired Server, as well as the containers on end users' mobile devices. The single hybrid app can be deployed to containers on mobile devices using the following platforms: iOS, Android, BlackBerry, and Windows Phone. Once the hybrid app is deployed, it runs on its own and interacts with the Unwired Server for its data needs.
- **Control**
  The management control center handles control over both hybrid apps and devices. The control center is a Windows-based app that runs on the Unwired Server. Devices must first be registered with this control center before they can receive or run hybrid apps. This control center handles the registration, including device IDs, activation codes, and passwords. The control center can also download apps directly to the end users' devices.

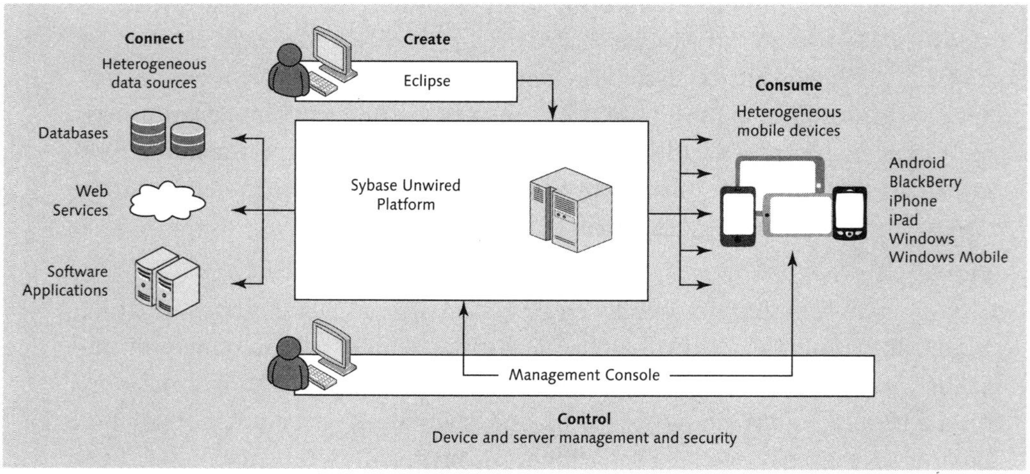

Figure 1.8 Sybase Unwired Platform

The SUP developer uses a set of tools called the Unwired Workspace. The Unwired Workspace is a perspective that is added to Eclipse, which is a commonly used public domain software design tool (for more details, see Chapter 4). The user downloads the Sybase Mobile SDK and then installs Eclipse, the Unwired Workspace, and a series of libraries in a single step. The Sybase Mobile SDK installs only on a Windows-based machine, just like the Unwired Platform runs only on Windows-based servers (for more details, see Chapter 3). Once the Unwired Workspace is installed, the final step involves connecting to the various backend data sources.

SAP Mobile Platform and SUP support the three different types of app development we introduced in Section 1.4, as we explain next.

Hybrid Apps

Hybrid apps provide a feature-rich, web-based user interface and the simplicity of a container-based development. MBOs are developed via the Mobile Application Diagram Editor, and the apps are created with the Hybrid App Designer (formerly called the Mobile Workflow Forms Editor) or by using third-party web tools. The code generated by these tools is standards-based HTML5 and JavaScript. Requirements for this type of apps include the following:

- Low data volume
- Simple, forms-based user experience involving notifications, lookup, and action-decision forms
- Lightweight persistence

With this approach, a container for each target operating system is developed as a native app and downloaded to the end user's mobile device. The same hybrid app can be deployed to the containers on multiple mobile platforms, with one app running on several mobile platforms. When the app is run, connectivity is provided between the app and the EIS by the Unwired Platform Runtime.

The strength of the SUP platform is for exactly this type of development. The technology behind this platform was originally developed by Sybase, which was later acquired by SAP and has since been integrated into SAP Mobile Platform. The designer, as shown in Figure 1.9, first develops MBOs for the required data and then develops the hybrid app. The design is generated by graphical tools, and while this involves specifying object properties, it does not require any programming because the system generates all the required code. The skill sets required of the designer are basic database concepts, such as understanding tables, attributes, and queries, and system design concepts, such as screen flow. Screen design skills are helpful but not required.

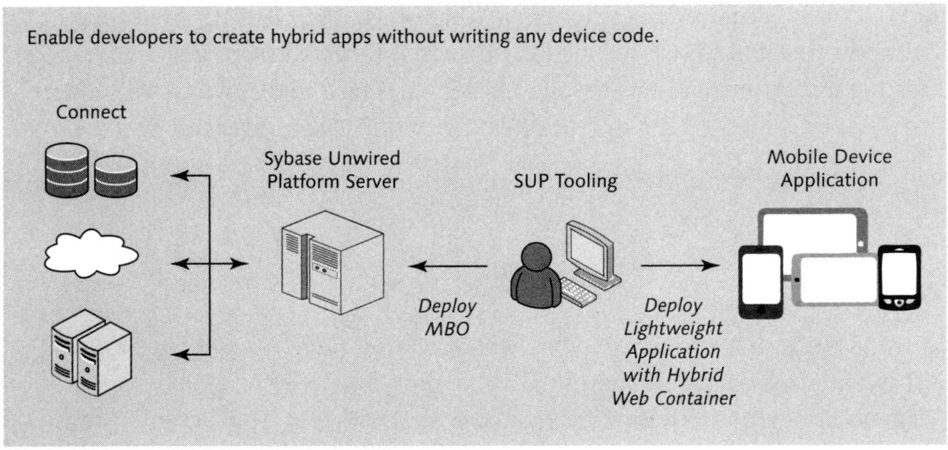

Figure 1.9 Designing Hybrid Apps

An important characteristic of developing hybrid apps using SUP is that all design work is done using the simple graphically oriented tool. While no coding is required, designers with coding skills (HTML5 and JavaScript) can use them to customize the app to improve the appearance and user experience of the app (for more details, see Chapter 10 and Chapter 11). The developer does not require any device platform–specific knowledge. The same app runs on all platforms (for more details, see Chapter 5). The types of apps that this approach is best suited for are so-called "lightweight" apps.

Native Apps

The SUP platform also provides some support for developing native apps, as shown in Figure 1.10 (for more details, see Chapter 12 through Chapter 14). The first step in the development process is the same as with hybrid apps—that is, the designer develops the MBOs and then deploys them to the Unwired Server. These MBOs have the same functionality with native apps as they have with hybrid apps. Once the MBO development is complete, its details are saved to the Unwired Server and to the current project folder.

The next step is the selection of a platform on which the native app is to run. As was shown in Table 1.1, each platform requires its own IDE, and the developer requires knowledge of the device platform and the language used for the specific platform. In some cases, fees or memberships are required if you wish to utilize the platform's store.

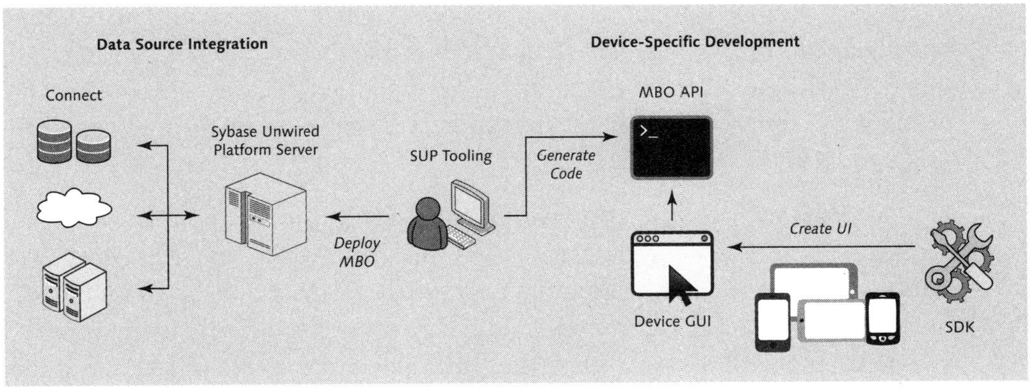

Figure 1.10 Native App Development—Object API Using MBO

Once you have selected and installed your native app development environment, the SUP tool can be used to export to that native app development environment all the necessary code to utilize those MBOs in the native app. At that point, the design process is separated from SUP. The screens, screen flow, and UI are developed using tools such as Xcode or ADT. Once the native apps are deployed to the end user's device, the end user can utilize the MBOs and the Unwired Server to access the backend data systems.

A specific type of native app that can be developed using SAP Mobile Platform is an OData app. The only component of the SUP environment that is used with this app type is the Unwired Server. This approach uses its own development tool, the OData SDK, for describing its data requirements, as shown in Figure 1.11.

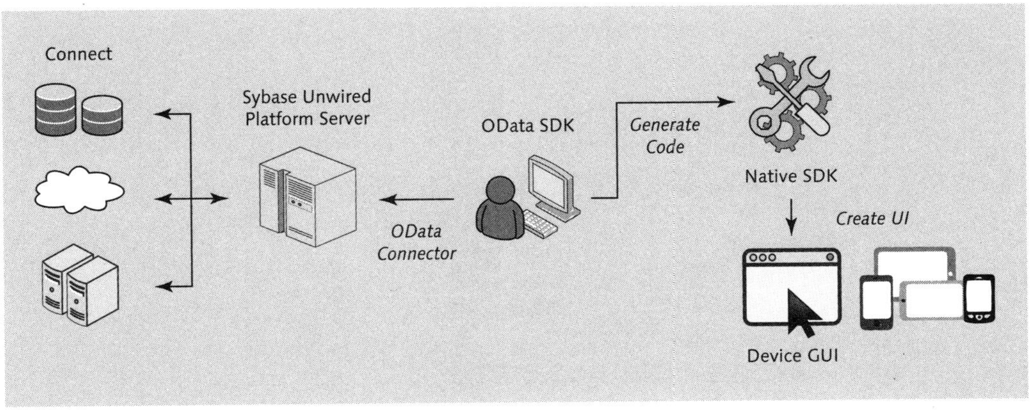

Figure 1.11 Native App Development—OData SDK

Open Data Protocol (OData) is a standard web protocol for querying and updating data. This protocol is currently supported by a number of different apps, including browser-based apps. The data structures supported in OData are considerably simpler and easier to use than the data structures supported by the traditional SAP RFCs and BAPIs.

SAP NetWeaver Gateway, as shown in Figure 1.12, provides powerful and convenient tools to generate OData Services for access to SAP systems. It enables users with a less technical background to create OData Services based on the SAP Business Suite. This OData Service can then be connected directly to the Unwired Server using the OData connector. The remainder of the design is the same as that for native app development. As stated earlier, this approach does not use any of the tools provided by the Sybase Mobile SDK, including MBOs and hybrid apps.

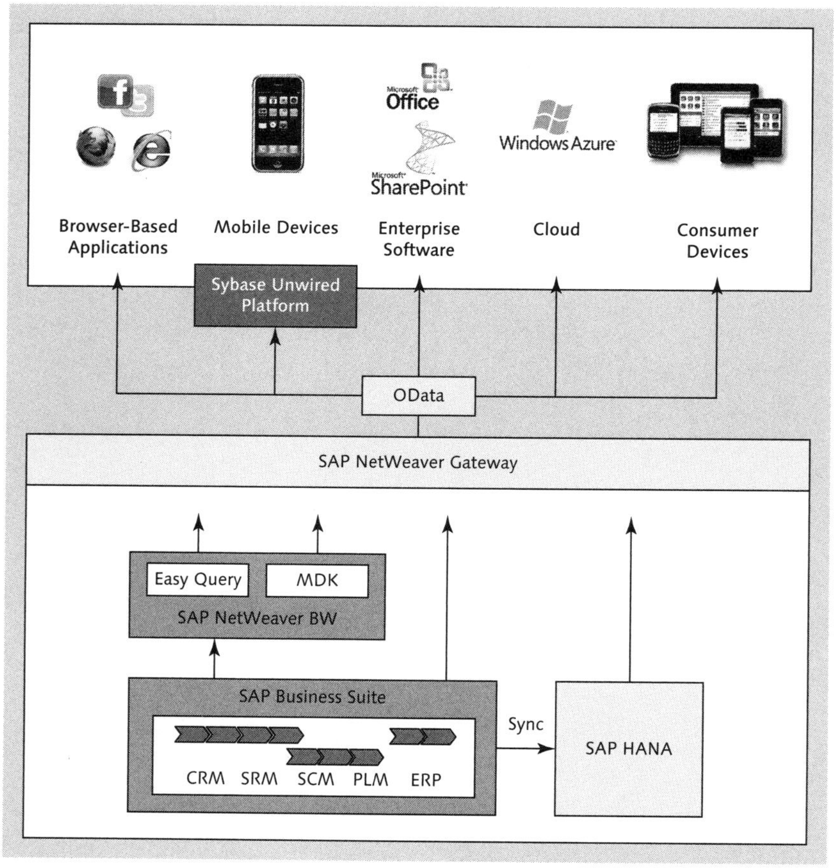

Figure 1.12 OData Services Model

Web Apps

SAP has entered into agreements with a couple of organizations that have products designed to produce web-based apps; these include the PhoneGap Touch and the Appcelerator Titanium products. Both of these products produce web-based user interfaces for mobile devices that utilize HTML5, CSS, and JavaScript (for more details, see Chapter 16). The current release of SUP includes portions of these products integrated as library components of the Sybase Mobile SDK.

Again, web app development is outside the scope of this book; we focus specifically on hybrid and native development.

Choosing a Development Approach

While SAP Mobile Platform supports both API-based native and hybrid apps, it is not possible to state that one approach is superior to the other. The choice of which approach to use depends on the type of app being developed and the skill set of the developer. One important variable is whether you app needs to function when connectivity is not present.

For online apps, data is available only when connectivity is available. Requirements for this type of app include the following:

- Request/reply sent directly from the EIS.
- Lightweight editing of form-based information.
- Few screens and simple entries.

Online apps can be developed with either the hybrid approach or the API-based native approach.

It is important for an offline mobile app to remain operational even if connectivity is not possible. The data synchronization between the data on the device and the EIS takes place the next time connectivity is available. Requirements for this type of apps include the following:

- Ability to store data locally on the device if the app is offline.
- More complex task orientation, data input, and screen navigation.
- Fully offline apps can be developed only with the API-based native approach, but the hybrid approach can support limited offline apps—it has the ability to save data updates to be applied when connectivity returns.

If you are developing a fully offline app, then your choice would be a native app. An app with fully offline capabilities can store data in a local database and then create, update, and delete data without connectivity. Once connectivity is restored, the changes to the data can be synchronized with the backend systems.

The following is a summary of some the advantages and disadvantages of native versus hybrid apps.

API-Based Native Apps

Disadvantages of API-based native apps include:

- Specific programming language skills are required.
- Complexity and development time are increased.
- Separate code bases must be maintained for each supported platform.

Advantages of API-based native apps include:

- Native apps allow more complex interfaces, logic, and data handling.
- Native apps have full access to device resources, such as location services.
- Native apps allow fully offline operation and data synchronization.

Hybrid Apps

Advantages of hybrid apps include:

- A single app runs on all supported platforms.
- Limited programming skills are required of the developer.
- Development is simple, straightforward, and quick.
- Server notifications are queued and transmitted when a device has connectivity.
- Submit hybrid apps can be executed when the device has no connectivity and are queued to be transmitted when connectivity to the Unwired Server is restored. The results of an object query can be cached on the device for a time interval.

Disadvantages of hybrid apps include:

- There is limited support for offline apps.
- Choice of user interfaces is limited.

In summary, hybrid apps are easier and quicker to develop, require fewer technical skills of the developer, and can be used for online and limited offline apps.

API-based native apps can be used to support offline apps and provide access to more device-specific features.

> **Note**
>
> We'll go into a bit more technical detail about some of these advantages and disadvantages in Section 6.1 of Chapter 6, where we discuss the mobile architectures of hybrid, native, and web apps in more detail.

1.6 Summary

This chapter presented a discussion of the rapid growth of smart phones and smart tablets and the roles they currently and will continue to play in most organizations, as well as some of the basic technical concepts related to mobility. An introduction to SAP Mobile Platform was presented, along with an introduction to the Sybase Unwired Platform, which is the focus of this book.

This chapter provides a description of Sybase Unwired Platform, which is used to create, manage, and secure mobile apps.

2 Introduction to Sybase Unwired Platform

Sybase Unwired Platform (SUP) provides a development and deployment platform that addresses many of the technical challenges associated with making an organization truly mobile. The purpose of SUP is to extend enterprise applications out to the mobile users. During development time, the platform provides a series of graphical tools for connecting to enterprise information systems (EISs), defining the data requirements in terms of Mobile Business Objects (MBOs), and developing apps. During runtime, the platform acts as a hub between the EISs and the mobile devices. The SUP platform, as shown in Figure 2.1, consists of two major components:

- **Sybase Mobile SDK**
  The Sybase Mobile SDK is the tool set used to develop mobile apps. It is installed on the developer's machine and contains the following components:
  - The Sybase Unwired Workspace, which is an Eclipse plug-in that provides the Integrated Development Environment (IDE) for modeling, designing, and deploying mobile apps.
  - Libraries to support building native apps (Object API), hybrid apps (HTML5/JavaScript), and OData apps (OData SDK).
- **Unwired Platform Runtime**
  The role of the Sybase Unwired Platform Runtime is to run and mange mobile apps. The platform consists of two major components, along with some additional servers discussed later in the chapter:
  - The Unwired Platform. This provides integration with the EIS, and server-side runtime services, such as security, caching, and synchronization to mobile apps.

- The Sybase Control Center (SCC). This is a web-based console that provides platform management and control. The SCC provides monitoring and management of both mobile devices and mobile apps.

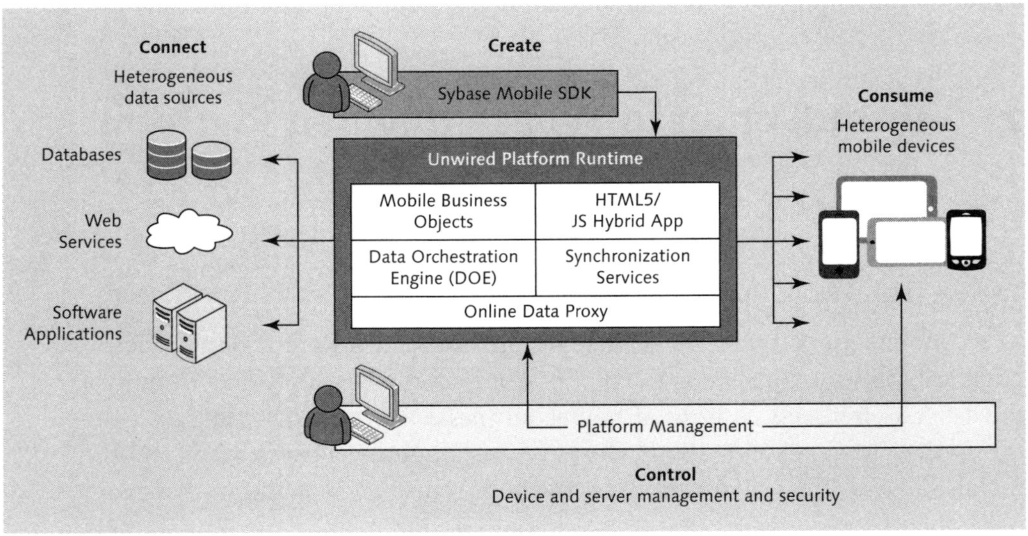

Figure 2.1 Sybase Unwired Platform

This chapter is devoted to explaining these two components, an understanding of which is crucial to mobile app development.

2.1 Sybase Mobile SDK

The Sybase Mobile SDK is installed on the developer's machine and includes both the Eclipse plug-in, which is called the Unwired Workspace, and a series of libraries. Some of the features of the Mobile SDK include the following:

- **Development tools that minimize total cost of development**
  - Features development tools that leverage existing developer expertise, technologies, and commonalities, such as data models and objects, to effectively produce mobile apps for a variety of device types and device platforms.

- **Design requirements for end-user devices**
  - Device choice. Allowing the end user to choose the device. More organizations now support this, as well as the ability to configure a personal device for work productivity and vice-versa.
  - Ease of use. Devices and apps that are easy to configure and use. App interfaces must be intuitive and integrate with device features, such as automatic layout, touch screen or trackball navigation.
  - Flexibility. The ability to load and upgrade apps, receive notifications, and perform work when connectivity is unavailable.

The following is a description of the Unwired Workspace and the libraries included in the Sybase Mobile SDK.

2.1.1 Unwired Workspace

A conceptual view of the development process for a mobile app using Unwired Workspace is shown in Figure 2.2.

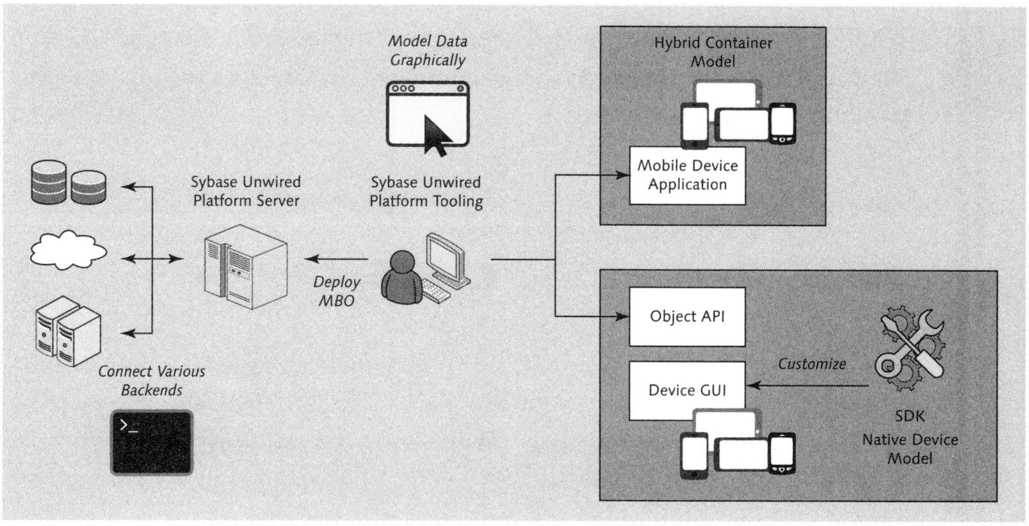

Figure 2.2 Developing Mobile Apps Using the Sybase Unwired Platform

The Unwired Workspace consists of a series of graphical editors, each of which performs a task in the development of a mobile app:

- **Enterprise Explorer**
  The Enterprise Explorer editor assists with identifying the backend systems required and connecting to those systems. For the example shown in Figure 2.3, connections have been made to a database by the name of SAMPLEDB, an SAP system by the name of RD1, and a web service by the name of WEATHERWS. A connection is also required for the Unwired Server.
- **Workspace Navigator**
  The Workspace Navigator editor, also shown in Figure 2.3, assists with the creation of a project. The project (in this example, named TRAVEL_REQUEST) has its own folder (sub-directory) on the developer's machine. All artifacts associated with this project are saved in that folder.
- **Mobile Application Diagram Editor**
  The Mobile Application Diagram Editor assists with the design of the data model. In this example, as shown in Figure 2.4, a single MBO has been created by the name of TRAVEL_REQUEST. Once the design is completed and saved in the project folder, the designer deploys it to the Unwired Server.
- **Hybrid App Designer (formerly called the Mobile Workflow Forms Editor)**
  If you are developing a hybrid app, the Hybrid App Designer is used to define all screens, as well as the flow between those screens. In the example shown in Figure 2.4, a series of screens has been designed to allow for creating, updating, and deleting travel requests.

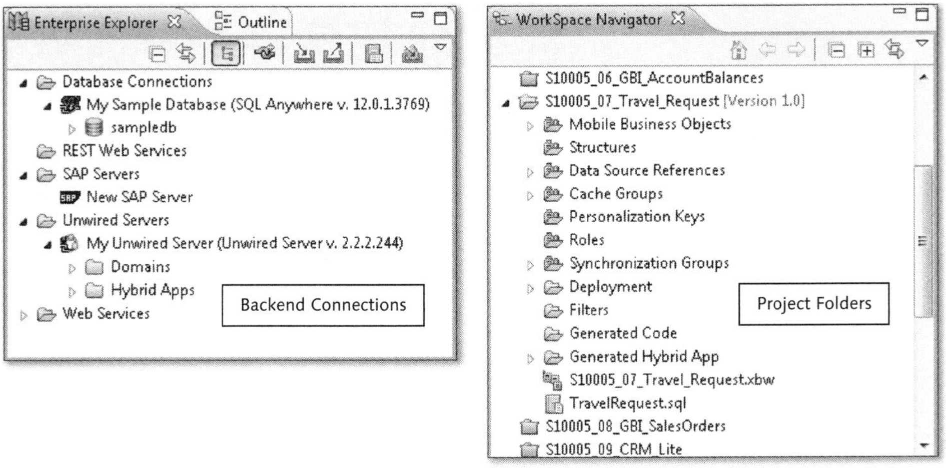

Figure 2.3 Establishing Backend Connections and Creating a Project Folder

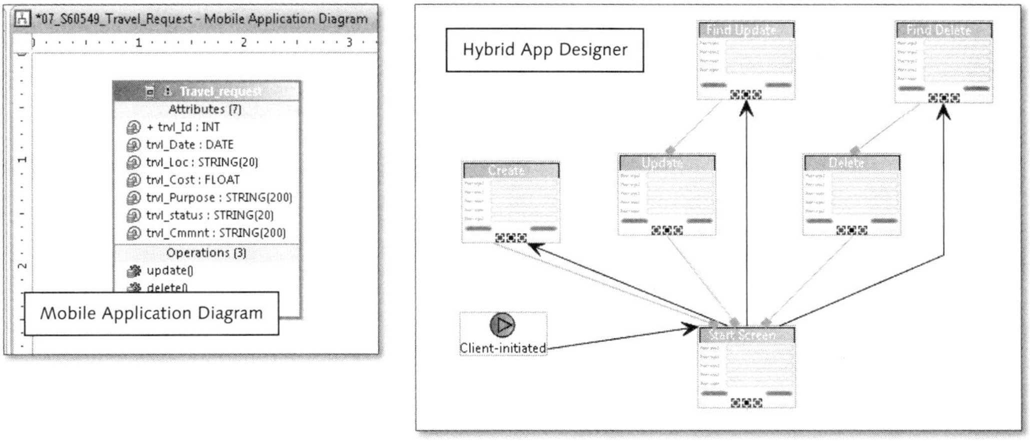

Figure 2.4 Designing an MBO and a Hybrid App

2.1.2 Mobile SDK Library: Object API

Native Object API apps are customized, full-featured mobile apps that use mobile data model packages, using Mobile Business Objects (MBOs), to facilitate connection with a variety of enterprise systems and leverage synchronization to support offline capabilities.

The Object API app model enables developers to write custom code— such as C#, Java, or Objective-C, depending on the target device platform—to create native apps. Development of Object API apps provides the most flexibility in terms of leveraging platform-specific services, but each app must be provisioned individually after being compiled, even for minor changes or updates.

Development involves both server-side and client-side components. The Unwired Server brokers data synchronization and transaction processing between the server and client components:

- Server-side components address the interaction between the EIS data source and the data cache. EIS data subsets and business logic are encapsulated in artifacts called *Mobile Business Object packages* that are deployed to Unwired Server.
- Client-side components are built into the mobile app and address the interaction between the data cache and the mobile device data store. This can include synchronizing data with the server, offline data access capabilities, and data change notification.

These apps have the following capabilities:

- They allow users to connect to data from a variety of EISs.
- They build in more complex data handling and logic and support for offline apps.
- They leverage data synchronization to optimize and balance the device response time versus the need for real-time data.
- They ensure secure and reliable transport of data.

2.1.3 Mobile SDK Library: OData

OData stands for *Open Data Protocol* and is a resource-based web protocol for querying and updating data. It is released by Microsoft under the Open Specification Promise to allow anyone to freely interoperate with OData implementations. OData defines operations on resources using HTTP verbs (`GET`, `PUT`, `POST`, and `DELETE`) and identifies those resources using a standard URI syntax. Data is transferred over HTTP using the Atom or JSON format.

In order to enable SAP to consume OData services, SAP offers a standalone product known as SAP NetWeaver Gateway, which creates OData services that expose SAP data. These services can then be used by a number of different apps, including the OData SDK. The OData SDK, which is included with the Mobile SDK library, is an alternate approach for building native mobile apps. It consists of a collection of runtime libraries and classes. The OData SDK supports Android, BlackBerry, and iOS platforms and is based on the native device SDKs of the platforms. There is an implementation for each platform, and native apps installed on the devices allow the client app to leverage the support provided by the given platform. For example:

- Adapt to each device's form factor (e.g., automatic layout)
- Exploit different input methods (e.g., touch screen, keyboard, or trackball)
- Cache data in native device data stores for better performance
- Tightly integrate with the features of the device

2.2 Unwired Platform Runtime

The Unwired Platform Runtime, as shown in Figure 2.5, provides the platform that allows the developer to deploy and manage mobile apps. The following are some of the features provided by the Unwired Platform Runtime:

- **Connectivity**
  - Provides the middleware platform where users connect to and interact with various enterprise information systems and apps that exist in any large or medium size corporate IT infrastructure that supports customers, employees, sales, partners, suppliers and executives. This includes SAP backend systems such as SAP Business Suite as well as other corporate databases.
- **Security**
  - Ensures that connections, apps, and data objects are deployed and managed securely.
  - Partitions data for mobile apps, securing data while in transport and while accessed by the app.
  - Ensures that data access and app functionality are available only to authorized mobile users.
- **Device and app management**
  - Provides a common runtime infrastructure that supports deployment and management of a variety of mobile app types.
  - Supports the rapidly expanding and changing selection of devices and device operating systems on the market.
  - Implements a data model that fits the app requirements, physical restrictions, and capabilities of the mobile device, such as network bandwidth, storage capacity, connectivity, security, and availability of consistent data.
  - Reliably propagates user transactions initiated on a device that may or may not have network connectivity.
  - Manages device and app life cycles—including device upgrades and turnover, as well as deployment of app updates and new apps—with limited interruption to productivity.

The components that make up the Unwired Platform Runtime include the following:

- DMZ (includes Relay and BES server)
- Unwired Server (includes a data tier)
- SAP Afaria Server
- Sybase Control Center

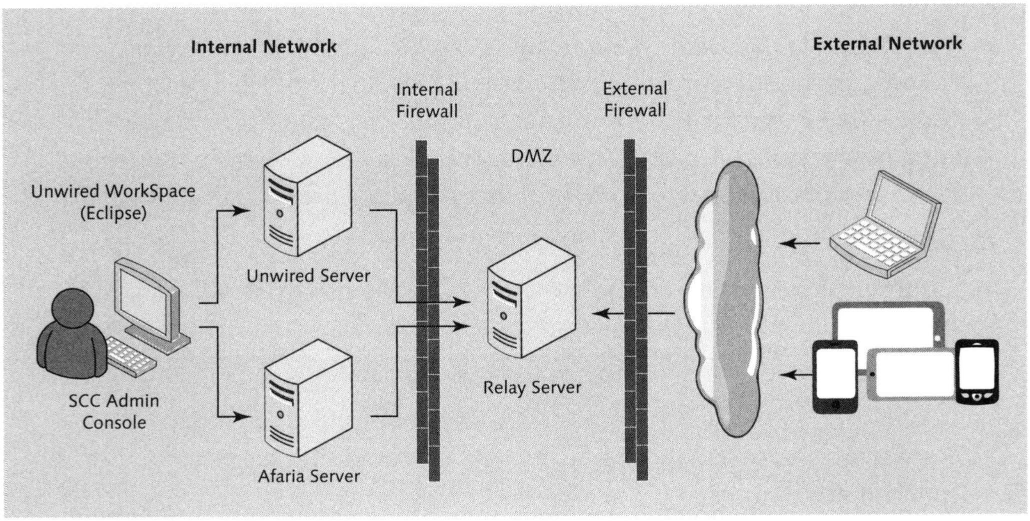

Figure 2.5 Unwired Platform Runtime

We discuss each of these in more detail next.

2.2.1 DMZ

The demilitarized zone (DMZ) provides for communications between the Unwired Server and mobile devices, eliminating the need to open any internal firewall ports. The standard port upon which this communication takes place is port 5100, using either HTTP or HTTPS. The communications that flow through the DMZ at runtime include the following:

- **Device registration**
  Each mobile device must be configured using its settings feature to be recognized by the Unwired Server. Once registered, secure communications can take place between the specific mobile device and the Unwired Server.
- **App deployment**
  Mobile apps are deployed from the Unwired Server to the mobile devices.

- **Runtime data access**
  The data tier supports data operations between the backend devices and the mobile apps.
- **Messaging**
  The Unwired Server sends messages to specific mobile devices. (BlackBerry handles this uniquely. This is discussed separately.)

The DMZ contains a server that functions as a reverse proxy, or Relay Server. This server is used to securely integrate mobile devices into your system landscape. This allows the installation of the Unwired Server completely within the corporate firewall. This Relay Server does the following:

- Provides a single point of contact for devices and is a specialized reverse proxy that avoids opening inbound ports in the firewall to Unwired Server.
- Supports load balancing.
- Accepts and forwards requests from remote clients to Unwired Platform components.
- Has the ability to be implemented as a pair of web extensions that runs in a web server (IIS on Windows or Apache on Linux).
- Integrates with existing security for web and enterprise infrastructure, eliminating the need for changes to existing corporate firewalls and IT policies.

When there is a single Unwired Server and no load balancing is required, this Relay Server is optional under any of the following conditions:

- All the mobile devices are going to connect to the network inside the corporate firewall (devices would only work onsite).
- There is a willingness to open the corporate firewall to communications that use port 5001.
- The devices used are only BlackBerry devices (in which case, you need a BES).
- As mentioned previously, the Relay Server supports messaging for all the mobile devices except the BlackBerry devices. The BlackBerry operating system uses its own proprietary technology to provide messaging to its mobile devices. If messaging to BlackBerry devices is required, it is necessary to install BlackBerry Enterprise Server (BES) as a component of the web tier. The BES requires an ongoing licensing fee.

2.2.2 Unwired Server

The Unwired Server stores and supports various mobile app components, and also handles enterprise data source and app access and communication between the EIS data source and the mobile device. It provides support for the following:

- **MBOs**
  Once the MBOs are completed for a project, they are deployed and saved on the Unwired Server as a package. The package can contain one or more MBOs. In order to separate multiple projects within an organization (or possibly multiple organizations) sharing a common server, the Unwired Server supports the creation of domains. When the MBOs are deployed, the designer selects the domain in which they should be saved. As you can see in Figure 2.6, a DEFAULT domain exists, and in this case, a series of additional domains have also been added. In domain DM101, we can see a package with the name S10102_EMPLOYEE, which contains an MBO with the name EMPLOYEE.

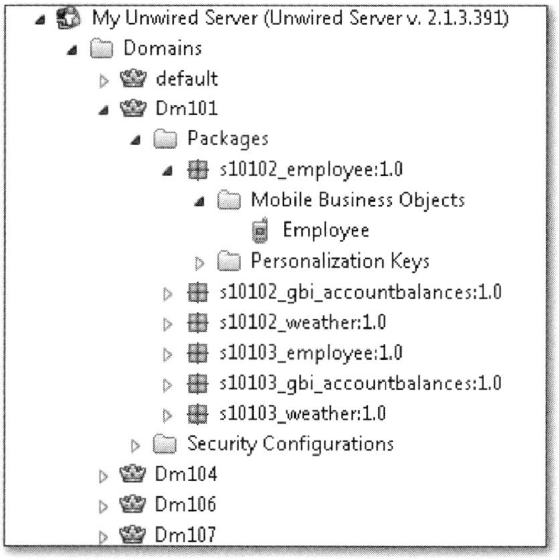

Figure 2.6 Deployed Packages (MBOs) to an Individual Domain

- **Hybrid apps**
  Once the hybrid app is designed, it is also deployed to the Unwired Server. These hybrid apps are all stored in a common folder location called HYBRID APPS. As you can see in Figure 2.7, a hybrid app with the name C08_SALES_ORDERS has been saved in the HYBRID APPS folder. Once the mobile app is deployed to

the Unwired Server, it can be deployed to a mobile device or provisioned to multiple devices using Afaria.

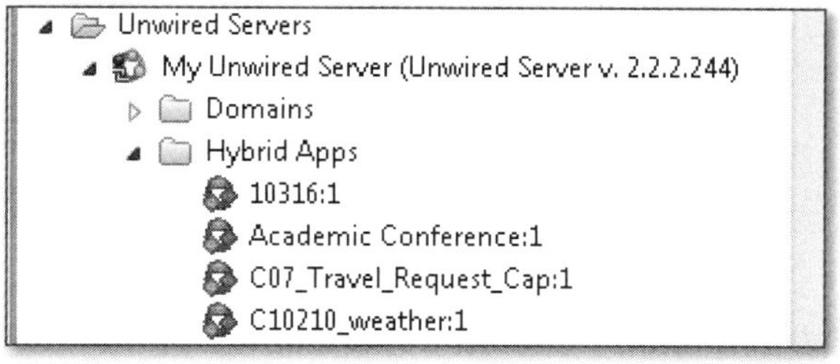

Figure 2.7 Deployed Hybrid Apps

- **Data Orchestration Engine**
When the backend enterprise system includes the Data Orchestration Engine (DOE) component of SAP NetWeaver Mobile, DOE data objects define the mobile data model for Object API apps. DOE consolidates data from diverse SAP backend sources and distributes it to apps via Sybase Unwired Platform messaging.
- **OData**
SAP NetWeaver Gateway exposes OData with extensions specific to SAP. Service documents that describe service interaction and data allow users to interact with the SAP app.
- **Synchronization services**
The goal of synchronization is to maintain data consistency among the client, data tier, and enterprise system. When changes occur on one tier, synchronization mechanisms ensure that the changes are replicated on the other tiers.

The Unwired Server serves as the communications channel between mobile devices and backend data systems. It is responsible for optimized access to backend systems, messaging, security services, multi-tenancy capabilities, and monitoring and development activities. It provides data delivery to and from device applications by applying transactions to the Cache Database and propagating transactions to backend systems.

The data tier, as shown in Figure 2.8, comprises multiple databases installed on and used by the Unwired Server and the management console within the runtime landscape. The databases in the data tier include the following:

- **Cache Database (CDB)**
  Maintains records of device synchronization activity, along with any cached data that may have been retrieved from the backend or pushed from the device. Each MBO has one or more tables in the CDB, depending on whether or not it is partitioned.
- **Cluster Database**
  Maintains knowledge of and coordination between Unwired Server clusters in an environment with multiple Unwired Servers.
- **Monitoring and Domain Log Database**
  Stores monitoring, tracking, and log messages for app activities.
- **Messaging Database**
  For apps that employ messaging services, HTML5/JavaScript hybrid apps, the OData SDK, and apps that use notification mechanisms, this database stores in-transit asynchronous messages until they are processed by the mobile app infrastructure layers. Message data is encrypted on the device and in transit.

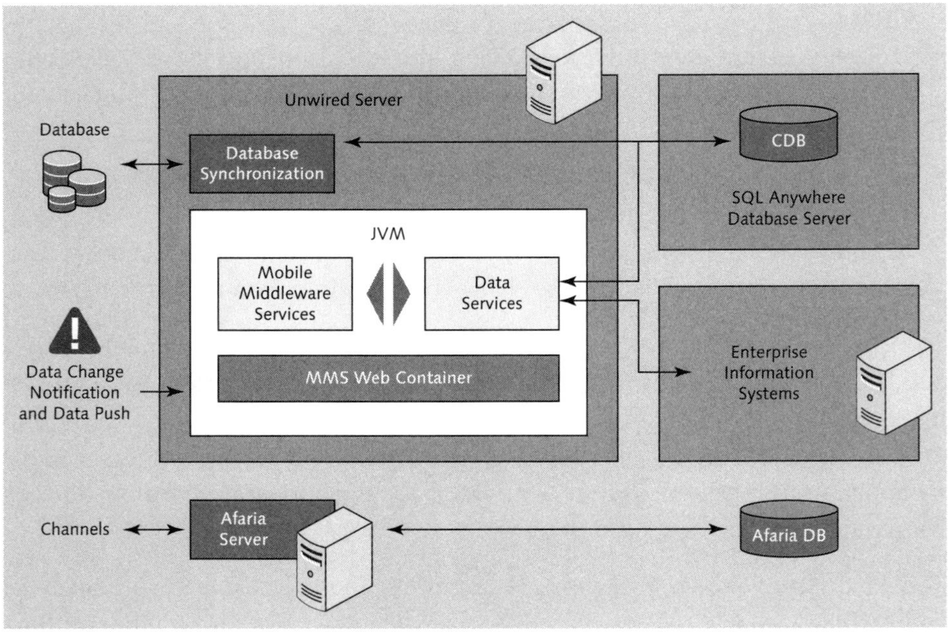

Figure 2.8 Data Tier

2.2.3 SAP Afaria Server

While not a requirement for using SUP, SAP Afaria, as shown in Figure 2.9, provides comprehensive device management, security, and application management for all mobile devices. Specific features include the following:

- **Configuration of mobile devices**
  Allows the developer to centrally define and maintain systems attributes, preferences, and settings for remote devices. These settings can include synchronization options for email, calendar, and contact information. It can be applied to both corporate and BYOD devices.
- **Management of mobile apps**
  Makes it easy for administrators to centrally distribute, install, maintain, and support mobile apps. Can be used for corporate apps and apps from stores such as the Apple App Store or Google Play.
- **Management of enterprise security**
  Enables IT to centrally manage security requirements, including enforcing power-on password, encryption of data, and updating signature files and antivirus engines. It also provides for automated data backup and the ability to remotely wipe a mobile device if it were to be stolen or lost.

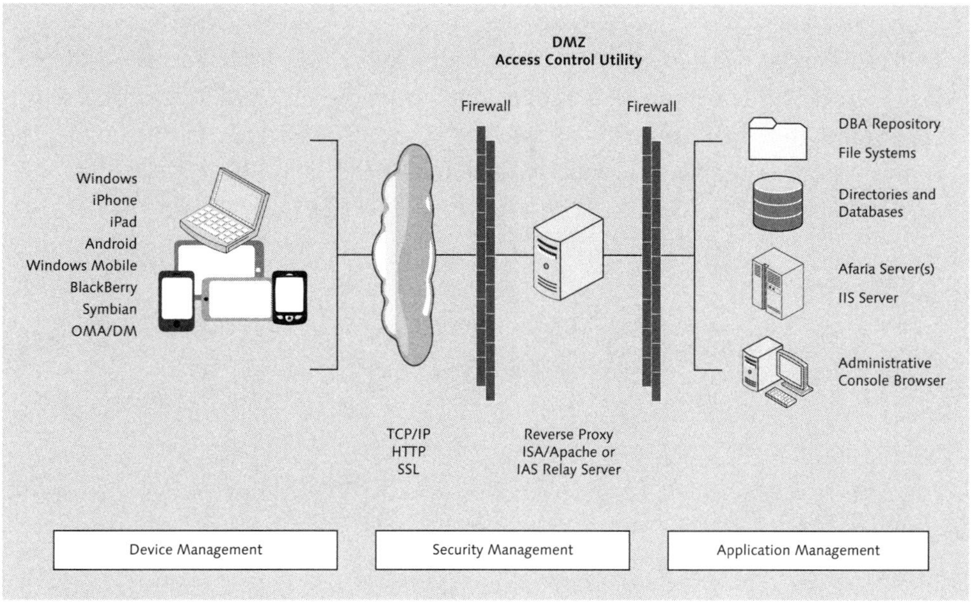

Figure 2.9 Basic Architecture of the On-Premise SAP Afaria

2.2.4 Sybase Control Center (SCC)

Sybase Control Center (SCC) for Sybase Unwired Platform offers integrated, web-based access to all Unwired Servers running in the corporate network. This console enables administrators to do the following:

- View server status.
- Perform start, stop, and restart operations.
- Configure server settings.
- Deploy packages.
- Configure package settings (including cache group schedule and synchronization settings).
- Manage subscriptions (including unsubscribe, recover, suspend, and resume).
- Set push notifications.
- Perform role mapping, cluster management tasks, and device lifecycle and licensing tasks (including registration, unregistration, locking, and unlocking, among others).

2.3 Summary

This chapter has presented an overview of the Sybase Unwired Server (SUP). SUP provides for the development of mobile apps using the Unwired Workspace, which supports both hybrid and native apps. The Sybase Unwired Platform provides the runtime support for these mobile apps and their access to EISs.

This chapter presents the SAP Unwired Platform Runtime, reviews the multiple kinds of licensing supported, and presents an installation procedure for a development system.

3 Installing the Sybase Unwired Platform Runtime

This chapter teaches you what you need to know in order to install the Unwired Platform Runtime. We start by presenting a discussion of the server landscapes that can be used for the Unwired Platform, then move on to the rather complex licensing options for the Unwired Server. In the last and most important section of the chapter, we walk you, step by step, through a typical installation.

3.1 Choosing the Unwired Platform Landscape

An important step in installing the Unwired Platform Runtime is the design of the server landscape. When choosing this design, SUP offers four possible options: single-server installation, a simple load-balancing cluster, a standard Microsoft failover cluster, and a Microsoft failover cluster with shared hosts.

The simplest, least expensive, and easiest-to-maintain landscape is a single-server installation. The disadvantages are that it cannot be scaled up when growth takes place and it does not support any load-balancing or failover mechanisms. The solution to the scaling and load balancing is the addition of load-balancing clusters. The solution to the failover issues is the data tier with Microsoft failover clusters.

In this section, we describe the four landscape options for the SUP Runtime.

3.1.1 Single-Server Installation

A single-server installation places the complete Unwired Platform on a single server. This environment offers the lowest complexity in terms of administration and security. The downsides are that the server must be of sufficient size to support all the required activities and that it is not easily scalable. It is also designed more for small development teams as opposed to large enterprise development. The basic configuration is shown in Figure 3.1.

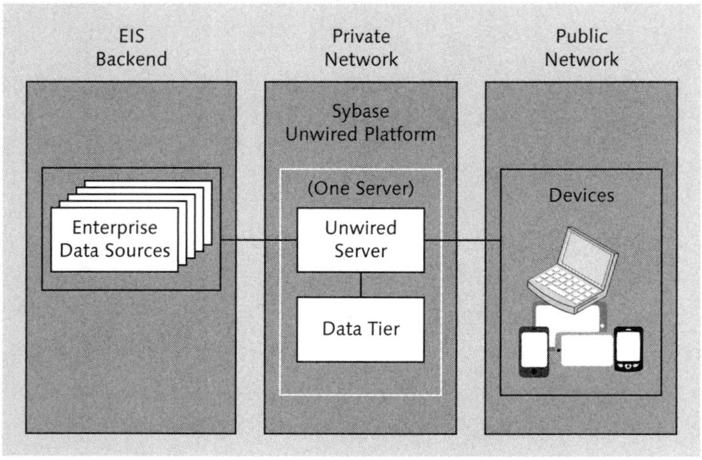

Figure 3.1 Install on a Single Server

3.1.2 Simple Load-Balancing Cluster

The simple load-balancing cluster consists of two Unwired Servers that are installed as the app server nodes and any number of additional Unwired Servers as scale-out nodes, as shown in Figure 3.2. If the primary server fails, the secondary server takes over. Each Unwired Server is installed on its own machine, and the single data tier is installed on its own machine. A proxy server farm provides load balancing by distributing requests to different Unwired Server nodes. This approach works well for multi-user development and test environments, as well as low-volume production environments. However, there is no failover support for the database, so it is a single-point-of-failure risk. The major benefits are that it is scalable and provides improved performance.

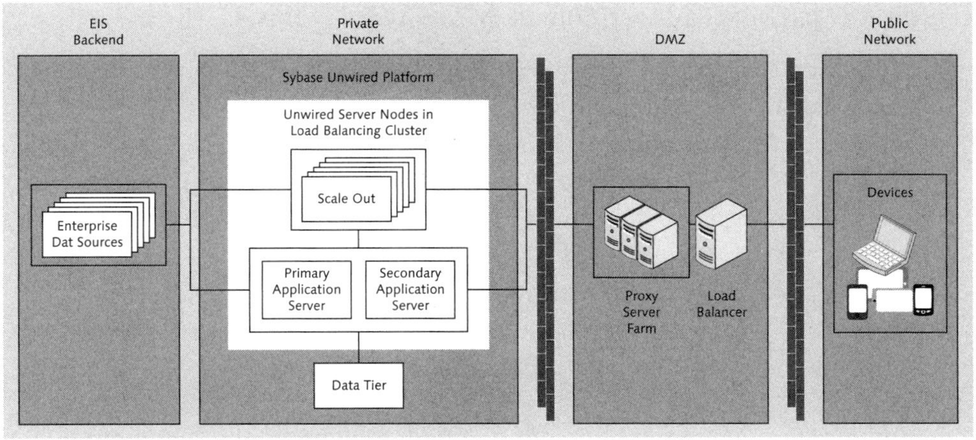

Figure 3.2 Simple Load-Balancing Cluster

3.1.3 Standard Microsoft Failover Cluster

The standard Microsoft failover cluster, as shown in Figure 3.3, adds two data tier nodes in a Microsoft failover cluster to the previous load-balancing cluster. This ensures that if the active data tier fails, you can continue the passive data tier. This design functions well in a multiple-user development and test environment, as well as in high-volume production environments. This architecture provides improved performance and has no single point of failure. The major disadvantages are cost and possible increased response times.

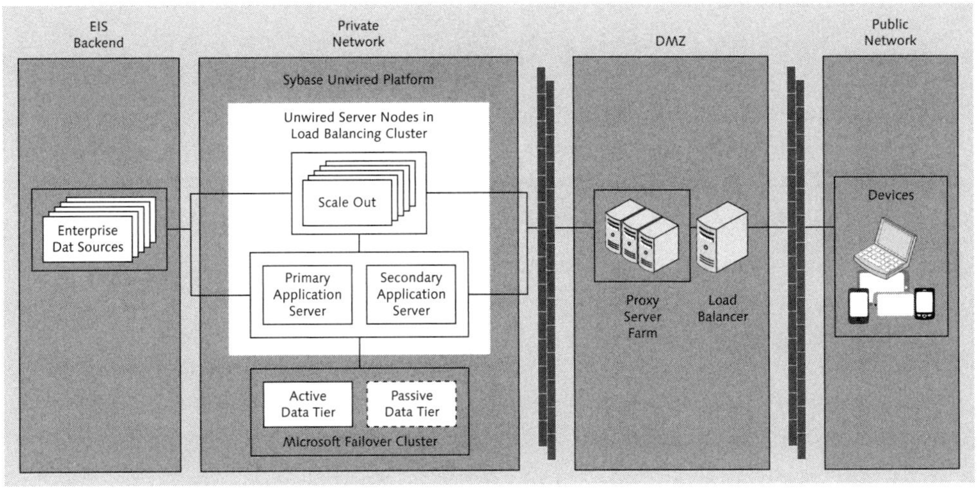

Figure 3.3 Standard Microsoft Failover Cluster

3.1.4 Microsoft Failover Cluster with Shared Hosts

A Microsoft failover cluster with shared hosts (see Figure 3.4) resembles two single-server installations. Each of the servers has both an Unwired Server and a data tier installed. The data tiers are also installed in a Microsoft failover cluster. This framework works well for multi-user development and test environments of larger scale and supports moderate volume production environments that require some high availability. It also leads to some increase in response times, yet provides an opportunity to economize on hardware costs.

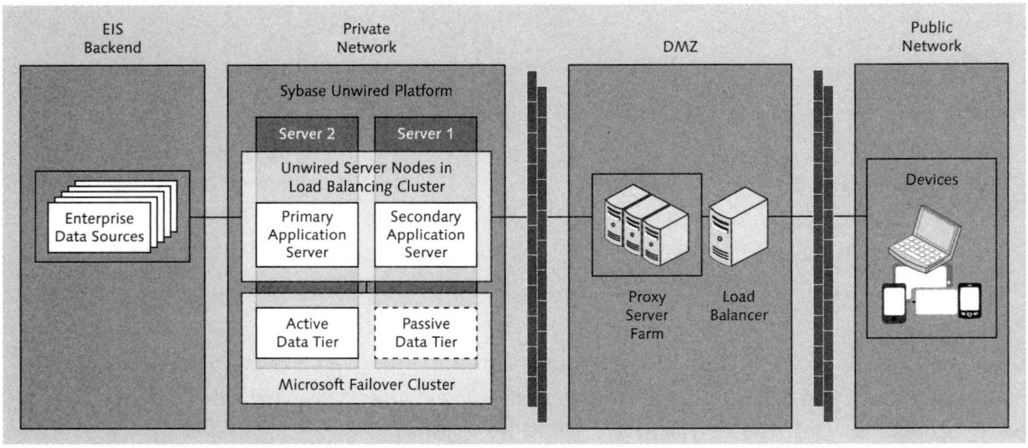

Figure 3.4 Microsoft Failover Cluster with Shared Hosts

3.2 Choosing Licensing Options

The licensing process for Unwired Platform Runtime components offers a complex range of options. When you select the appropriate licensing option, there are three things to consider: the type of environment in question, possible product editions, and possible license types. In this section, we walk you through these elements of licensing, and then show you how to map your environment to the best product edition and license types. At the end of the section, we walk you through two typical licensing examples.

3.2.1 Types of Environments

The type of environment helps you to determine the product edition and license type. The following environments are supported by the Unwired Platform Runtime:

- **Development**
  This is a preproduction environment where apps are developed on a single host using a personal development server license. A simple load-balancing clustering using an enterprise development server license may be necessary if app performance is to be tested. For this environment, purchase a Sybase Mobile SDK license for each developer workstation.
- **Qualification**
  This is another preproduction environment used to test apps and runtime properties. If your budget allows, qualification environments should replicate production environments as closely as possible. A Microsoft failover cluster is used, with Unwired Platform installed, using an enterprise development server license. Any developer workstation involved requires its own Sybase Mobile SDK license.
- **Production**
  This is a live runtime environment that uses a Microsoft failover cluster, with Unwired Platform installed, using an enterprise server license. This environment requires Unwired Platform Runtime that is licensed according to the number of CPU cores, clients, and developer workstations used. If any developer workstations are involved, each requires its own Sybase Mobile SDK license.

3.2.2 Product Editions

Unwired Platform Runtime components are licensed according to the product edition. The following describes the three product editions:

- **Personal Development Server Edition**
  - Allows use in development systems and testing systems only; not for use in production systems
  - Requires all Unwired Platform server components to be installed on the same single-user host with Sybase Mobile SDK
  - Allows a maximum of five mobile client devices

- **Enterprise Development Server Edition**
  - Allows use in development systems and testing systems only; not for use in production systems
  - Allows each installable component to be located on a separate host
  - Allows clustered systems
  - Allows a maximum of 20 mobile client devices
- **Enterprise Server Edition**
  - License type determines allowed use (production only, or development and testing only)
  - Allows each installable component to be located on a separate host
  - Allows clustered systems
  - Requires separate license for mobile client devices (production)
  - Allows unlimited mobile client devices (development and testing)

3.2.3 License Types

Each license type is associated with one or more Unwired Platform product editions. The following describes the four license types:

- **Standalone Seat license**
  - Unwired Platform Runtime components must be installed on the same host as Sybase Mobile SDK
  - Available only with Personal Development Server Edition
- **Development and Test License**
  - Servers licensed for development and testing use only; no production use allowed
  - No limit on CPU/cores or host configuration (single or multiple hosts, clusters, and so on)
  - Available with Enterprise Development Server Edition and Enterprise Server Edition
- **CPU/Core License**
  - Servers licensed by CPU/cores and for production use only; no development or testing use allowed

- No limit on host configuration (single or multiple hosts, clusters, and so on)
- Available only with Enterprise Server Edition
- **OEM License**
  - Unwired Platform servers to be bundled with packaged apps and redistributed
  - Servers licensed for production use only; no development or testing use allowed
  - Unserved License deployment model only (see below); same license on all server hosts (no host ID required); no limit on CPU/cores or host configuration (single or multiple hosts, clusters, and so on)
  - Only clients are counted for licensing/royalties (various terms)
  - Available only with Enterprise Server Edition

Software licenses for Unwired Platform are provided via two different license deployment models:

- **Unserved License**
  Each license is granted for one specific host. The license file must be stored locally, on the Unwired Platform host. The license cannot be transferred to another host. This is the easiest to use if you have a small number of servers.
- **Served License**
  A license is granted for a number of hosts. The license file is stored on a SySAM license server, and the license can be automatically acquired (checked out) by any Unwired Platform host. If you choose this, you must install a SySAM license server. This approach is used more often when a large number of Unwired Servers are used.

3.2.4 Mapping Environment to Product Editions and License Types

Once you have selected your environment, the first step is to select the available edition. The options are presented in Table 3.1. As shown in the table, there are two editions that will support development and two editions that will support production.

	Enterprise Development Server Edition	Personal Development Server Edition	Enterprise Server Edition
Development	No	Yes	Yes
Qualification	Pilot	No	Pilot; QA test
Production	Yes	No	Yes

Table 3.1 Choosing a Product Edition for the Environment

The second option, as shown in Table 3.2, is to choose a license type given the product edition selected.

	Standalone Seat License	Development and Test License	CPU/Core License
Personal Development Edition	Yes	No	No
Enterprise Development Edition	No	Yes	No
Enterprise Server Edition	No	Yes	Yes

Table 3.2 Choosing a License Type for the Selected Product Edition

3.2.5 Examples: Single-Developer Development and Team Development

The number of different licensing choices can be confusing. In hopes that this will help simplify your choices, let's discuss two typical examples that would apply if your objective is to develop mobile apps. Our first example assumes a single-developer situation, and the second example assumes a team development situation.

Having a single developer gives you the opportunity to adopt the simplest architecture for developing hybrid apps. This example uses the Personal Development Edition with the Standalone Seat License. The limitations require that all of the software, including the Unwired Server and Sybase Mobile SDK, must be installed on the same physical box. This configuration, as shown in Figure 3.5, delivers each developer their own development workstation. A series of virtual servers could be within a physical server environment. A limitation of this approach is that you can have a maximum of only five client devices.

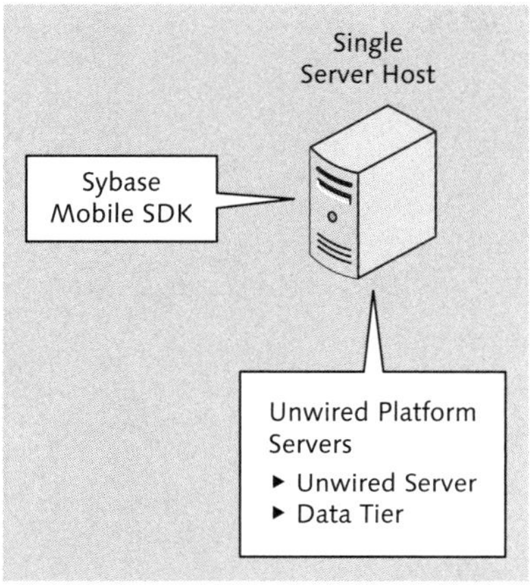

Figure 3.5 Single Server with Personal Edition Standalone Seat License

Team development, on the other hand, supports a development team whereby all the developers share a common Unwired Server, yet each developer has their own copy of the Unwired Workspace on their workstation. In this example, we are using the single-server installation landscape with the Enterprise Server Edition and the Development and Test License, as shown in Figure 3.6.

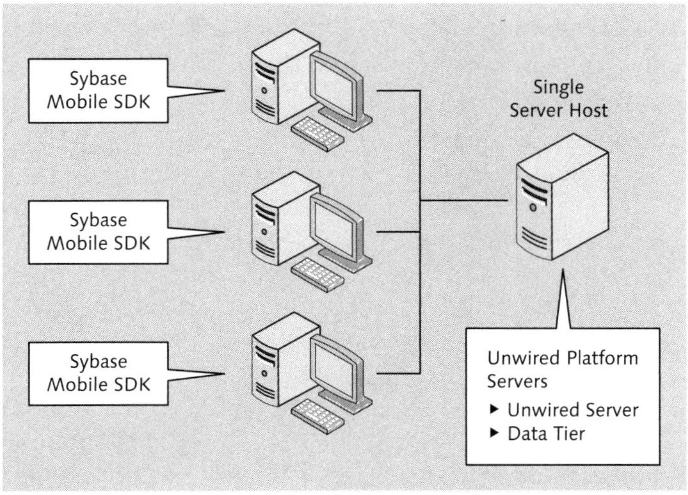

Figure 3.6 Single Server with Enterprise Edition and Development and Test License

Cloud-Based Access to Sybase Unwired Platform

A developer version of the Sybase Unwired Platform is available from Amazon Web Services (AWS). Discussion of its use is presented in the next chapter.

3.3 Typical Installation Steps

The following is a typical installation of the Unwired Platform Runtime. For this installation, we are using the following configuration, which is the same as shown in Figure 3.6:

- Landscape: single-server installation
- Edition: Enterprise Edition
- License: Development and Test License
- Licensing Model: Unserved License

In this section, we start by discussing the hardware and software requirements you should be familiar with and then go on to describe the major steps involved in the installation process: generating license keys, downloading installation files, and, finally, actually installing the Unwired Platform Runtime.

3.3.1 Hardware and Software Requirements

The only supported operating system for the Unwired Server and data tier is Microsoft Windows. The supported versions are shown in Table 3.3. The Enterprise Server Edition can be installed only on 64-bit operating systems. The Personal Development and Enterprise Development Editions can be installed on either 32-bit or 64-bit operating systems. The Unwired Server requires a minimum of 1.2 GB, the data tier requires a minimum of .72 GB of disk space, and 4 GB of RAM is recommended.

Operating System	Service Pack
Microsoft Windows XP: Professional (32-bit only)	SP2, SP3
Microsoft Windows Vista: Business, Enterprise, or Ultimate	SP2
Microsoft Windows 7: Professional, Enterprise, or Ultimate	SP1
Microsoft Windows Server 2003: Standard or Enterprise	SP2
Microsoft Windows Server 2008: Standard or Enterprise	SP2
Microsoft Windows Server 2008 R2: Standard or Enterprise	SP1

Table 3.3 Host Operating Systems

3.3.2 Generating License Keys

Now that the type of license has been determined, the next step in the process is to download the license file. Since the unserved license model is being used in this example, it is necessary to download a license file onto the new server being built. To obtain this file, go the SAP Service Market Place and create the new system, as shown in Figure 3.7. It is important to select DEVELOPMENT SYSTEM for the system type.

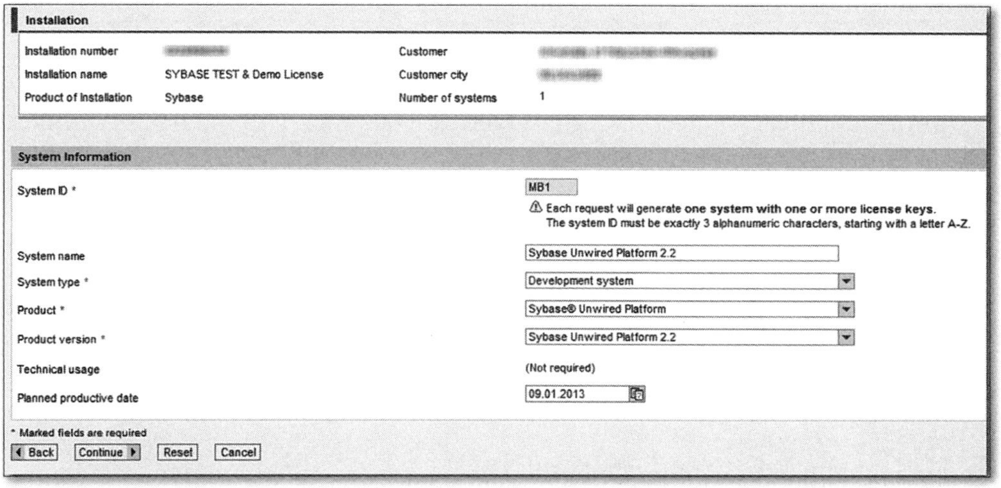

Figure 3.7 Creating a New Development System

In the next step, select the radio button to generate the license. Make sure that you select ENTERPRISE SERVER and DEVELOPMENT AND TESTING, as shown in Figure 3.8.

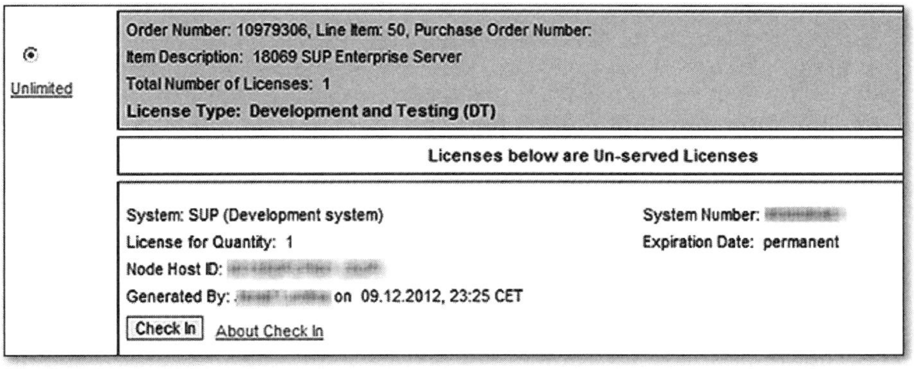

Figure 3.8 Selecting Enterprise Server and Development and Testing

The next step is to select UNSERVED LICENSE as the license model, as shown in Figure 3.9.

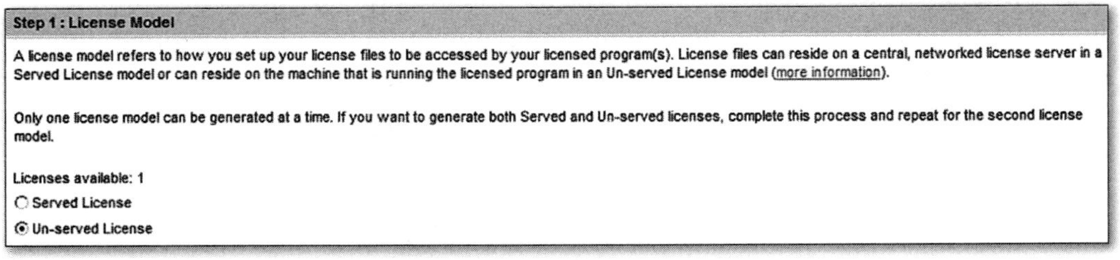

Figure 3.9 Selecting Unserved as the License Model

In this example, we have a single server, so you enter 1 as the number of host machines, as shown in Figure 3.10.

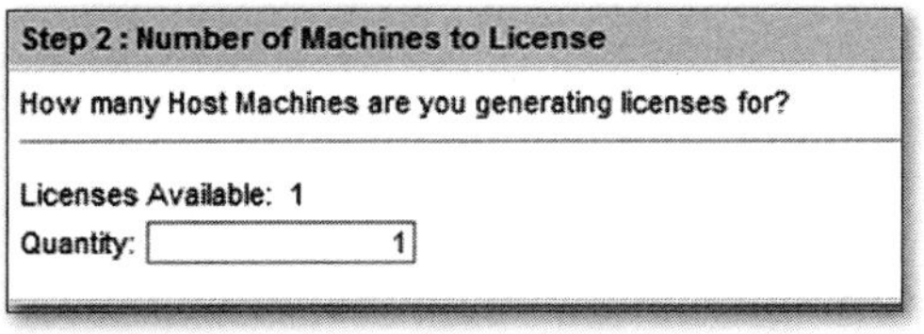

Figure 3.10 Specifying the Number of Machines

For the next step, you need to determine the 12-digit physical address for the server upon which the Unwired Server will be installed. To obtain this, you need to do the following:

1. Log on to the server.
2. Open a command prompt.
3. Enter the command `ipconfig/all`.

The physical address is the six pairs of digits separated by dashes, as shown in Figure 3.11.

Enter the System ID (SID) for the server along with this physical address (minus the dashes), as shown in Figure 3.12.

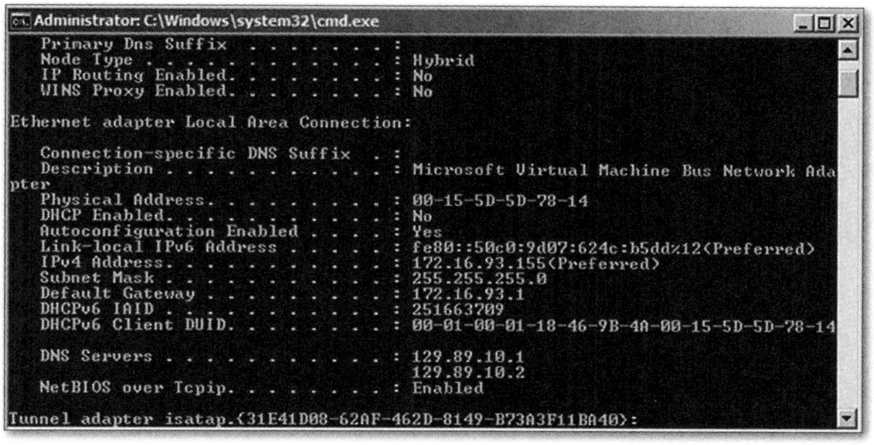

Figure 3.11 Determining the System's Physical Address

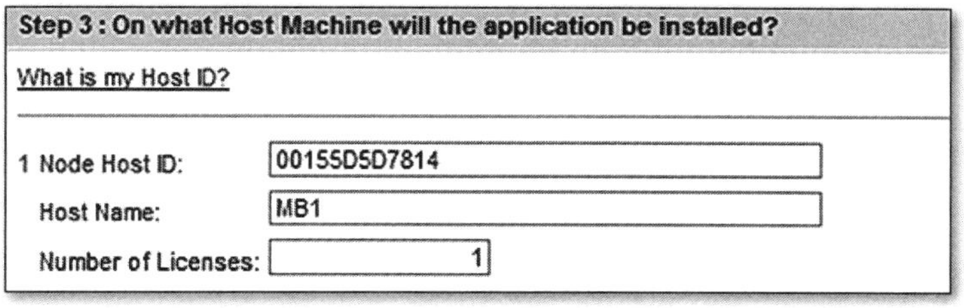

Figure 3.12 Entering the Host ID and Name

Click the Download License File button, as shown in Figure 3.13.

Figure 3.13 Downloading the License File

3.3.3 Downloading Installation Files

The installation files are also downloaded from the SAP Service Marketplace. It is important that you select the Sybase Unwired Platform Runtime 64-bit server DVD and download these files to the new Unwired Server machine.

3.3.4 Installation of the Unwired Platform Runtime

There is a series of requirements that must be satisfied before you can begin the installation:

- You need to download the license file on this server.
- You need to download and expand the installation files on this server.
- You must have administrative rights.
- You must disable WINDOWS USER ACCOUNT CONTROL (UAC) prior to starting the installation. To do this, search for the term UAC in the WINDOWS START menu and then click CHANGE USER ACCOUNT CONTROL settings when the results appear. Pull the slider all the way down to NEVER NOTIFY, and click OK.
- Disable the built-in Windows Firewall.
- Install the latest ADOBE FLASH PLAYER used by SCC.
- Restart the server.

Once the server reboots, start the installation by clicking the *setup.exe* file in the installation source directory, as shown in Figure 3.14.

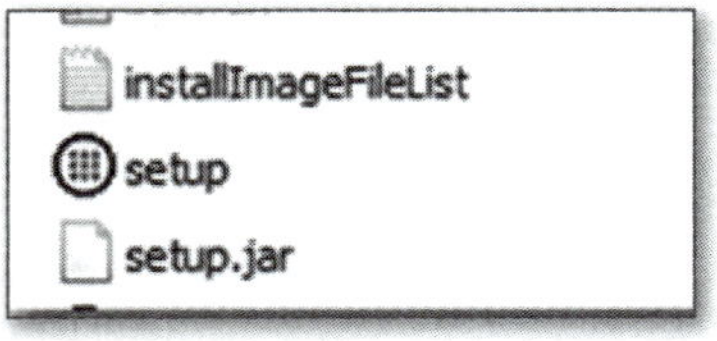

Figure 3.14 Clicking setup.exe to Start Installation

Select UNSERVED (LOCAL) LICENSE and click NEXT, as showed in Figure 3.15.

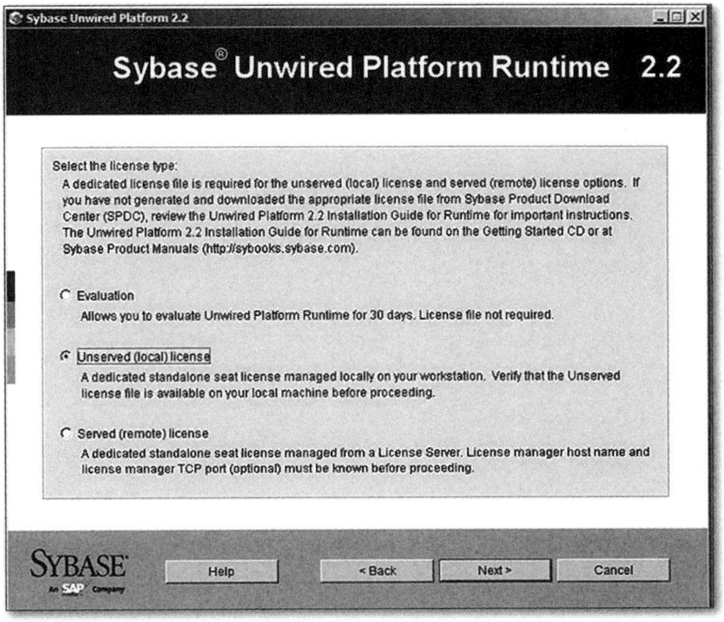

Figure 3.15 Selecting Unserved (Local) License

Select Enterprise Server for License Product Edition, select Development and Test License for License Type, and enter 0 for an unlimited license, as shown in Figure 3.16.

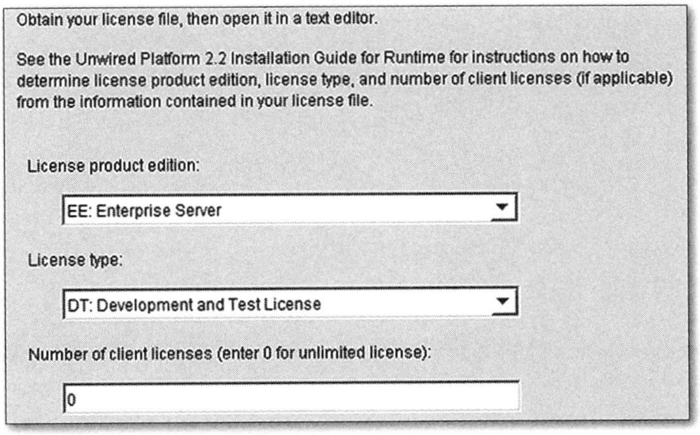

Figure 3.16 Specifying License Information

Browse to the location of the license you generated earlier, as shown in Figure 3.17.

Figure 3.17 Browsing for the License File

Select SINGLE SERVER as the installation type, as shown in Figure 3.18.

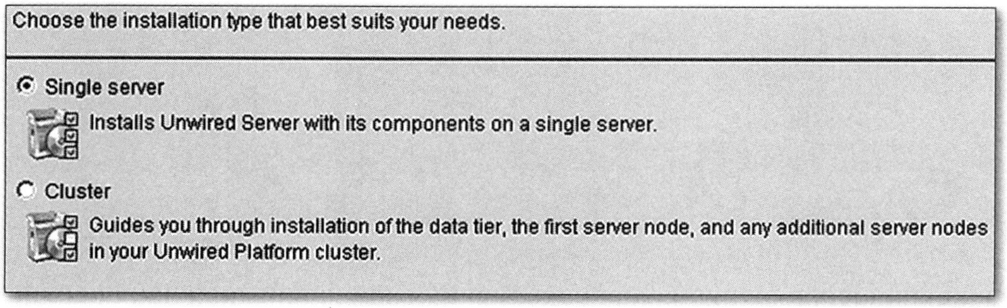

Figure 3.18 Selecting Single Server

If the message shown in Figure 3.19 appears, simply click NEXT.

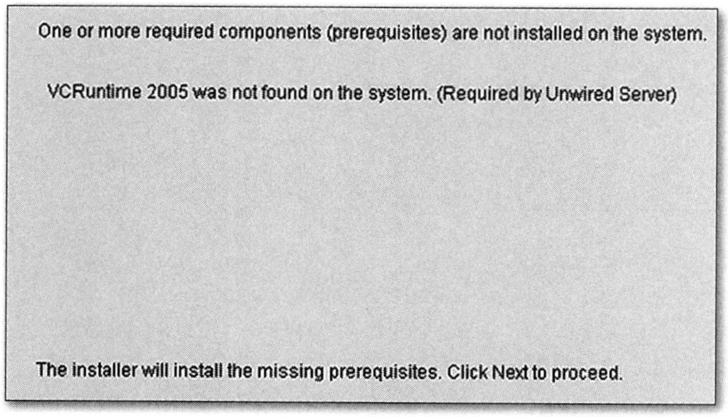

Figure 3.19 Ignoring the Warning

Check the CONFIGURE UNWIRED PLATFORM COMMUNICATIONS PORTS and SET SYBASE UNWIRED PLATFORM SERVICES TO START AUTOMATICALLY, as shown in Figure 3.20.

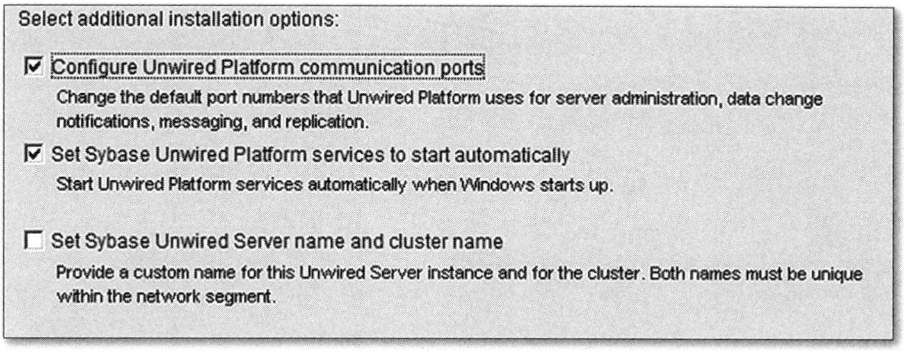

Figure 3.20 Selecting Installation Options

The next screen, as shown in Figure 3.21, presents a series of default ports to be used by the Unwired Server and provides the opportunity to change any of those ports. The default port for Server Administration is 2000. This port plays a major role in the communications between the Unwired Server and the Unwired Workspace. Most organizations tend to block communications on all ports below 4000. While it may be easy to open port 200 at the server location, it is not as easy to open port 2000 for every location where a developer may be using the Unwired Workspace. The solution is to change the server administration port to, for example, 4500. This change is also required when you configure the Unwired Platform, as discussed in the next chapter.

More on Ports

Various components of SUP use a number of different port numbers to communicate among themselves. The complete list of those ports used is shown in Table 3.4. Experience has indicated that opening these ports on the server firewall leads to a more successful test of the system.

Device	Type	Default	Secure
Unwired Server	Administration	2000 (4500)	2001
	HTTP Listener	5001	N/A
	HTTP Listener	8000	8001
	Message services admin	5100	N/A
	Replication	2480	2481

Table 3.4 Ports Used by Various SUP Components

Device	Type	Default	Secure
Data Tier	CDB	5200	N/A
	Cluster DB	5300	N/A
	Monitor DB	5400	N/A
	Domain log DB	5400	N/A
SCC	RMI service	9999	N/A
	Message service	2100	N/A
	SCC repository DB	3638	N/A
	Web container	8282	8283
Sample Database	Database	5500	N/A

Table 3.4 Ports Used by Various SUP Components (Cont.)

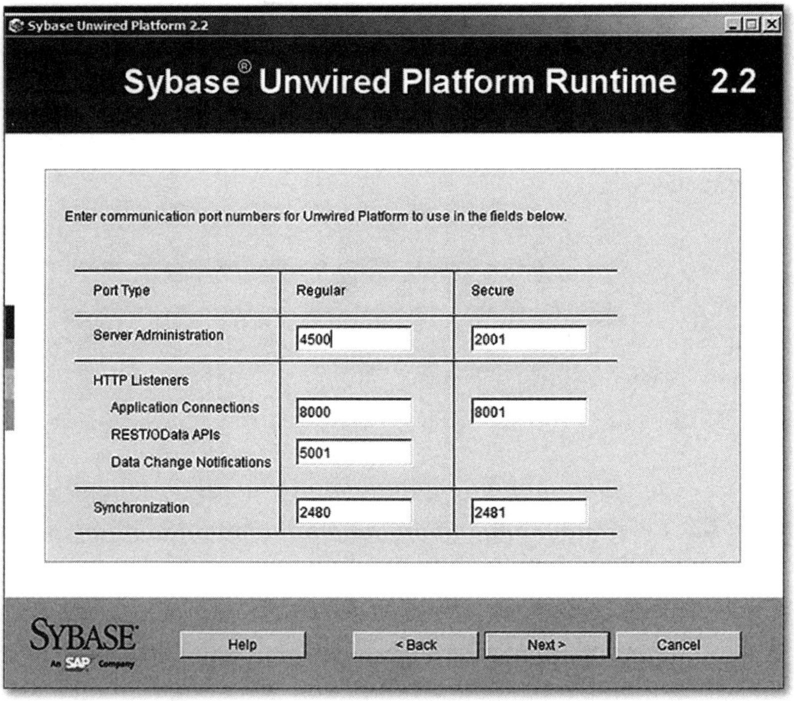

Figure 3.21 Entering Communication Ports for the Unwired Server

The next screen asks you to create an SUPADMIN PASSWORD. As shown in Figure 3.22, enter this value twice.

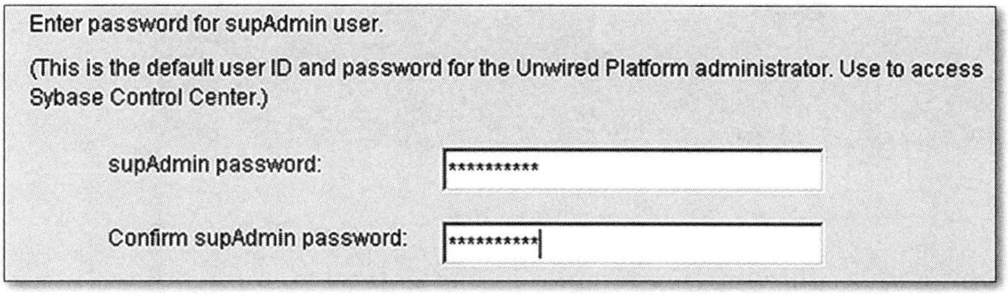

Figure 3.22 Entering the supAdmin Password

Verify that all the information is correct, and click INSTALL, as shown in Figure 3.23.

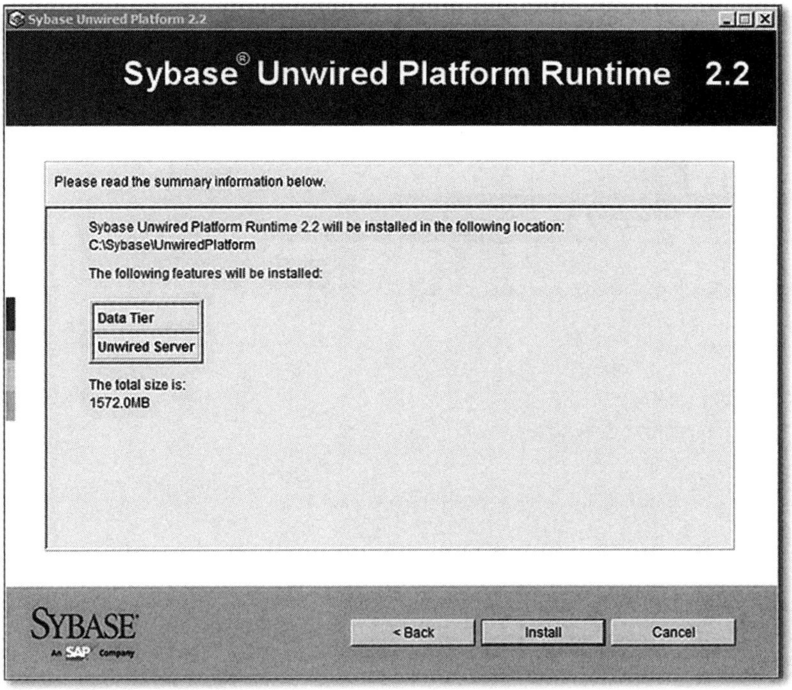

Figure 3.23 Clicking Install

The installation takes about 10 minutes and generates the screen shown in Figure 3.24 when complete.

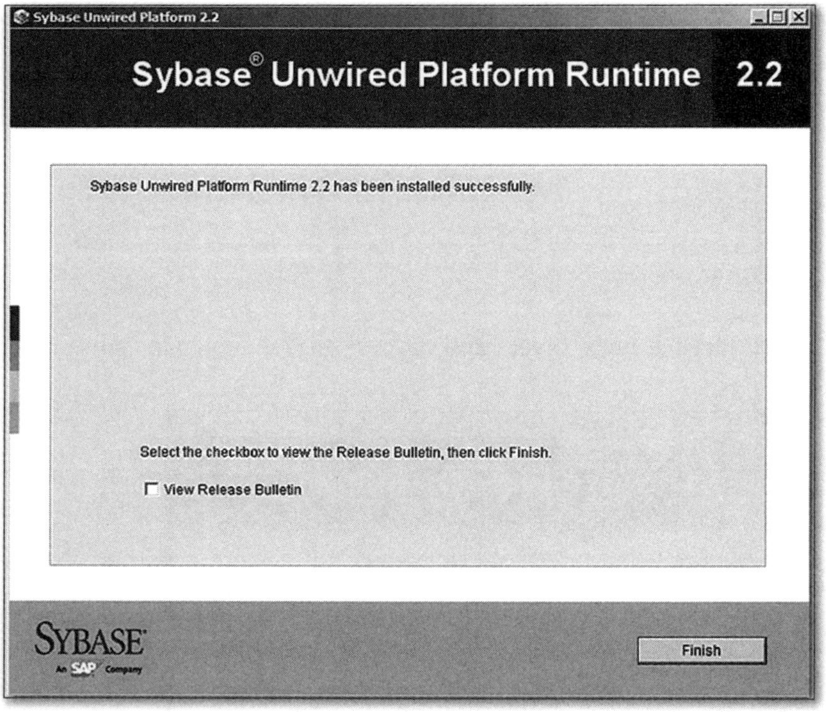

Figure 3.24 Successful Installation

3.4 Sybase Control Center

The Sybase Control Center (SCC) is a web interface that provides a console to configure and manage the Unwired Platform. The SCC is created as part of the standard Unwired Platform installation. The user interface for the SCC is shown in Figure 3.25.

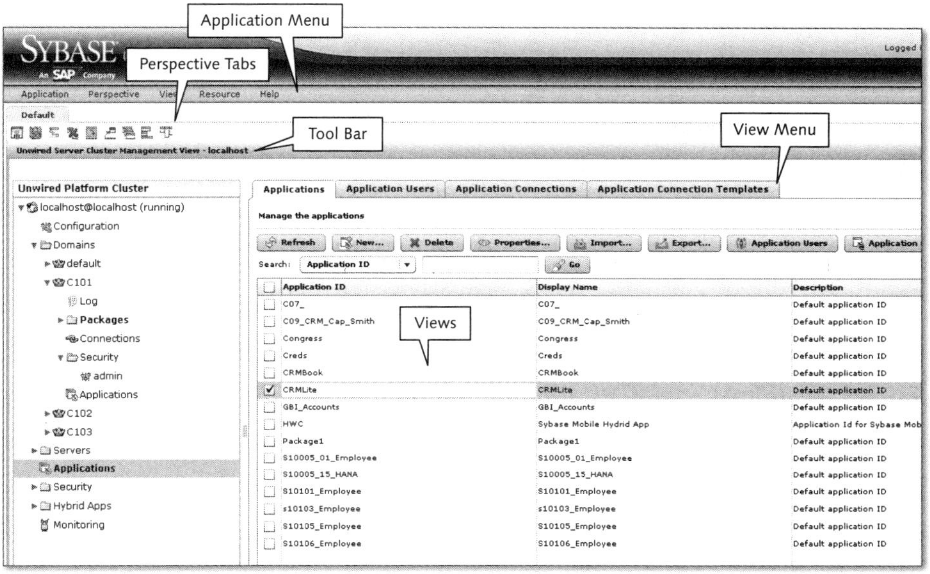

Figure 3.25 SCC User Interface

When the ADMINISTRATIVE application is selected, the left navigational pane in the SCC console displays a tree of administrative functions in the form nodes. Clicking these nodes allows you to administer the functions display shown in Table 3.5. You can create administrative users who have one or more of these administrative functions. For example, the SCC shown in Figure 3.26 has access to only DOMAINS, SERVERS, SECURITY, and HYBRID APP.

Node	Purposes
CLUSTER	View general cluster and server node properties
DOMAINS	Add, delete, enable, and disable domains
SERVERS	Configure JVM properties of each server
APPLICATIONS	Add, view, delete, and edit apps, app users, and app connections
SECURITY	Add, view, edit, and delete domain administrators
HYBRID APP	Deploy and manage hybrid apps
MONITORING	Create and manage settngs for monitoring security

Table 3.5 Administrative Nodes

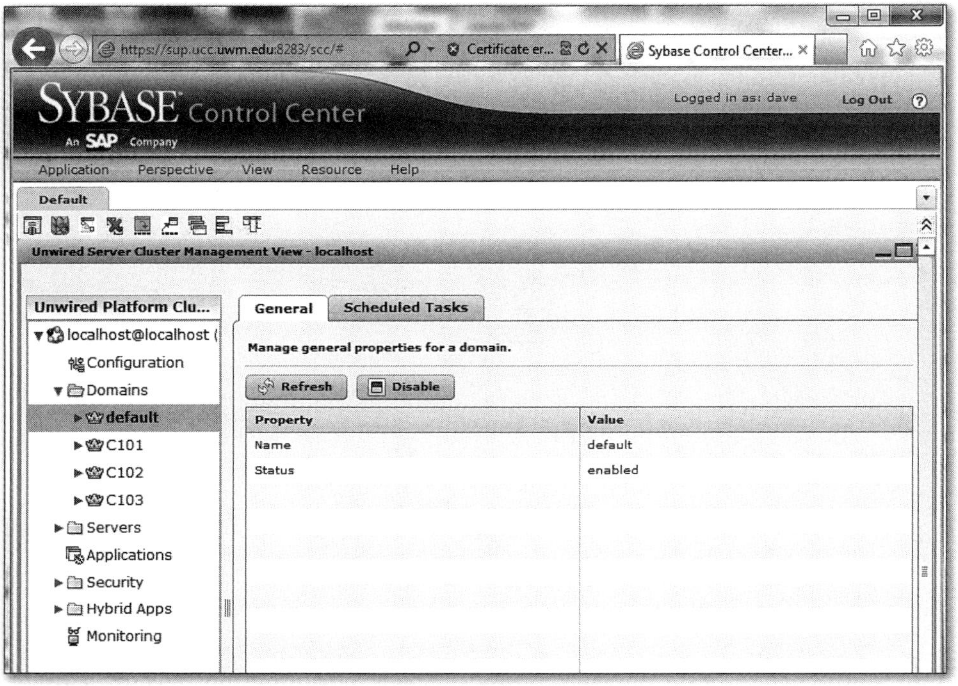

Figure 3.26 Administrator with Restricted Functions

In this section, we show two examples of how the SCC is used to manage the Unwired Platform.

3.4.1 Creating a New Administrative User

To create a new administrative user, click the security node and select Login, as shown in Figure 3.27.

Click Create Login and enter the Login Name (Figure 3.28), and then click Next.

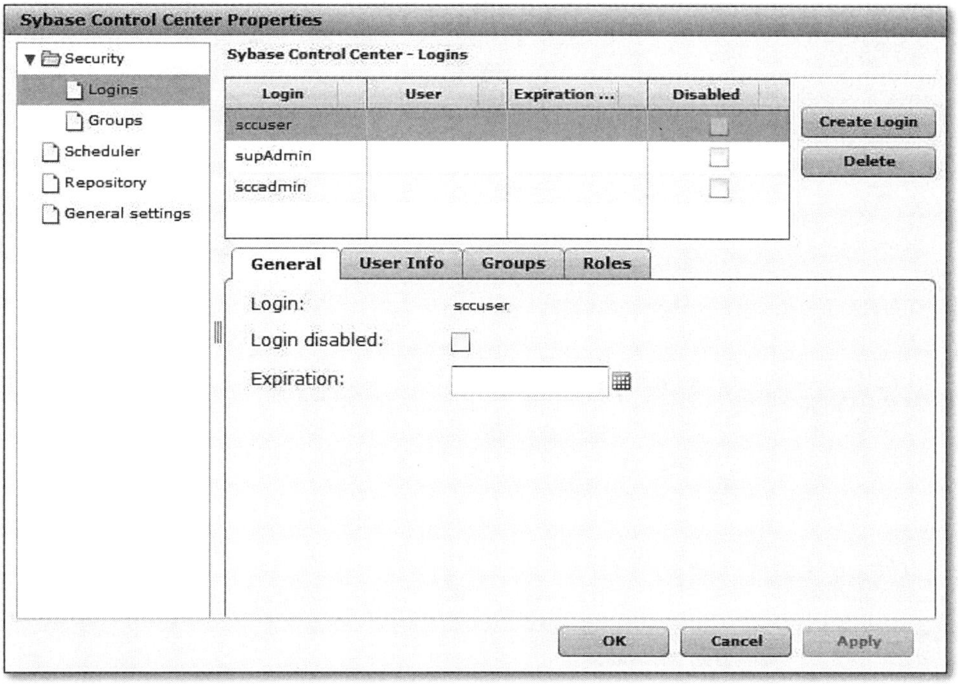

Figure 3.27 Selecting Login under Security

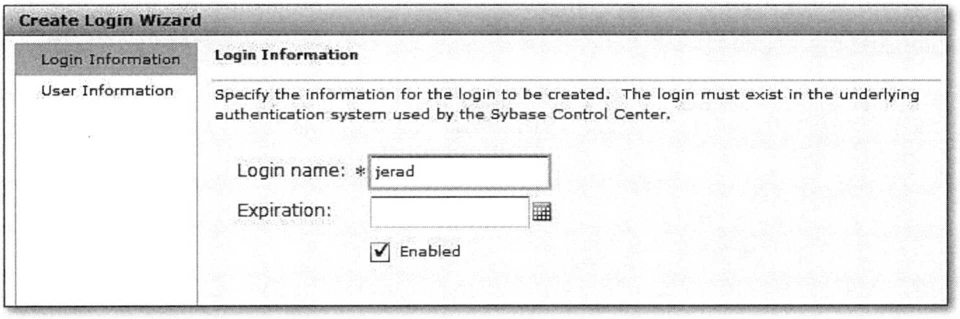

Figure 3.28 Entering the Login Name

Enter the additional user information and then click FINISH, as shown in Figure 3.29.

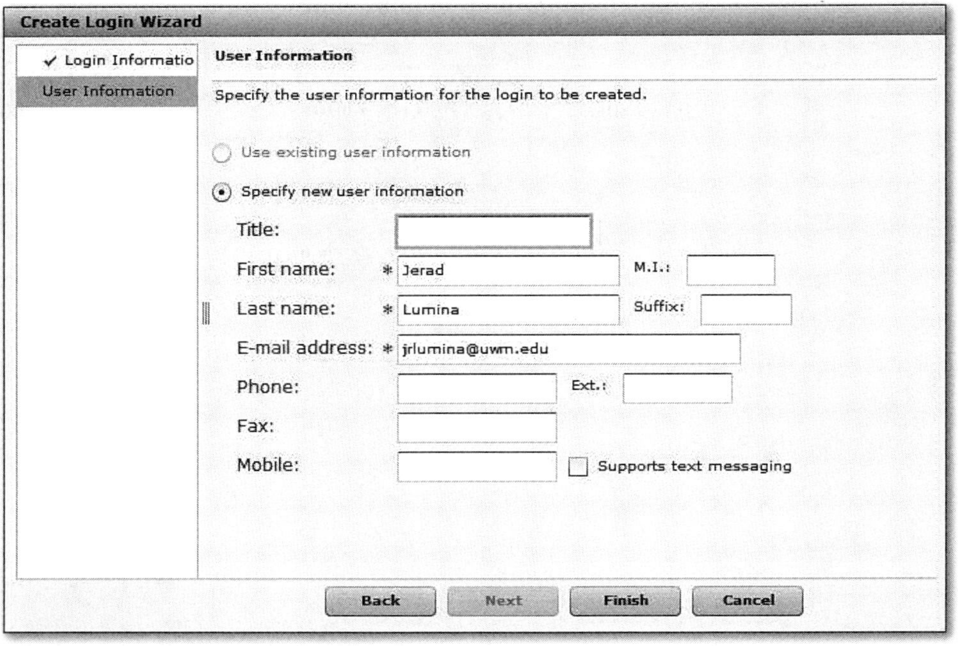

Figure 3.29 Entering User Information

To assign a password, you must create an AUTHENTICATION PROFILE for this user. Select the SECURITY node and then the ADMIN sub-node. Select the AUTHENTICATION tab, and then click NEW, as shown in Figure 3.30.

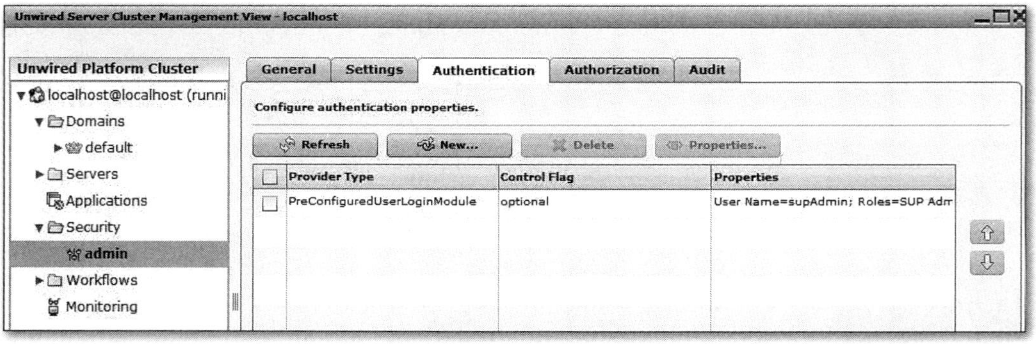

Figure 3.30 Creating a New Authentication Profile

Enter the following values, as shown in Figure 3.31:

- Implementation Class: *com.sybase.security.core.PreConfiguredUserLoginModule*
- Roles: SUP Administrator
- User Name: <new user>
- Password: <new password>

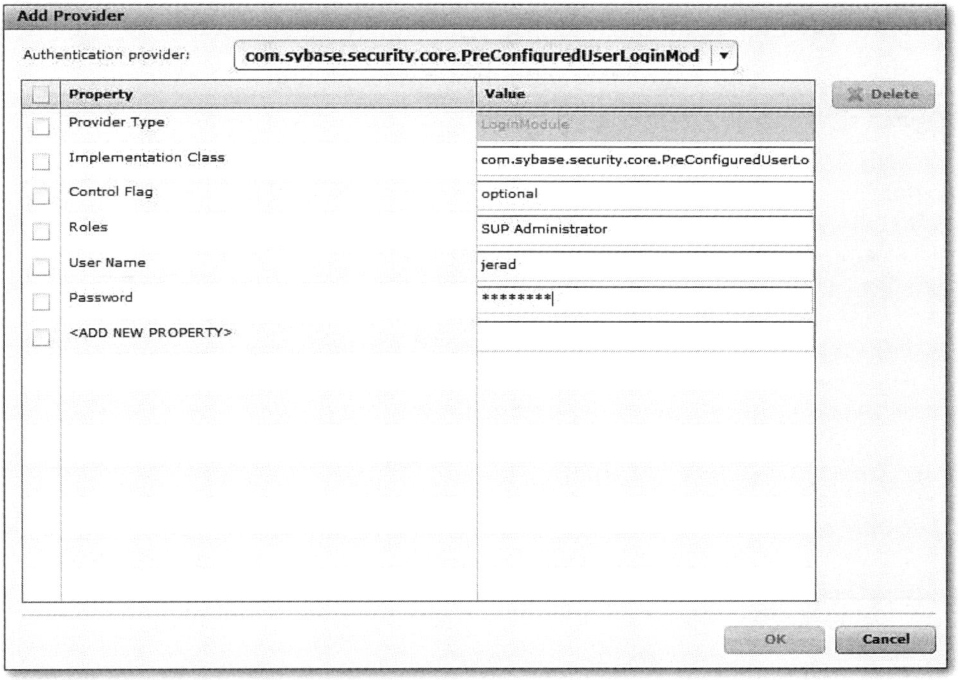

Figure 3.31 Providing Authentication Information

An information window appears; click Yes. Select the General tab, and click the Apply button. Log out and test your new login.

3.4.2 Creating a New Domain

Select Domains and click New, as shown in Figure 3.32.

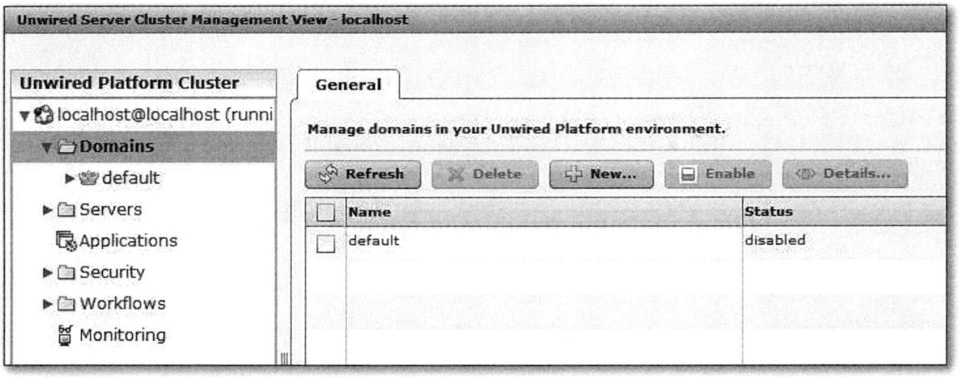

Figure 3.32 Creating a New Domain

Enter the new domain name and click NEXT, as shown in Figure 3.33.

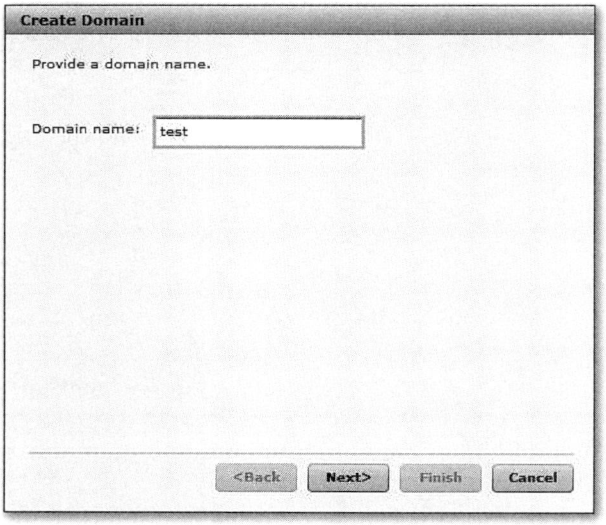

Figure 3.33 Entering the Domain Name

Assign the DOMAINADMIN security configuration, and click NEXT, as shown in Figure 3.34.

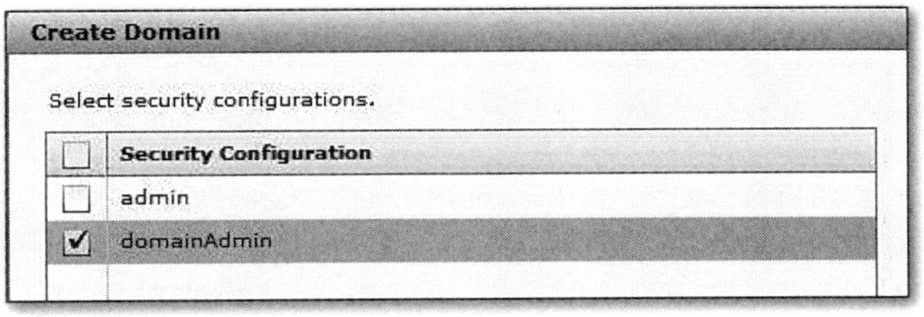

Figure 3.34 Assign Security Configuration

3.5 Summary

This chapter presented the installation of the Unwired Server Platform Runtime. This included a discussion of the server landscapes and the license options associated with them. A typical landscape that focuses on development of mobile apps and downloading of a license was selected. The installation of the Unwired Platform was then presented. A description and an example of using the SCC were also presented.

The Sybase Mobile SDK is the development platform for container apps. This chapter describes the process for installing the Sybase Mobile SDK and configuring connections to the Unwired Server and various data sources.

4 Installing and Connecting the Sybase Mobile SDK

This chapter describes the installation and connection of the Unwired Workspace, which is the development platform for SAP Mobile Platform. The Unwired Workspace and a set of libraries are delivered as a single package referred to as the Sybase Mobile SDK. The Unwired Workspace is based on the integrated development environment Eclipse, which is extensible, open-source software that supports multiple development environments through the use of *perspectives*. A perspective consists of a number of views that provide the editors and other tools required for a given task. The Unwired Workspace includes the MOBILE DEVELOPMENT perspective, which is used to develop hybrid apps.

This chapter provides the information required to install the Sybase Mobile SDK and configure connections to the Unwired Server and various data sources. If you want to experience SAP Mobile Platform but do not have access to the software, the last section describes how you can access a personal instance of the platform on Amazon Web Services.

4.1 Getting Started

Before you start your installation, there are a few things you should know. In this section, we tell you what they are.

4.1.1 Hardware and Software Requirements

The only operating system supported by the Unwired Workspace is Microsoft Windows. The supported versions are shown in Table 4.1. The install requires 1.4 GB of disk space, and 4 GB of RAM is recommended.

Operating System	Service Pack	CPU
Microsoft Windows XP: Professional	SP2, SP3	32-bit only
Microsoft Windows Vista: Business, Enterprise, or Ultimate	SP2	32- and 64-bit
Microsoft Windows 7: Professional, Enterprise, or Ultimate	SP1	32- and 64-bit
Microsoft Windows Server 2003	SP2	64-bit
Microsoft Windows Server 2008	SP2	64-bit
Microsoft Windows Server 2008R2	SP1	64-bit

Table 4.1 Operating System Support for the Unwired Workspace

4.1.2 Backward Compatibility

Beginning with version 2.1.3, installing a new version of the Unwired Workspace does not conflict with previously installed versions. The installation folder of versions prior to 2.1.3 was named MOBILESDK, but beginning with versions 2.1.3, the installation folder name includes the version number. For example, the installation folder for version 2.2 is MOBILESDK22. This allows multiple versions to exist in parallel. You must use the Windows Control Panel to remove a Sybase Mobile SDK installation.

Eclipse stores project files in a workspace folder. If you open a new version of the Unwired Workspace using a workspace that contains projects from an older version, the Unwired Workspace automatically converts open projects to the new version. To be safe, you should use a new workspace folder with each version. You can then manually import the projects you want to convert individually.

4.2 Steps for Installation and Connection

Before beginning the installation, make sure the JAVA_TOOLS_OPTIONS environment variable is not set on the system. In Windows 7, right-click My Computer and select Properties. Select Advanced System Settings, and then click the Environment Variables button. If JAVA_TOOLS_OPTIONS appears in either the User Variables or the System Variables, select and delete it.

This section describes the process used to install the Unwired Workspace and to connect to an Unwired Runtime server, as well as various data sources.

4.2.1 Installing the Unwired Workspace

If you are installing from a network share, you must use a mapped drive, not a UNC path. To begin the installation, locate and execute the *setup.exe* application. Click OK to acknowledge the message about User Access Control (UAC) and then click Next on the welcome screen (Figure 4.1).

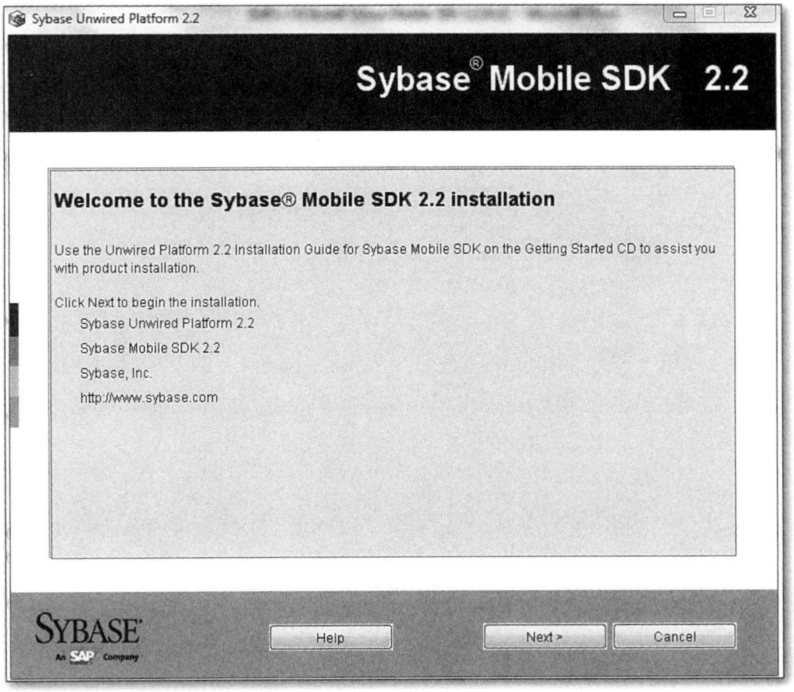

Figure 4.1 Installation Welcome Screen

The screen shown in Figure 4.2 requires you to choose the country-specific end user license agreement. The choice of country has no impact on the appearance or function of the Unwired Workspace. Select your country, click to agree to the terms of the agreement, and then click NEXT.

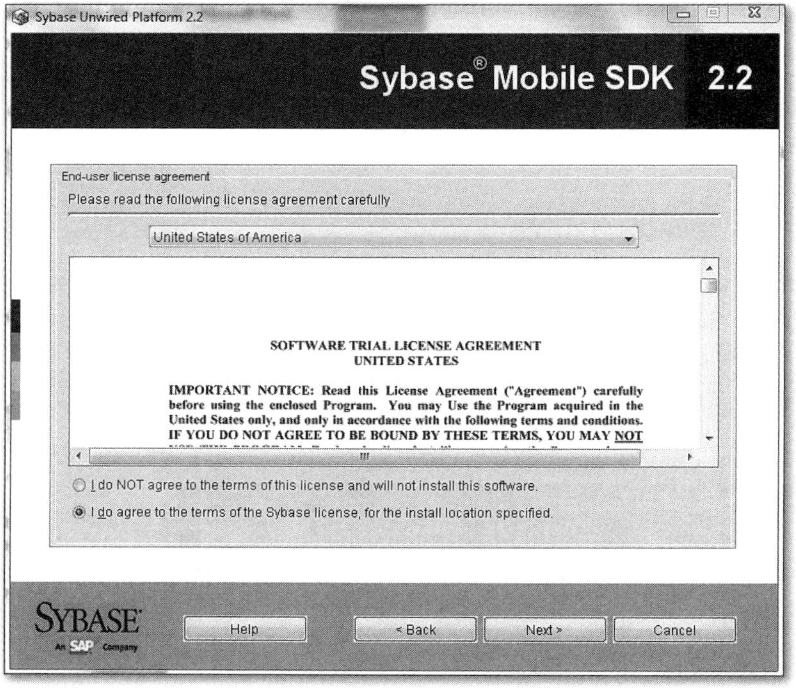

Figure 4.2 Country-Specific License Agreement

On the screen shown in Figure 4.3, enter the path to the installation folder. The path must be no more than 30 characters and can contain only ASCII alphanumeric characters, underscores, and hyphens. Click NEXT and then INSTALL, and the installation proceeds.

Warning

The Eclipse components included with this installation have been tested together and work well together. Even though they may not be the latest versions, SAP suggests that you not use the Eclipse update site to update any Eclipse plug-ins or contributed plug-ins like GMP, GEF, DTPs, etc.

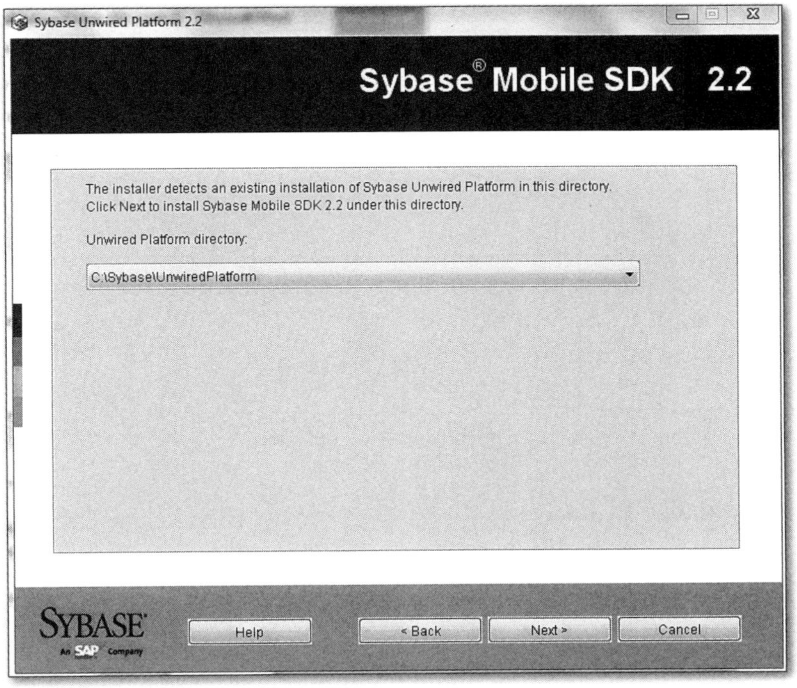

Figure 4.3 Choose the Installation Path.

4.2.2 Connecting the Unwired Workspace

This section describes how to connect to the Unwired Runtime server and to data sources. Use either the desktop icon or the Windows menu path START • ALL PROGRAMS • SYBASE • UNWIRED PLATFORM • UNWIRED WORKSPACE <VERSION> to start the Unwired Workspace. The first time you start the program, you see the WELCOME view. Close the window to see the MOBILE DEVELOPMENT perspective. (You can reopen the WELCOME view by using the menu path HELP • WELCOME.)

In the MOBILE DEVELOPMENT perspective, connections are managed using *connection profiles*, which you can see in the ENTERPRISE EXPLORER view on the lower-left corner of the Unwired Workspace. When you first install the Unwired Workspace, the Enterprise Explorer appears as it does in Figure 4.4. Sample connection profiles to the sample database and the Unwired Server are created, but the connection parameters must be edited before they can be used.

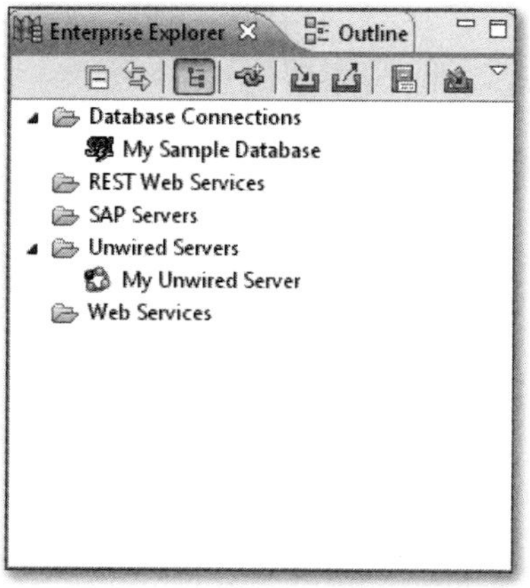

Figure 4.4 Enterprise Explorer View before Connections Are Configured

Connecting to the Unwired Server

In the Enterprise Explorer, expand the Unwired Servers node, right-click My Unwired Server, and click Properties. If you want the connection to the Unwired Server to open automatically each time you start the Unwired Workspace, select Connect every time the workbench is started.

In the left panel, click Unwired Server Connection Properties. The following fields can be configured, as shown in Figure 4.5:

- Host
  The host name on which server resides. By default, this is the local machine name.
- IP domain
  The IP domain name of the server. Leave this field blank if you entered a fully qualified address in the host field.
- Port
  The port on the Unwired Server to which you are connecting. By default, this is 2000 for HTTP and 2001 for HTTPS.
- User Name
  The user name you use to connect to the Unwired Server. This is "supAdmin" by default.
- Password
  The password on the Unwired Server associated with the user name. The default is "s3pAdmin."

- Save Password
  Select this if you do not want to enter the password each time you connect to the Unwired Server.
- Secure
  If you have enabled the secure management port on the Unwired Server, checking this box enables secure connections to the server.
- Cluster information
  These are read-only properties of the Unwired Server.

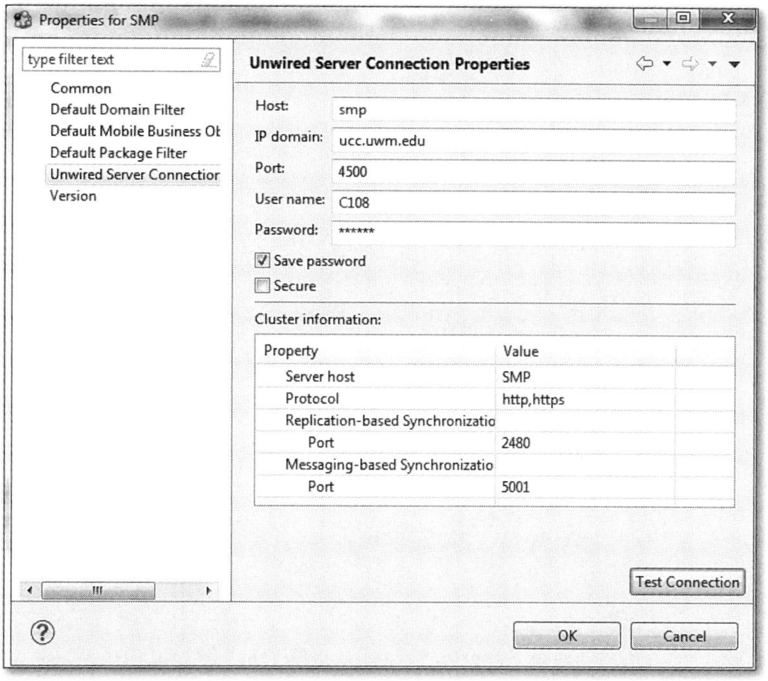

Figure 4.5 Unwired Server Connection Properties

Click Test Connection. If the connection information is correct, you see a dialog that indicates that the ping succeeded. If the ping failed, recheck your connection parameters. A common problem is that the port number is incorrect or that a firewall on your machine or in your organization's network is blocking the port.

If the ping succeeds, click OK to close the connection dialog. To connect to the Unwired Server, right-click My Unwired Server in the Enterprise Explorer, and select Connect. Once the connection is established, the two folders shown in

Figure 4.6 appear. Drilling down into these folders allows you to view and delete packages and hybrid applications deployed on the Unwired Server.

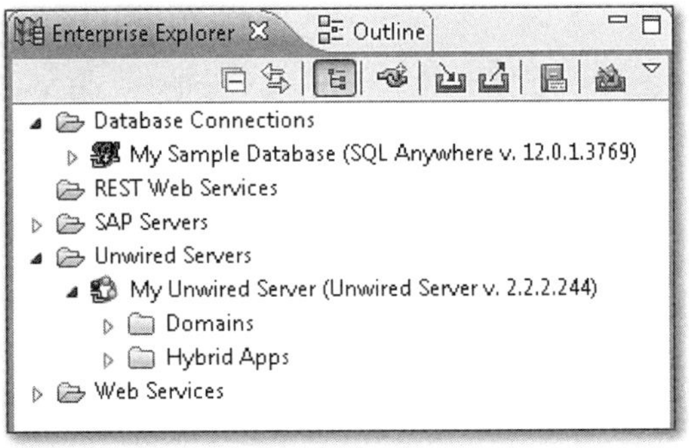

Figure 4.6 The Domains and Hybrid App Folders

Connecting to the Sample Database

In the Enterprise Explorer, expand the Database Connections node, right-click My Sample Database, and select Properties. If you want to connect to the database automatically each time you start the Unwired Workspace, select Connect every time the workbench is started.

In the left pane, select Sybase ASA Connection Properties. The following fields can be configured, as shown in Figure 4.7:

- Drivers
  Choose the driver for the version of the Sybase Adaptive Server to which you are connecting.
- Host
  The host name of the machine where the Database Server is installed. This is "localhost" by default.
- Port
  The port used to connect to the Database Server. This is "5500" by default.
- Database name
  The database to which you wish to connect. The sample database is called "sampledb."
- User name
  The user name you use to connect to the Database Server. This is the database administrator, "dba", by default.

- PASSWORD
  The password associated with the user name used to connect to the Database Server.
- SAVE PASSWORD
  Select this option if you do not wish to enter the password each time you connect to the Database Server.

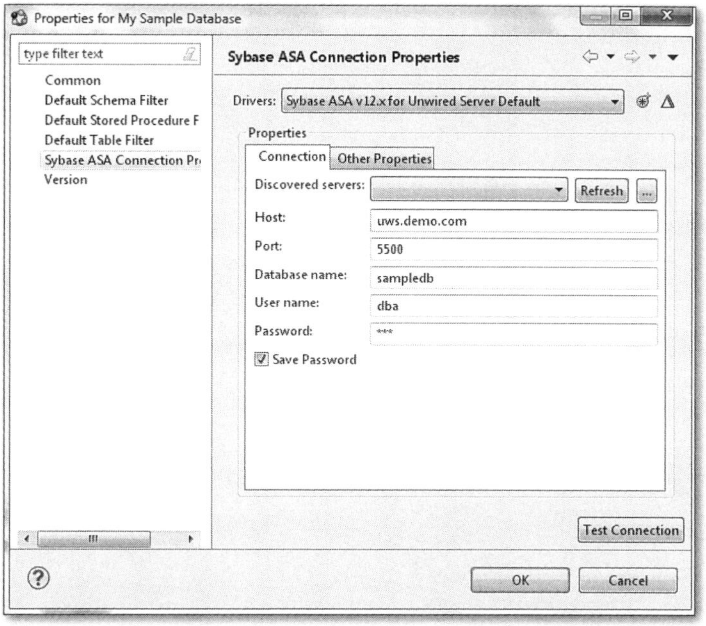

Figure 4.7 Connection Properties for the Sample Database

Click TEST CONNECTION to confirm that the information is correct. If the connection information is correct, you see a dialog indicating that the ping succeeded. If the ping fails, recheck your connection parameters. A common problem is that the port number is incorrect or that a firewall on your machine or in your organization's network is blocking the port.

If the ping succeeds, click OK to close the connection dialog. To connect to the sample database, right-click the connection profile in the Enterprise Explorer, and select CONNECT.

As shown in Figure 4.8, you can drill down into the database in the Enterprise Explorer to view and edit various objects in the database. To edit data in a table, right-click the table name and select Edit Data. The Table Data Filter dialog shown in Figure 4.9 allows you to select which data is retrieved from the database.

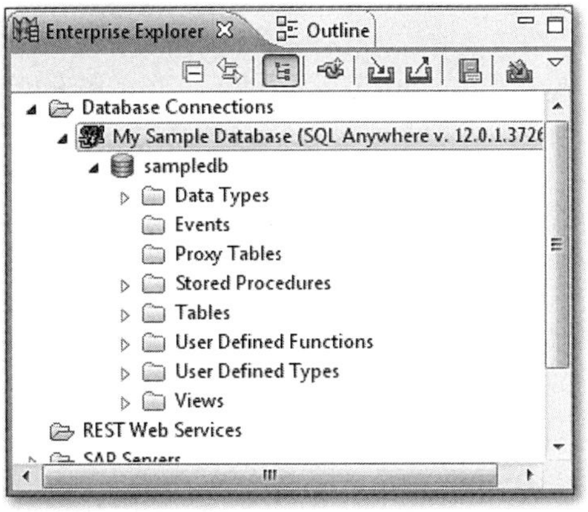

Figure 4.8 Viewing and Editing Details of a Database from within the Unwired Workspace

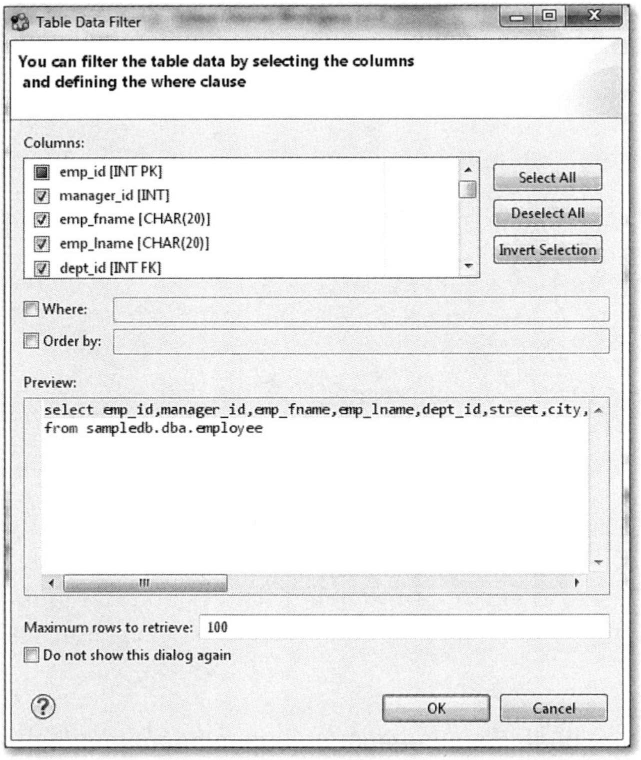

Figure 4.9 Editing the SQL Used to Retrieve the Data from the Database

Creating a New Database Connection

The Unwired Workspace can connect to the following databases:

- DB2 on Linux, Unix, and Windows
- DB2 for i5/OS
- DB2 for z/OS
- Oracle
- SQL Server
- Sybase ASA
- Sybase ASE

You must install the correct drivers and JAR files before you can create connection profiles to SQL Server, DB2, or Oracle databases. Table 4.2 shows where to obtain the drivers.

JDBC driver	URL
Oracle	*http://www.oracle.com/technology/software/tech/java/sqlj_jdbc/index.html*
DB2	*http://www-306.ibm.com/software/data/db2/express/download.html*
SQL Server	*http://msdn.microsoft.com/en-us/data/aa937724.aspx*

Table 4.2 Locations for Downloading the JDBC Drivers for Supported Databases.

Copy the JDBC JAR files into the following folders:

- **On the machine with the Sybase Mobile SDK**
  <Unwired Workspace installation root r>\Unwired_Workspace\Eclipse\sybase_workspace\mobile\eclipse\plugins\com.sybase.uep.com.<database>_<version>.<plugin version number>\lib. The placeholder for *<database>* in the path will be Oracle, DB2, or SQL Server.
- **On the machine with the Unwired Server**
  <Unwired Server installation root>\Servers\UnwiredServer\lib\3rdparty

Once the files are copied, you must restart the Unwired Workspace and Unwired Platform Services.

To create a new database connection profile in the Unwired Workspace, right-click DATABASE CONNECTIONS in the Enterprise Explorer and select NEW. Select a CONNECTION PROFILE TYPE, enter a name for the connection and an optional description, and then click NEXT. On the next screen, click the NEW DRIVER DEFINITION icon shown in Figure 4.10.

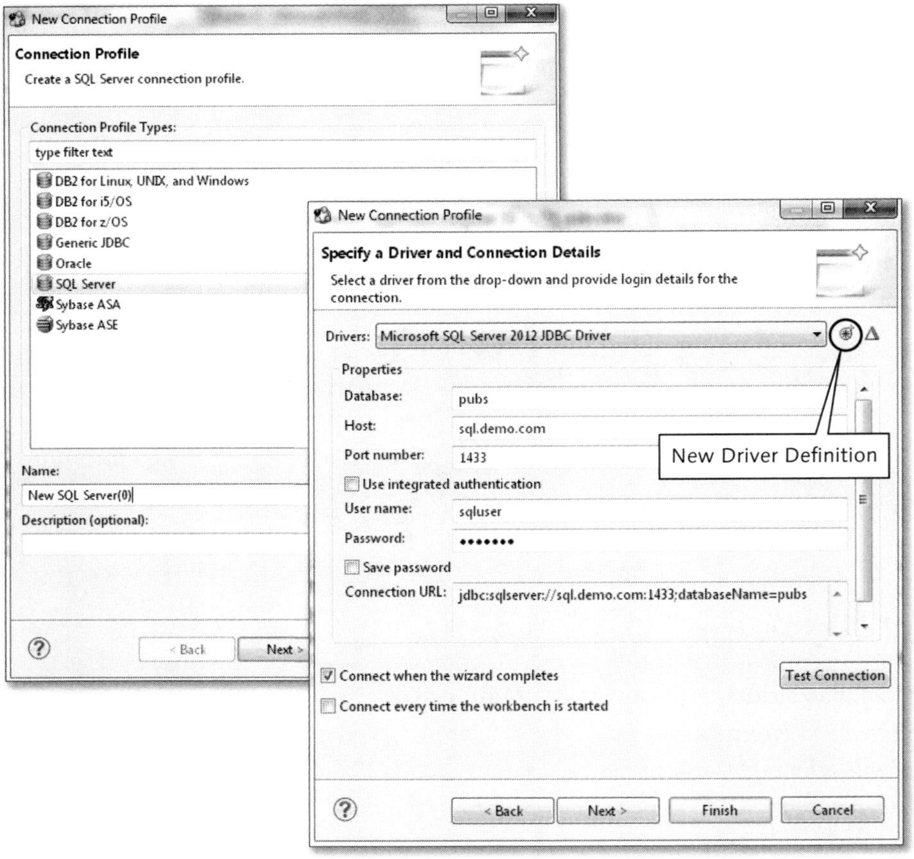

Figure 4.10 Creating a New Connection Profile Using a Non-Sybase Database.

On the NEW DRIVER DEFINITION dialog, select a profile and enter a name. Select the JAR LIST tab. If an entry already exists, delete it and add an entry that points to the JAR file you copied in the previous step. You can use the PROPERTIES VIEW to edit the connection properties or edit those properties on the NEW CONNECTION PROFILE dialog (Figure 4.10 and Figure 4.11).

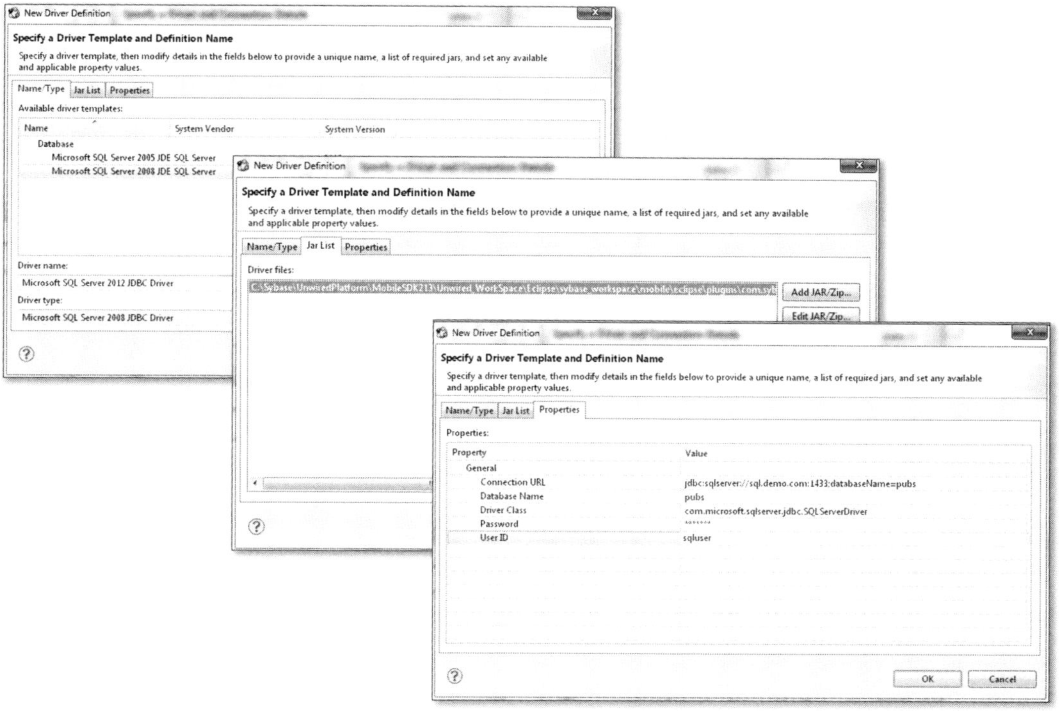

Figure 4.11 Defining the Properties of the New Database Driver Definition

The integration with non-Sybase databases is not as tight as it is for Sybase databases. As a result, some of the functionality is limited. For example, it is not possible to view or edit data from a non-Sybase database from within the Unwired Workspace. Also, when you create MBOs, you see a dialog with a warning that the Unwired Workspace cannot guarantee valid SQL syntax for non-Sybase databases and that manual editing of the SQL may be required.

Connecting to an SAP Server

To create a connection to an SAP server on a Windows 7, Windows Vista, or Windows Server 2008 machine, the Microsoft *msvcr71.dll* and *msvcp71.dll* files must be installed in either the <WINDOWS_HOME>\SYSTEM32 (32-bit) or <WINDOWS_HOME\SYSWOW64 (64-bit) folder, depending on the version of the operating system you have.

Right-click the SAP SERVERS node in the Enterprise Explorer and select NEW. In the dialog that appears, enter a name and an optional description and click NEXT.

On the next screen, either enter the connection information manually or enter the location of a properties file that contains the information. Property files are text files with a .properties extension. The following code shows an example of the contents of a properties file:

```
jco.client.client=758
jco.client.user=sapuser
jco.client.passwd=password
jco.client.ashost=sap.demo.com
jco.client.sysnr=00
```

Enter the absolute path to the properties file or use the BROWSE button. If you use a properties file, the connection properties are assigned according to the values in the file, even though they do not populate the fields in the dialog box. Alternatively, you can manually enter the values in the fields, as in Figure 4.12.

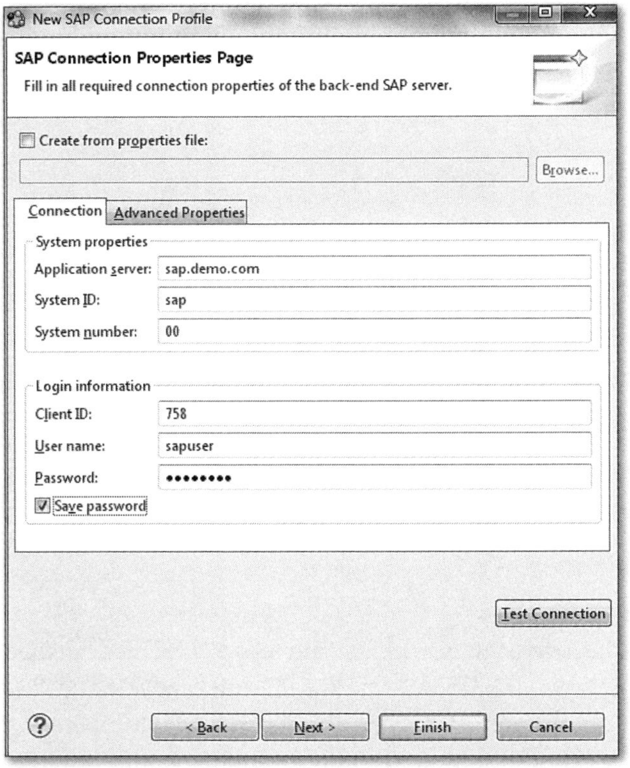

Figure 4.12 Entering the Connection Properties for an SAP Connection Profile

Click TEST CONNECTION to confirm that the information is correct. If the connection information is correct, you see a dialog indicating that the ping succeeded. If the ping fails, recheck your connection parameters. Close the connection dialog. To connect to the SAP server, right-click the connection in the Enterprise Explorer and select CONNECT.

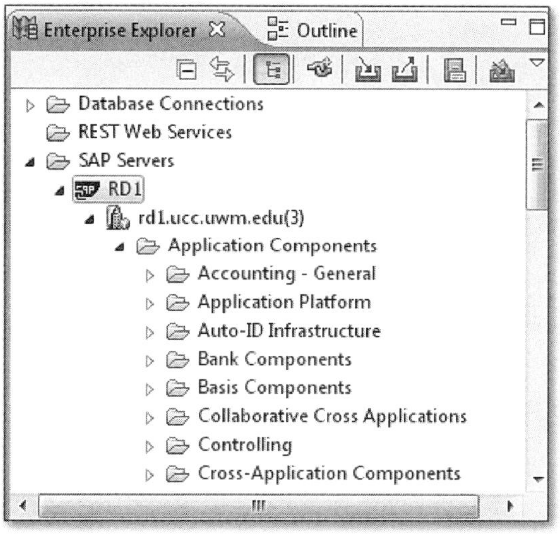

Figure 4.13 Browsing the Application Components from within the Enterprise Explorer

Connecting to a Web Service

To connect to a web service, right-click WEB SERVICES in the Enterprise Explorer and click NEW. After entering a name and clicking NEXT, enter the location of the WSDL file for the web service. The file can be accessed from a local drive, from within the project, or from a URL (see Figure 4.14).

As before, you can test the connection using the TEST CONNECTION button. To connect to the web service, right-click it and select CONNECT (Figure 4.15).

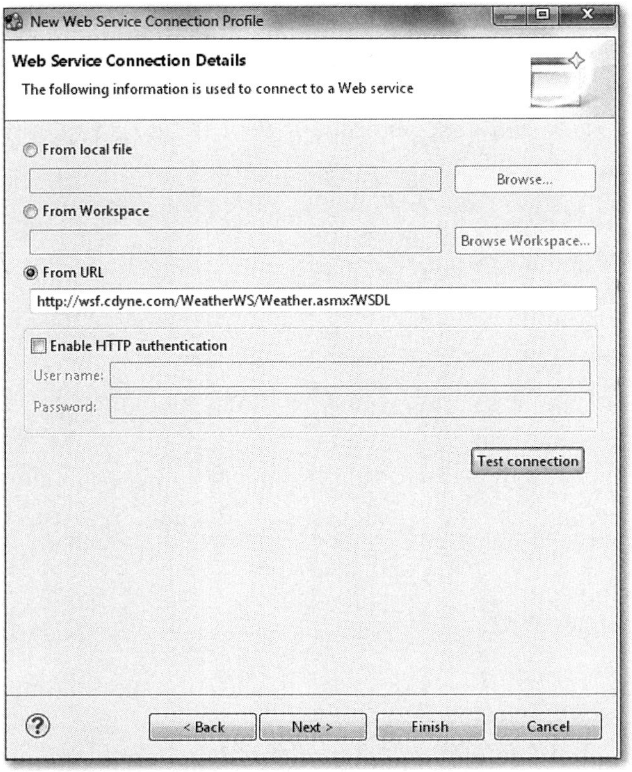

Figure 4.14 Entering the Location of the WSDL File to Connect to a Web Service

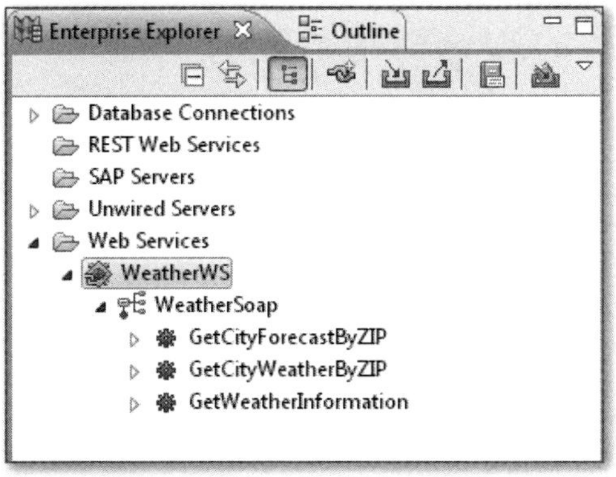

Figure 4.15 Web Service Connection

4.3 Installing and Connecting a Cloud-Based Sybase Unwired Platform

A developer version of the Sybase Unwired Platform is available from Amazon Web Services (AWS). You have access to all the features of SUP server, as well as the Unwired Workspace. Details can be found at *http://scn.sap.com/docs/DOC-30502*.

While there is no charge for the Sybase license, you are subject to monthly fees from AWS. You can estimate the cost under various conditions at *http://calculator.s3.amazonaws.com/calc5.html?key=calc-F45DEB05-9160-418A-A1D0-508C462803A9*. This tool allows you to test various parameters (see Figure 4.16).

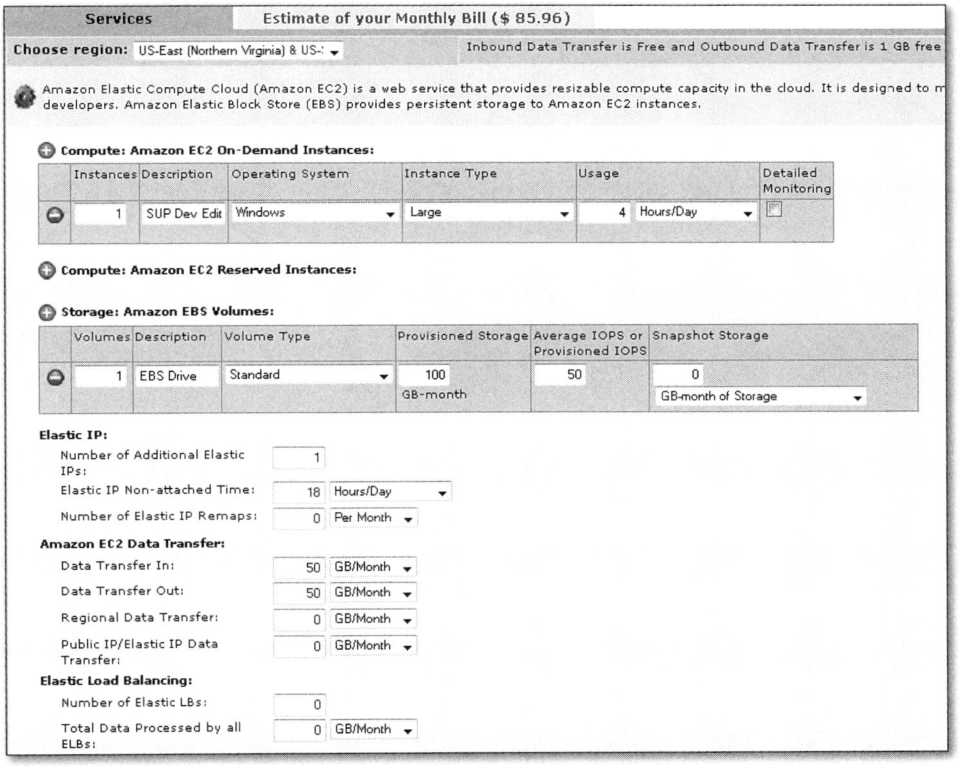

Figure 4.16 AWS Cost Estimator

In this section, we walk you through the steps of installing and connecting the cloud-based version of SUP. To start, you must have an account on the SAP Community Network (SCN) (*https://www.sdn.sap.com/irj/scn/register*) and an account on AWS (*http://aws.amazon.com*). After you create your AWS account, you must register to use the AWS CloudFormation tool. This requires entering a credit card, although you are not charged until you create a server instance. To complete the process, you receive a phone call from an automated system and are required to enter a PIN provided during the registration process.

4.3.1 Creating the SUP Server Instance

Make sure you are logged on to SCN, and then navigate to *http://www.sdn.sap.com/irj/scn/index?rid=/webcontent/uuid/206dbd82-ca9c-2f10-69a4-a5cfcb902540*. Once you click I HAVE READ AND AGREE, you are forwarded to the AWS website to complete the process.

On the next screen (shown in Figure 4.17), enter your AWS account number (the instructions on the page describe how to determine your account number), select the nearest AWS region, and click SUBMIT.

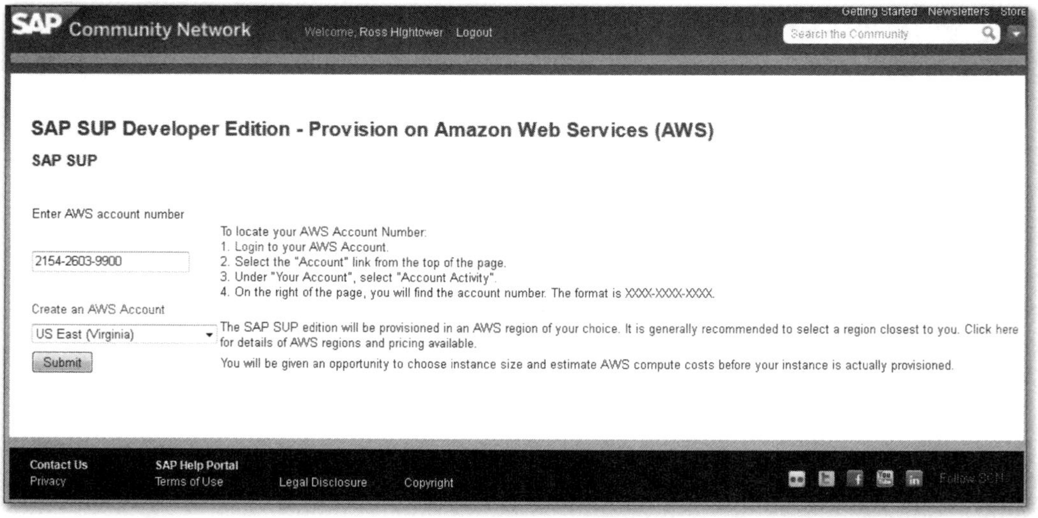

Figure 4.17 Page Used to Create an SUP Instance in the AWS Cloud

The next screen you see may require you to log on to your AWS account if you are not already logged on. Once you log on, you see the Stack Instance Creation Wizard shown in Figure 4.18. The wizard provides a default stack name and template URL. You can change the name if you like, but do not change the template.

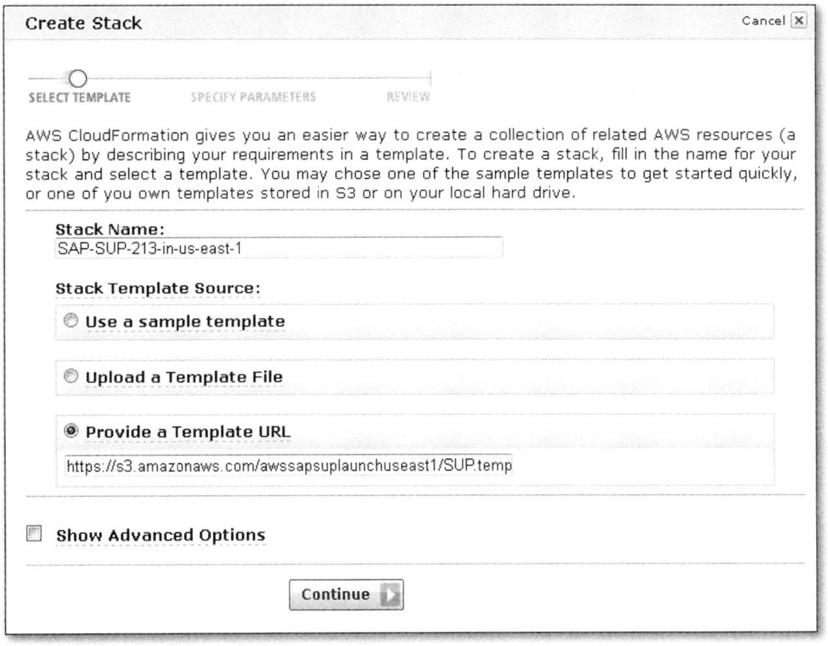

Figure 4.18 Stack Creation Wizard

On the next screen, leave the default values and click CONTINUE. Once you click CLOSE on the last screen, you see the stack in the AWS Management Console. The status should be CREATE_IN_PROGRESS. The creation process takes a few minutes. Click the REFRESH button on the upper-right side of the page until the status changes to CREATE_COMPLETE. This completes the process of creating the SUP server instance.

4.3.2 Installing the Unwired Workspace

If you do not already have the Unwired Workspace installed, you can download it from *http://www.sdn.sap.com/irj/scn/go/portal/prtroot/docs/webcontent/uuid/*

f00164c9-bcb7-2f10-d2b3-ac79a6faf833. Once the file downloads, install the Unwired Workspace according to the instructions earlier in the chapter (Section 4.2.1).

4.3.3 Managing the Server

The SUP server instance must be running before you can connect to it. You manage the instance from the EC2 Management Console, which you can reach from *https://console.aws.amazon.com/console/home* (Figure 4.19). Click EC2.

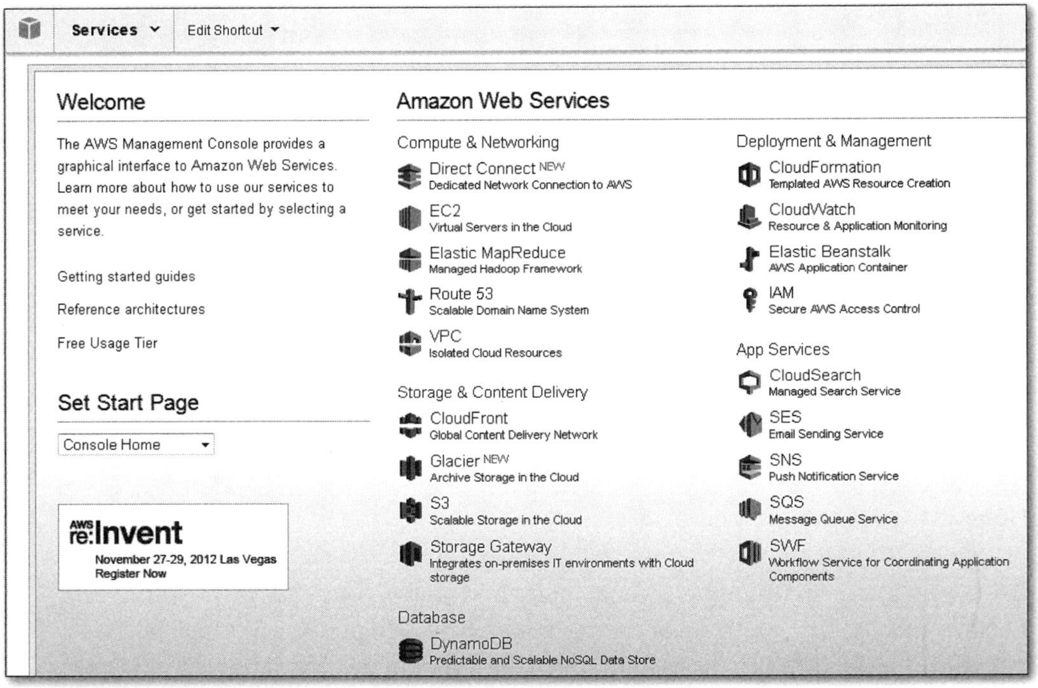

Figure 4.19 Amazon Web Services Management Console

In the EC2 Management Console (Figure 4.20) on the upper-left side, select the region in which you created your instance.

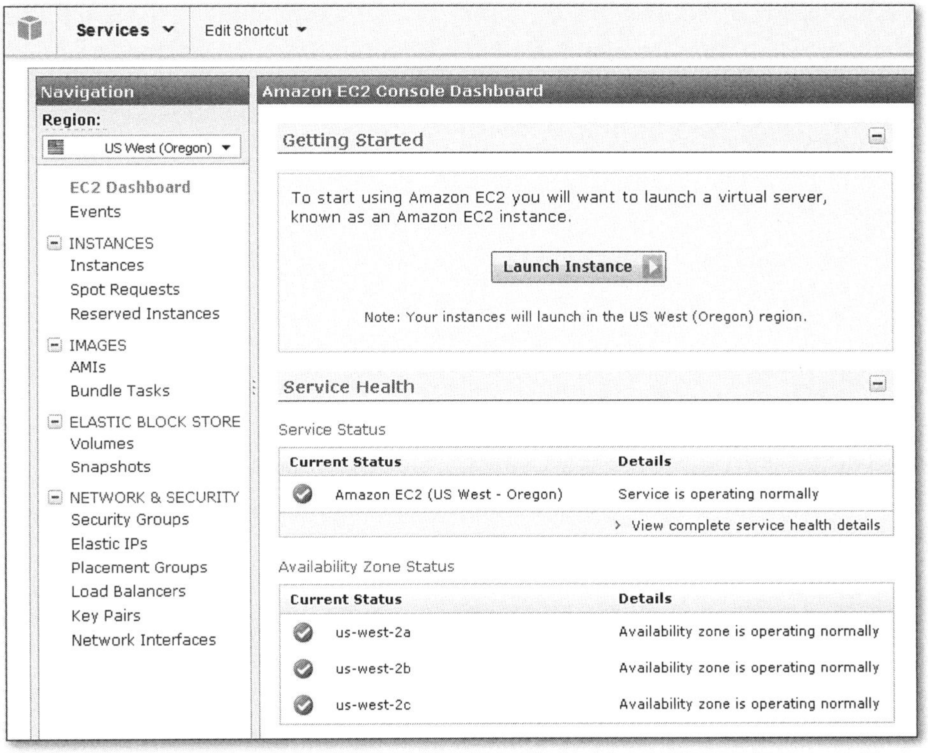

Figure 4.20 Amazon Web Services EC2 Management Console

The My Resources section on the right side of the screen shows how many instances are running. If you just created your SUP server instance, it should already be running (see Figure 4.21).

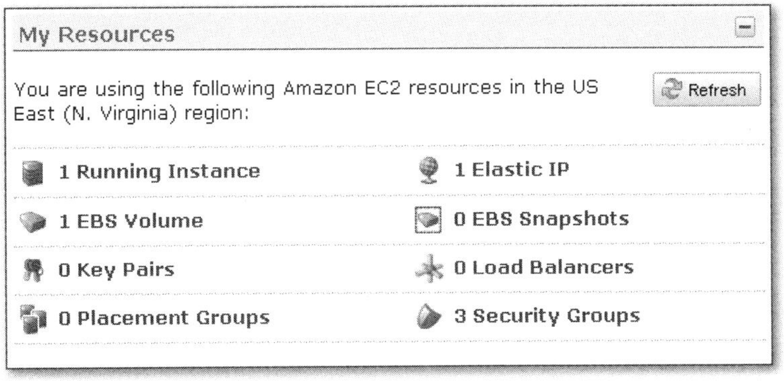

Figure 4.21 AWS EC2 Management Console

You can start the instance by clicking INSTANCES • INSTANCE in the navigation frame on the left. Select your instance and click the LAUNCH INSTANCE button.

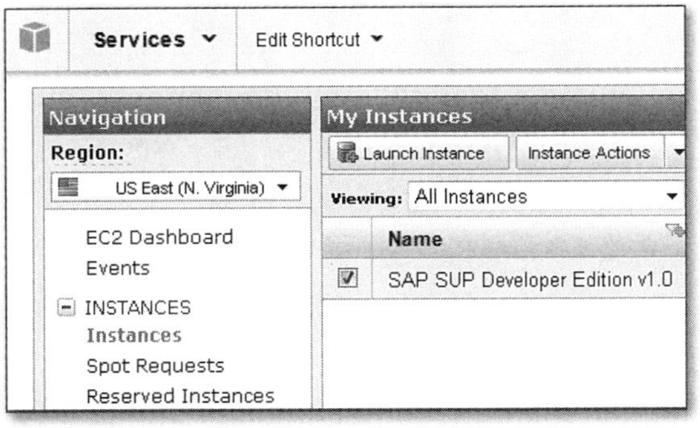

Figure 4.22 Launching an Instance in the AWS EC2 Management Console

To connect to the instance from the Unwired Workspace, you need the IP address of the instance. AWS instances use an elastic IP address, which means that a new IP address is assigned each time the instance starts but does not change as long as the instance is running. You can find the IP address assigned in the details section of the MY INSTANCES page after you launch the instance (see Figure 4.23).

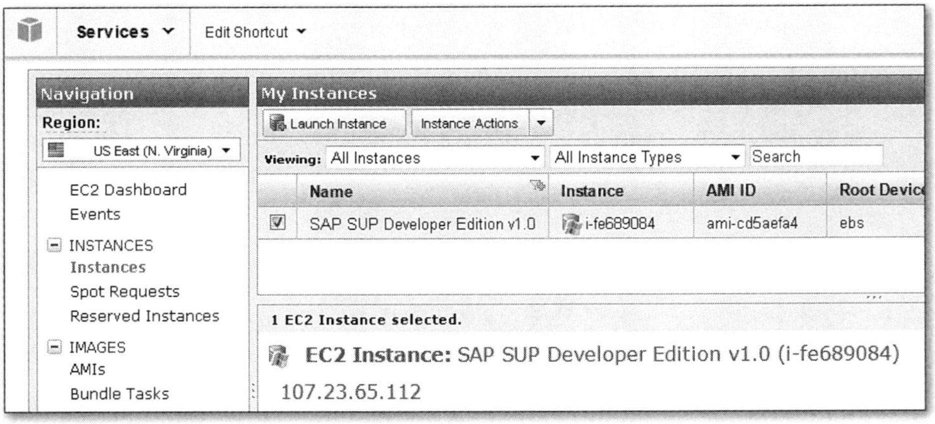

Figure 4.23 The IP Address Found in the Details Section of the My Instances Page

To manage the server, create a remote desktop connection to it. To configure this connection, right-click the instance and select CONNECT.

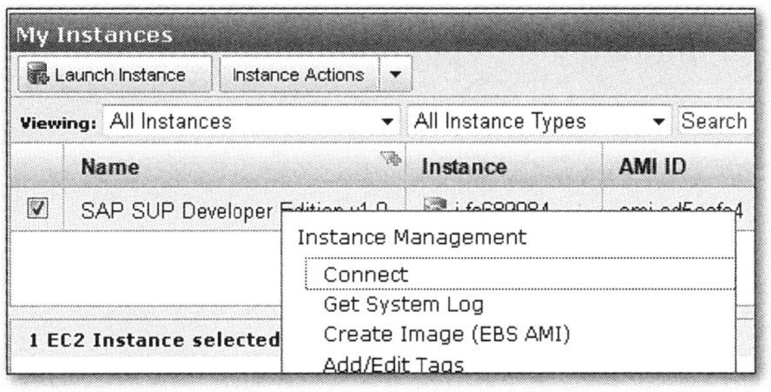

Figure 4.24 Connecting to the Instance

The dialog that appears gives you the option of downloading a shortcut file you can use to open the remote desktop connection (see Figure 4.25). Download this file and double-click it to open the connection.

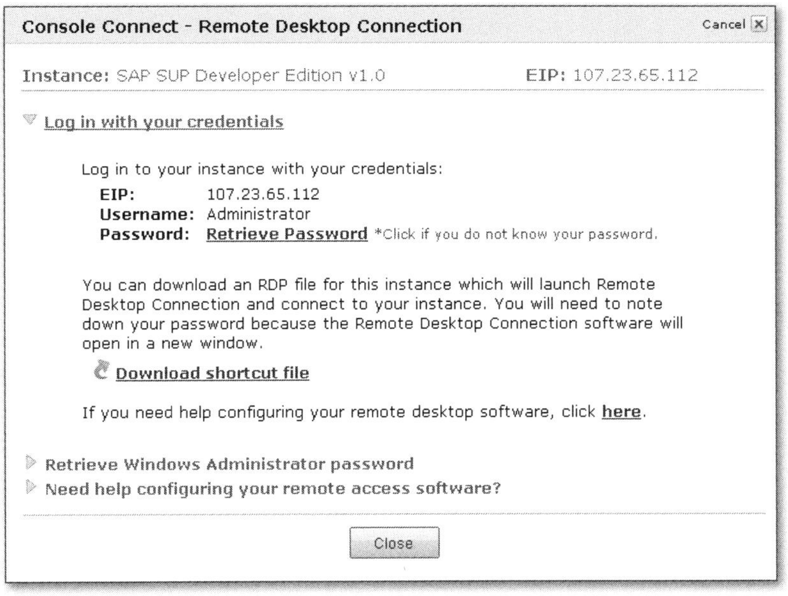

Figure 4.25 Downloading a Shortcut for the Remote Desktop Connection to the Instance

The initial user ID is "Administrator" and the initial password is "Admin1234." You are prompted to create a new password the first time you log on. The remote computer includes the SUP server and the Sybase ASA database server, as well as the Unwired Workspace. The server should start automatically when you start the instance, but if you need to start or stop the server, use the icon on the desktop.

From your own computer, you can connect to the Sybase Control Center using *https://<your-elastic-ip-address>:8283*. Skip over the certificate warning and log on using the user ID "supAdmin" and the password "Admin1234."

4.3.4 Configuring Connections in the Local Unwired Workspace

This section describes how to connect to the SUP server instance and the database server.

Connecting to the SUP Server Instance

The procedure for connecting the SUP server instance is the same as discussed in Section 4.2.2 (Connecting to the Unwired Server). The connection parameters are similar to those shown in Figure 4.26. Enter the elastic IP of your SUP server instance for the Host, "supAdmin" for the user name, and the password. If you did not change the password in the SCC, the password is "Admin1234."

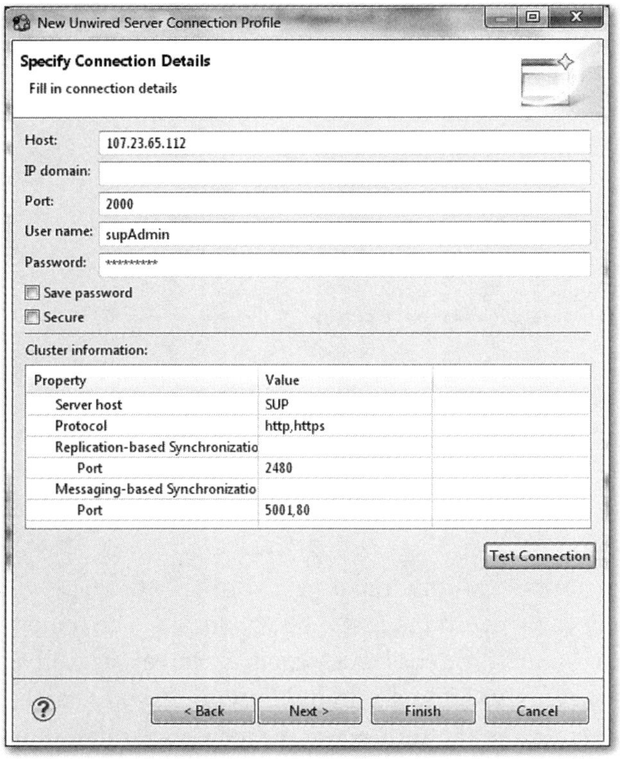

Figure 4.26 Properties for Connecting to the AWS SUP Server Instance

Connecting to the Database Server

Connecting to the database server follows the same procedure described in Section 4.2.2 (Connecting to the Sample Database). The connection parameters are similar to those in Figure 4.7. Select SYBASE ASA V10.X FOR UNWIRED SERVER DEFAULT in the DRIVERS dropdown list, and enter your elastic IP as the host and "5500" for the port. Do not change the user name or password.

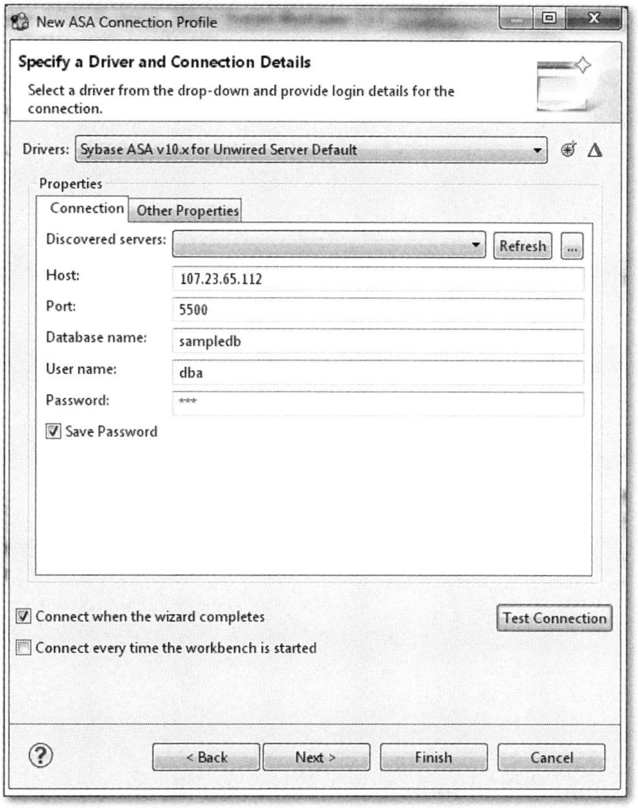

Figure 4.27 Properties for Connecting to the Database Server on the AWS SUP Server Instance

Once these connections are created, you are ready to begin developing apps.

4.4 Summary

After Chapter 3 and Chapter 4, you are almost ready to begin developing hybrid apps. The last piece of the puzzle is to install the container app in an emulator or device so that you can run the apps you write. Chapter 5 covers this process.

This chapter describes the installation and configuration of the Hybrid Web Container application, including the emulators and simulators required to run it.

5 Installing the Hybrid Web Container Application

The *Hybrid Web Container application* is a native app that includes an embedded web browser. Hybrid apps, which we briefly introduced earlier and will spend the rest of the book discussing, run inside the container. The container allows you to develop cross-platform apps using web technologies, while also providing some features of native apps. (More on this in Chapter 6.)

In this chapter, we explain how to install the Hybrid Web Container application on Android and iOS devices and emulators. We walk you through these processes for each of the platforms.

Note

Emulators/simulators are software simulations of mobile devices. Emulators allow you to test apps on multiple platforms without the expense of having multiple devices. From the point of view of the Unwired Server, there is no difference; the process of creating users and registering the device with the user is the same. However, there are some differences in how apps perform on different platforms. Apps should be tested on each of the platforms in the target user community.

5.1 Android Devices

In this section, we walk you through the steps involved in getting the Hybrid Web Container application running on Android.

5.1.1 Installing Android Emulators

The Android emulator consists of a system image, which includes the full Android system stack down to the Linux kernel. As a result, the emulator allows you to test most features, except phone calls, of an Android device. This chapter serves as a starting point for taking advantage of the emulators' capabilities. Complete documentation for using the emulators can be found at *http://developer.android.com/tools/devices/emulator.html*.

Development for Android is more challenging than it is for iOS. There are a wider variety of Android devices from different manufacturers, and because the Android operating system is open source, manufacturers are free to customize the operating system. For example, the Kindle Fire operating system is based on Android but is significantly different from the Android found on many other platforms. In addition, there are more versions of the Android API than there are for iOS. Fortunately, because hybrid apps run within the container, hardware and operating system differences are less relevant than for native apps. As long as the container application runs correctly, the hybrid app should run consistently within it. Customizations that make use of the device's native capabilities are more problematic. In any case, testing on multiple devices or emulators is recommended. While the number of Android emulators available is not exhaustive, there is a variety available for different manufacturers and different Android API levels.

In order to install the Android emulator, you must have the Java SE Development kit (JDK) installed. Because the Android SDK tools installer is 32 bit, you must install the 32-bit JDK. Once this is installed, you're ready to get started; in this section, we walk you through the steps.

Installing the Android SDK Tools

The first step in installing the Android emulator is to install the Android SDK tools. These include the Android SDK Manager and the Android Virtual Device (AVD) Manager, which is used to manage emulators.

The installation files for the SDK tools can be found at *http://developer.android.com/sdk/index.html*. Click Use an Existing IDE if you have Windows or Download for other Platforms to find links for Mac OS X and Linux (see Figure 5.1). Next, click Download the SDK Tools for Windows.

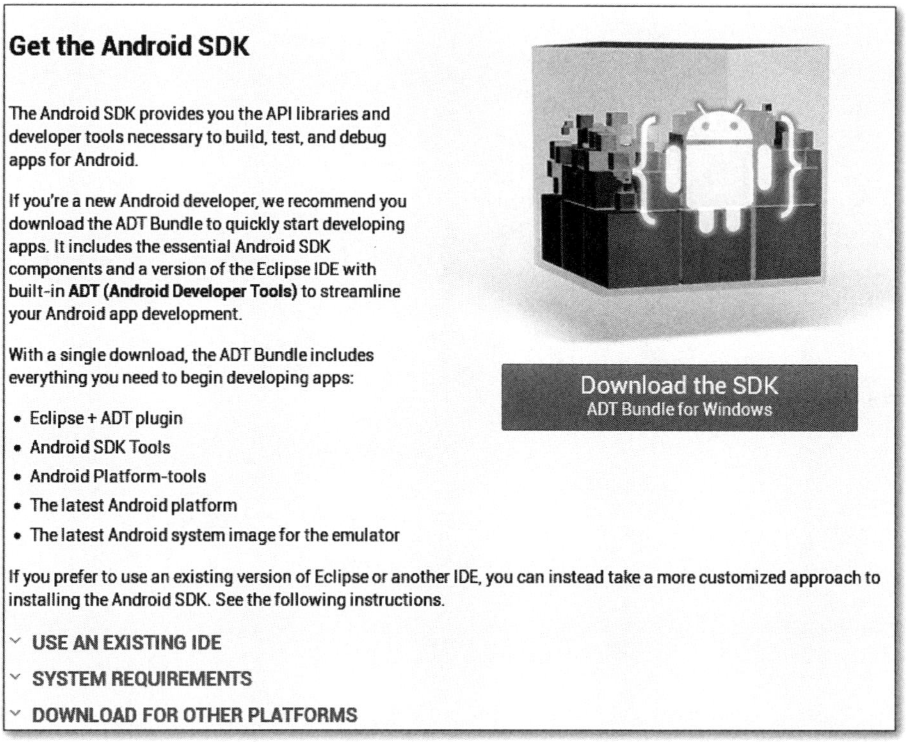

Figure 5.1 Downloading the Android SDK Tools

Once the installer downloads, execute it and click Next. The installer checks that the Java SDK is present. If you see the screen shown in Figure 5.2, the installer is unable to find a 32-bit Java JDK. Clicking Visit java.oracle.com opens a browser to a page that allows you to install the JDK. Once the JDK installs, click Back on the screen shown in Figure 5.2 and then resume the installation.

When your installation completes, the Android SDK Manager, shown in Figure 5.3, appears. If you need to start the Android SDK Manager from the Start Menu, use Start • All Programs • Android SDK Tools • SDK Manager.

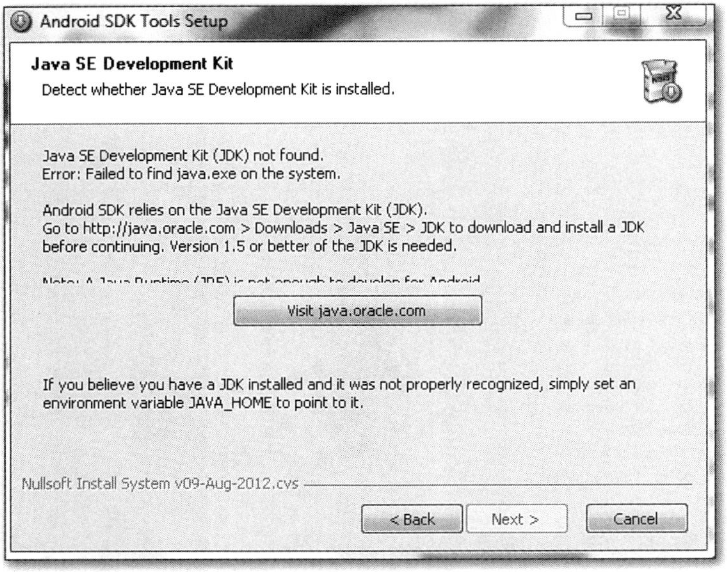

Figure 5.2 Installer Unable to Find a Java JDK

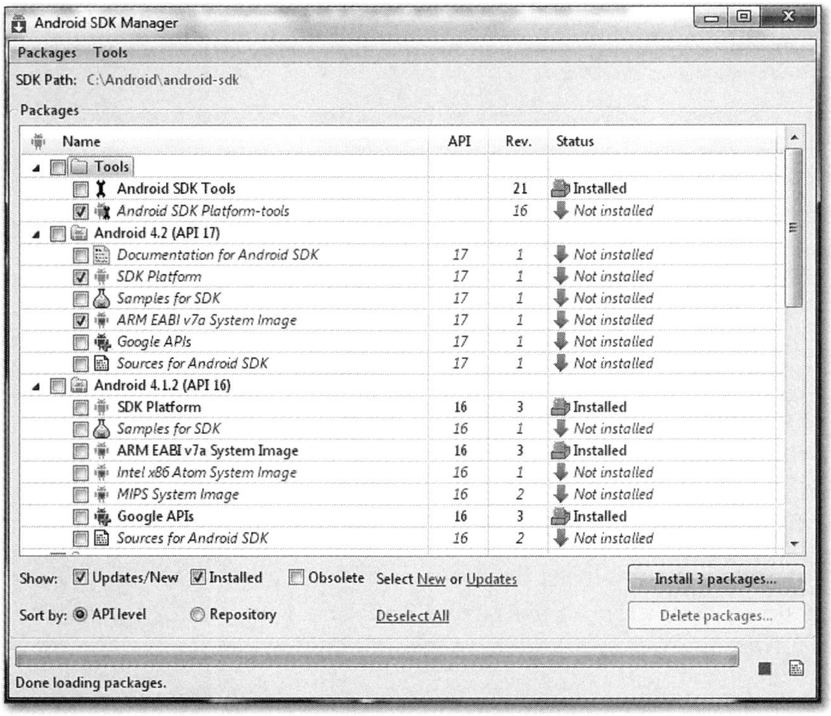

Figure 5.3 Android SDK Manager

Installing the Android SDK and System Images

With the Android SDK Manager, you can install emulators for each version of the Android API. To install an emulator, select the following three packages in the SDK Manager (see Figure 5.4):

- Android SDK platform tools
- SDK Platform (from whichever API level you wish)
- Emulator system image associated with the API level you chose

When you have selected the options you require, click INSTALL X PACKAGES (where *x* equals the number of packages you selected). In the next screen, accept the license agreements and then click INSTALL to begin the installation. The packages must be downloaded and decompressed, so the installation may be lengthy.

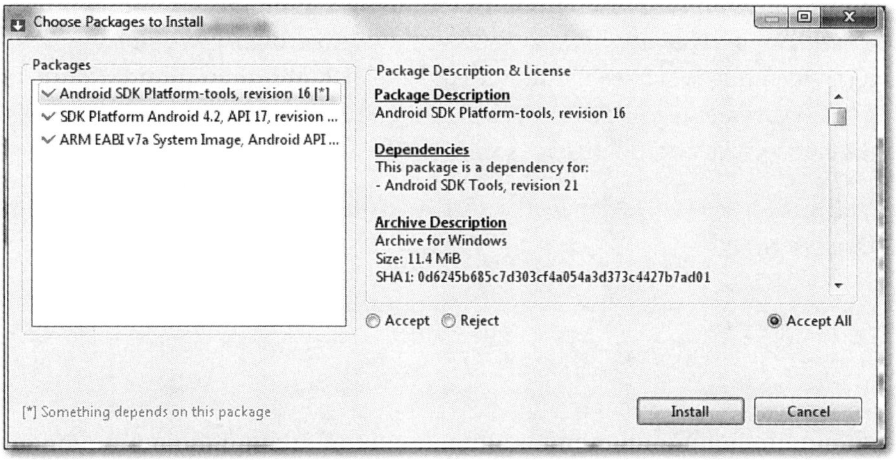

Figure 5.4 Choosing Packages to Install

The log window, shown in Figure 5.5, displays the progress of the installation. If errors occur, the log can be used to troubleshoot the problem. When the installation completes, click CLOSE.

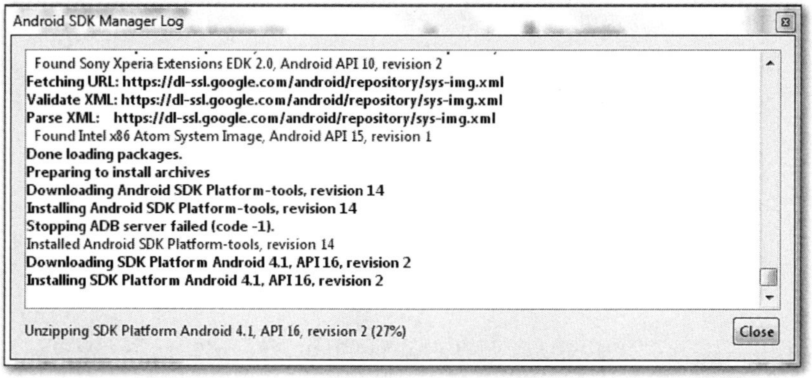

Figure 5.5 Installation Log

Creating an Android Virtual Device

The emulator uses an Android Virtual Device (AVD) configuration, which specifies the emulator's hardware options and skin. AVDs are managed using the AVD Manager, which can be started within the Android SDK Manager using the TOOLS • MANAGE AVDS... menu path. Or, in Windows, you can use the menu path START • ALL PROGRAMS • ANDROID SDK TOOLS • AVD MANAGER.

Figure 5.6 shows the AVD Manager before a virtual device is created. To create an AVD, click NEW.

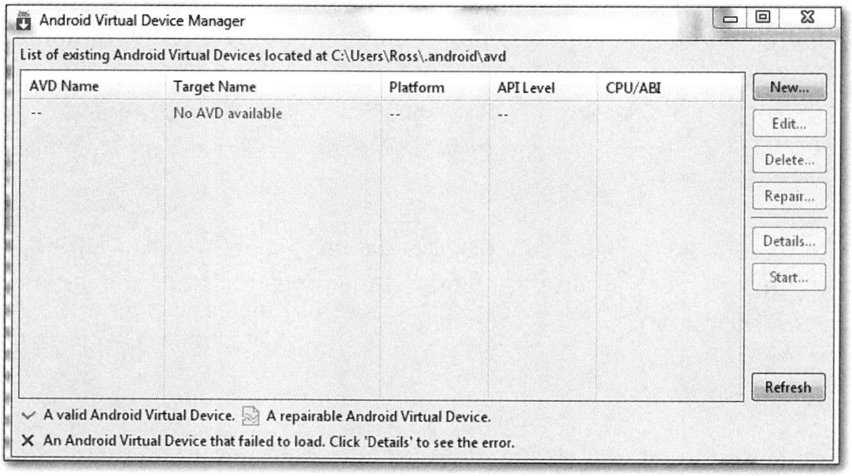

Figure 5.6 Android AVD Manager before Virtual Devices

Figure 5.7 shows the dialog used to create the AVD. The fields in the dialog are as follows:

- AVD NAME
  The name can have alphanumeric characters, periods, underscores, and hyphens, but no spaces.
- DEVICE
  The specific device you wish to emulate.
- TARGET
  The dropdown list includes an item for each installed API level.
- KEYBOARD
  Allows you to use the machine's keyboard rather than relying on the emulator's soft keyboard.
- SKIN
  The emulator includes simulated hardware controls in addition to the soft controls.
- MEMORY OPTION
  Allows you to set the amount of RAM. If you encounter a memory allocation error, reduce this amount.
- INTERNAL STORAGE
  Memory available for file storage.
- SD CARD SIZE
  You can specify the amount of memory in a simulated SD card.
- SNAPSHOT
  If you enable snapshots, the emulator's state can be saved when the emulator is closed. Snapshots speed the process of starting the emulator considerably and allow you to return to a working state if a problem occurs.
- USE HOST GPU
  Uses the machine's graphic processing unit, which may increase the emulator's graphical performance. You cannot choose to use both snapshots and the GPU.

Click OK to create the AVD.

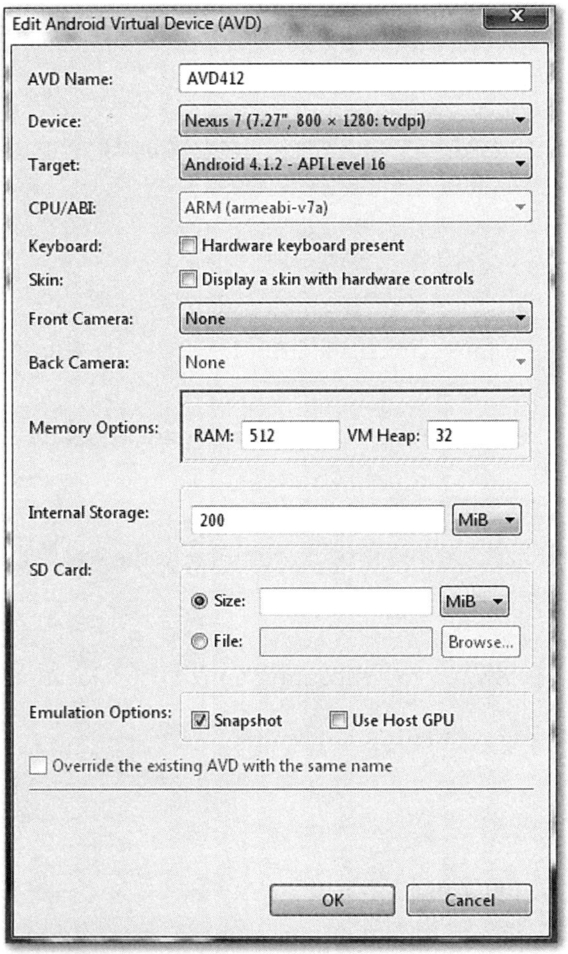

Figure 5.7 Creating an Android AVD

Starting the AVD

To start the AVD, select it in the AVD Manager and click START (see Figure 5.8).

Figure 5.9 shows the dialog used to start the AVD. The following fields can be configured:

- SCALE DISPLAY TO REAL SIZE
  This can be used to adjust the size of the emulator. If you do not specify a size, the size is determined by the skin applied when the AVD is created.

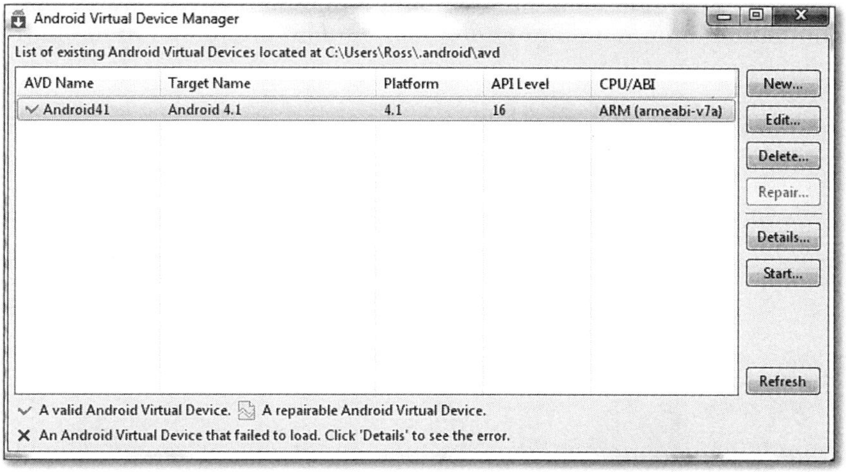

Figure 5.8 Starting the AVD

- WIPE USER DATA
  This option erases all data previously saved by the emulator when it is run. If you have started the emulator before and registered it with the Unwired Server, choosing WIPE USER DATA prevents the Unwired Server from recognizing the emulator. You then have to create a new device user in the SCC and register the emulator with the new user. (There are few circumstances that require you to use this option.) If the emulator becomes corrupted and does not start, simply deselect the option to LAUNCH FROM SNAPSHOT. This usually allows it to start without deleting the device ID register with the SCC.
- LAUNCH FROM SNAPSHOT
  If you have a saved snapshot, this option causes the emulator to start using the snapshot. The emulator starts much more quickly and retains all the settings you configured on the emulator.
- SAVE TO SNAPSHOT
  This option causes the emulator to save a snapshot when it closes. You should select this option only if you expect to make changes to the emulator's configuration. Once you are satisfied with the emulator's configuration, you should leave this option unselected. This reduces the possibility of corrupting a clean snapshot.

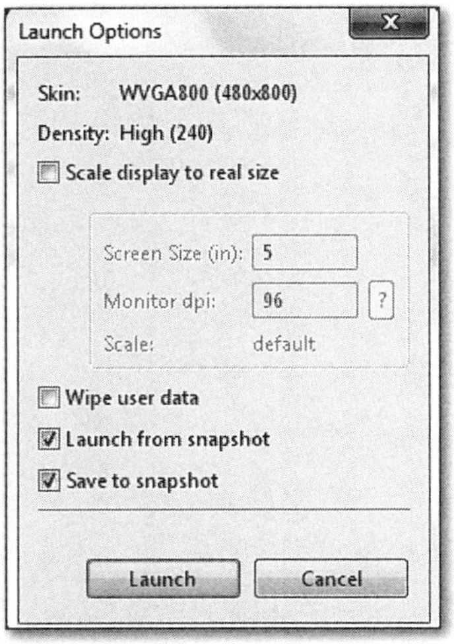

Figure 5.9 Launching from a Snapshot

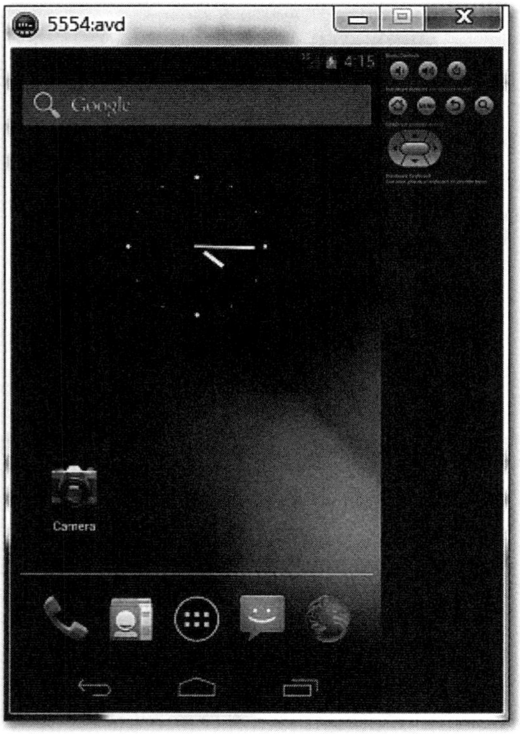

Figure 5.10 The Android API 4.1.2 Emulator

Without a snapshot, the emulator can take several minutes to start. Once the emulator starts, you see a help message, which you can dismiss by clicking OK. The emulator is shown in Figure 5.10. Note that other system images likely look different. At this point, you can use many of the preinstalled apps on the emulator, and if the computer on which the emulator is running has network access, the emulator can access the network, as well. Before the emulator can be used to test hybrid apps, the container application must be installed and registered with the SCC.

5.1.2 Installing and Configuring the Container Application on the Emulator

For an Android device, the container application can be downloaded from Google Play; although emulators can access the Internet, they cannot access Google Play, so apps must be installed from a command prompt. The container apps for all four supported platforms are located in folders under the <MOBILE SDK INSTALLATION FOLDER>\HYBRIDWEB folder. The Android container application is called HYBRIDWEBCONTAINER.APK and is located in the Android folder (see Figure 5.11).

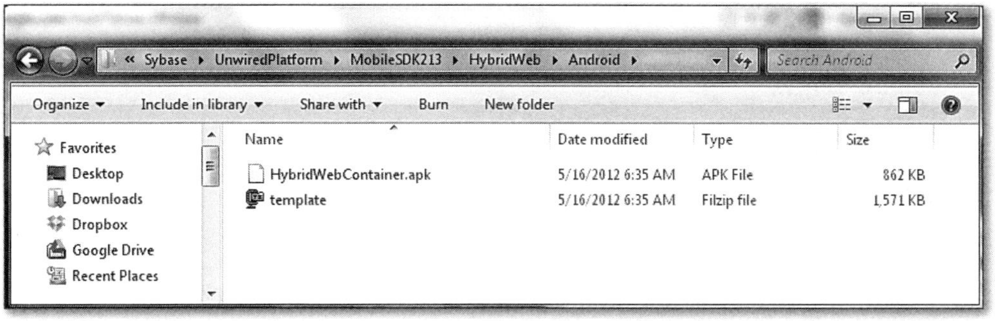

Figure 5.11 The Mobile SDK Directory Structure

Copy this file to the Android platform tools folder located at *<Android installation folder>\Android-sdk\platform-tools*. To install the container in an emulator, make sure the emulator is running and open a command prompt. Change the folder to the platform tools folder. You can ensure that the emulator is running using the command ADB DEVICES. If you don't see the emulator, close the emulator and start it again from the AVD Manager.

To install the container application, use the command ADB INSTALL HYBRIDWEBCONTAINER.APK. Figure 5.12 shows the results of a successful installation.

Figure 5.12 The ADB Command

In the emulator, click the APPLICATIONS icon to display the installed apps. Depending on the emulator you are using, you may have to drag the screen to show the container application (see Figure 5.13).

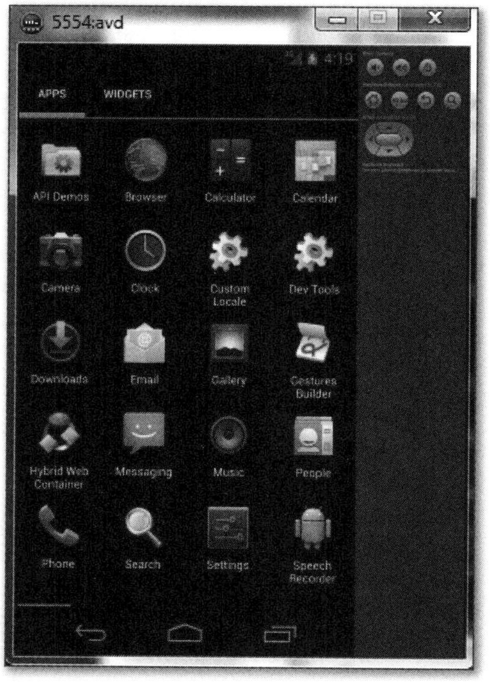

Figure 5.13 The Container Application in the Android Emulator

Click the icon to open the container application. The first time you open the app, you are prompted to create a password. This password is used to access the container application, not the Unwired Server. The password must be at least six non-sequential characters long.

After you click OK, you see a screen that allows you to enter the connection parameters for the Unwired Server. Note that you must have a user configured on the server to complete this step. Figure 5.14 shows the settings after they are configured. The following properties are configured:

- Enable Automatic Registration
  This is only used if a PreconfiguredUserLoginModule has been created in the SCC.
- Server Name
  The host name or IP address of the Unwired Server. If using a Relay Server, this is the host name or IP address of the Relay Server.
- Port
  The port used for messaging connections on the Unwired Server or the Relay Server, if one is being used.

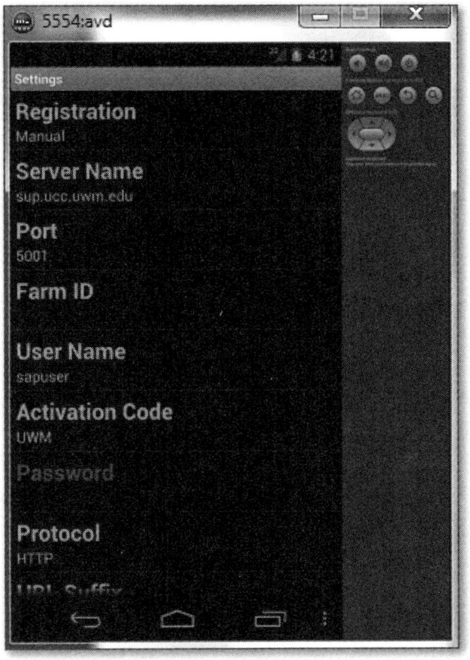

Figure 5.14 User Configurations on the Unwired Server

- FARM ID
  A string associated with a Relay Server. If no Relay Server is used, this is ignored.
- USER NAME
  The user name configured on the Unwired Server.
- ACTIVATION CODE
  The activation code if one has been configured for the user on the server. This is not used if automatic registration is used.
- PROTOCOL
  The protocol used to communicate with the server for messaging connections.
- PASSWORD
  The password is used only if automatic registration is used.
- URL SUFFIX
  This is used with a Relay Server or proxy server.

Once the settings are entered, click the MENU button and select SAVE. As soon as the settings are saved, the container application attempts to connect to the server and register the emulator with the user. You can use the SHOW LOG menu option to view the connection log (see Figure 5.15). The log shows you whether the emulator registered successfully on the Unwired Server. If the user name or activation code is not correct, the emulator connects and disconnects repeatedly. If the emulator's device ID doesn't match one previously registered with the user, the log displays a message that says WRONG DEVICE FOR USER. The log also shows when hybrid apps are downloaded from the server.

If the process is successful, the emulator is registered with the user on the Unwired Server. Figure 5.16 shows an example of the APPLICATION CONNECTION screen in the SCC. The DEVICE TYPE and DEVICE ID columns show whether a device or emulator is registered with the user. In this case, the values indicate that an Android emulator is registered. The device ID is unique to this specific incarnation of the emulator. If you start the emulator by selecting the WIPE USER DATA option, the device ID is regenerated and the SCC does not recognize the emulator. The PENDING column indicates whether the Unwired Server has anything in the queue to download to the emulator or device. This value must be greater than 0 before the emulator or device registers the first time or new hybrid apps are assigned to the user. The STATUS column shows whether the device is currently online.

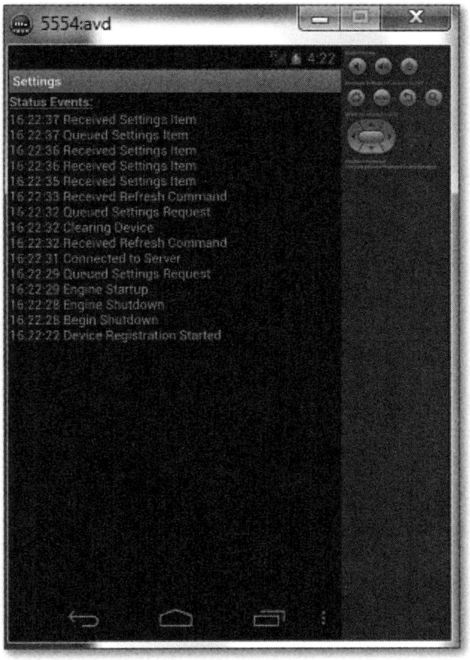

Figure 5.15 The Connection Log

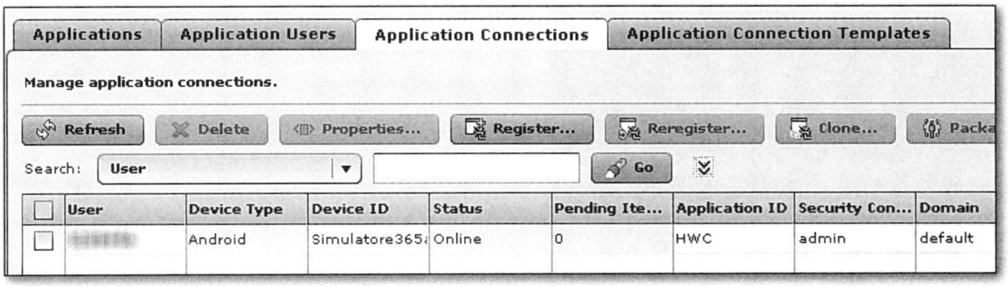

Figure 5.16 The Application Connection Screen

In the emulator, click the back button on the keypad until you see the screen shown in Figure 5.17. This screen shows messages that are notifications sent by server-initiated apps. To see client-initiated apps that have been assigned to the user, click the MENU button and select HYBRID APP.

Figure 5.17 The Messages Screen Showing Server Notifications

Figure 5.18 shows the emulator with two hybrid apps. Click an app name to start one.

Figure 5.18 The Container Application with Two Hybrid Apps

5.1.3 Installing the Container Application on the Device

This section describes the process for installing the container application on Android devices. The process for configuring and using the container application is the same whether it is installed on an emulator or device. The difference is the method used to install it.

The easiest way to install the container application on an Android device is via Google Play. You can also install the container application on a physical device using the method described for emulators. This is sometimes necessary, as Google Play does not recognize some devices as compatible with the application and does not allow you to install the application on them.

Find the Google Play icon on your device's APPLICATIONS screen and activate it. In Google Play, search using the term *Sybase* (see Figure 5.19). Select SYBASE HYBRID WEB CONTAINER and then click INSTALL to install it.

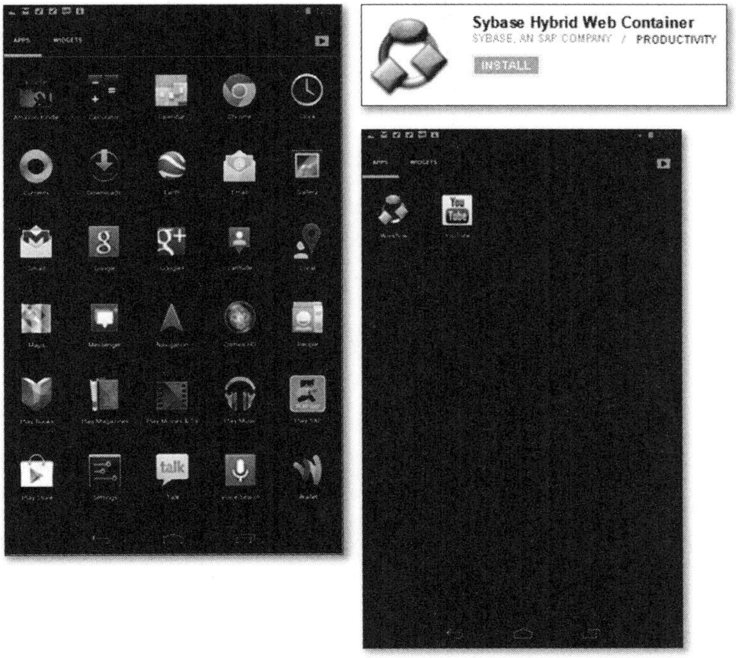

Figure 5.19 The Container Application Found in Google Play

Once the installation is complete, locate the app on your APPLICATIONS screen and configure it as described in Section 5.1.2.

5.2 iOS Devices

In this section, we walk you through the steps involved in getting the Hybrid Web Container application running on an iOS device.

Note

In the iOS world, there are a limited number of operating system versions and only two devices. However, it is still a good idea to test apps on multiple platforms.

5.2.1 Installing the Simulator

While the Unwired Workspace runs only in Microsoft Windows, the iOS simulator runs only in the Mac OS. Fortunately, it is relatively easy to install both the Mac OS and Windows on a Mac, and even run Windows apps and Mac OS apps side by side. This makes it easy to use the iOS simulator, which is significantly faster and more stable than the other options.

To install the iOS simulator, you must first install the Xcode developer tools from the Apple App Store (see Figure 5.20). You can access the Apple App Store from APPLICATIONS in Finder.

Once Xcode is installed, open it and select PREFERENCES from the XCODE menu (Figure 5.21). Select DOWNLOADS, and then click INSTALL next to the simulator you wish to install.

You are prompted to log on using an Apple developer ID. If you do not already have one, you can register for one. It's free to register for the ID and to run hybrid apps. If you want to run native iOS apps on a physical device or submit apps to the Apple App Store, you must pay an annual fee.

To start the simulator, use the XCODE • DEVELOPER TOOLS • iOS SIMULATOR menu option (see Figure 5.22). You can change the type of simulator using the simulator's HARDWARE • DEVICE menu.

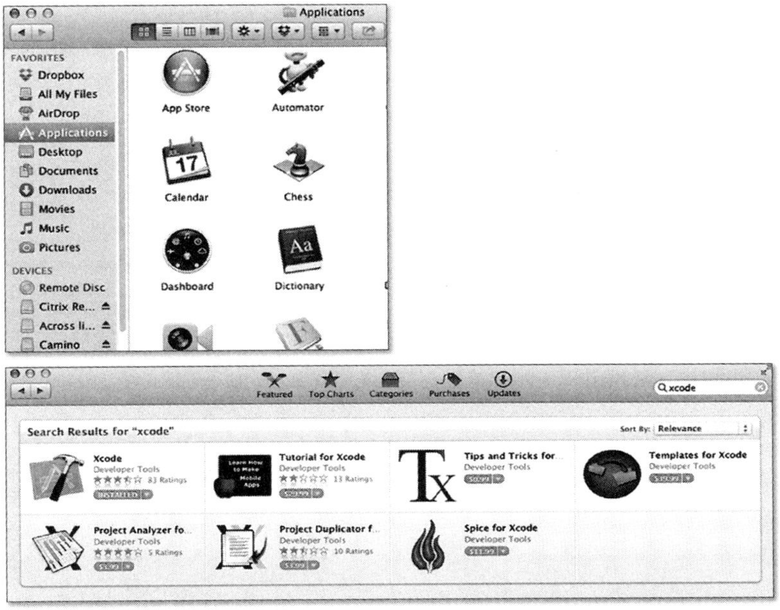

Figure 5.20 The Xcode Developer Tools

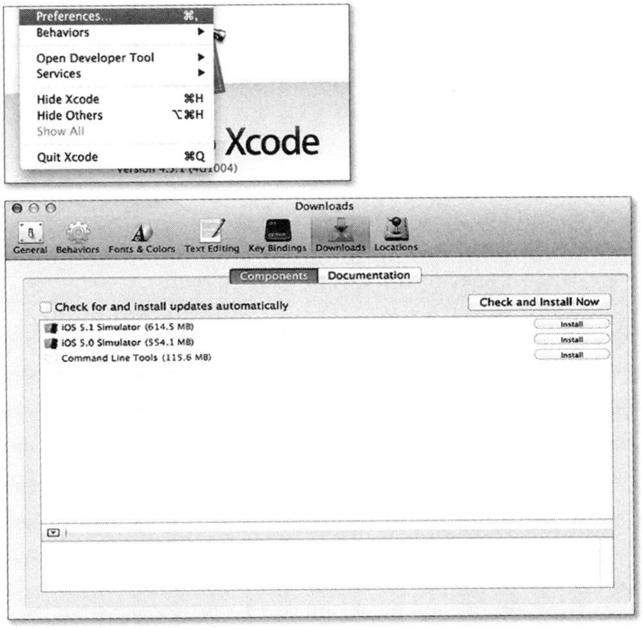

Figure 5.21 The iOS Simulators

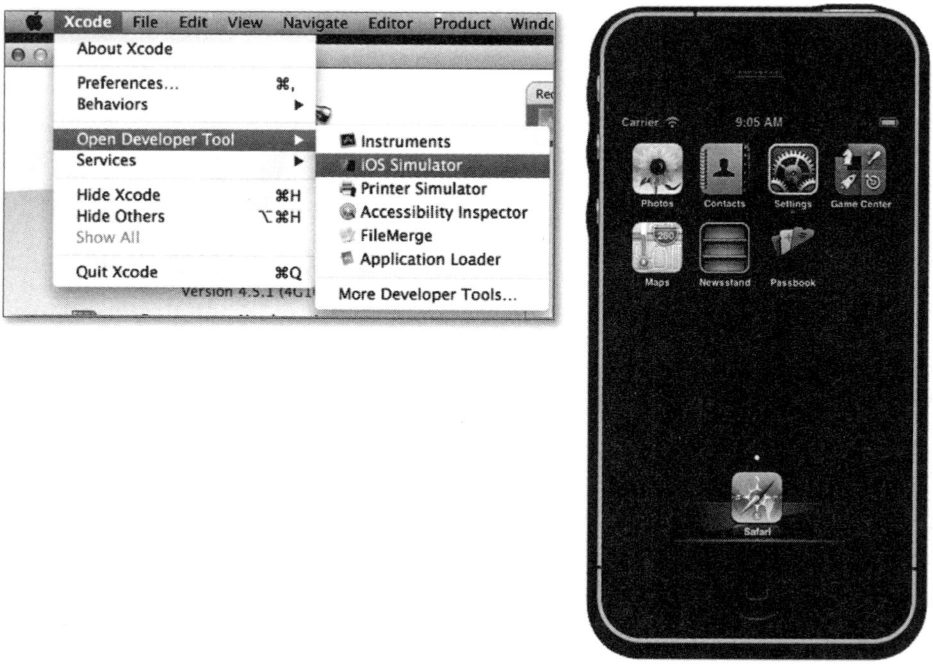

Figure 5.22 Starting the iOS Simulator

5.2.2 Installing and Configuring the Container Application on the Simulator

The source code for the container application is installed with the Unwired Workspace in the <Installation folder>\HybridWeb\iOS folder. Copy the file *MobileWorkflow-2.1.3.tar.gz* to the computer on which the iOS simulator is installed and expand it to create a folder called Workflow. In Xcode, use the File • Open menu and then navigate to the Workflow folder to open the project (see Figure 5.23).

You can select the type of device using the dropdown list on the top left, and then click the Run button to run the simulator (see Figure 5.24).

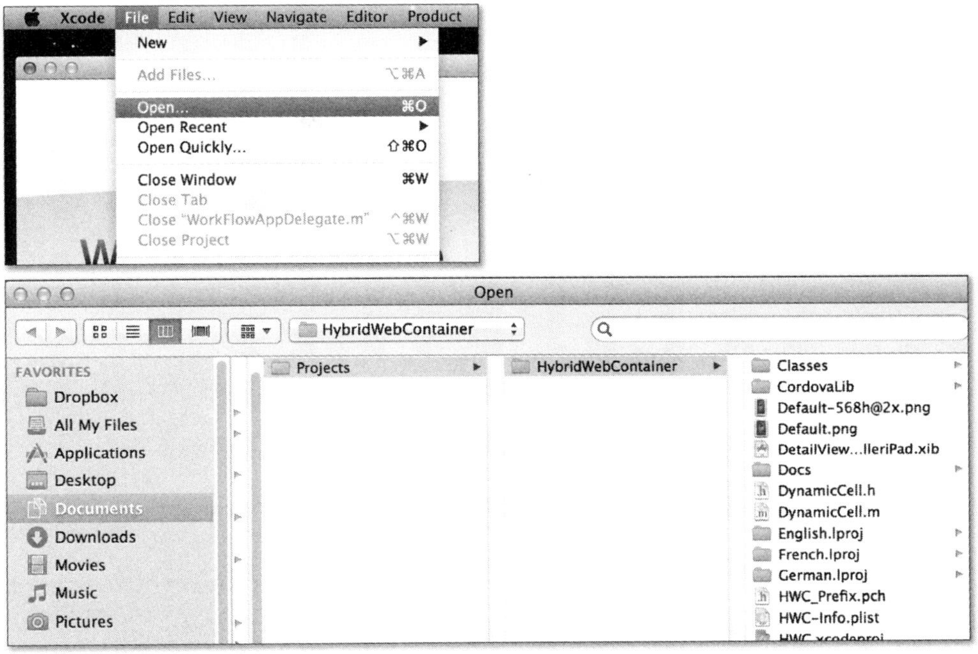

Figure 5.23 Loading the Project

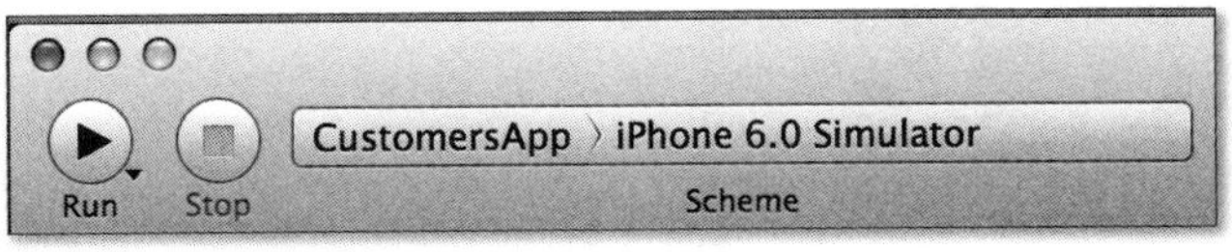

Figure 5.24 Running the Project to Start the Simulator

Once the simulator starts, you can use the HARDWARE menu to control it. For example, you can select the type of device, navigate to the device's home screen, or rotate the device (see Figure 5.25).

To configure the container application, navigate to the HOME screen of the simulator using the HARDWARE menu or the simulator's controls. Start the SETTINGS application, choose the WORKFLOW application, and complete the fields (see Figure 5.26).

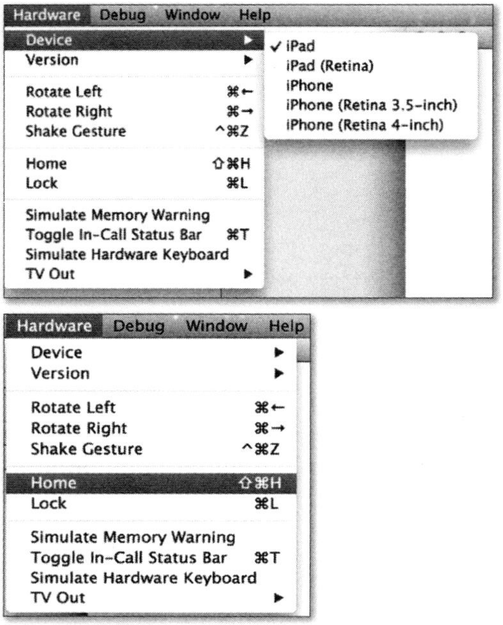

Figure 5.25 The Hardware Menu

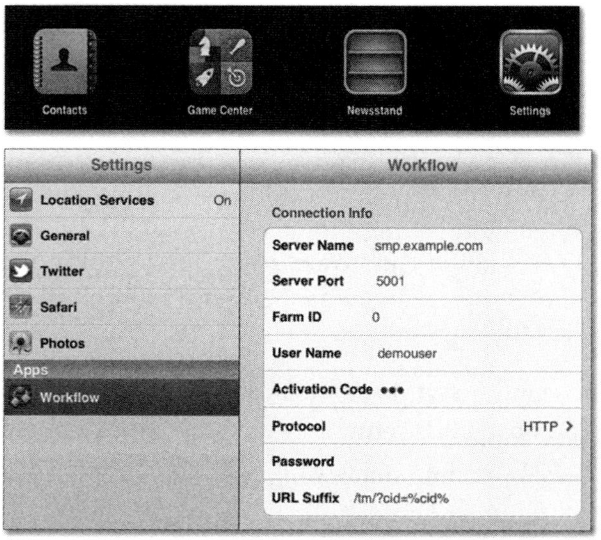

Figure 5.26 The Settings App

The connection settings are the same as those described for the Android emulator. However, unlike the Android emulator container application, the iOS container application has no option to view the connection log.

5.2.3 Installing the Container Application on the Device

The container application for iOS devices can be downloaded from the Apple App Store. Start the App Store app on the device and search using the keyword *Sybase* (see Figure 5.27). Install the application and then configure the settings using the procedure described earlier in this chapter.

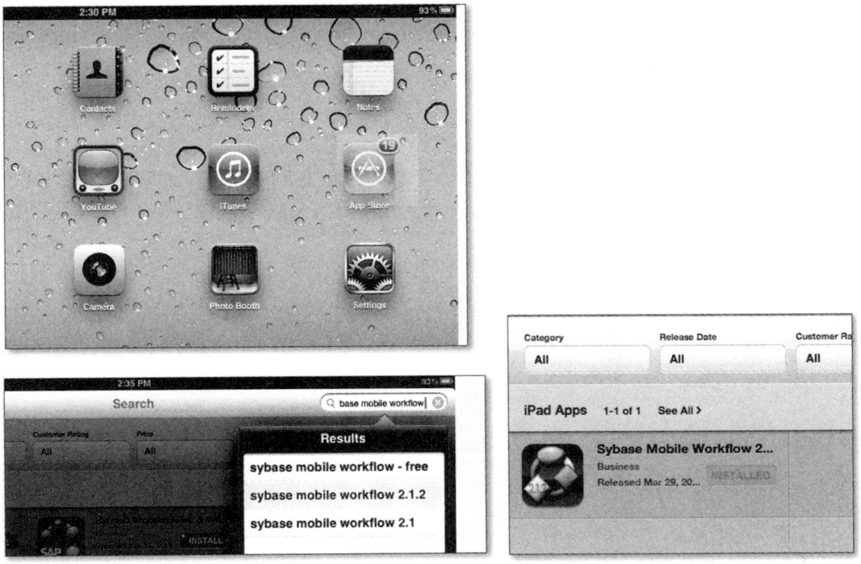

Figure 5.27 The Container Installed from the Apple App Store

5.3 Summary

After Chapter 5, all the pieces are in place to develop hybrid apps. You will find that testing apps on multiple platforms is a good idea because, although hybrid apps are cross platform by design, there can be subtle differences between the different platforms and between different screen sizes.

PART II
Building Hybrid Apps

One of the desirable objectives of mobile app design is the ability to develop a single mobile app that can run on different mobile platforms. In this chapter, you learn how Sybase Unwired Platform uses the concept of a container to accomplish this objective.

6 Working with the Hybrid Web Container

This chapter introduces the *Hybrid Web Container*, or simply *the container*. This container is a runtime application that is installed on the mobile device and executes Sybase Unwired Platform (SUP) hybrid apps. The containers are native apps that depend upon their mobile operating system (OS). A different container has been developed for each of the four major mobile platforms (Android, iOS, BlackBerry, and Windows Phone), and these apps can usually be downloaded from their associated app stores.

This chapter explains what you need to know to work with the Hybrid Web Container: its architecture, hybrid app design, and hybrid app patterns. Before we dive into that, though, we start the chapter with an overview of the different mobile development architectures, which is necessary to understand how the Hybrid Web Container works, and why a developer would use it.

6.1 Mobile Architectures

In Chapter 1, we defined the three types of mobile app development: native development, mobile web development, and hybrid development. In this section, we'll go into a bit more detail about the mobile architectures of the apps that are created as a result of these three different types of development, including their advantages and disadvantages.

6.1.1 Native Apps

A native app is an executable file that is typically downloaded from an app store, saved on the mobile device as a file, and run by the operating system on that device. Smart devices come from the manufacturer with several native apps already installed; these are referred to as *built-in apps*. The built-in apps are used for standard activities, such as making phone calls, browsing the Internet, sending and receiving email, looking up contacts, taking and saving pictures, etc. Once a native app is saved to the mobile device, it simply appears as an icon and functions very similarly to the built-in apps.

A native app has the ability to access the mobile OS functions through what are referred to as *Application Program Interface (API) calls*. These APIs provide access to both hardware and software features of the smart device. Also provided are APIs that give access to the data associated with the built-in apps. Examples of APIs that are common on today's mobile devices are shown in Table 6.1.

Standard Hardware APIs	Advanced Hardware APIs	Built-in App APIs
Keyboard	Network access	Mail
Touch screen	Vibration	Contact list
Graphics	Compass	Calendar
Microphone	GPS	Phone
Speakers	Accelerometer	Music
Storage	Future	Text messaging

Table 6.1 APIs Commonly Provided on Mobile Devices

Native apps also have access to the mobile device's GUI toolkit. This supports the development of native apps that have a look and feel similar to that of the built-in apps. The final capability offered by APIs is access to the features provided by the built-in apps, such as contact lists and the browser. As shown in Figure 6.1, a native app has access to all the APIs for a given mobile operating system.

Some of the major advantages offered by native apps are that they are self-contained, are downloaded from an app store, respond quickly, and utilize all the features of the mobile device. Additionally, many have the same look and feel of the built-in apps. These advantages have led to the adoption of the native approach for many apps. The Apple App Store alone offers hundreds of thousands of native apps.

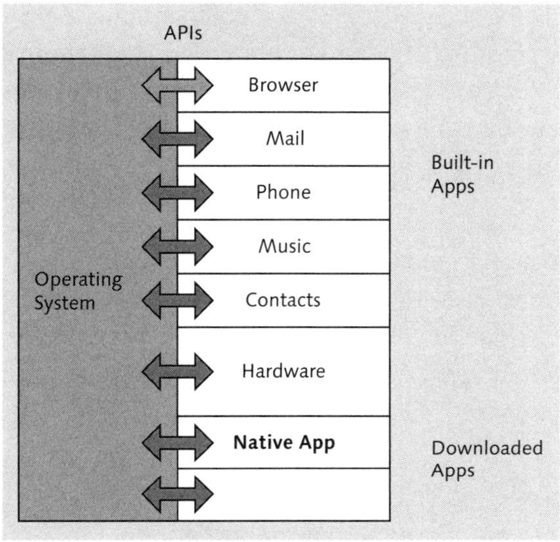

Figure 6.1 A Native App's Access to Hardware, Software, and GUI Toolkit via APIs

With all these advantages and the current community of native app developers, it is important to look at why other approaches have been developed. The primary disadvantage of the native app approach is native apps' direct dependence on the mobile operating system itself. Each major mobile application platform utilizes its own proprietary operating system, which support different languages and have their own sets of APIs. They also require the use of their own SDKs and utilize different Integrated Development Environments (IDEs). Table 6.2 presents the four major platforms, the programming languages they use, and the IDEs that are normally used for developing their native apps.

Platform	Programming Language	IDE
Apple iOS	Objective-C	Xcode (Apple only)
Android	Primarily Java	Android Development Tools Plug-in for Eclipse
Microsoft Windows Phone	C# or VB.Net	Visual Studio
BlackBerry	Java	BlackBerry Plug-in for Eclipse

Table 6.2 Programming Languages and IDEs Used for Developing Native Apps for the Major Mobile Platforms

When developing a native app, it is necessary to first decide which mobile platform is to be supported. If a single platform is selected, the second step requires that all end users acquire a mobile device that utilizes that operating system. For example, if the native app will be developed for iOS, then all the users must have access to an iPhone or iPad. In today's world of bring your own device (BYOD), this is becoming an unlikely scenario.

If, instead, the user base has devices from more than one mobile platform, it is necessary to develop several different native apps, one for each platform using its programming language and development environment. It is also necessary to support and maintain all these different apps. As shown in Figure 6.2, a different app is required for each mobile platform.

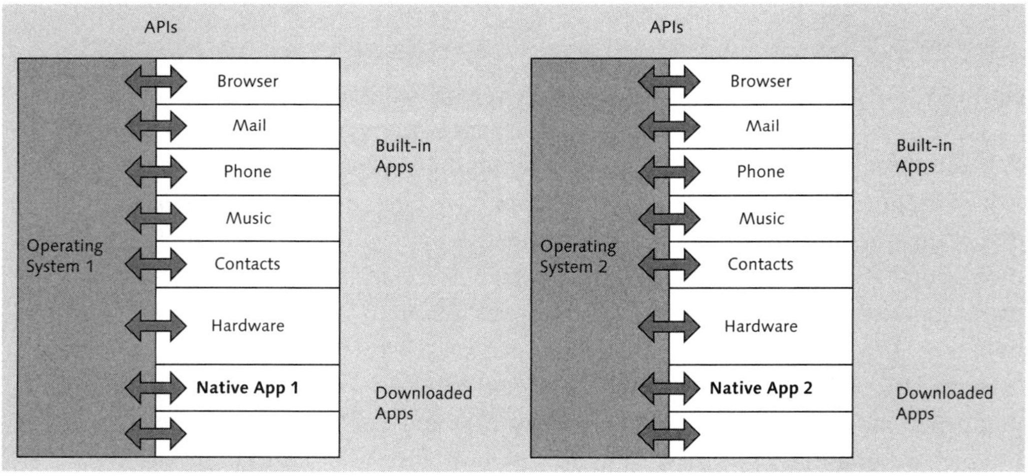

Figure 6.2 Different Native Apps as Required for Each Mobile Operating System

Since the APIs are different and unique to each platform, it is not financially feasible to write the native app in one language and then automate the conversion to other languages. This leads to the major disadvantage of native apps: it is very expensive to use the native app approach when trying to support multiple device platforms.

A second disadvantage, which is important for enterprise applications and which is addressed in more detail later, is that, since native apps are self contained, they do not easily support access to backend data sources. A final disadvantage is that the developer skill required for each platform is different, which has an impact on the availability of developers, most of whom know only one platform.

6.1.2 Mobile Web Apps

A mobile web app is actually a website stored on a web server. The web server delivers web pages to the mobile web browser—a built-in app on the mobile device. These web pages are typically written using HTML5, CSS3, and advanced JavaScript. A brief description of these languages is presented in Table 6.3.

Language	Description
HTML5	HTML is a hypertext markup language used for presenting content on the World Wide Web. While previous versions were primarily static, version 5 has added multimedia, graphics features (bitmapped and vector), GPS, and support for limited APIs. Presentation features, now provided by CSS, have been dropped.
CSS3	CSS controls the style (the look and feel) and formatting of web pages written using HTML.
JavaScript	JavaScript (JS) provides client-side scripting for enhanced user interfaces and dynamic websites. JS is processed by the browser. JS is a function-based scripting language that allows controls on the page to react to inputs (keyboard, screen touches) without the need to make a round trip to the server, making it more responsive to the end user.

Table 6.3 A Description of the Languages Used to Build a Mobile Web App

These are some of the same languages commonly used to develop traditional websites, and some websites even support both traditional and mobile browsers. A large number of developers are familiar with these languages. The newer versions of these languages provide the ability to create high-quality web pages that appear similar to the screens of native apps.

One of the unique characteristics of a mobile web app is that the code is stored on an external web server, as opposed to the mobile device, and is displayed on the mobile device using only the browser. This provides the major advantage of mobile web apps: a single app can run on more than one platform. Since most mobile devices use a similar webkit for displaying pages, most mobile web apps can run on almost all major platforms. As shown in Figure 6.3, the only component that these apps interact with is the browser. As mobile browsers improve and add new features, the quality of mobile web apps will also improve.

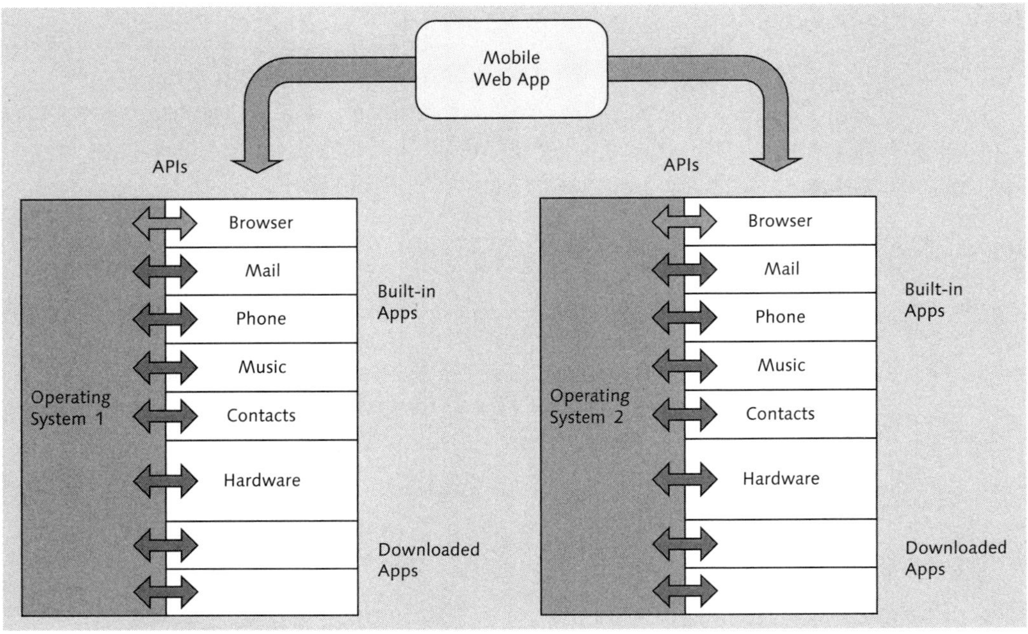

Figure 6.3 A Single Mobile Web App's Ability to Run on Multiple Platforms

An important impact of serving pages from a web server is that slower performance can be viewed as a sluggish interface because delays can take place while the browser is waiting for the next page to be delivered from the server. The result is that most end users view the app experience as more like browsing than like using a native app. Some tricks exist to save the URL link to look like an icon, and in some cases, it is possible to eliminate some of the traditional web browser frames, but it is difficult to completely eliminate the browser feel.

There are also some new techniques used to help with the delay time that is associated with downloading pages. The most popular is Asynchronous JavaScript and XML (AJAX). The AJAX technology allows web apps to send and receive data in the background without interfering with the current display. For example, AJAX is being used when only a small area of a screen needs to be modified; only that modified area is downloaded and displayed, as opposed to the entire page.

Another technique being used to improve performance is executing more server-side code in the browser using advanced JS, as compared to making decisions on the server, which would require an additional round trip to the server. A number of third-party advanced JS toolkits that make this coding easier, many of which specifically focus on mobile apps, have been released. Most of these toolkits focus on interfacing with the webkit, but others are starting to add some of the features that require API access. This is an area where functionality will improve as the browsers improve. Examples of some of the leading libraries or toolkits are shown in Table 6.4.

JS Kit or Library	Function
Sencha Touch	A JavaScript library built specifically for providing user interfaces for mobile web apps. It is used to develop user interfaces that have the look and feel of native apps.
jQuery Mobile	A touch-optimized web framework that is compatible with a wide variety of smart phones and tablet computers.
Dojo Mobile	A mobile framework that enables rapid development of mobile apps with a native look and feel.
MAKit	A mobile analytics kit that provides graphical controls with data visualization and analytics capability for mobile devices.

Table 6.4 List of Example JS Toolkits for Developing Mobile Web Apps

Sencha Touch and MAKit

In April of 2012, SAP and Sencha announced a collaboration that would focus on the integration of the Sencha Touch library into SAP Mobile Platform. jQuery is a component of Sybase Unwired Platform and MAKit has also been introduced as a standard component. We introduce you to MAKit in Chapter 15 of this book.

Another limitation of the mobile web app is that it interacts with only the mobile platform's browser. While a few of the operating systems' hardware APIs are available to the browser (such as touch screen and keyboard), most are not. Few, if any, of the APIs associated with built-in apps are available through the browser; as such, mobile web apps are very limited in their ability to access the OS APIs. A final limitation of the mobile web app is that it can be used only when the mobile device is actively connected to the network. These apps do not work in a disconnected environment.

6.1.3 Hybrid Apps

The comparison of the native app with the mobile web app clearly indicates that both have their own advantages and disadvantages. Mobile web apps can utilize the same app for all platforms and are implemented using commonly known web-development languages; however, they lack the ability to utilize most of the APIs provided by the OS. Native apps can access all OS APIs but require the development of a different app for each platform.

A third type of app has been introduced that combines native apps and mobile web apps with the goal of trying to capture the advantages of both while eliminating their disadvantages. This third type of app is referred to as the *hybrid app*. The hybrid app contains both a web portion and a native app portion. The web portion utilizes the HTML5/CSS3/JS languages, which provide the primary advantage of the mobile web approach. A native portion of the app can access most OS APIs, including hardware, built-in apps, and the user look and feel. The hybrid web app includes architecture that allows these two portions to communicate. Conceptually, the hybrid web app has all the advantages of both of the other approaches.

Two different techniques have been developed to bring about this communication between the two portions. The first is closer to the native approach, and the second is closer to the mobile web approach. The first technique defines an abstract set of APIs that can be mapped to the APIs supported by the operating system. The developer is able to create a single app utilizing HTML5/CSSS3/JS for the code portion. Within the code, the developer includes calls to the abstract APIs. (An example of such an abstract API is one that utilizes the camera to take a picture. The developer does not need to know the specific camera API for any of the operating systems—just the abstract camera API.)

This approach includes a set of plug-ins for each API and the operating system it supports. In the case of the abstract camera API, a plug-in is provided for iOS that is written in Objective-C and maps the abstract camera API to the iOS camera API. A second plug-in is developed for Android that is written in Java and maps the abstract camera API to an Android camera API. The systems come with a number of these abstract APIs and their associated plug-ins for all the supported operating systems. Tools are also available for developers to create their own plug-ins.

The final step with this approach involves the creation of an app using the common code portion and the plug-ins for the desired operating system. The creation of this app requires using the IDE for the desired operating system to generate a native app. For example, to generate an iOS app, the Xcode IDE (Mac only) is used; for an Android app, the Android IDE for Eclipse is used.

This approach uses a single tool for building hybrid web apps with the common code base and abstract API calls. A separate IDE for each supported OS is then required. Some examples of leading products using this approach include Adobe PhoneGap and Appcelerator Titanium. This procedure is shown in Figure 6.4.

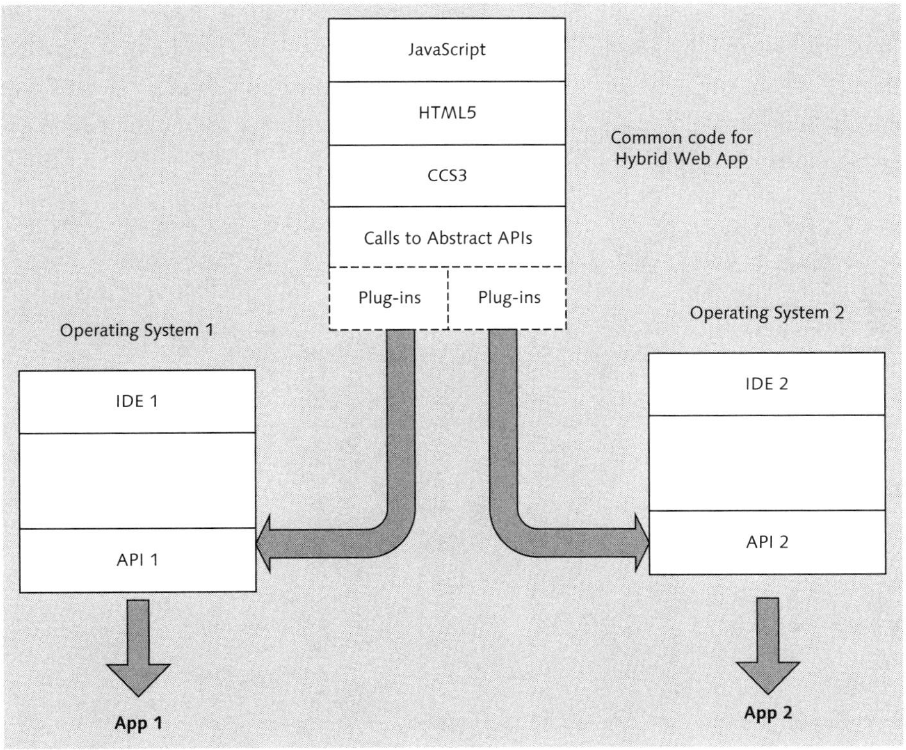

Figure 6.4 Hybrid Web Apps Built from a Common Code Base with Abstract APIs and OS-Specific Plug-ins

PhoneGap, Titanium, and SAP Mobile Platform

PhoneGap was developed by Nitobi and then purchased by Adobe Systems and donated to the public domain. It is currently the focus of an open source project. In April of 2012, SAP announced a relationship with both Adobe and Appcelerator, with the intention of integrating their frameworks with Sybase Unwired Platform. Starting with release 2.1.3, PhoneGap has been integrated into Sybase Unwired Platform.

A second technique for developing hybrid apps involves the use of a container application. The first component of this framework is the development of a native app (the container) for each of the mobile platforms to be supported. While these container applications could have access to all the APIs, they may not offer all of them to the developer. These apps are typically available in the app stores for the end user to download and configure.

The second component of this framework is the set of tools (i.e., Sybase Mobile SDK) that are used by the developer. These tools are used to develop a hybrid app that runs inside the container application. The single app can be downloaded and run in any of the containers. This hybrid approach allows a single hybrid app to run on every mobile platform for which a container has been developed. This approach is shown in Figure 6.5.

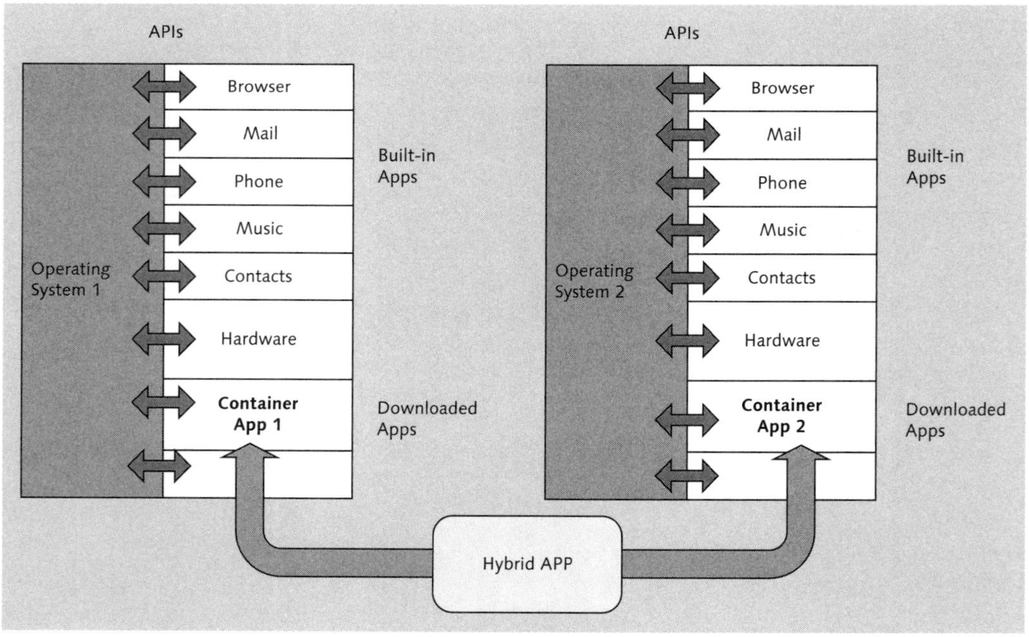

Figure 6.5 The Hybrid Approach of Running a Single App to Run within a Container on Any Mobile Platform

While the container applications can be distributed from a store, these individual hybrid web apps are not standalone apps and cannot be distributed from a store. A separate component of this framework is required to download them into the container on the end users' devices. Some of the leading examples of products that develop hybrid mobile apps utilizing this approach include Antenna Software, Pyxis Mobile, and Sybase Unwired Platform (SUP).

6.2 Hybrid Web Container Architecture

Sybase Unwired Platform (SUP) supports the development of hybrid apps utilizing the container approach. These apps run in these containers. Containers have been developed for the four major platforms: Apple iOS, Android, Microsoft Windows Phone, and BlackBerry. Most of these containers can be downloaded directly from the respective app store. An example of this process for the iOS container is shown in Figure 6.6.

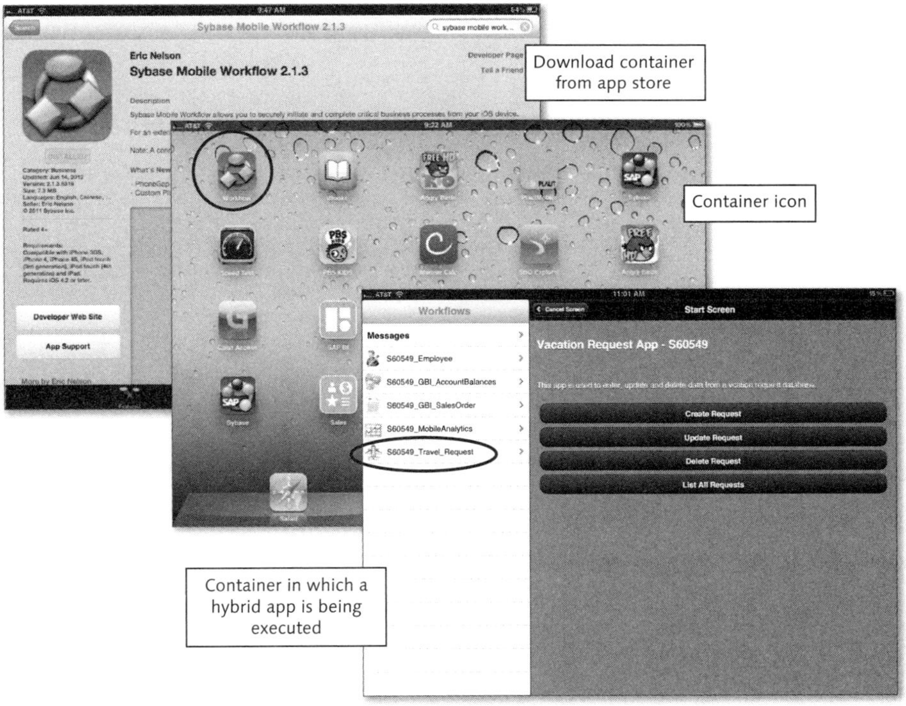

Figure 6.6 Installing an iOS Container and Running a Hybrid App

The end user firsts needs to install this free app from the Apple App Store to their device. Some configuration is then required to connect the device to the Unwired Server. (This was discussed in Chapter 5.) Once installed, the user simply clicks to start the app. The final step involves the selection of which hybrid app to execute. A similar process is used for the other three platforms.

While container applications have been written for the four major operating systems, these apps do not support all versions of the operating systems or, in some cases, all devices. This is particularly true for mobile devices that utilize the Android OS. A list of OS versions and devices currently supported is shown in Table 6.5.

Platform	Supported Versions	Tested Devices
Apple iOS	Apple iOS 4.3 Apple iOS 5.0 Apple iOS 5.1 Apple iOS 5.1.1 Apple 6.0	iPod Touch iOS 5.0 iPad 2 on iOS 6.0 iPad 3 on iOS 6.0 iPhone 4 on iOS 6.0 iPhone 5 on iOS 6.0
Google Android	Android 2.2 Android 2.2.1 Android 2.3.2 Android 2.3.3 Android 2.3.5 Android 3.0.0 Android 3.1 Android 3.2 Android 4.0 Android 4.0.4	Samsung Galaxy SCH-1800 Samsung Galaxy GT-1900T Samsung Galaxy SPH-D700 Samsung Galaxy GT-i9100G HTC Desire HTC Desire HD Motorola XOOM
Microsoft Windows Mobile	WM 6.0 Standard WM 6.0 Professional WM 6.1 Standard WM 6.1 Professional WM 6.5 Standard WM 6.5 Professional	Palm Treo HTC Tilt HTC 8900 HTC Tilt II HTC Snap S521 HTC HD2
BlackBerry	BlackBerry OS 5.0.x BlackBerry OS 6.0.x BlackBerry OS 7.0.x BlackBerry OS 7.1.x	Torch 9700 (only 5.0.x) Torch 9800 Bundle 2647 Torch 9810 Torch 9810 Bundle 998

Table 6.5 Operating Systems and Devices Supporting the Current Containers

An overview of the architecture used by SUP is shown in Figure 6.7. This view represents the runtime component of the architecture and consists of two major components. The left side of the drawing focuses on the container, while the right side focuses on the Unwired Server.

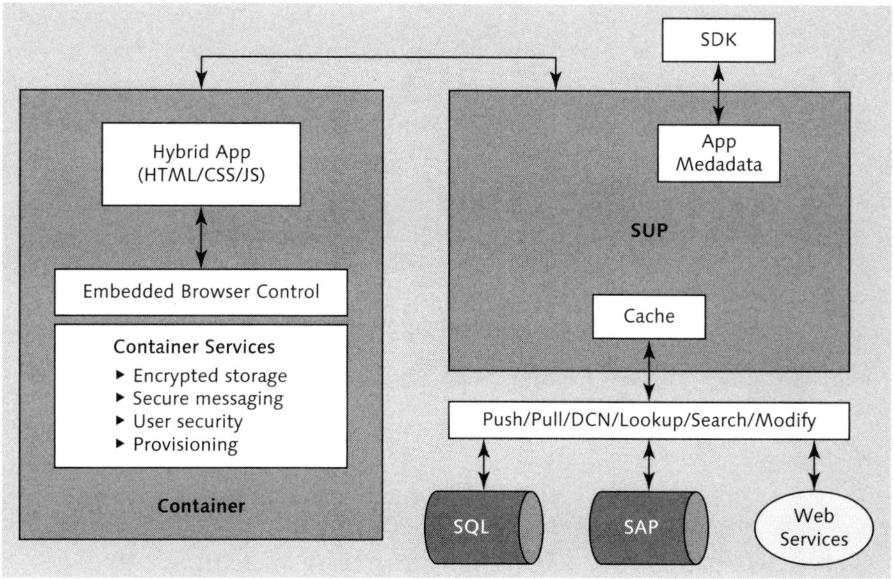

Figure 6.7 Runtime Architecture for Sybase Unwired Platform

The Unwired Server (on the right side) is a critical component of the runtime environment. This Unwired Server provides connections to all the enterprise information systems (EISs) and has the ability to search, access, and update, as well as pull and push data from these EISs. This is accomplished utilizing the metadata provided by the hybrid app. The Unwired Server also has a data cache for storing transient data. In addition to data, the Unwired Server can communicate with the container via messages with XML packets. The Unwired Server can also be used to download the hybrid apps to specified mobile devices.

Hybrid apps are stored and executed in the container (on the left side). The container has a directory service for tracking hybrid apps and an embedded browser within which the hybrid app code is executed. The container provides several services critical for apps used in the enterprise world, which are described in Table 6.6.

Service	Description
Encrypted storage	The container uses a small database manager on the device for storing local data. When encryption is enabled for this data, the user needs to supply a user name and password before running the app.
Secure messaging	Messaging between the device and Unwired Server is compressed and can optionally use HTTPS.
User security	A user name and password are required when starting the container and following a few minutes of inactivity.
Provisioning	Secure, over-the-air provisioning is provided between the Unwired Server and the device. A device must first be registered utilizing the unique device hardware key, and then the hybrid app needs to be provisioned to that specific device.
Native functionality	With version 2.1.3, PhoneGap is supported for the iOS and Android platforms.

Table 6.6 Services Provided by the Container

6.3 Hybrid App Design for the Hybrid Web Container

Apps created for the Hybrid Web Container are referred to as *hybrid apps* and are developed using the Unwired Workspace. The Unwired Workspace, a component of the Sybase Mobile SDK is a perspective that has been added to the Eclipse IDE (for more details, see Chapter 4). The Unwired Workspace needs a connection to the Unwired Server and to all the EISs. An overview of the steps required to develop a hybrid app is show in Figure 6.8. The following is an introductory overview, with details of many of these steps presented in the following chapters.

The first step is the creation of a named project, which is a container in which all the code and artifacts associated with the hybrid app are saved. This container is saved as a folder in the developer's defined workspace directory on the developer's machine.

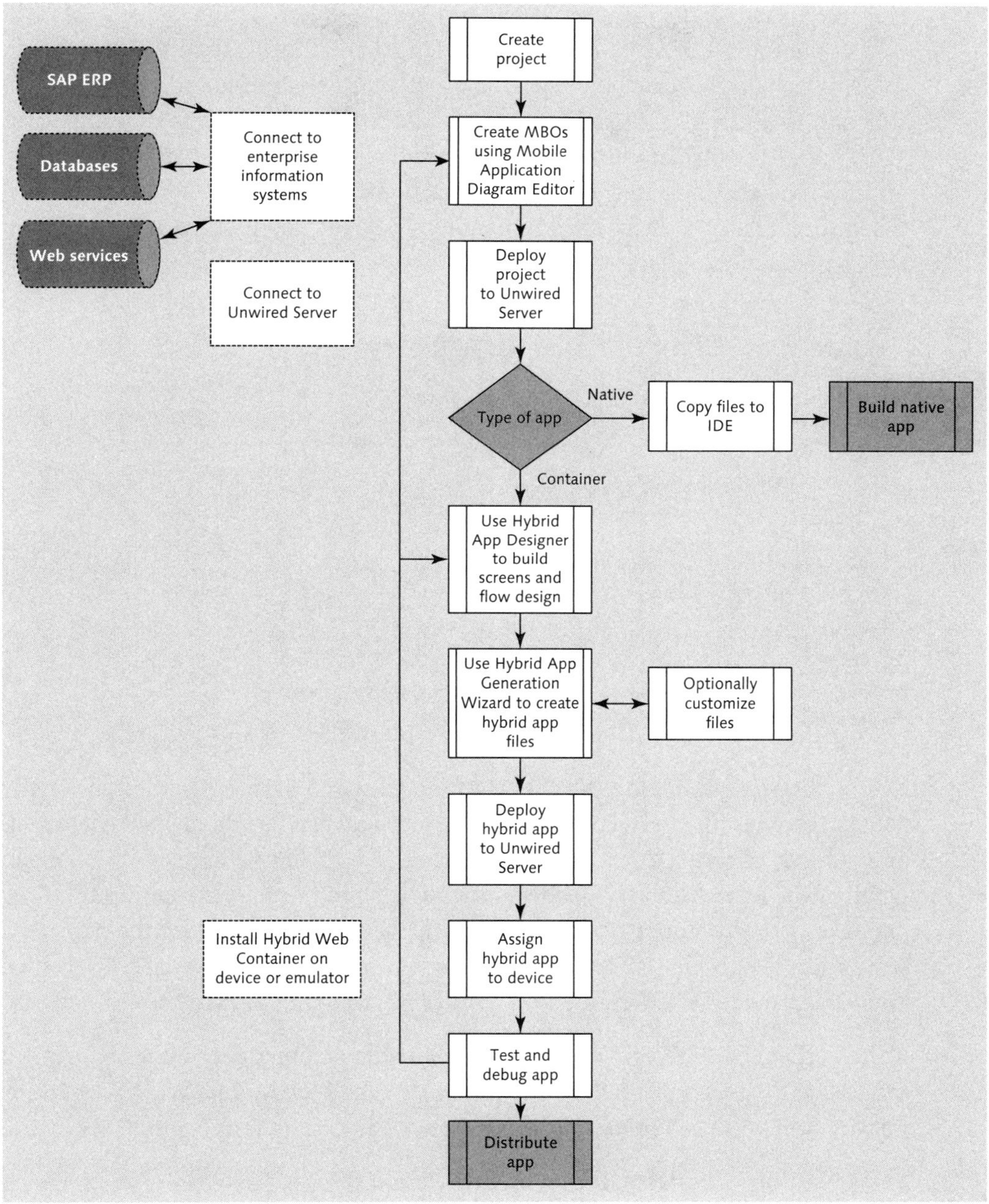

Figure 6.8 Steps Required for the Development of a Hybrid App

The second step is the creation of one or more Mobile Business Objects (MBOs). One of the tools, the Mobile Application Diagram Editor, is used to construct an MBO, which encapsulates the business process objects and their data into a reusable unit that can be accessed by multiple clients. The first step in developing an MBO is locating the data source from the available EISs. The next step is to specify the specific data attributes, operations, relationships, and parameter definitions. (The process is covered in detail in Chapter 7.) An example of a project with two MBOs is shown in Figure 6.9.

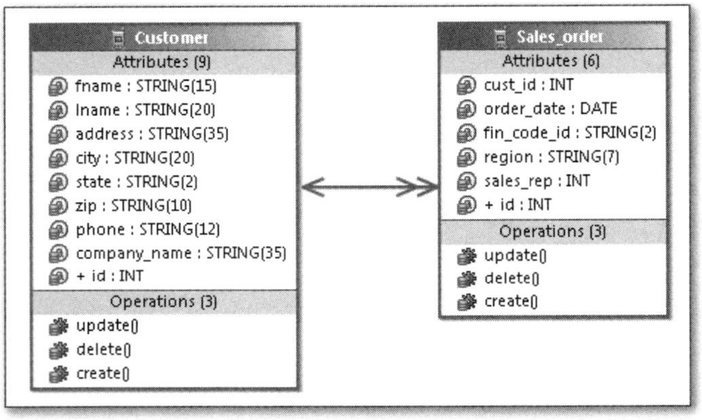

Figure 6.9 Two Related MBOs

The next step is to deploy the MBOs to the Unwired Server. The design for the MBO is saved in the project (in the workspace) and can be edited and redeployed any number of times. Once the MBO is deployed, the decision needs to be made whether the application will be developed as a hybrid or native application. If the decision is to develop a native application, the project (including the MBOs) is copied to the specific IDE for the development of that native application. (Native application details are presented in Chapter 12 through Chapter 14.)

If the decision is to develop a hybrid app, the development continues by using a tool called the Hybrid App Designer (formerly called the Mobile Workflow Forms Editor), which is used to design the screens, such as shown in Figure 6.10.

The Hybrid App Designer is also used to design the flow between these screens. An example flow design is shown in Figure 6.11. (The Hybrid App Designer is described in more detail in Chapter 8.)

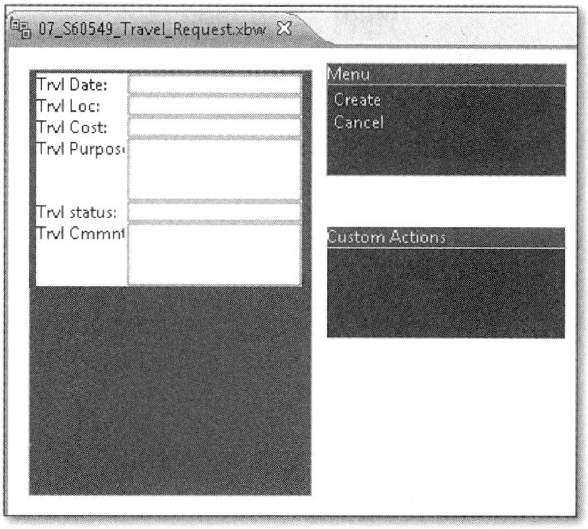

Figure 6.10 Using the Hybrid App Designer to Create Screens

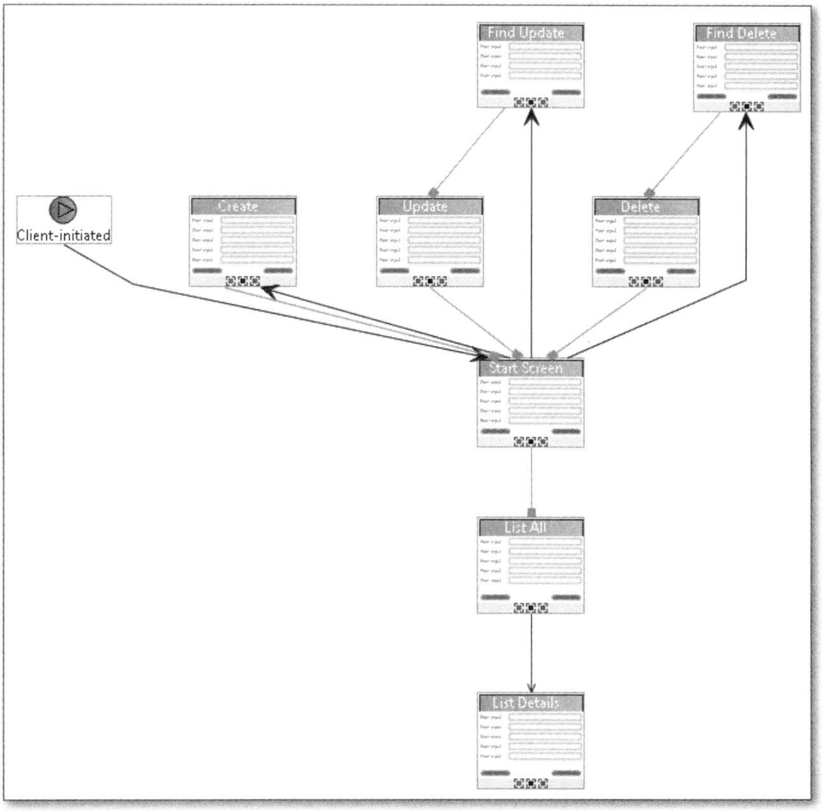

Figure 6.11 Using the Hybrid App Designer to Create Flow Design

Once the design is complete, the next step involves the generation of the hybrid app package. First, the model is saved to the workspace; then the deployment of that hybrid app can be made to the Unwired Server; and, finally, the hybrid app can be assigned to the selected devices. These options, along with a description of each one, are described in Table 6.7.

Option	Description
Generate hybrid app	Builds all the code for the hybrid app and saves it in the workspace
Deploy hybrid app	Deploys that generated hybrid app to the Unwired Server
Assign hybrid app to device(s)	Assigns (i.e., downloads) the generated hybrid app to one or more devices

Table 6.7 Hybrid App Deployment Options

The final step is testing and debugging. This process may include modifications to both the MBOs and the flow design. Once the process is complete, the resulting hybrid app can then be assigned to the user community.

6.4 Hybrid App Patterns for the Hybrid Web Container

The container is used to implement a number of lightweight hybrid apps for solving a number of different business problems. Some of the common patterns (i.e., models) developed for the container include online lookup, server notification, and cached data. These basic patterns can be used as building blocks for more complex patterns. The following is a brief overview of these patterns.

6.4.1 Online Lookup

The online lookup pattern is one of the most common applications and is triggered when the end user requests data. The first screen is used to specify the parameter values for the desired data. The steps in the online lookup process pattern shown in Figure 6.12 are as follows:

❶ The end user provides the values for the required parameters and then clicks a button or menu item, which initiates an online request utilizing an SQL query.

❷ Assuming that the MBO data is not cached on the Unwired Server, the Unwired Server places a request for the required data to be retrieved directly from the EIS.

❸ The EIS passes the data back to the Unwired Server.

❹ The Unwired Server passes the requested data back to the mobile application.

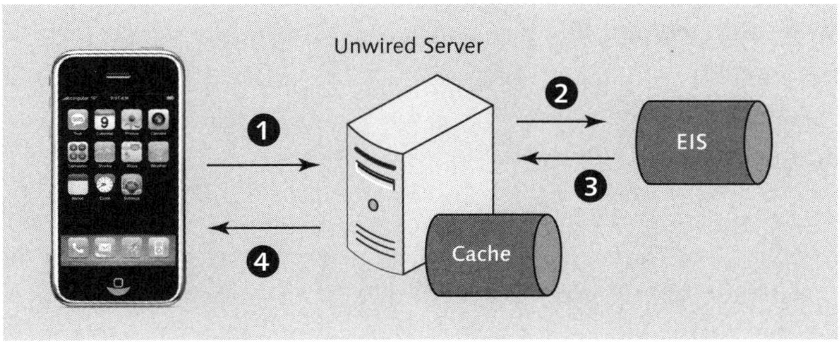

Figure 6.12 Steps Required for an Online Lookup Pattern

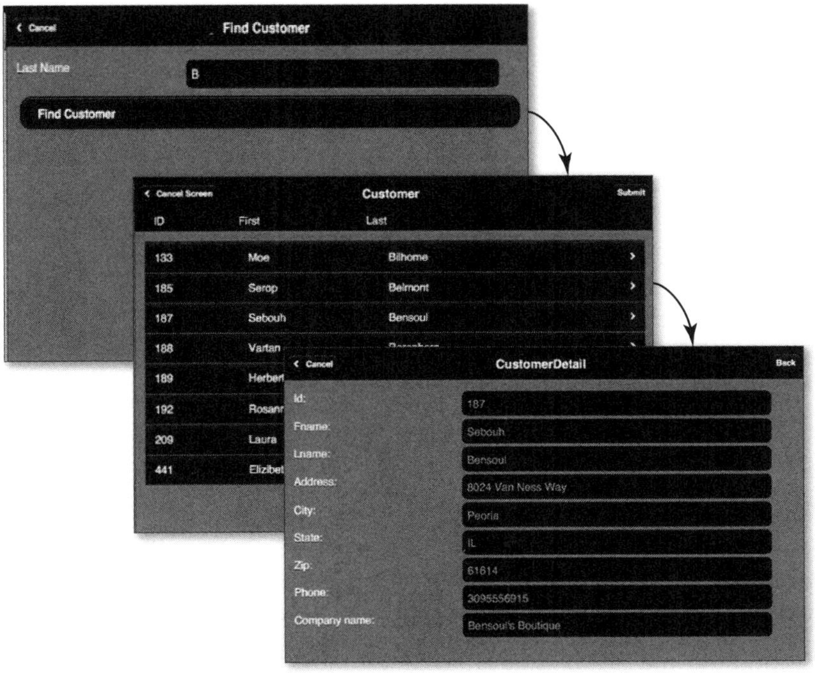

Figure 6.13 Example Hybrid App Using the Online Lookup Pattern

Figure 6.13 shows the screens associated with an example of an online lookup hybrid app. The example application is used to retrieve information about customers. The first screen narrows the retrieved customers by requesting the first character or two of the desired customer's last name. In this example, the letter *N* is entered, and the FIND CUSTOMER button is clicked. This button causes an online request to take place using an SQL query to retrieve all the customer records that satisfy the criteria. The second screen shows a list of those customers. The final step is to click the desired customer, and the third screen, which contains the customer's detail information, appears.

6.4.2 Server Notification

The sever notification pattern is a model in which the execution comes from some external activity that takes place related to the EIS. A component of the application includes a set of data matching rules for MBO data. If any changes to this data match these defined rules, then the application is triggered, and a message can be generated. This message includes a fixed, formatted string in the subject that can be used to pass parameter data. In this pattern, the application and its MBO are stored only on the Unwired Server and in the container on the mobile device. Figure 6.14 shows the steps required in a typical server notification pattern:

❶ A change takes place in the EIS that matches a predefined rule (for example, a new sales order is created).

❷ The cached data in the MBO is updated on the server.

❸ The Unwired Server sends a notification to the mobile client.

❹ The end user opens this message and sees the resulting data change.

An example server notification application is shown in Figure 6.15. Note that end users do not select and run this app from the list—they can only respond when a message is received. The first screen shows that one message is currently available and that several hybrid apps have been downloaded to the container. These hybrid apps in the container have nothing to do with a sales order application. When the end user clicks the message, a second screen that provides details about the message, including who it is from, the date received, and the subject, appears. Finally, clicking this message displays a sales order screen, from the application, that is saved on the Unwired Server. The subject of the message has been inspected, and the value "2001" has been used as a parameter for the SQL query. The information displayed on this third screen is for sales order 2001.

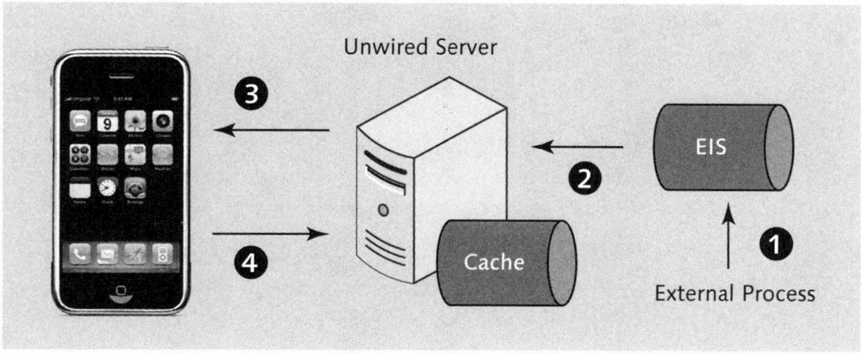

Figure 6.14 Steps Required for a Server Notification Pattern

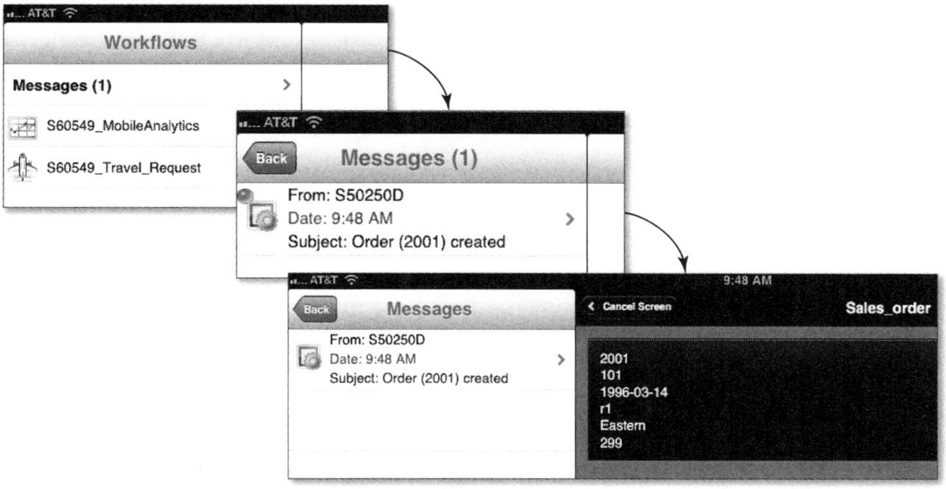

Figure 6.15 Example Application Using the Server Notification Pattern

Several different supported server notification scenarios are shown in Table 6.8.

Source	Description
Mobile app	A message sent by another mobile application. For example, a user may require an approval from a manager. This scenario requires an email account to route messages.
Data change notification (DCN)	A data change notification (DCN) sent by an EIS. For example, a sales order is created in the system, and a program on the EIS sends a DCN to the Unwired Server, which sends a notification to an application used by a customer service representative.
Scheduled load	A scheduled data load between the EIS and the Unwired Server. This requires a cache group on the Unwired Server with a scheduled update policy.

Table 6.8 Server Notification Scenarios

6.4.3 Cached Data

The cached data pattern takes full advantage of the data cache that is a component of the Unwired Server. One of the steps in the MBO design involves selecting the cache policy. An example policy might be to update the cache from the EIS every 20 minutes. With this policy, the data returned to the hybrid app is always data that is saved in the Unwired Server cache. A separate component of the Unwired Server is responsible for periodically refreshing that data in the cache on the Unwired Server. The advantages of this pattern are a quicker response to the queries to the mobile device and less overhead on the EIS and Unwired Server.

The data flow for this data cache pattern is shown in Figure 6.16:

❶ The Unwired Server periodically updates the data stored in the cache as an independent process.

❷ The user requests data from the Unwired Server via an SQL query.

❸ The Unwired Server returns the current data values stored in the Unwired Server cache.

Except for the cache policy and performance issues, the example app could appear to the end user to be the same as the hybrid app shown in Figure 6.13.

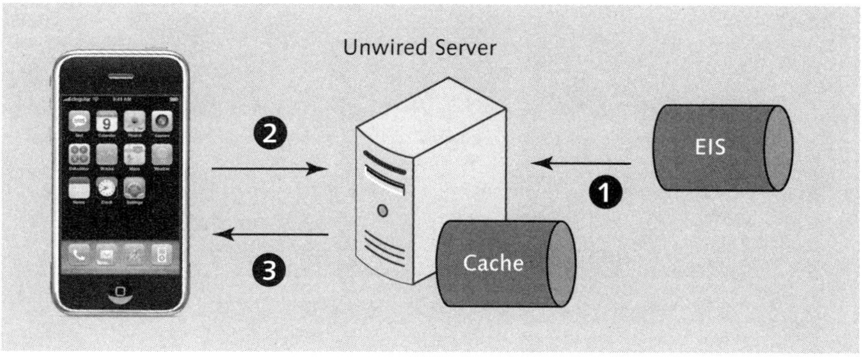

Figure 6.16 Steps Required for the Cached Data Pattern

6.5 Summary

The focus of this chapter has been the Hybrid Web Container. The chapter first presented an overview of the different approaches used to develop the three most common mobile applications: native applications, mobile web applications, and hybrid web applications, setting the stage for our focus on the Hybrid Web Container. This was followed by a more in-depth look at the container architecture used by Sybase Unwired Platform. The process of developing a hybrid app that runs in those containers, as well as three of the most common patterns used to develop such hybrid apps, were presented.

The Mobile Business Object (MBO) forms the business logic for the defining data required from your backend systems. This chapter focuses on the role of the MBO and the steps required to build MBOs.

7 Building Mobile Business Objects

One of the characteristics of SAP Mobile Platform is the ability to develop mobile apps that can utilize data from existing enterprise information systems (EISs). The cornerstone of this functionality is based on the MBO and its use by the Unwired Server and the mobile app. This chapter introduces the concept of the Mobile Business Object (MBO), discusses the importance of the MBO, and provides a detailed description of how to use the Unwired Workspace to build an MBO.

Building MBOs is one of the first steps in the creation of hybrid apps—however, it's important to note that MBOs are actually used in both hybrid *and* native apps. While the examples being discussed in this chapter often refer to hybrid apps, the development procedure is identical for both. In fact, the MBO is the one single development process that is common between these two approaches.

This chapter starts out with an introduction to the concept of MBO, followed by a description of the Unwired Workspace tools that are used to build the MBOs. We then go into the technical details of building MBOs, which can be divided into five steps:

1. Create the MBO based on the backend system data (Section 7.3).
2. Control data flow and performance (Section 7.4).
3. Create database operations (Section 7.5).
4. Create object queries (Section 7.6).
5. Create relationships between MBOs (Section 7.7).

7.1 Introduction to Mobile Business Objects

MBOs help form the business logic for the development of mobile apps. The following is a more detailed definition of an MBO, followed by a brief introduction to the concepts involved in developing an MBO.

7.1.1 Defining MBOs

An MBO provides the logical description of the portion (or slice) of the backend EIS data required for the mobile app. EISs are large data stores consisting of many tables, each of which contains a number of attributes and many rows of data. It is not possible to replicate these large data stores into the Unwired Server or onto the mobile device. The primary role of the MBO is to provide the business logic to filter (i.e., simplify and limit) the EIS data to that which will be exposed to the mobile app.

The creation of an MBO is a logical data modeling process. The first step involves identifying the backend data source and connecting to it using an existing connection profile. The second step involves using the Unwired Workspace to create a project and then define the MBOs. Once the MBOs are defined and customized, the MBO package is both saved locally in the project folder and deployed to the Unwired Server. The MBO package saved on the Unwired Server manages the synchronization of data between the EISs, Unwired Server, and user's device. The MBO artifacts saved in the project folder are used to develop the hybrid app that generates the code required to interact between the Unwired Server and the database on the mobile device.

A given hybrid app can utilize one or more MBOs, and a given MBO can be reused by more than one hybrid app. The EIS data sources that can be used to generate MBOs include SAP systems, database systems, and web services. The MBO plays the role of data abstraction among the EIS, Unwired Server, and mobile device. This abstraction is shown in Figure 7.1.

The MBOs include the following information:

- Implementation details, which is metadata that includes information about the data from the data source
- Abstract-level details, including the following:

  - List of the desired attributes from the selected data source, excluding all attributes that are not required by the mobile app
  - Parameters used to select desired rows of data
  - MBO relationships that are used to link MBOs with other MBOs based on common attributes
  - Inclusion of predefined data operations such as create, update, and delete
- Additional details, including about the following:
  - Data caching on the Unwired Server and the mobile device
  - Data synchronization among the EIS, Unwired Server, and mobile device

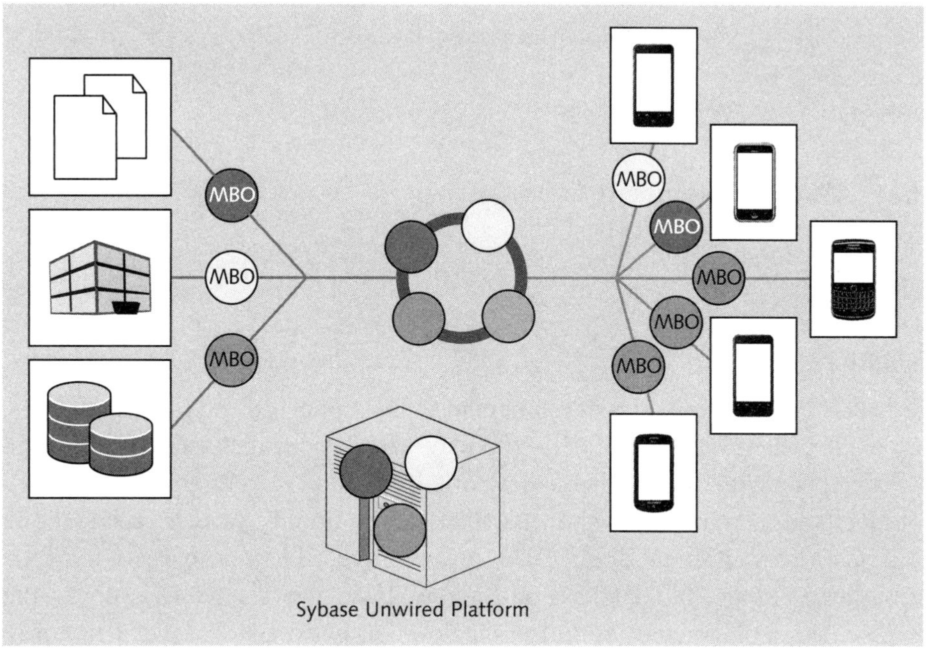

Figure 7.1 The Abstraction Role of the MBO

The abstract flow of data is shown in Figure 7.2.

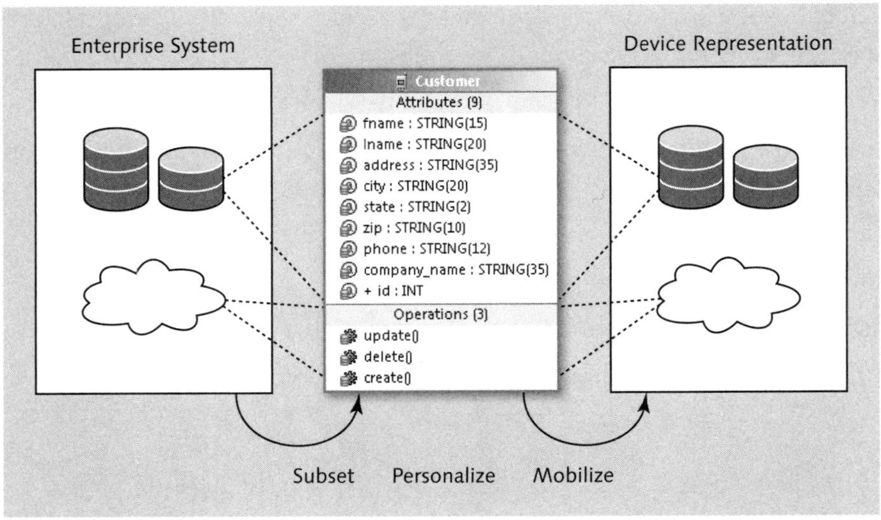

Figure 7.2 Data Abstraction Model

7.1.2 MBO Development Concepts

The following is a description of the major concepts involved in the development of an MBO.

Mobility Patterns

The MBO allows mobile app developers to build support for mobility patterns—both data virtualization and operation replay. The data virtualization presents the app developer with a normalized view of all data subsets, eliminating the need to understand the unique physical characteristics required to access each EIS. The MBO provides a Cache Database (CDB) on the Unwired Server that lays between the backend EIS and the database on the mobile device. The MBO, utilizing the specified design parameters, optimizes the flow data between these two data stores. Operation replay provides support for data flowing from the mobile device back to the EIS.

Attributes

One common characteristic of the EIS is that their data structures can be viewed as a table whose columns are referred to as attributes and whose rows are referred to as *tuples*. A tuple contains attribute values (i.e., information) about a specific object.

The attributes selected for the MBO define the data structure for the mobile device (i.e., device-side data store). These attributes are usually a subset of all the attributes in the EIS data structure with the goal of limited data transfer to the mobile device.

MBO Performance

The design of the MBO includes several options that can be used to improve the performance of the mobile app. These options include load parameters, caching policy, and synchronization parameters, and are designed to decrease the time required for mobile apps to access the required data. The MBO includes parameters for portioning the CDB data to limit download size to mobile apps and rules for when and how often to update the CDB data. For example, all customer data stored in the CDB may be updated from the EIS on an hourly basis and partitioned by customer region (referred to as a *load parameter*). When the mobile app is executed, it syncs its data store (all or a subset of) on just the desired CDB partition and then presents this data to the user.

MBO Operations

The design of the MBO may include predefined operations used to modify the EIS database. The standard database operations are referred to as *create, update, and delete* (CUD). The update and delete operations normally follow a find or load operation. For example: find Customer X, change some of the attributes, and then update that record in the EIS. When the EIS is a database, the logical CUD operations are performed by system-generated SQL that can be modified by the developer. These operations support validation and concurrent changes that can be detected when the original EIS data has been modified since the original data was also retrieved.

MBO Relationships

Relationships can be defined between two MBOs. These relationships involve the linking of attributes to those from another MBO. The relationships can be one-to-one, one-to-many, or many-to-many, and both bi-directional and composite relationships are supported. The linking together of two or more MBOs assures that their processing is handled consistently.

7.2 MBO Development Tools

The Unwired Workspace (which, as you'll recall, is a component of the Sybase Mobile SDK) is a set of graphical tools used to develop both MBOs and hybrid apps. These tools have been developed as an add-on perspective (MOBILE DEVELOPMENT) for Eclipse, a well-excepted multi-language software development environment. In Chapter 4, the installation of the Sybase Mobile SDK was presented. Eclipse, with the MOBILE DEVELOPMENT perspective loaded, is shown in Figure 7.3

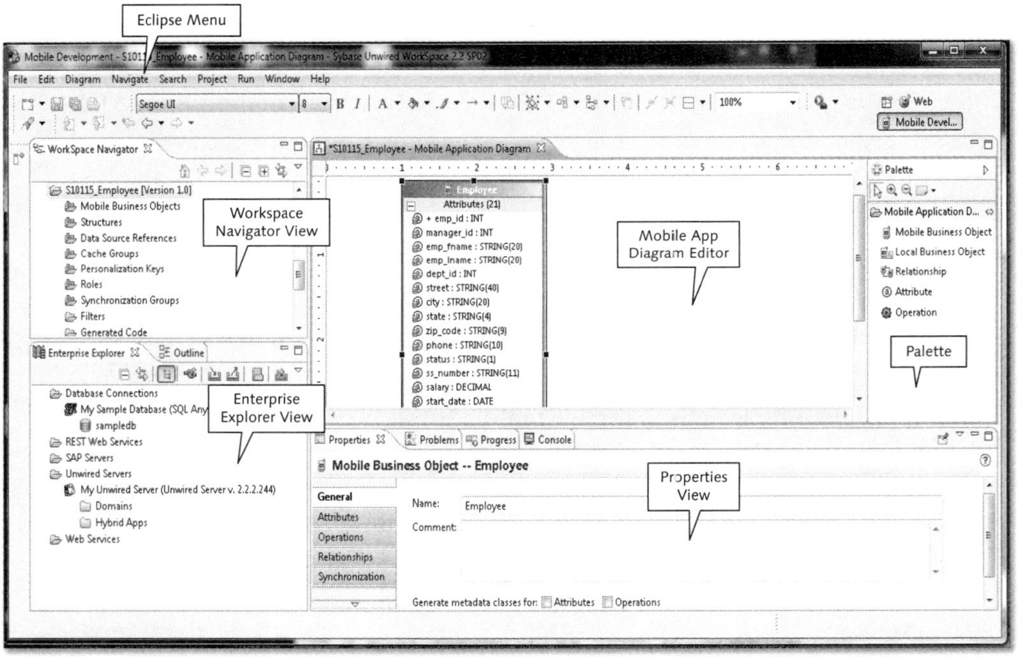

Figure 7.3 Eclipse, with the Mobile Development Perspective Open

The following is a description of the major components of Eclipse with the MOBILE DEVELOPMENT perspective.

7.2.1 Eclipse Menu

The Eclipse menu appears on the top of the window and provides a large number of commands. A few of these commands are required for developing MBOs and hybrid apps. The first menu item is WINDOW • OPEN PERSPECTIVE • MOBILE DEVELOPMENT. As shown in Figure 7.4, it is used to open the MOBILE DEVELOPMENT perspective.

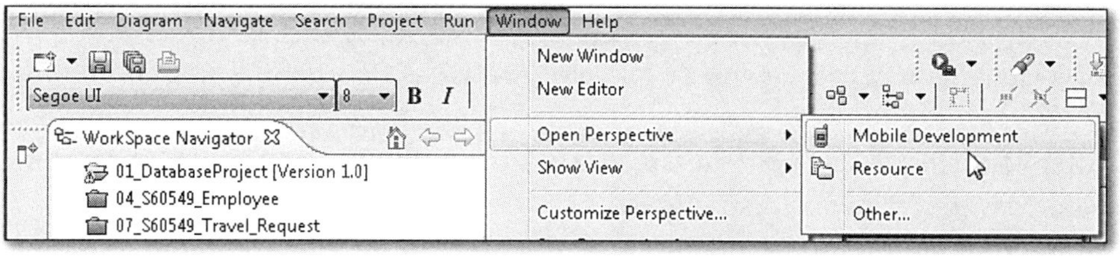

Figure 7.4 Menu Path for Opening Mobile Development Perspective

One of the challenges with using the Eclipse tool is that, often, several views open at the same time. The user may accidently close one of these views, and then the user is unable to continue developing the project. This problem can be solved with the second menu, with the path WINDOW • RESET PERSPECTIVE. The RESET perspective, as shown in Figure 7.5, resets the window and all its views back to its starting format.

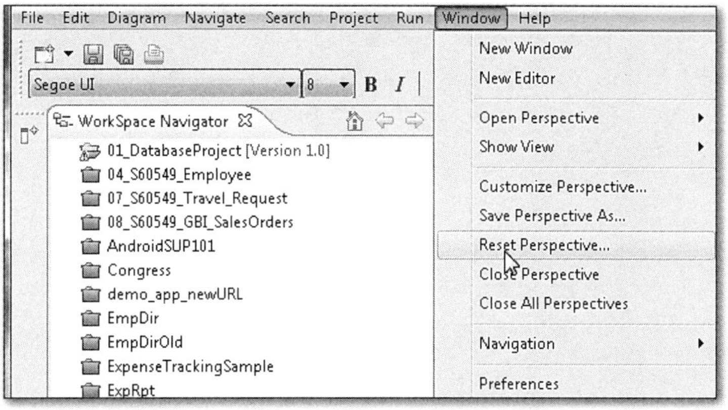

Figure 7.5 Menu Path for Resetting the Perspective

The third menu item is used to create a new project. This project consists of a folder (with the same name as the project) created in the workspace folder. All the artifacts for all the MBOs and the hybrid app are stored in this project folder. The menu path FILE • NEW • MOBILE APPLICATION PROJECT, as show in Figure 7.6, can be used to create (and open) a new project.

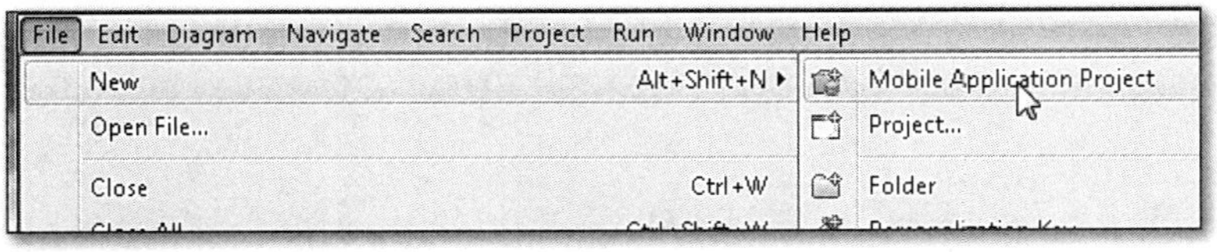

Figure 7.6 Menu Path for Creating a New Mobile Application Project

After you select the menu item, a dialog box requests a PROJECT NAME. The project name has the following restrictions:

- Must be unique within the workspace
- Cannot contain a space, #, %, or other non-ASCII character

The location can be changed from the default workspace. After a valid name is entered, the FINISH button is enabled, as shown in Figure 7.7. Click this button to create the new project.

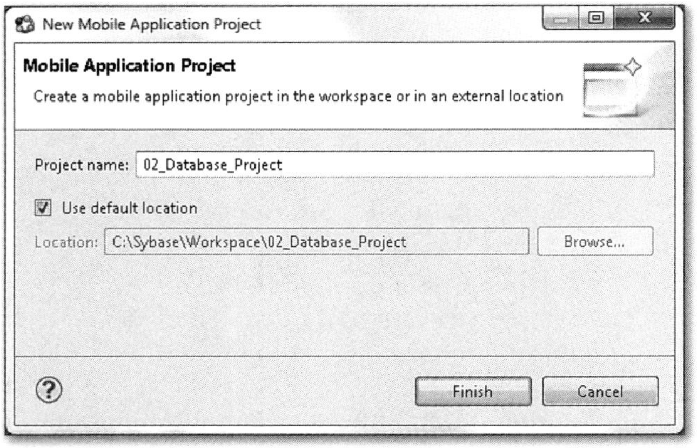

Figure 7.7 Dialog Box for Project Name

7.2.2 Workspace Navigator View

The WORKSPACE NAVIGATOR view is located in the upper left-hand corner and displays the list projects that have been created in the current workspace directory. The icon for each project is a closed folder. When a project is opened, the icon changes to an open folder. While multiple projects can be open at the same time, you can save time and confusion by closing a project when work is complete and having only one project open at a time. Right-clicking the folder provides several options, including OPEN PROJECT and CLOSE PROJECT. Once a project is open, clicking it opens the folder and shows its contents. A new project starts with a number of sub-folders used to save project artifacts. Each step of the design process for both MBOs and hybrid apps creates more artifacts that are saved in these folders or the additional sub-folders that may have been created. The initial folders created for the 01_DATABASEPROJECT is shown in Figure 7.8.

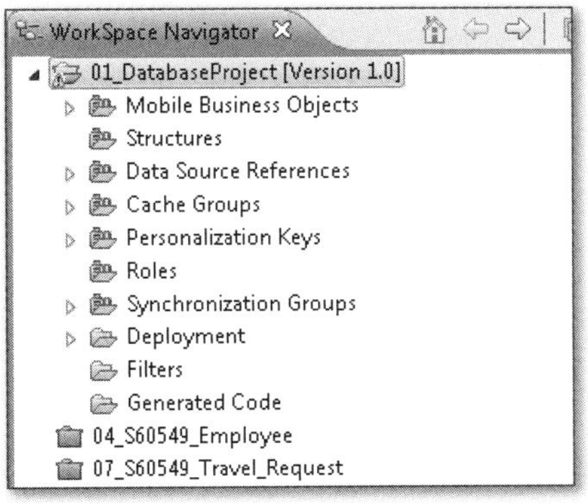

Figure 7.8 Workspace Navigator with an Open Project Folder

7.2.3 Enterprise Explorer View

The ENTERPRISE EXPLORER view is located in the lower-left corner and displays the connections to the Unwired Server and the various EISs. The connection properties need to be defined for each of these systems, and then the connection must be open before they can be used. (A discussion of defining and opening these connections is presented in Chapter 4.) The ENTERPRISE EXPLORER view is shown in Figure 7.9.

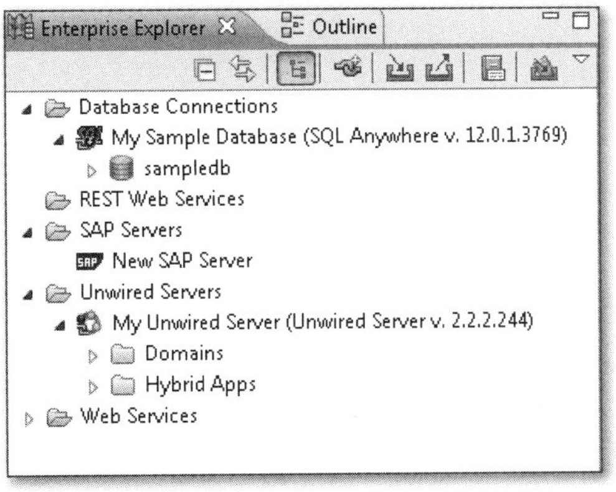

Figure 7.9 Enterprise Explorer View

7.2.4 Mobile Application Diagram Editor

The PRIMARY WORK view for Eclipse is located in the upper-middle area of the window. This view is used for the various graphical development tools. When more than one tool is in use, the tools each appear as a separate tab within this view. The tool used for developing MBOs is the *Mobile Application Diagram Editor*. This tool is used to created and customize the MBO. When a new project is created, the Mobile Application Diagram Editor is automatically started, and can be used to create multiple MBOs for the same app. When more than one MBO is present, you must select the desired MBO by clicking it before modifications can be made to it. The current or selected MBO appears with a frame around it. When more than one tool is currently open (for example, editing both MBOs and a hybrid app), each tool appears in a separate tab. Selecting the tab brings forward that tool. The Mobile Application Diagram Editor for a pair of MBOs is shown in Figure 7.10. The unselected tab is a tool for building a hybrid app.

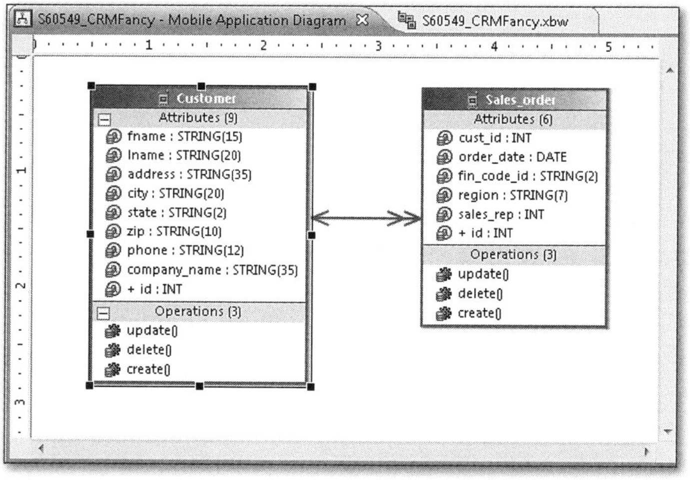

Figure 7.10 Mobile Application Diagram Editor

7.2.5 Properties View

The PROPERTIES view appears in the lower right-hand corner of the screen and contains several tabs. While the PROPERTIES view is used most often, other tabs exist—for example, for displaying error messages and showing deployment progress. Within the PROPERTIES view, there are a number of choices on the left-hand side, and in some cases, each of these choices has multiple tabs across the top of its window. The PROPERTIES view is used to enter and modify properties associated with the object that is currently selected in the WORK AREA view. When the Mobile Application Diagram Editor is the current tool and a specific MBO is selected, the PROPERTIES view provides many options for modifying the MBO. These options are presented in detail later in this chapter. The PROPERTIES view shown in Figure 7.11 is for the previous MBO example shown in Figure 7.10.

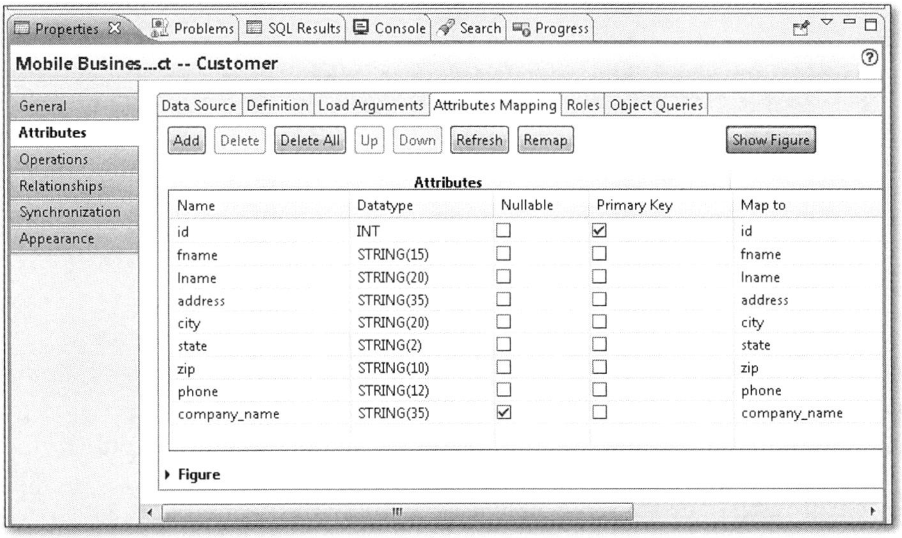

Figure 7.11 Properties View

7.2.6 Palette

The Mobile Application Diagram Editor provides two different approaches for creating an MBO. The first is the drag-and-drop approach—desired data is first located by browsing through the EIS in the ENTERPRISE EXPLORER view and then dropped onto the workspace. The second approach uses a menu approach, which utilizes the *Palette*.

The Palette appears in the upper right-hand corner of the window and provides operations that can be performed based on the current tool. These functions, or wizards, are discussed in detail in Section 7.3.3. The Palette is shown in Figure 7.12.

Figure 7.12 Palette

7.3 Creating MBOs

The remainder of this chapter offers a detailed description of the MBO build process, which begins after the creation of a named project. This starts the Mobile Application Diagram Editor and creates a folder named after the project in the Unwired Workspace. This folder is used to store all source components (artifacts) required for building the app, including those related to the MBO. Right-clicking the empty screen provides a dialog box for switching between the two profiles within the Mobile Application Diagram Editor:

- BASIC
  This profile displays only a subset of the features available and allows you to deploy and build MBOs.
- ADVANCED
  This profile includes all the features, wizards, and properties that enable additional MBO customization not offered in BASIC, and is the recommended option.

The following is an overview of the steps for creating an MBO:

1. Locate and connect data. Identify the required data and then connect to the backend EIS that can provide that data.
2. Filter or limit. Eliminate attributes and rows of data that are not required for the application.
3. Improve performance. Customize the MBO to improve performance, including issues such as caching and synchronization.

4. Deploy. Deploy the project (MBO) to the Unwired Server. The Unwired Server uses the MBO to coordinate the flow of data between the EIS(s) and the mobile devices.
5. Save the project. Save the MBO artifacts in the project folder. These artifacts are used in the development of the data flow portion of the application.

In this section, we explain the initial creation of an MBO. This first step depends somewhat on the EIS that is selected. In Section 7.3.1, we present the creation of an MBO that uses a database for its data source. In Section 7.3.2 and Section 7.3.3, we show the differences that are involved in creating an MBO when it uses a web service and when is uses an SAP system as its data source.

7.3.1 Database MBO

For this example, we assume we are developing a hybrid app that will allow the end user to request a sub-list of the company's employees and then select detailed information about a specific employee on that list. We create the MBO using a database table and the drag-and-drop procedure.

Exploring the EIS Database Table

The EIS that contains the necessary data for the app has been identified, and it is an SQL Anywhere database with the name SAMPLEDB. The data is in a table with the name EMPLOYEE. The connection properties for the database are defined, and the database is connected. The next step, as shown in Figure 7.13, is to locate the table folder within that database.

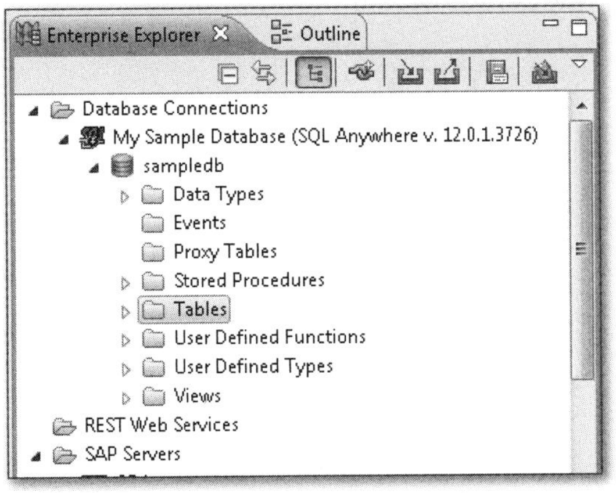

Figure 7.13 Opening Table Folder in the Sampledb Database

You can open this table folder by right-clicking it and selecting OPEN. (If the folder does not have the caret in front of it, it may be necessary to disconnect and reconnect the EIS.) As you can see, a number of different tables exist in this database. The first limiting step is to select one table and eliminate the remainder of the database; in this case, select the EMPLOYEE table. Before creating the MBO, it is often useful to view the table structure and some of the data in the table. You can accomplish both of these, as shown in Figure 7.14, by right-clicking the table name and selecting EDIT DATA.

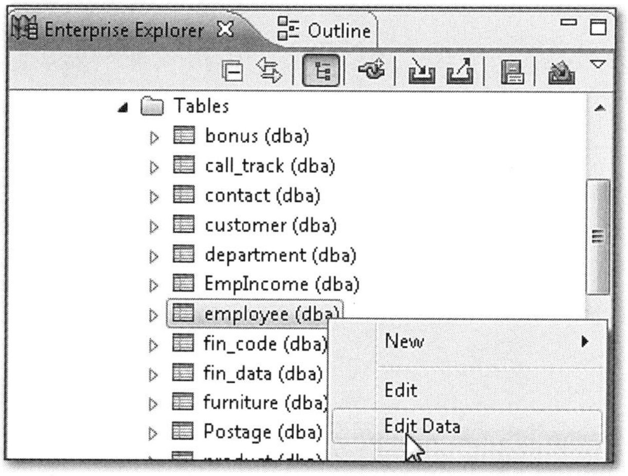

Figure 7.14 Requesting Display of the Data Structure and Sample Data

The first dialog that appears provides a list of all the data columns (i.e., attributes) in the table, as well as the data type and length information. It also identifies the primary key (PK), as shown in Figure 7.15. To close this dialog box, click the OK button.

The second dialog box displays the sample data, as shown in Figure 7.16. To close this box, click the X on the EMPLOYEE tab.

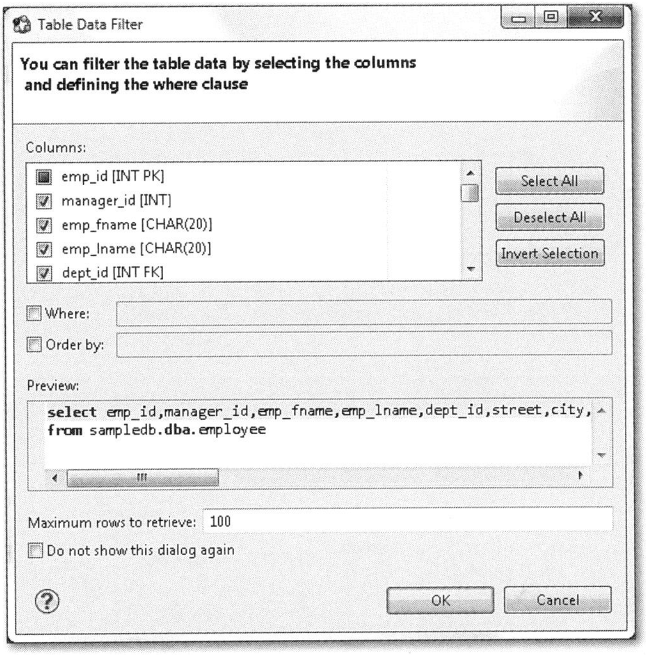

Figure 7.15 Data Columns in the Employee Table

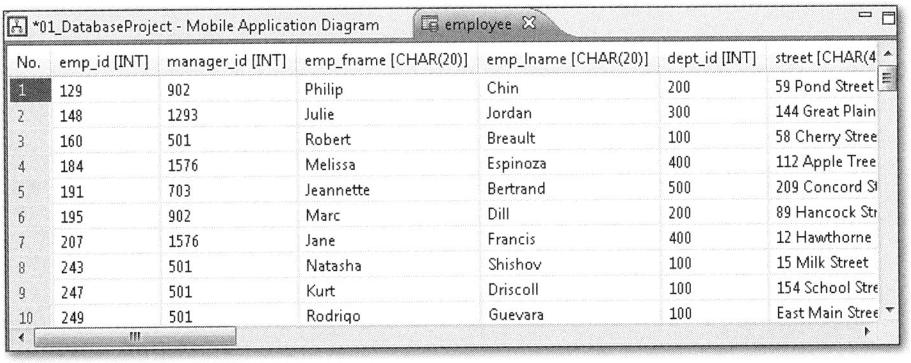

Figure 7.16 Sample Data from the Employee Table

Dragging and Dropping the Table

To start the MBO design, select the EMPLOYEE table, hold down the left mouse button, drag the EMPLOYEE table to the Mobile Application Diagram Editor, and then release the key to drop it. A Quick Create Wizard, as shown in Figure 7.17, appears, showing the default MBO name, the SQL that selects all columns in the table, and the option to select the three CUD operations. For this initial MBO, uncheck all three of the CUD operations. Once this is complete, click the OK button, and the wizard proceeds with creating the MBO.

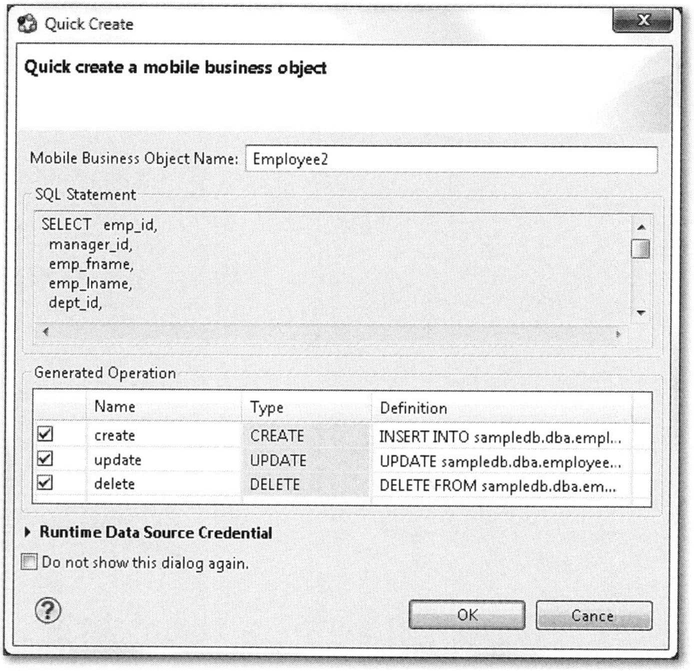

Figure 7.17 Creating MBO with Drag and Drop Using Quick Create

The initial version of the MBO contains all the attributes that were in the original data table. This MBO can now be customized by changing the MBO's properties, such as eliminating some of these attributes. To view the properties of the MBO, select the specific MBO by left-clicking. A frame appears around the selected MBO, as shown in Figure 7.18, and then you can view and modify properties in the PROPERTIES view.

Removing Attributes

To open the ATTRIBUTES MAPPING dialog box, use the path PROPERTIES • ATTRIBUTES • ATTRIBUTES MAPPING. This dialog box is shown in Figure 7.19. The box displays a list of the attributes currently making up the MBO, as well as their one-to-one mapping to the database table. You can remove attributes from the MBO by selecting an unwanted attribute and clicking the DELETE button. Repeat this process until all the unwanted attributes have been removed from the MBO. Changes that have been made to the MBO are not immediately saved to the project folder; the developer is responsible for determining when to save changes. In order to do this, click the icon that looks like a stack of disks, just below the Eclipse menu.

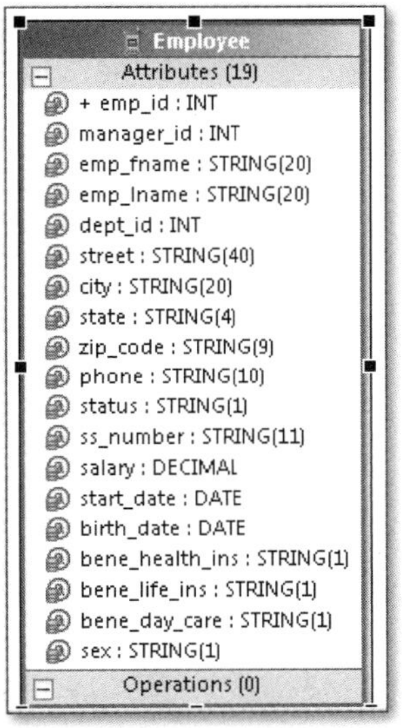

Figure 7.18 MBO with the Same Attributes as the Database Table

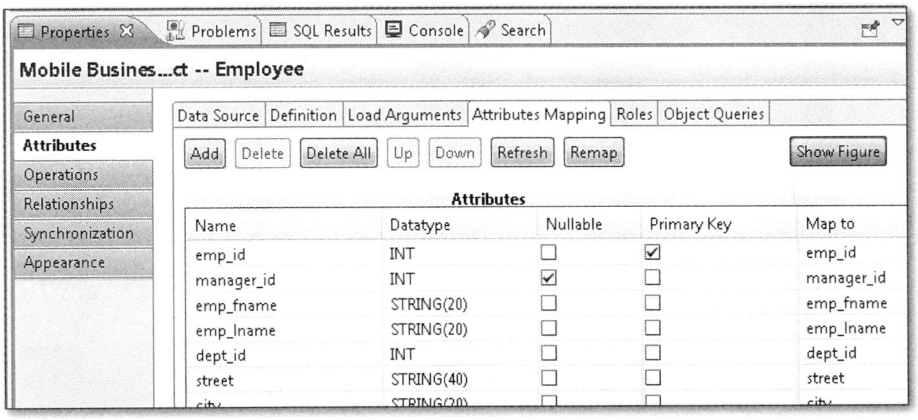

Figure 7.19 Attribute Mapping Tab

The modified MBO, with fewer attributes, is shown in Figure 7.20.

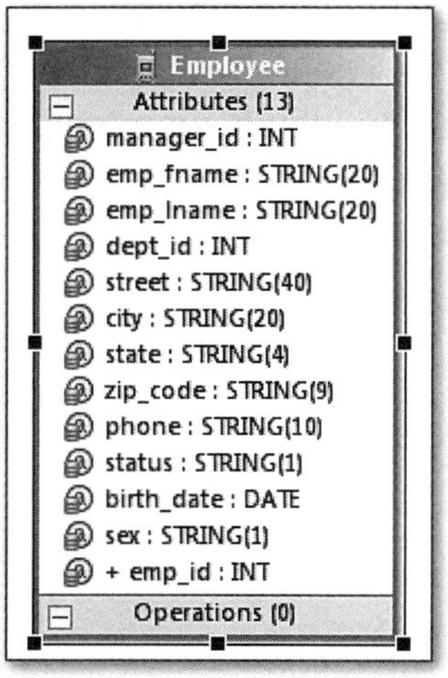

Figure 7.20 Revised MBO with Unwanted Attributes Removed

Deploying the Project

The final step is to deploy the package (which, at this point, is only this single MBO) to the server. Right-click anywhere in the white space in the Mobile Application Diagram Editor and then select Deploy Project, as shown in Figure 7.21.

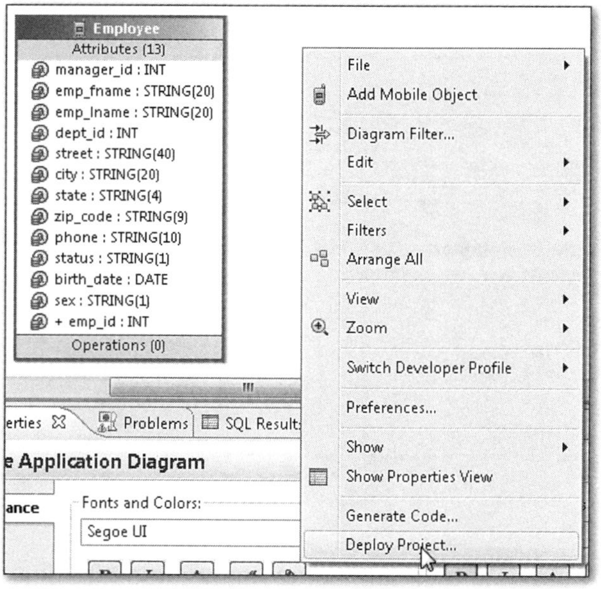

Figure 7.21 Deploying the Project

The first of a series of dialog boxes appears. For the first dialog box, select the UPDATE option and then enter a name for the project. (The project name must start with a letter and cannot contain spaces, "#,""%," or other non-ASCII characters.) Then, click the NEXT button, as shown in Figure 7.22.

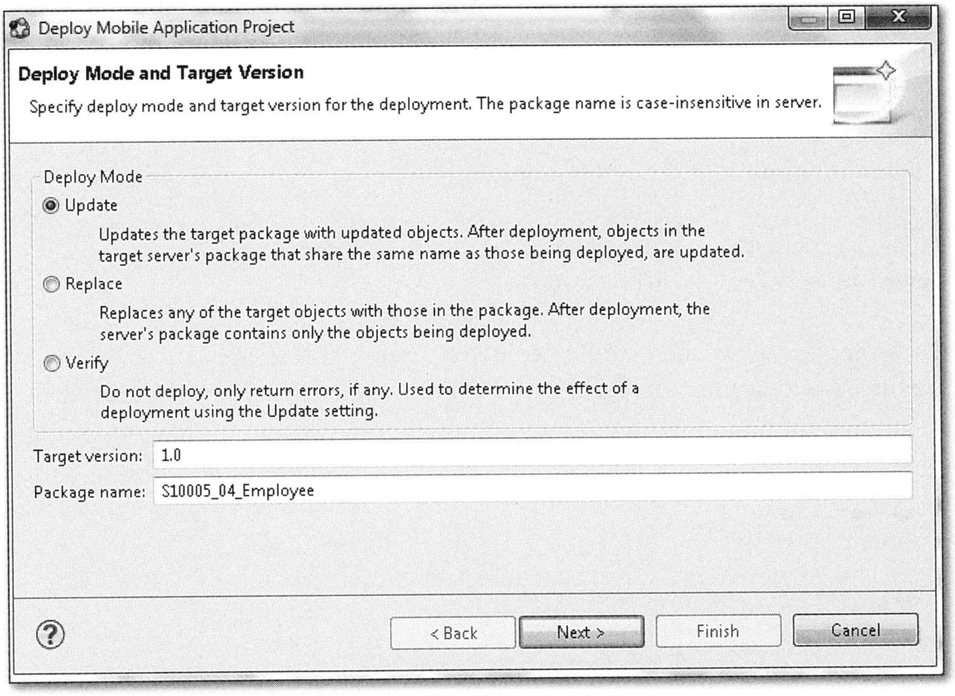

Figure 7.22 Selecting the Deployment Mode and Naming the Package

Accept the contents in the next four dialog boxes by clicking NEXT each time. The final dialog box requires the selection of the server connection mapping for the EIS. Click the dropdown menu under SERVER CONNECTION and select the database name (in this example, SAMPLEDB), as shown in Figure 7.23. Once the connection is selected, complete the dialog by clicking FINISH.

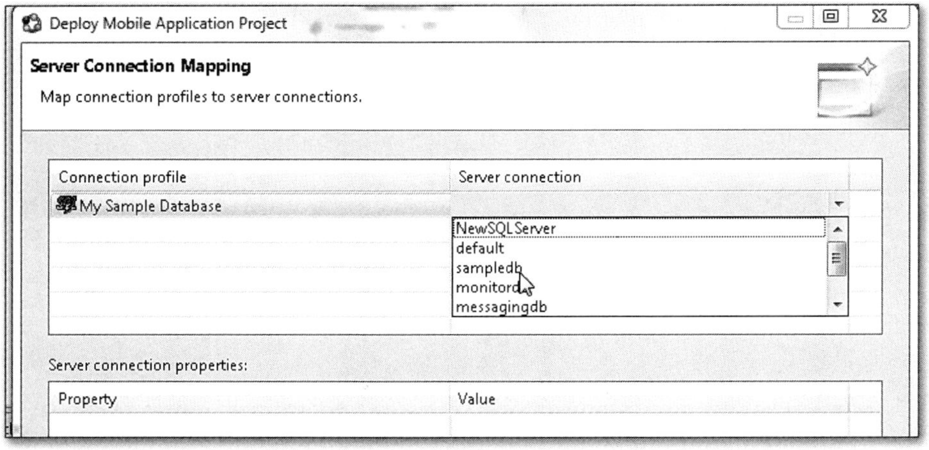

Figure 7.23 Selecting the Database Mapping

Once the package is successfully deployed, a message box appears, as shown in Figure 7.24, indicating that the deployment was either successful or unsuccessful.

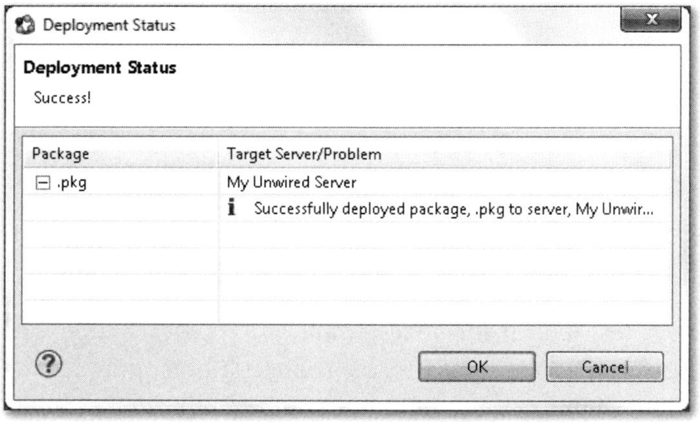

Figure 7.24 Dialog Box Indicating that Deployment to the Unwired Server Was Successful.

You can confirm deployment of the project and MBO to the Unwired Server by locating the package on the Unwired Server in the ENTERPRISE SERVER view, as shown in Figure 7.25.

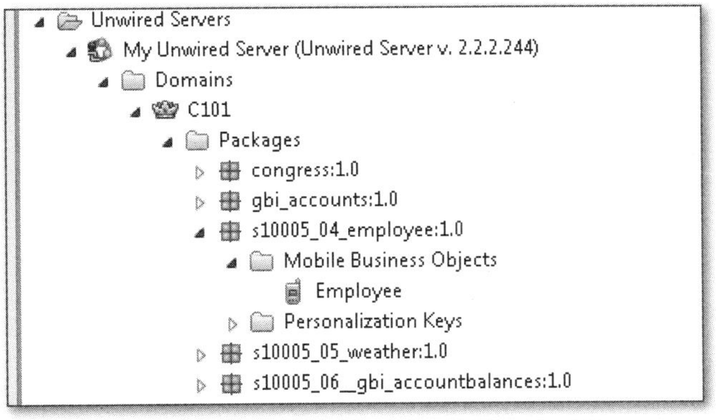

Figure 7.25 Confirming Deployment to the Unwired Server

7.3.2 Web Service MBO

A web service is designed to support machine-to-machine interaction over a network; that is, machines pass data without a human in the loop. A web service describes and embeds its data into a flat file using Extensible Markup Language (XML). A Web Services Description Language (WSDL) is an XML-based language used to describe web services and how they can be accessed. The protocol used to transfer data packets is referred to as Simple Object Access Protocol (SOAP). Recently, a lightweight alterative approach called Representational State Transfer (REST) has begun to be used. REST uses the HTTP protocol, along with CRUD operations. Both SOAP and REST EISs are supported.

The example web service we will demonstrate provides weather information (i.e., current weather conditions and weather forecasts by zip code) and sample weather drawings. While web services are designed for computer-to-computer communications, a browser interface is often also provided to demonstrate the web service. That is the case for this web service, which can be viewed at the following address: *http://wsf.cdyne.com/WeatherWS/Weather.asmx*.

This browser interface appears, as shown in Figure 7.26. This interface describes the three operations supported by the web service.

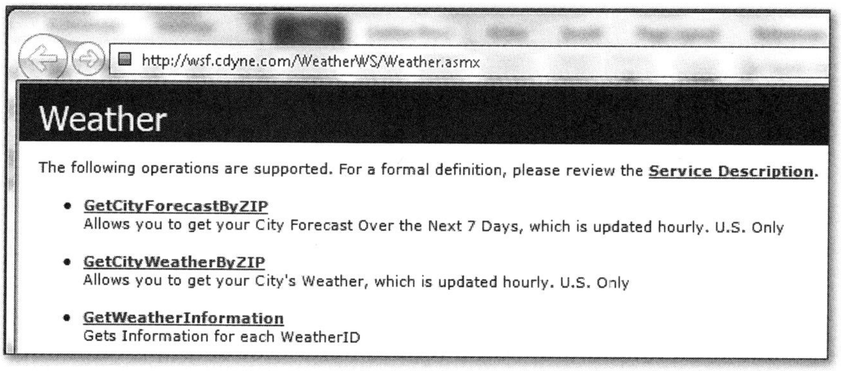

Figure 7.26 User Interface for Weather Web Service

The current weather operation can be demonstrated by the following steps:

1. Click the operation name, for example GETCITYWEATHERBYZIP.
2. Enter a valid US zip code.
3. Click INVOKE.

What is returned is the XML SOAP message, which is shown in Figure 7.27. As you can see from this XML, the temperature in Milwaukee was 47 degrees at the time the web service was called.

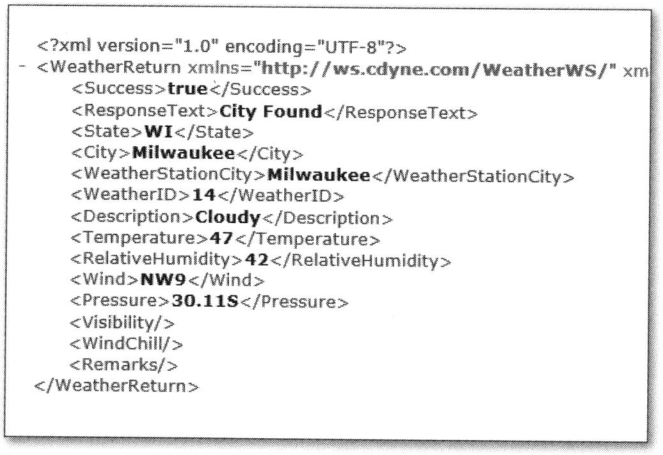

```
<?xml version="1.0" encoding="UTF-8"?>
- <WeatherReturn xmlns="http://ws.cdyne.com/WeatherWS/" xm
    <Success>true</Success>
    <ResponseText>City Found</ResponseText>
    <State>WI</State>
    <City>Milwaukee</City>
    <WeatherStationCity>Milwaukee</WeatherStationCity>
    <WeatherID>14</WeatherID>
    <Description>Cloudy</Description>
    <Temperature>47</Temperature>
    <RelativeHumidity>42</RelativeHumidity>
    <Wind>NW9</Wind>
    <Pressure>30.11S</Pressure>
    <Visibility/>
    <WindChill/>
    <Remarks/>
  </WeatherReturn>
```

Figure 7.27 XML Containing Current Weather Conditions

The Unwired Server is able to communicate with this web service and retrieve the required data from the XML. The first step involves connecting to the web service, as described in Chapter 4. With this connection made, as shown in Figure 7.28, you then see that this is a SOAP web service and that it supports three operations.

Figure 7.28 Connecting to a Web Service

The first step is to drag and drop the CITYWEATHERBYZIP operation onto the Mobile Application Diagram Editor. In the first dialog box, select ATTRIBUTES and click OK. In the second dialog box, click the PREVIEW... button, as shown in Figure 7.29.

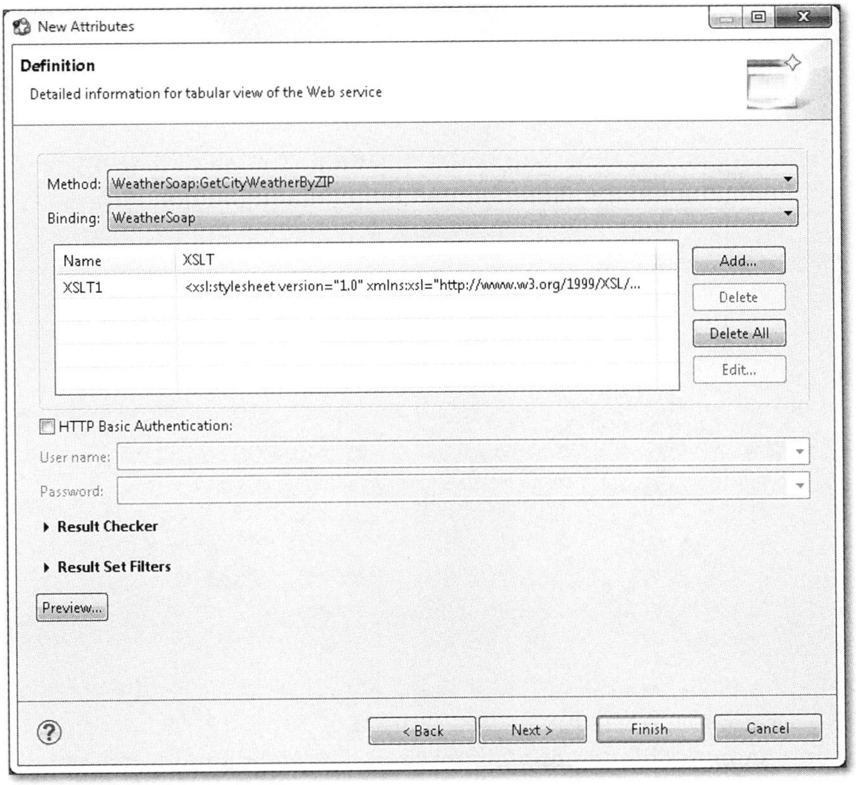

Figure 7.29 Select Preview of Web Service Operation

In the PREVIEW dialog box, enter an example US zip code (such as "53209"), select SAVE AS DEFAULT VALUES, click PREVIEW, and ignore the warning dialog box. The data returned from the web service then appears, as shown in Figure 7.30.

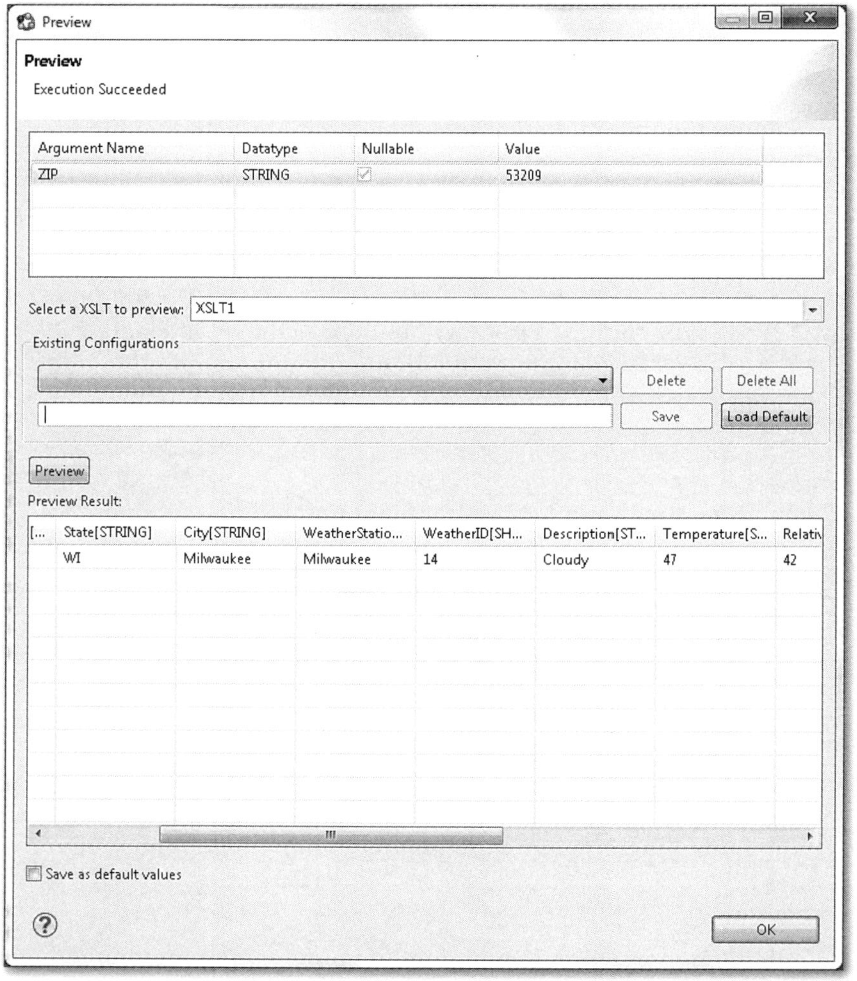

Figure 7.30 Current Weather Data Returned from the Web Service

After viewing the data, click OK to return to the XSLT. The XSLT for the web service MBO is similar to the SQL for the database MBO. The XSLT can be edited to eliminate unnecessary attributes. To accomplish this, highlight XSLT1 and click the EDIT button, as shown in Figure 7.31.

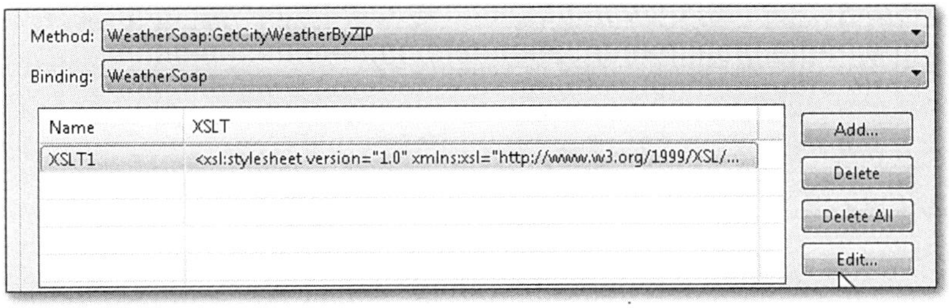

Figure 7.31 Editing the XSLT

When you click GENERATE XSLT FROM ... and accept the warning message, a list of the attributes appears. You can remove any attributes from the MBO by deselecting them, as shown in Figure 7.32. When you're finished, click OK.

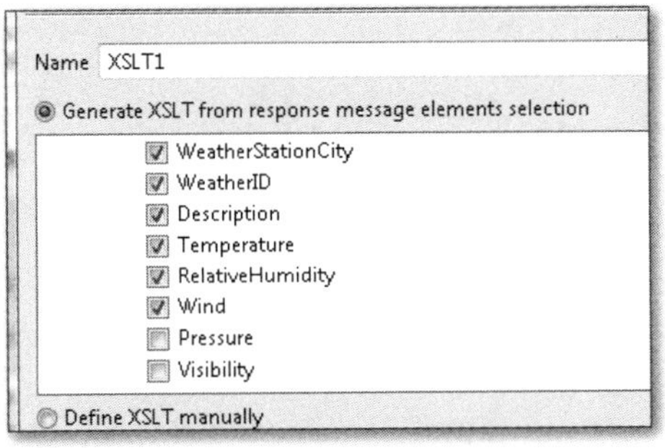

Figure 7.32 Deselecting Attributes in the Web Service MBO

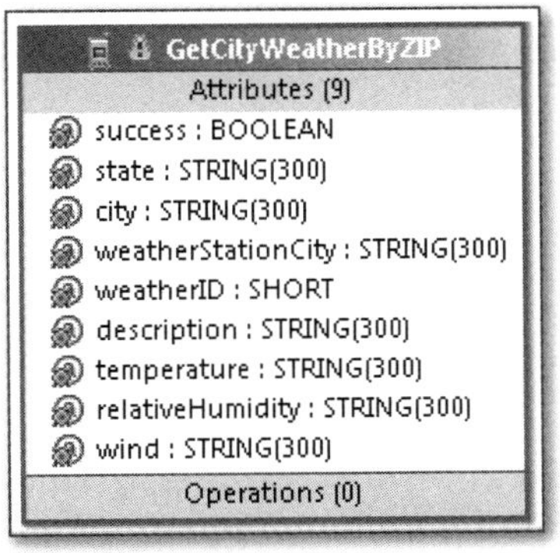

Figure 7.33 Completed Web Service MBO

The MBO then appears in the Mobile Application Diagram Editor, as shown in Figure 7.33. Once this step is complete, the MBO can be customized and used in the same way as the database MBO. This includes adding cache group policies, synchronization, and load and personalization keys and object queries. The web service MBO can also be used in the same way in the development of the mobile apps.

7.3.3 SAP MBO

The third type of EIS supported is SAP. MBOs that interact with SAP systems using Business Application Programming Interface (BAPIs) are created. A *BAPI* is a set of Application Program Interfaces (APIs) that supports both SAP and non-SAP software integration with the proprietary SAP system. A number of these interfaces that support both data retrieval and data creation, update, and delete are defined. Next, we show how to access the SAP data of interest for this particular example, and then discuss building MBOs that access their data from SAP systems using both the drag-and-drop approach and the Palette approach.

Accessing SAP Data via the SAP Easy Access Menu

Anyone who has worked with SAP is likely very familiar with the SAP Easy Access menu. For those who are new to SAP, though, it's worth a brief introduction.

The traditional "human interface" method for access to data within an SAP system—referred to as SAP GUI—is through the use of a thin client loaded on the user's computer. Once a user has provided the required security information, the user is given access to a proprietary interface referred to as the SAP Easy Access menu. This menu system allows users to drill down through the hierarchical menu structure until they locate the specific transaction they wish to execute.

For this example, we want to look up general ledger account balances. To access this data using the SAP Easy Access menu, the following path would be selected: SAP Easy Access menu • Information Systems • Accounting • Financial Accounting • General Ledger • Information system (New) • Account Balances.

The SAP Easy Access menu would appear as shown in Figure 7.34.

Clicking the transaction brings up the parameter screen. This screen requires the chart of accounts (such as "GL00"), the company code (such as "US00"), and the physical year (such as "2012"). The screen with these example values is shown in Figure 7.35.

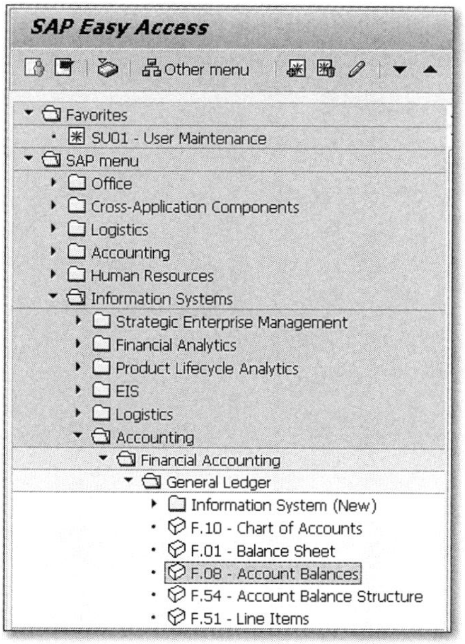

Figure 7.34 Using SAP Easy Access Menu to Run Account Balances Transaction

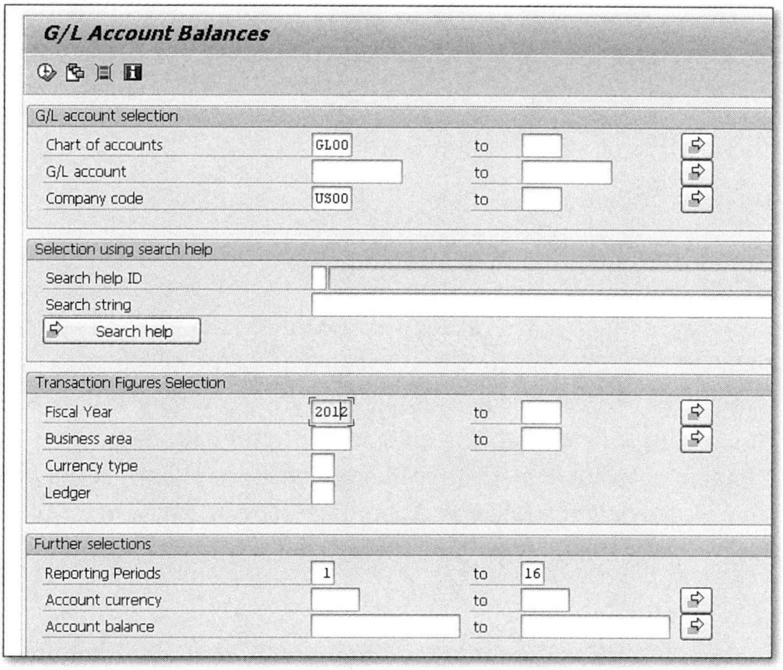

Figure 7.35 Parameters Required for Transaction F.08

Clicking the clock executes this transaction, which produces a listing of the general ledger accounts and their balances for a specific period. A subset of that report is shown in Figure 7.36.

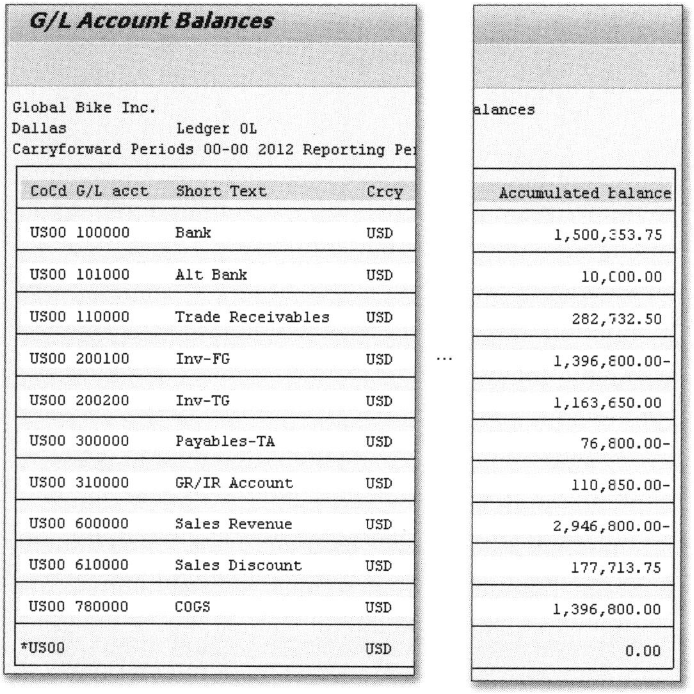

G/L Account Balances

Global Bike Inc.
Dallas Ledger 0L
Carryforward Periods 00-00 2012 Reporting Pe

alances

CoCd	G/L acct	Short Text	Crcy	Accumulated balance
US00	100000	Bank	USD	1,500,353.75
US00	101000	Alt Bank	USD	10,000.00
US00	110000	Trade Receivables	USD	282,732.50
US00	200100	Inv-FG	USD	1,396,800.00-
US00	200200	Inv-TG	USD	1,163,650.00
US00	300000	Payables-TA	USD	76,800.00-
US00	310000	GR/IR Account	USD	110,850.00-
US00	600000	Sales Revenue	USD	2,946,800.00-
US00	610000	Sales Discount	USD	177,713.75
US00	780000	COGS	USD	1,396,800.00
*US00			USD	0.00

Figure 7.36 General Ledger Account Balances

Using Drag and Drop

The first approach for creating an SAP MBO is to expand the connected system and search through the available app components. One of the problems with this approach is that not all the BAPIs have been implemented as components. In fact, the percentage that is currently available is quite small. In the example shown in Figure 7.36, a component is not provided for looking up account balances in the general ledger. In order to demonstrate this approach, we create a different MBO. We will come back to the general ledger account example in the next section.

The MBO we are going to demonstrate will be used to retrieve a list of banks for a given country. The bank object can be located by the following path: SAP SERVERS • GB4 • APPLICATION COMPONENTS • CROSS-APPLICATION COMPONENTS • BANK • BANK.

The result of this path is the object `Bank`, as shown in Figure 7.37.

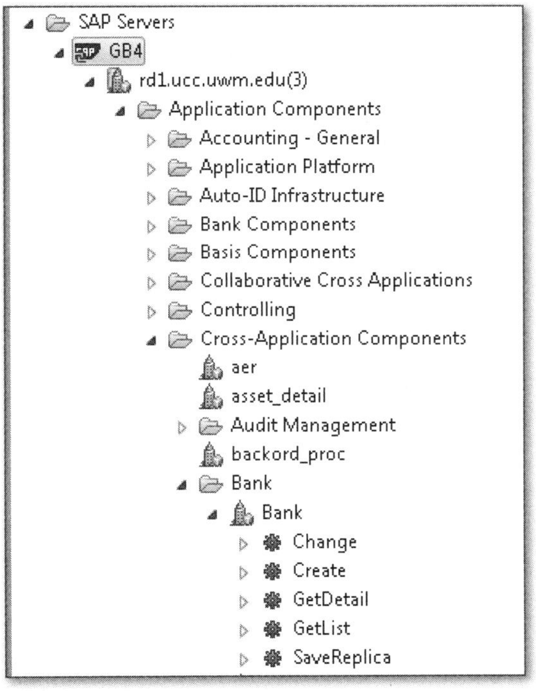

Figure 7.37 Path to the Object Bank

The BAPI functions that are supported for the `Bank` object include `Change`, `Create`, `GetDetail`, `GetList`, and `SaveReplica`. Let's assume our mobile app will be used to list banks for a given country; therefore, we use the `GetList` function. You can drag and drop this function onto the Mobile Application Diagram Editor. In the first dialog box, select ATTRIBUTES and click OK, and then an SAP definition appears. In this dialog box, select `Bank_City` as the input field and `Bank_List` as the output, and click the PREVIEW button, as shown in Figure 7.38.

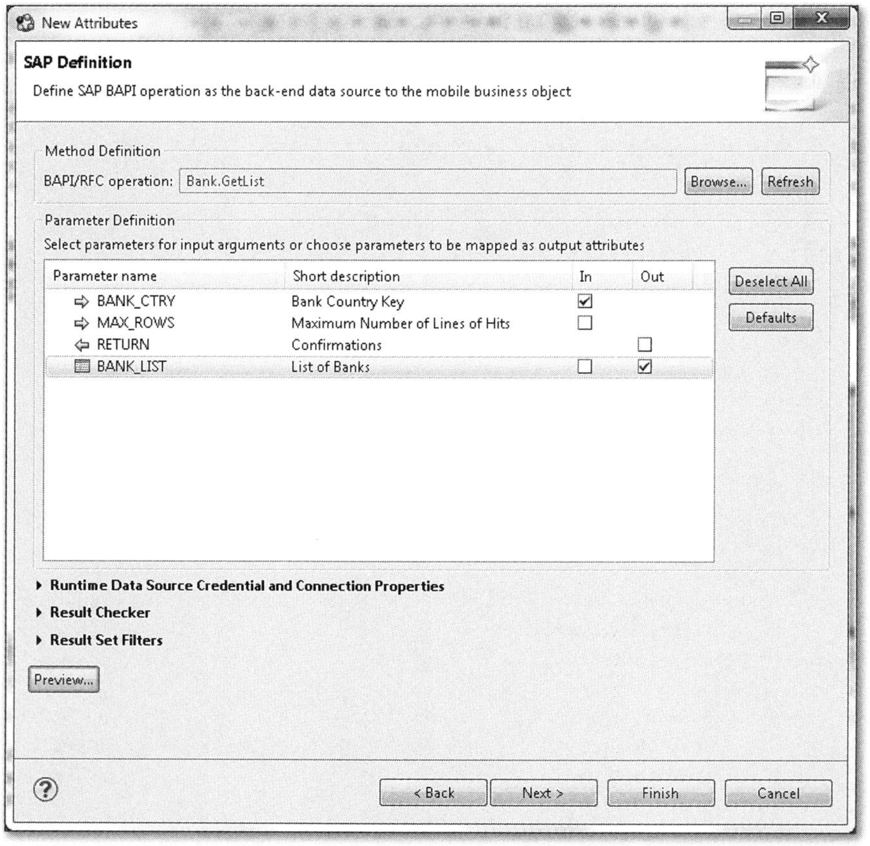

Figure 7.38 SAP MBO Definition

Then, enter an example country (such as "US"), click the SAVE AS DEFAULT VALUES, and then click the PREVIEW button. The list of banks is then displayed, as shown in Figure 7.39.

To close these last two dialogs, click OK and then FINISH. The MBO is created, as shown in Figure 7.40.

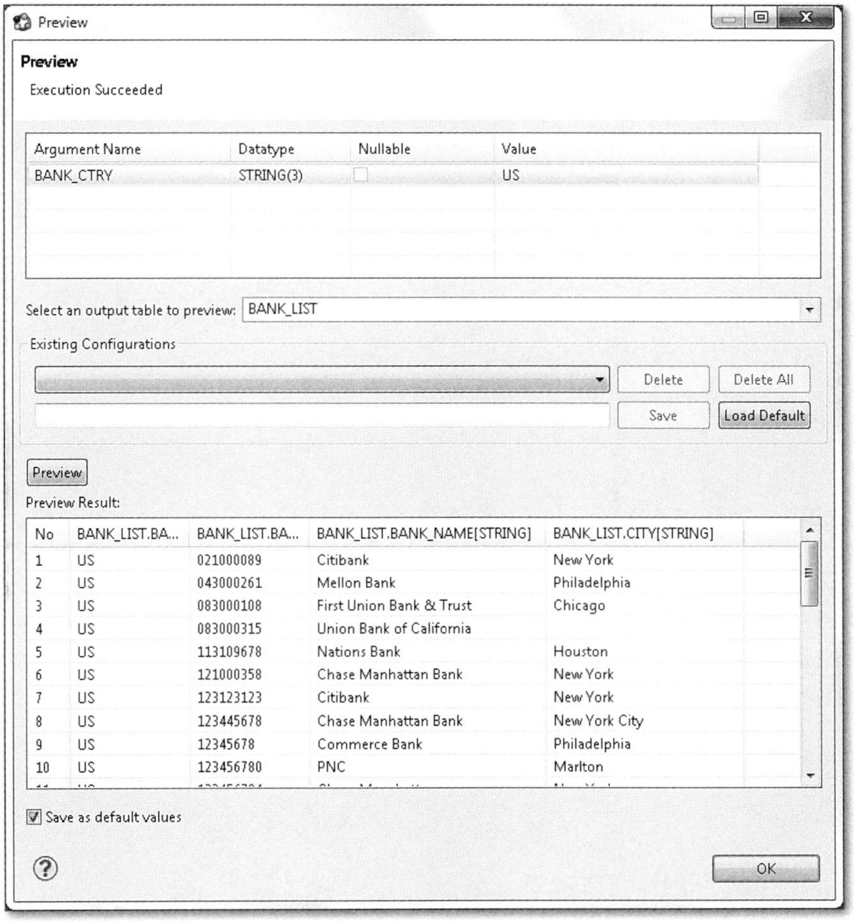

Figure 7.39 List of Banks for the US

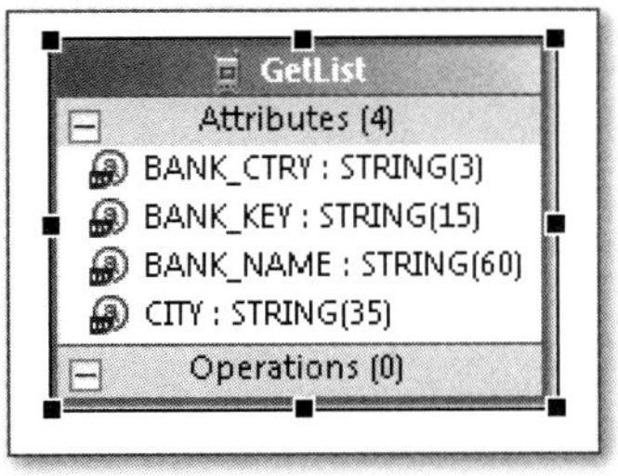

Figure 7.40 Completed SAP MBO

Using the Palette

An alternative approach to the drag-and-drop approach is to use the option to create an MBO that can be found in the Palette, as shown in Figure 7.41.

Figure 7.41 Creation Using Mobile Business Object in the Palette

To use this approach, double-click the MOBILE BUSINESS OBJECT menu item in the Palette, enter the MBO name (such as "Accounts"), and click the NEXT button, as shown in Figure 7.42.

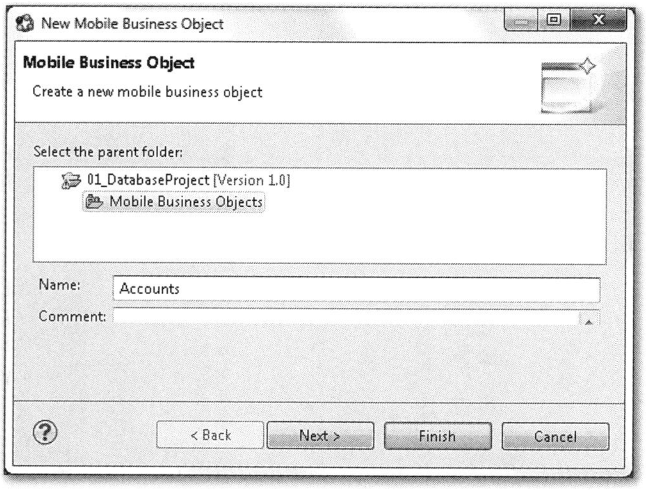

Figure 7.42 Naming the MBO

The next dialog box requests the data source to be used to create the MBO. The data source type (in this case, SAP) is selected from the dropdown list, and then the connection profile (in this case, GB4) is selected, as shown in Figure 7.43. Then, click the NEXT button.

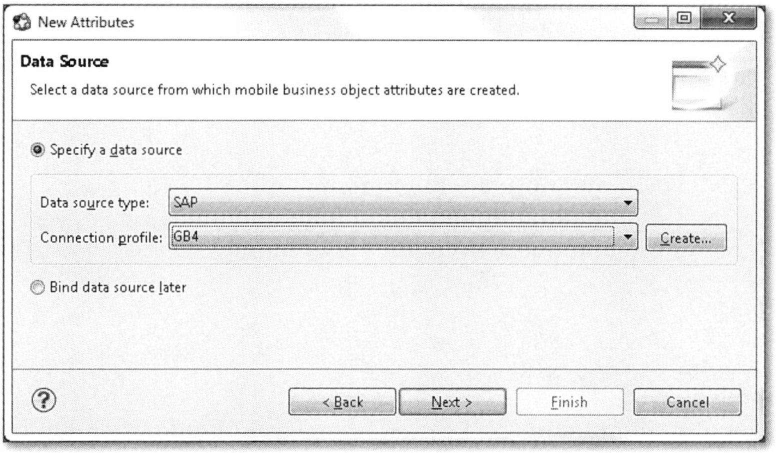

Figure 7.43 Selecting the Data Type and Connection Profile

The next dialog box is for selecting how to search for the specific data of interest. In this example, we search for a BAPI by providing a sub-string that is contained within its name. Using this search approach, we can locate any of the BAPIs within the SAP system, and not just the objects and functions displayed in the Enterprise Explorer. To start this process, click the BROWSE button for the BAPI/RFC operation, as shown in Figure 7.44.

Figure 7.44 Selecting Browse for a BAPI/RFC

In the next dialog box, enter the sub-string "*GLACCPeriodBalances*" in the SEARCH BAPIs/RFCs textbox. The "*" represent a wild card in this search process, so this search looks for all BAPIs that have `GLACCPeriodBalances` in their names. Then, click the SEARCH BAPIs/RFCs button, as shown in Figure 7.45.

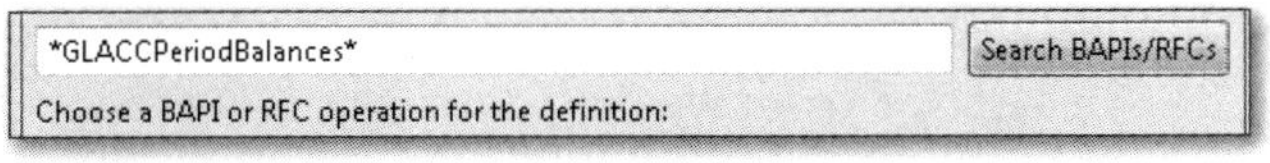

Figure 7.45 Searching for BAPIs with a Sub-String in Their Names

The result of this search is a single BAPI that contains this sub-string. Select the desired BAPI by clicking it, and then proceed to the next dialog by clicking OK, as shown in Figure 7.46.

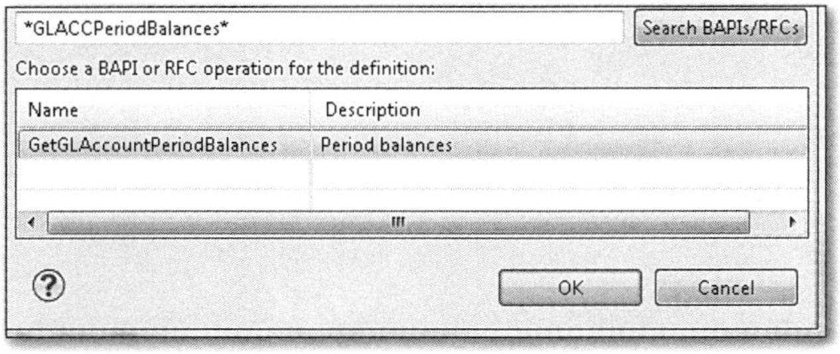

Figure 7.46 Selecting the Desired BAPI

The following dialog is used for selecting the required `Input` and `Output` fields. The dialog box appears as shown in Figure 7.47.

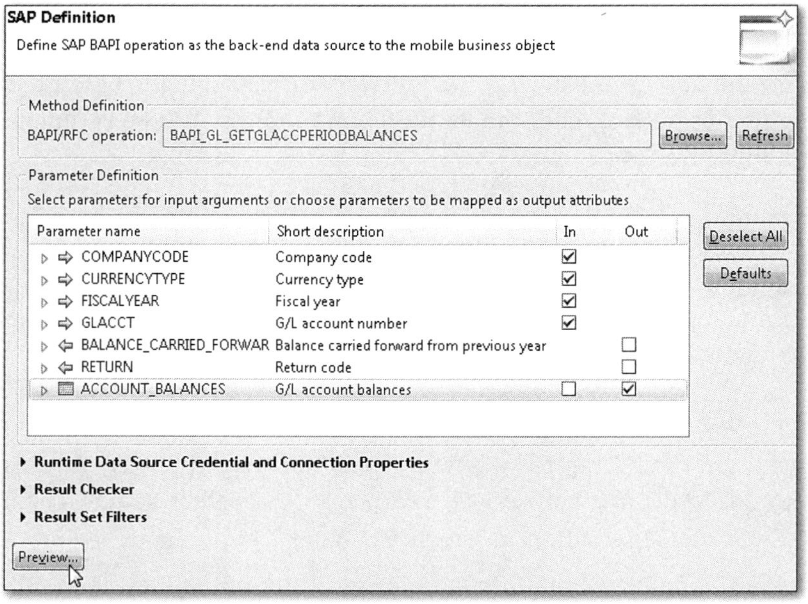

Figure 7.47 Selecting Input and Output Fields

Click the PREVIEW button to display the next dialog box, as shown in Figure 7.48.

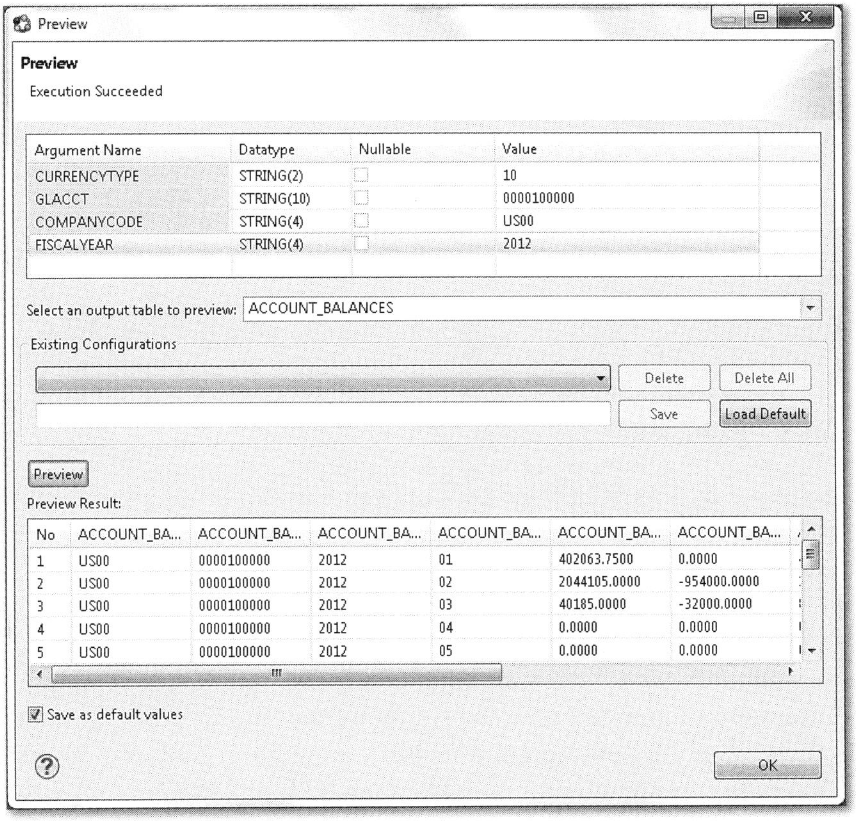

Figure 7.48 Example BAPI Data

Enter example data for all four arguments, check SAVE AS DEFAULT VALUES, and click PREVIEW to display some example data. Once you click OK and then FINISH, the MBO is created, as shown in Figure 7.49.

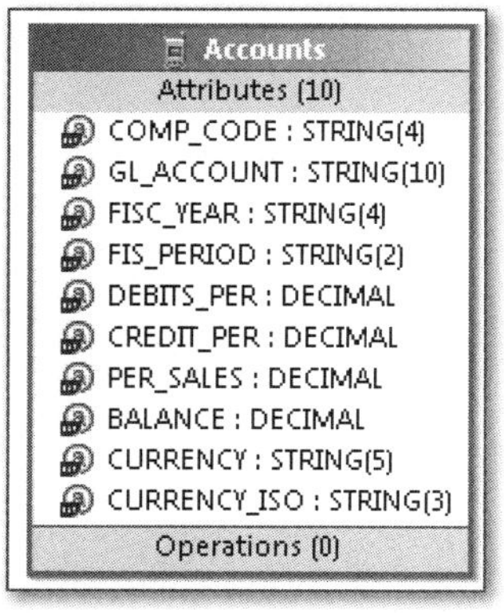

Figure 7.49 Completed MBO

7.4 Controlling Data Flow and Performance

The design element that has the most important impact on mobile app performance relates to those MBO properties that impact the quantity and frequency with which data is passed from the EIS to the Unwired Server and from the Unwired Server to the mobile device. The design goal should be to minimize the flow of this data while at the same time presenting the most accurate information to the user via the app. In many instances, these two requirements are conflicting. This issue becomes even more complex when the app moves from simply retrieving information to performing CUD operations. The final complication enters the picture when multiple users are simultaneously performing both simple access and CUD operations to the same EIS.

The data flow, as shown in Figure 7.50, involves (a), moving selected data to and from the EIS to a cache on the Unwired Server, which is referred to as the Cache Database (CDB); and (b), moving selected data to and from the CDB to a database stored on the mobile device.

There are several MBO properties that control which data is moved among the EIS, CDB, and mobile device and when this data movement takes place. A brief introduction to these properties is presented here, with detailed examples to follow:

- **Load arguments**
  The load arguments control which data is refreshed between the EIS and the CDB. A partition is created in the CDB for each data value for the load argument.

- **Synchronization parameters**
  The synchronization parameters filter the data (i.e., determine the amount) to be uploaded and downloaded from the CDB to the mobile device.
- **Personalization keys**
  Personalization keys allow mobile users to define their own values for keys that can be used by the mobile app to select data.
- **Cache and cache groups**
  Caching determines the data refresh behavior for MBOs between the EIS and the CDB. MBOs can be combined to form cache groups. A caching policy that controls the refresh behavior is created for each group.
- **Join between MBOs**
  Join allows you to logically connect two MBOs together using one or more common attributes. It works very similarly to the join used in the SQL data model.

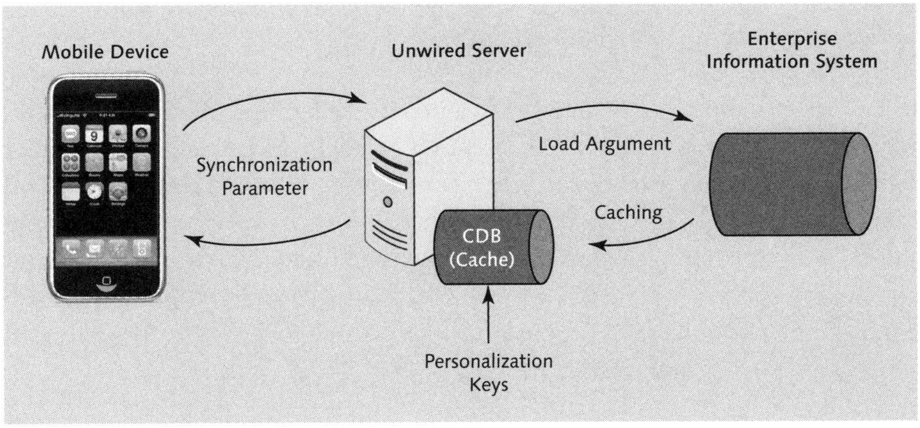

Figure 7.50 Data Flow among the EIS, CDB, and Mobile Device

The following is a more detailed discussion of how to create each of these properties and how they control the flow of data and interact with each other.

7.4.1 Load Arguments

The role of the load argument is to logically partition the data that is stored in the CDB based on the value of the load parameter. Given that a load argument has not been selected for the current EMPLOYEE MBO, a single partition that contains the entire data table exists. When the data in the CDB is updated, the entire MBO is downloaded from the IES to the CDB. This single large partition appears as shown in Figure 7.51 and contains all 75 rows of data. Any update to this MBO in the CDB requires a complete download of the entire table. This MBO is designed to be used by a mobile app that displays all the employees and then lets the user scroll through the data to find the employee of interest. Given this data requirement, a load argument is not required.

No.	emp_id [INT]	manager_id [INT]	emp_fname [CHAR(20)]	emp_lname [CHAR(20)]	dept_id [INT]
1	102	501	Fran	Whitney	100
2	105	501	Matthew	Cobb	100
3	129	902	Philip	Chin	200
4	148	1293	Julie	Jordan	300
5	160	501	Robert	Breault	100
6	184	1576	Melissa	Espinoza	400
7	191	703	Jeannette	Bertrand	500
8	195	902	Marc	Dill	200
9	207	1576	Jane	Francis	400

Figure 7.51 Employee MBO without a Load Parameter

A second mobile app could be developed and designed so that the user first enters a department ID (or saves the department ID on their device) and then reviews only employees within their department. For this app, selecting `Dept_Id` as a load parameter would cause the CDB to be partitioned based on the value of the department ID. With this approach, the EMPLOYEE MBO is broken into five separate partitions (current values for `Dept_Id` are 100, 200, 300, 400 and 500), whereby each partition contains all the rows that have the same value for `Dept_Id`. The partition for `Dept_Id` = 100 is shown in Figure 7.52.

No.	emp_id [INT]	manager_id [INT]	emp_fname [CHAR(20)]	emp_lname [CHAR(20)]	dept_id [INT]
1	102	501	Fran	Whitney	100
2	105	501	Matthew	Cobb	100
3	160	501	Robert	Breault	100
4	243	501	Natasha	Shishov	100
5	247	501	Kurt	Driscoll	100
6	249	501	Rodrigo	Guevara	100
7	266	501	Ram	Gowda	100

Figure 7.52 Employee MBO with a Load Parameter

To demonstrate the creation of a load property, a second MBO is created using the drag-and-drop approach. The default name for this second MBO is EMPLOYEE1.

More than one approach exists for defining a load parameter. One of the approaches is to modify the MBO's SQL definition. With the EMPLOYEE1 MBO selected, use the path PROPERTIES • ATTRIBUTES • DEFINITION, and then click the EDIT button, as shown in Figure 7.53.

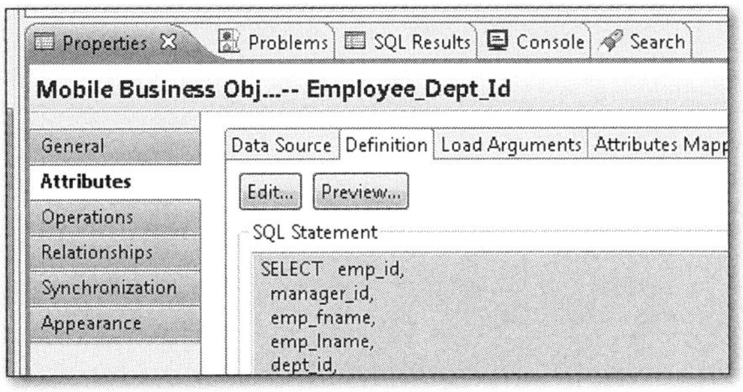

Figure 7.53 Modifying a Definition to Create a Load Argument

Add `WHERE dept_id="dept_id` at the bottom of the query and then click the OK button, as shown in Figure 7.54.

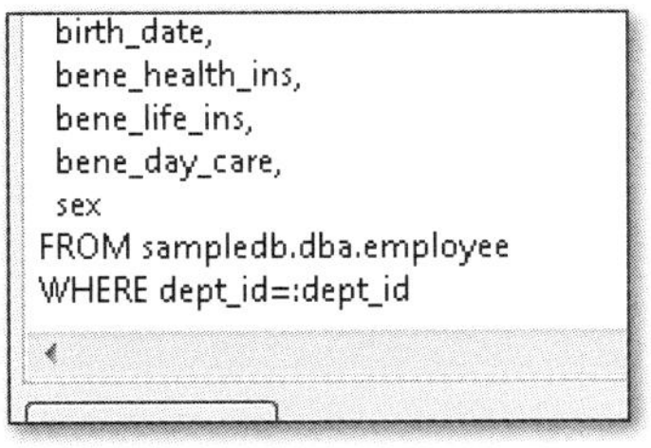

Figure 7.54 Adding Where Clause to Create a Load Parameter

Adding this condition to the MBO definition query creates a LOAD parameter for this MBO. A value for the parameter `Dept_Id` is required when this MBO is utilized. After adding this condition, you can remove the extra attributes by deleting them utilizing the ATTRIBUTES MAPPING tab, as with the original MBO. Selecting the LOAD ARGUMENT tab shows that this load argument has been created. Selecting this argument, the following changes need to be made:

1. Change the data type to `Int`.
2. Deselect NULLABLE.
3. Select `Dept_Id` as the propagating attribute.
4. Enter a default value, such as "100."

Following these changes, the load argument now appears as shown in Figure 7.55. The EMPLOYEE1 MBO has its data partitioned in the CDB based on the value of the `Dept_Id` attribute.

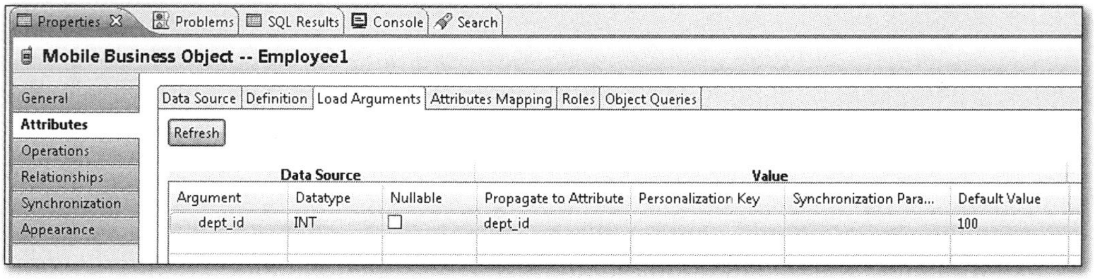

Figure 7.55 Connecting Load Parameter to Attribute

7.4.2 Synchronization Parameters

The synchronization parameter is used to restrict the rows that are transferred from the CDB to the mobile device. The only rows that are transferred are those that have an attribute value that matches the value of the synchronization parameter. These are useful when you have large amounts of data that does not change often. With the synchronization parameter, only the specific data of interest is downloaded to the mobile device. The mobile app does not have access to the EIS directly because all its data flows from the CDB. When data is downloaded to the mobile device, the entire partition is downloaded at once.

An MBO design that is often employed uses the same attribute for both the synchronization parameter and the load parameter. With this approach, each mobile user's device maps to one of the partitions in the CDB. With multiple users, one user may be accessing one partition while another user may be updating a different partition. This allows for parallel processing against a large data set. Note that the use of load and synchronization on the same attribute works best when several partitions are created with several records in each partition. As the number of partitions grows, and the size of each partition becomes smaller, the additional overhead tends to outweigh the advantages. The extreme case of selecting a primary key would yield a separate partition for each record.

To create a synchronization parameter, highlight the MBO and then use the path PROPERTIES • SYNCHRONIZATION and click ADD. Make the following entries to the dialog box:

1. Rename the parameter to, for example, "`deptParam`."
2. Change the data type to `Int`.
3. Deselect NULLABLE.
4. Enter "100" for the default value.
5. Using the dropdown, select the `Dept_Id` for the attribute to which this synchronization parameter is mapped.

A dialog box then appears, as shown in Figure 7.56.

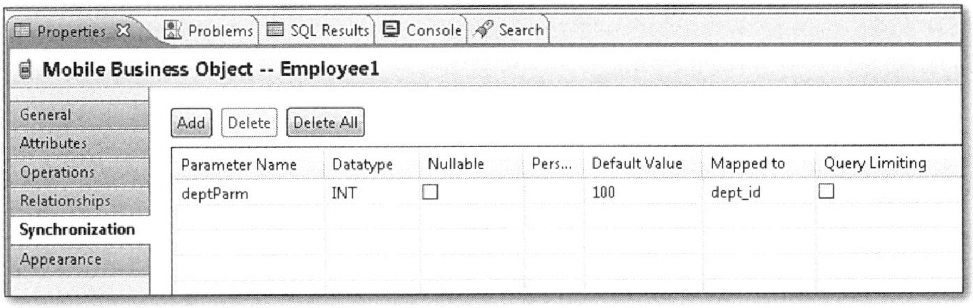

Figure 7.56 Creating a Synchronization Parameter

In this example, the same attribute is selected for both the load and synchronization parameters. Return to the LOAD ARGUMENT dialog box using PROPERTIES • ATTRIBUTES • LOAD ARGUMENTS, and then look up `DeptParm` as the synchronization parameter, as shown in Figure 7.57.

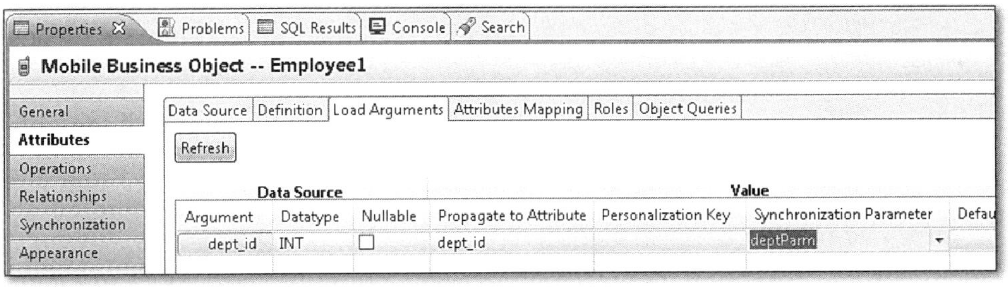

Figure 7.57 Adding a Synchronization Key to Load Arguments

7.4.3 Personalization Keys

In the previous example, a single attribute is selected to serve as both the load and synchronization parameters for the EMPLOYEE1 MBO. In order for these to function, a hybrid app needs a method by which the user can specify the value for this attribute that is used by the user's specific mobile device. Some users may want department 100 as their initial value, while others want department 300. Personalization keys can be created to provide this capability.

Personalization keys allow mobile users to personalize certain attribute field values that are used in the mobile app. These personalization keys allow the user to associate a named key with a data value. The model supports two types of personalization keys:

- **User defined**
  These keys are created as part of the MBO development. They can be used in the hybrid app, and each user can define their own specific (personalized) values for them. In this example, a personalized key is created for the `Dept_Id` value.
- **System defined**
  The system predefines two personalization key values called USERNAME and PASSWORD. These keys are read only and contain the values for accessing the EIS system.

A personalization key is created by the path [PROJECT NAME] • PERSONALIZATION KEYS • NEW • PERSONALIZATION KEY in the Workspace Navigator, as shown in Figure 7.58.

Figure 7.58 Creating a New Personalization Key

The following entries are required in the dialog box:

1. Enter a name for the key, such as "PK_Dept_Id."
2. Select the data type `Int`.
3. Deselect NULLABLE.
4. Enter a default value, such as "100."
5. Select TRANSIENT for the storage.

A dialog appears, as shown in Figure 7.59. Click the FINISH button.

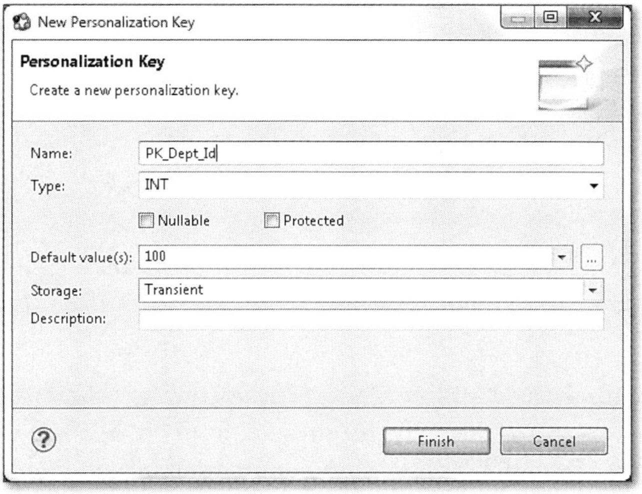

Figure 7.59 Creating a Personalization Key

You can then add this personalization key to the synchronization argument by following the path PROPERTIES • SYNCHRONIZATION • UPDATE. The personalization key is found in the dropdown menu, as shown in Figure 7.60.

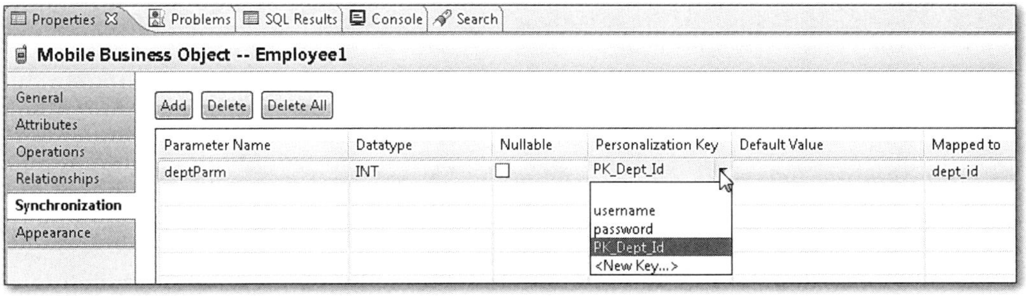

Figure 7.60 Adding the Personalization Key to the Synchronization Parameter

7.4.4 Cache and Cache Groups

As described previously, the data stored on the mobile device is updated using only the data stored in the cache on the Unwired Server in what is known as the CDB. In other words, all mobile devices using a specific MBO are synced to that MBO data in the CDB. If load arguments have been defined for a given MBO, then this data is partitioned into multiple tables, where all the rows in a given partition have a common value for the load parameter. With this as a background, this section focuses on the refresh/update pattern for the data flow between the CDB and the EIS.

A cache group can be created that contains one or more MBOs. A cache policy that describes how and when these refreshes take place can be created for each cache group. An MBO can belong to only one caching group. The reasons for including more than one MBO within the same caching group is that their data update characteristics might be similar in nature. The MBO data in the CDB can be refreshed only using these cache policy rules.

The data cache policy defines the refresh behavior for all MBOs within a cache group. The policy options include the following:

- **On demand**
  The cache expires after a fixed time limit, such as one hour. Once this limit has expired and a data request is made, the cache is refreshed from the EIS before the data is provided back to the mobile app. With this approach, the mobile app user may notice a delay in response.
- **Scheduled**
  The cache expires according to defined time(s). For example, at 6:00 am and 1:00 pm. This refresh is requested independently of whether a data request has been made. With this approach, the end user is less likely to see a delay, but there is a risk of receiving data that is not as current. (Note: transient personalized keys cannot be supported with this approach.)
- **DCN**
  The cache never expires with this policy. The data refresh is triggered by a data change notification (DCN) message sent from the EIS application. With this approach, the code used to update the EIS sends a message indicating that the cache needs to be updated. This requires code modifications within the EIS. An alternative approach is to have code that watches the EIS databases for changes.

When the cache expires, the following procedure is used to refresh the CDB data for all the MBOs in that group:

1. If the CDB is empty, all rows are inserted.
2. If a row exists in the CDB for a specific row in the EIS (based on the primary key value), a comparison takes place to determine whether all the attribute values match the EIS.
   - If all the attributes match, nothing happens.
   - If any of the attributes do not match, an update of the row takes place.
3. If a row does not exist, it is inserted.

7.4.5 Join within an MBO

In the case when the EIS is a database, the attributes contained in the MBO are defined by an SQL statement. This definition can be viewed via the path PROPERTIES • ATTRIBUTES • DEFINITION. In the previous example, the `Dept_Id` attribute was selected for the load and synchronization parameters, as well as for the personalization key. Assume that the developer of the app might also like to include some attributes about the department, such as the department name, in the MBO. A second table exists in the same database that contains this `Dept_Name` attribute, as shown in Figure 7.61.

No.	dept_id [INT]	dept_name [CHAR(40)]	dept_head_id [INT]
1	100	R & D	501
2	200	Sales	902
3	300	Finance	1293
4	400	Marketing	1576
5	500	Shipping	703

Figure 7.61 Department Table

The `Select` statement used to define the MBO can include any of the standard SQL clauses, including the `Join` clause. You can accomplish a `Join` between the employee table and department table by editing this MBO definition.

The modifications that need to be made to the SQL include the following:

- The `dept_id` attribute is qualified by the table name (`dept`).
- The attribute `dept_name` is added.
- The second table (department) is added to the `from` clause, and both tables are given aliases.
- The first condition in the `where` clause is added, which joins the two tables on the `dept_id` attributes.
- The `dept_id` attribute in the second condition is qualified by the table name (`dept`).

The modified SQL is shown in Figure 7.62.

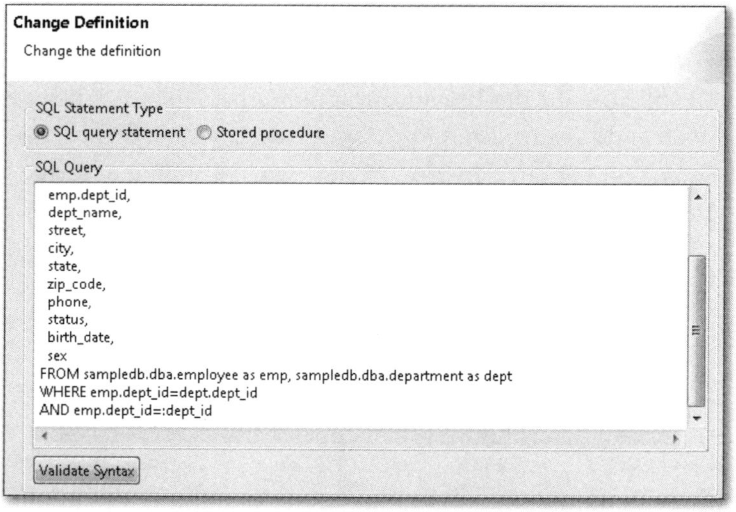

Figure 7.62 Modifying the MBO Definition by Adding a Join

You can verify this `Join` by right-clicking the EMPLOYEE1 MBO and selecting PREVIEW. Enter "100" as an example value for `Dept_Id` and then click the PREVIEW button. The data values for the `Dept_Name` attribute are shown in Figure 7.63. (Note that using a `Join` between two tables to define an MBO is very different from creating a relationship between two MBOs. This process is discussed later in the chapter.)

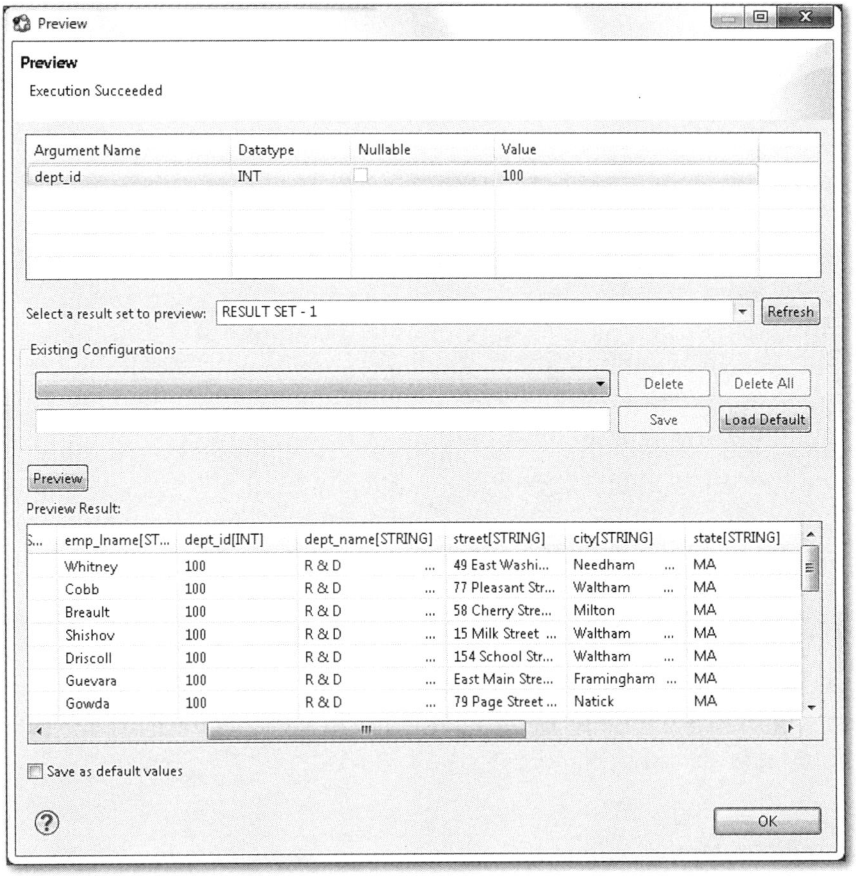

Figure 7.63 Preview MBO Data Showing dept_name

7.5 Creating Database Operations

A number of standard operations are often defined for database tables. The operations are often referred to as *create, update, and delete* (CUD). When you define an MBO from a database EIS, the system offers to create these operations. After selecting the table and dragging it onto the workspace, the wizard provides the option to generate or not generate these operations, as shown in Figure 7.64.

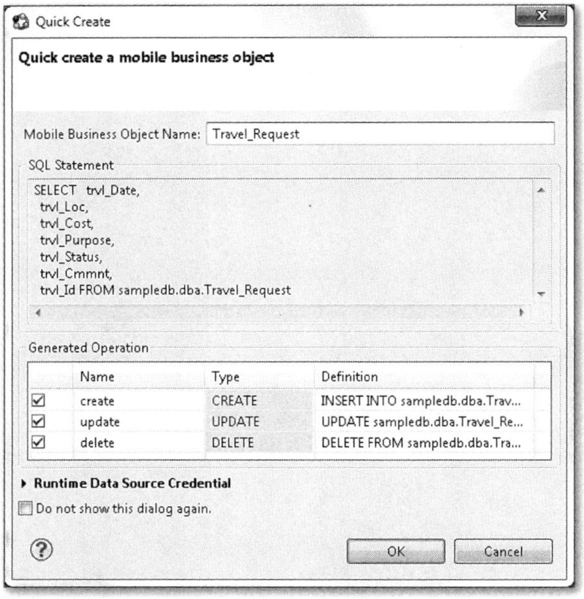

Figure 7.64 Option for the Generation of the CUD Operations

Leaving the box checked requests the generation of an MBO that includes these operations. In this case, the selected operations are displayed at the bottom of the MBO definition, as shown in Figure 7.65.

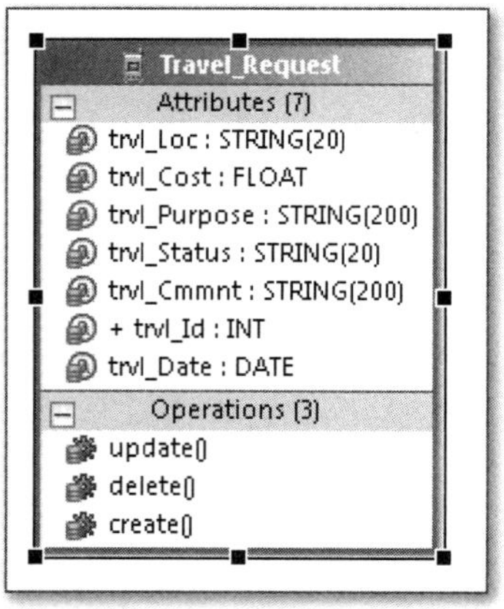

Figure 7.65 MBO with CUD Operations

You can view and edit the SQL generated for each of these operations using the path PROPERTIES • OPERATIONS • EDIT, as shown in Figure 7.66.

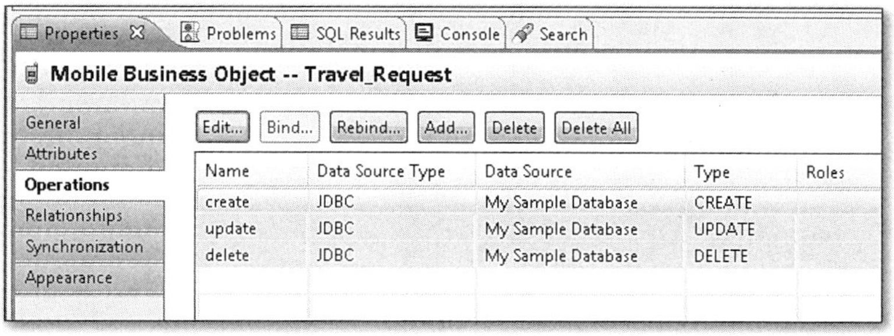

Figure 7.66 Editing the Operators

For example, selecting the `Create` operation and clicking EDIT displays the generated SQL, as shown in Figure 7.67.

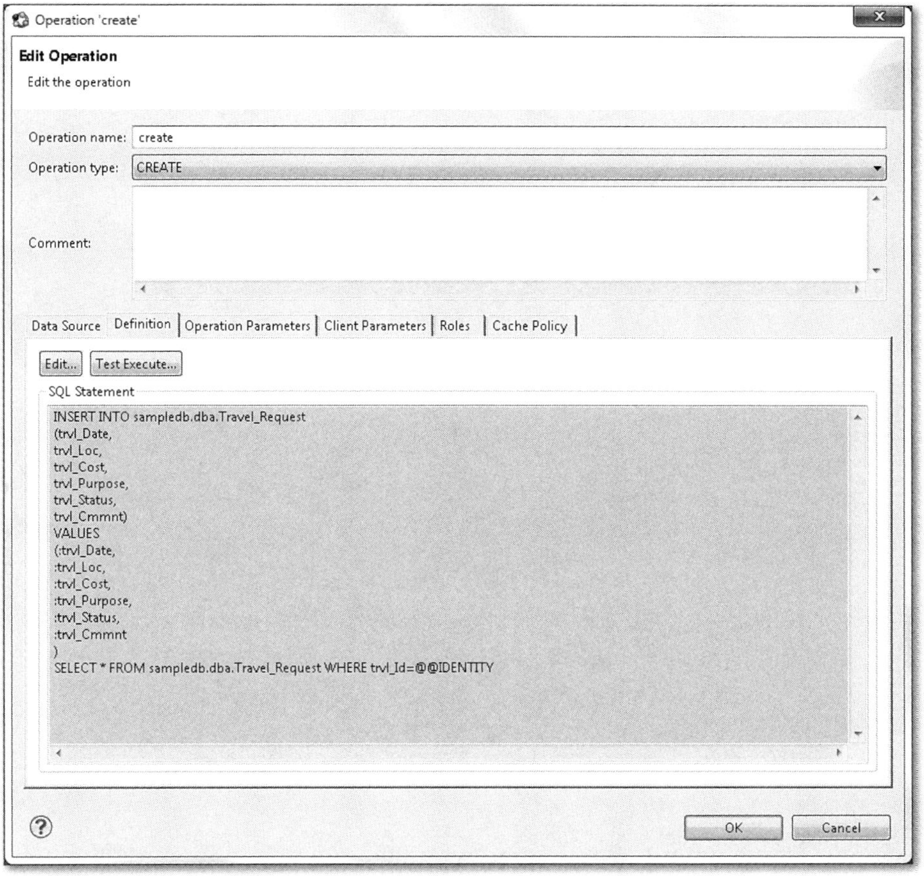

Figure 7.67 SQL Generated for the Create Operation

7.6 Creating Object Queries

Object queries that are used to run against the data cached on the Unwired Server can be created for an MBO. Their role is to process the data stored on the Unwired Server and provide those results to the screen controls. In the previous EMPLOYEE1 MBO example, the employee data for each specific department is a partition in the CDB, and only one partition (for example, 100) is downloaded to the mobile device. An object query could be created for the MBO that might select a single row, such as `Where Emp_Id = 105` or `Order by State, City`.

The containers installed on the different mobile operating systems all contain their own variants of compact database managers. A common SQL supported by all these database managers is used to construct object queries. The syntax is a subset of the SQL used by Ultralite. The restrictions to the SQL language are shown in Table 7.1.

SQL	Restrictions
`Select` statement	Supports: ▸ `Order by` Does not support: ▸ Bulk and math functions ▸ `Group by` ▸ `For` ▸ `Option` ▸ Row limitation ▸ `As`
`Input` parameter	Supports: ▸ `:name`
`Comparison` operators	Supports: ▸ Date format YYY-MM_DD and contained in single quotes ▸ Literal strings (must be in single quotes)
`From` clause	Supports ▸ Multiple MBOs

Table 7.1 SQL Limitations for Object Queries

The MBO wizard automatically creates some object queries, and the designer can add their own. The path to view these queries is PROPERTIES • ATTRIBUTES • OBJECT QUERIES, as shown in Figure 7.68.

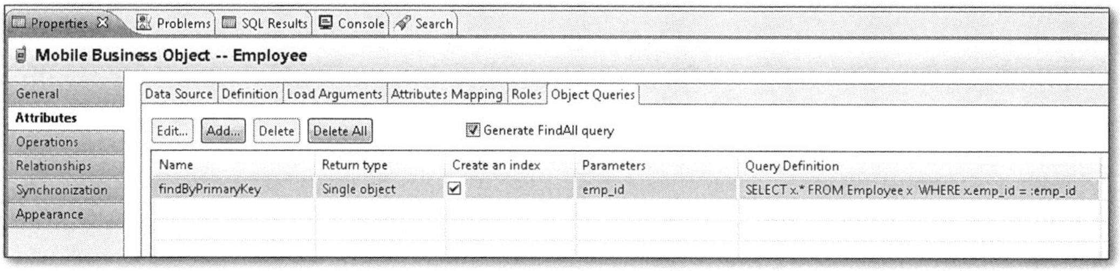

Figure 7.68 Object Queries

Next, we explain three queries that are required for this example. The first two queries are created by the system, and the third query is created by the developers. These queries are used to retrieve (or update) data that is used in the various app screens.

7.6.1 FindAll

Each MBO has a default query: the `FindAll` query. An example for this object query appears as follows:

```
SELECT x.* from Employee x
```

One of the requirements of an object query is it must use aliases. In this example, the MBO is EMPLOYEE, and the alias being created is `X`. The `X.*` indicates that all attributes for the MBO are selected. Since there is no `Where` clause, all rows in the MBO are selected. The `FindAll` query, therefore, returns the entire table.

The phrase "return" is important to understand here. The object queries are called in the mobile apps (as we discuss in the next chapter) to generate data to fill a control on the screen. In this case, the entire table is passed to an on-screen control that might display all the attribute values of the MBO, row by row. The object queries are used to propagate data from the mobile cache to the mobile apps and their logic controls. The developer can choose to exclude this query from the MBO by deselecting the box.

7.6.2 FindByPrimaryKey

A `findByPrimaryKey` object query is also generated for each MBO for which a primary key is defined. An example of this object query is as follows:

```
Select x.* from Employee x Where x.emp_id = :emp_id
```

This query returns zero rows or one row of data from the MBO, based on whether the provided value for the key field exists in the table. If a match is found, all attribute values for that row are returned.

7.6.3 Designer Developed Queries

The developer can generate additional object queries for the MBO. To add an additional query, follow the path Properties • Attributes • Object Queries and click the Add button. A wizard for creating the object query appears. For the Employee MBO, assume that the mobile app includes an option to find all employees who live in a particular location. This might be accomplished by finding employees with a specific zip code. The following entries are required:

1. Enter "findByZipCode" as the name.
2. Click to Add a parameter.
3. Change the parameter name to `Zip_Code`.
4. Change the Datatype to `STRING(9)`.
5. Deselect Nullable.
6. Select `Zip_Code` as the `mapped to` attribute.
7. Click Generate to create the query.
8. Select Return Multiple Objects.

The dialog box then appears, as shown in Figure 7.69. To complete this process, click the Finish button, and you then see the complete list of object queries for this MBO, including this `findByZipCode` query. Once this object query is generated, it is also necessary to create a personalization key for `Zip_Code`.

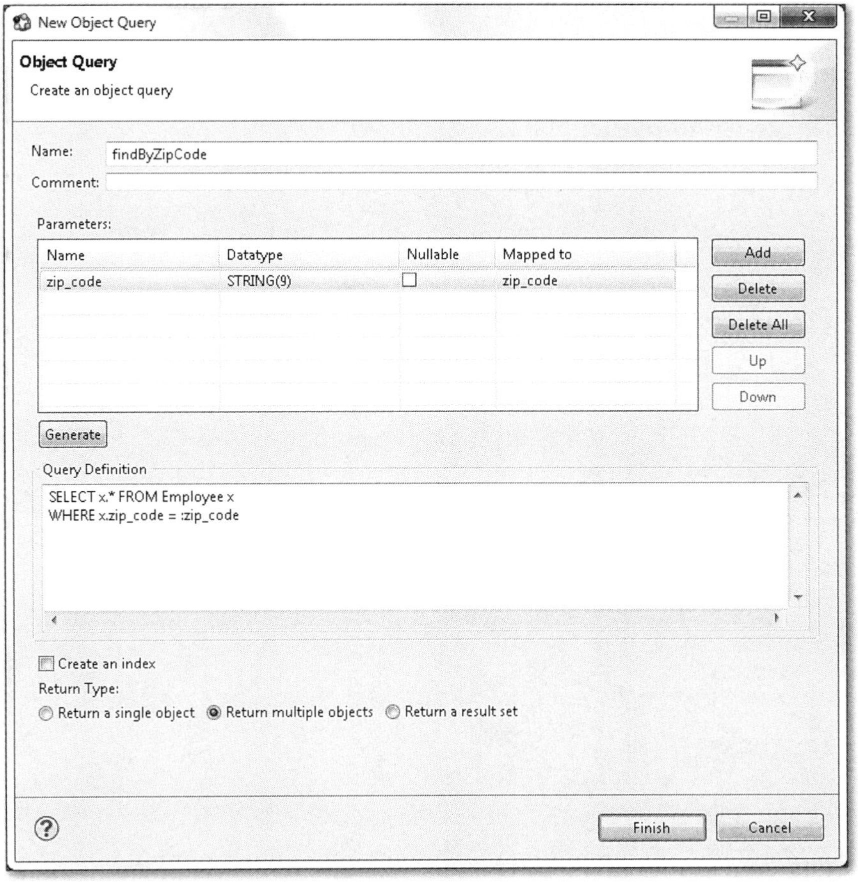

Figure 7.69 Creating a Custom Object Query

7.7 Creating Relationships between MBOs

Relationships involve linking together two MBOs using attributes that are common between the two MBOs. These relationships allow the developer to treat the results as a single object. The relationships that can be supported include one-to-many, many-to-one, and one-to-one relationships.

To demonstrate a relationship between two MBOs, we will create two new MBOs. The first MBO is entitled "Customer" and contains a row of data for each customer. The second MBO is entitled "Sales_Orders" and contains a row for each sales order. The CUSTOMER MBO contains the attribute named `ID`, which uniquely identifies the customer, and the SALES ORDER MBO contains the attribute named `Cust_Id` for the customer associated with the sales order. These two attributes can be used to form a relationship between the two MBOs. The following is description of the steps used to create this relationship:

1. Drill down into the database to open the TABLES folder.
2. Drag the CUSTOMER table onto the Mobile Application Diagram Editor.
3. Accept the defaults on the QUICK CREATE dialog and click OK.
4. The CUSTOMER MBO that includes all the attributes and the three operations is created, as shown in Figure 7.70.

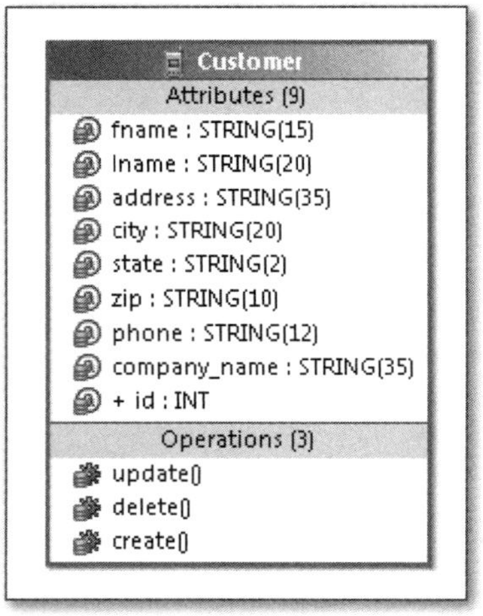

Figure 7.70 Customer MBO

Before creating the SALES ORDER MBO, it is necessary to create two object queries for the CUSTOMER MBO. After selecting the CUSTOMER MBO, select the path PROPERTIES • ATTRIBUTES • OBJECT QUERIES. The MBO has, by default, a `FindAll` query, and since the CUSTOMER MBO has a primary key, there is also a `findByPrimaryKey query`, as shown in Figure 7.71.

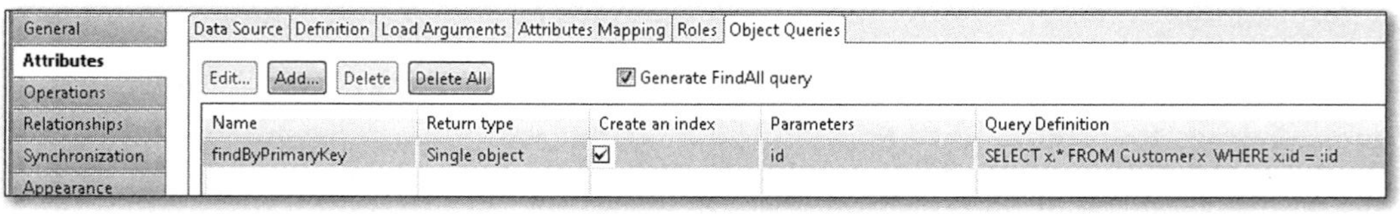

Figure 7.71 Customer MBO with FindAll and findByPrimaryKey Queries

The first additional object query to be created retrieves all the data, but unlike the `FindAll`, it orders the data by customer name. To create this query, click the ADD button and then make the following entries into the dialog box:

1. Enter “findAllSortedbyLName" for the name.
2. Click GENERATE.
3. Add the `Order By LName` clause to the end of the query.

The query appears, as shown in Figure 7.72.

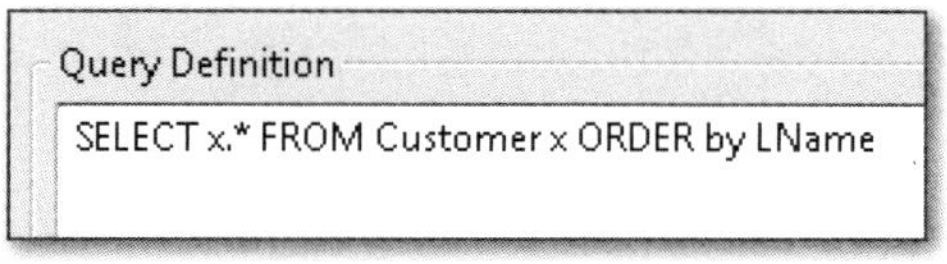

Figure 7.72 First Additional Object Query

Next, you must create the second additional object query, which finds a customer that has a character sub-string as the beginning of the customer's last name. Examples might be “Smith," “Jac," or even “H." To create this query, click the ADD button and then make the following entries into the dialog box:

1. Enter "findByLname" for the name.
2. Click ADD to add a parameter.
3. Rename the parameter "LnameParam".
4. Select `String(20)` for the data type.
5. Map the parameter to `lname`.
6. Click GENERATE.
7. Edit the generated SQL to look like the query shown in Figure 7.73. (The “%" is a wild card.)
8. Click FINISH to view three object queries.

The next step is to create the SALES ORDER MBO using the drag-and-drop procedure, accepting the three operations. The relationship between these two MBOs is created as follows:

1. Click RELATIONSHIP in the Palette, as shown in Figure 7.74.
2. Click the CUSTOMER MBO.

3. Drag and drop it onto the Sales Order MBO.
4. The New Relationship dialog appears.
5. Select Composite and One to many.
6. Click the yellow square next to `ID` in the Customer MBO.
7. Click on the yellow square next to `Cust_Id` in the Sales Order MBO, as shown in Figure 7.75.
8. Click Finish.

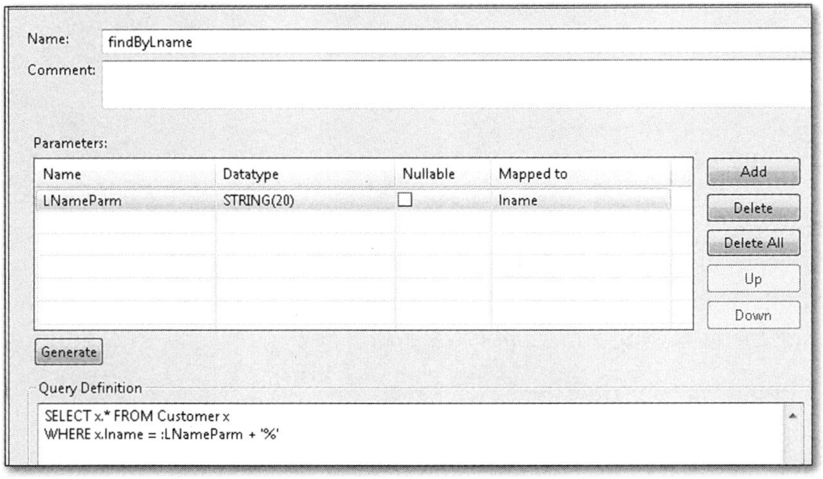

Figure 7.73 Second Additional Object Query

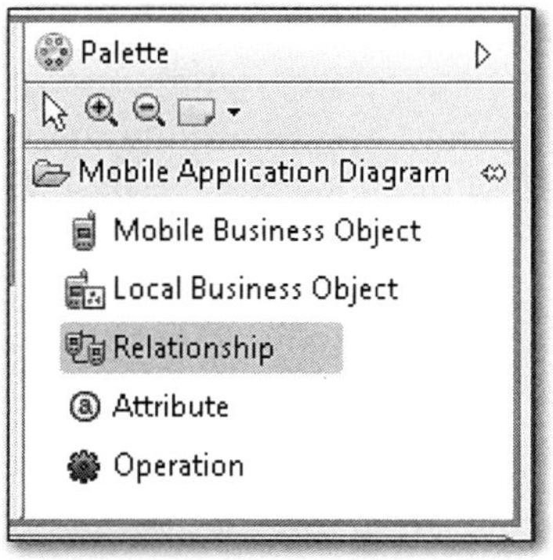

Figure 7.74 Selecting the Create Relationship Wizard

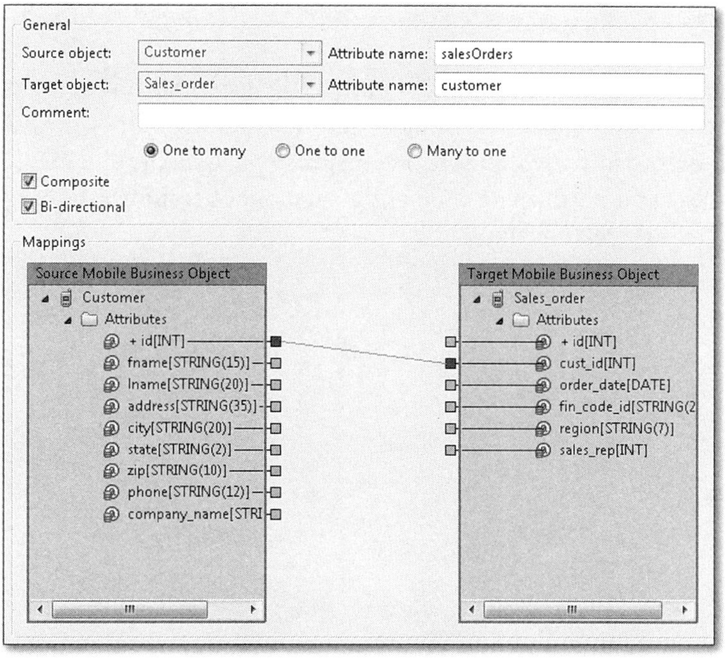

Figure 7.75 Creating the Relationship between Customer and Sales Order MBOs

The relationship between the two MBOs is created, as shown in Figure 7.76. This relationship is a one-to-many relationship between the CUSTOMER and SALES ORDER MBOs.

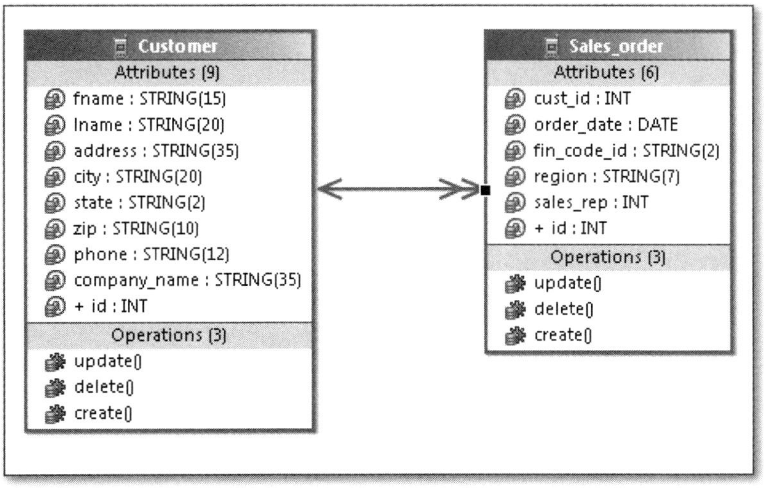

Figure 7.76 Relationship between Two MBOs

7.8 Summary

This chapter provided an explanation of MBOs, one of the essential components of a mobile app for SAP. You should now understand the entire process of MBO building, which can be broadly divided into five main steps: creating the MBO, controlling its data flow and performance, creating database operations, creating object queries, and creating relationships.

Hybrid app design involves a number of steps. We'll walk you through these steps in this chapter, and specifically focus on the most complicated: the design of the user interface.

8 Creating Hybrid Apps

Once you have created your MBOs, the next step is for you to create your hybrid app. In this chapter, we will walk you through what this entails. In Section 8.1, we'll explain the general process for creating a hybrid app, which involves five steps (creating the app, designing the screen flow of the app, designing the screens themselves, generating mobile packages, and testing the app). In subsequent sections, we'll go into more detail about the more complicated steps in this process. In Section 8.2 and Section 8.3, we'll explain screen flow and screen design, which together compose the design of the *user interface*, one of the most important steps in creating a hybrid app. Screen flow and screen design are controlled using the *Hybrid App Designer* (formerly called the Mobile Workflow Forms Editor), which is one of the graphical tools provided with the MOBILE DEVELOPMENT perspective of the Unwired Workspace. The tool is used for developing hybrid apps that will run on mobile devices that utilize one of the four supported platforms: Apple iOS, Android, BlackBerry, or Windows Mobile Professional. Finally, in Section 8.4, we'll explain how to generate mobile packages.

8.1 Designing a Hybrid App: Process Overview

Prior to the creation of the hybrid app, the project needs to be created, and the MBOs need to be built. Once those steps are complete, the high-level steps required to design a hybrid app are as follows:

1. Create the hybrid app.
2. Design the flow of screens.
3. Design the screens themselves.
4. Generate mobile packages.
5. Test the hybrid app.

8.1.1 Creating Hybrid Apps

To create a hybrid app, use the path [RIGHT CLICK ON PROJECT NAME] • NEW • HYBRID APP DESIGNER, as shown in Figure 8.1.

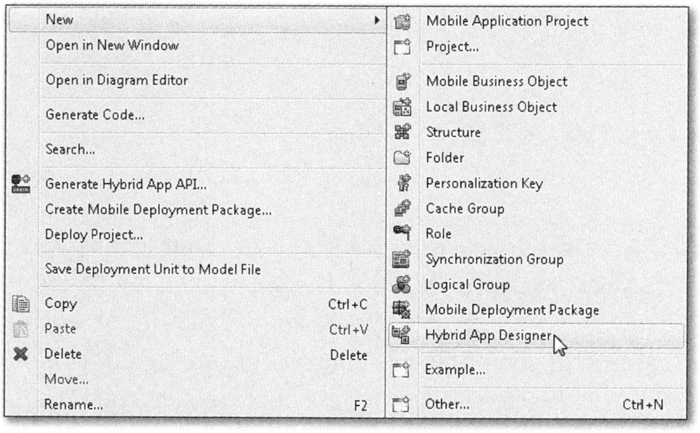

Figure 8.1 Creating the Hybrid App and Starting the Hybrid App Designer

The first dialog box requests the name for the hybrid app. This name cannot contain spaces or non-ASCII characters. The name also has to be unique among all hybrid apps deployed to the Unwired Server. After entering the name, as shown in Figure 8.2, click NEXT.

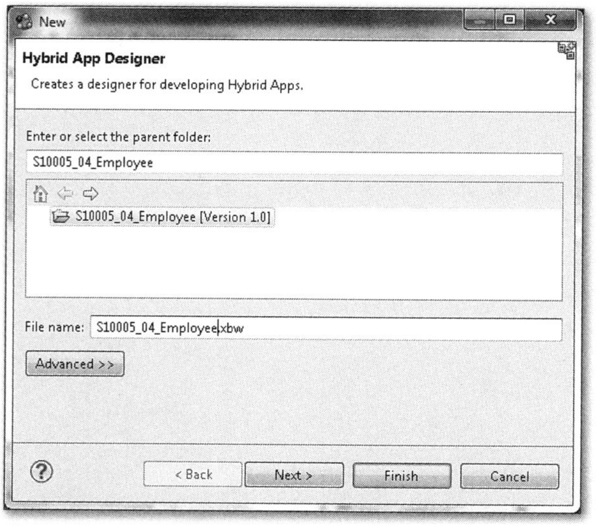

Figure 8.2 Naming the New Hybrid App

The next dialog box asks how this new hybrid app will be started. The apps can be started by the end user, the server, a first-time option, or a credential-requesting screen. In this example, the hybrid app is started by the user. Once this option is selected, as shown in Figure 8.3, click FINISH.

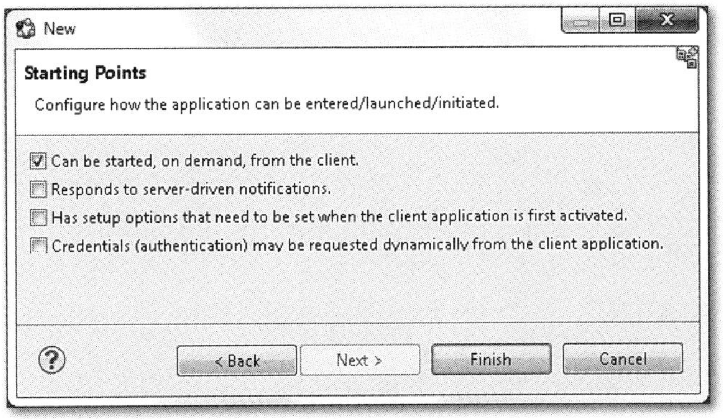

Figure 8.3 Selecting the Client-Initiated Starting Point

The HYBRID APP DESIGNER INTRODUCTION screen is displayed in the EDITOR area. This optional splash screen displays a link to START, which brings up the Flow Design Editor and a link to the HELP system. The tabs in the lower left-hand corner of this view provide the option to move between this introduction screen, the Flow Design Editor, and the Screen Design Editor, as shown in Figure 8.4.

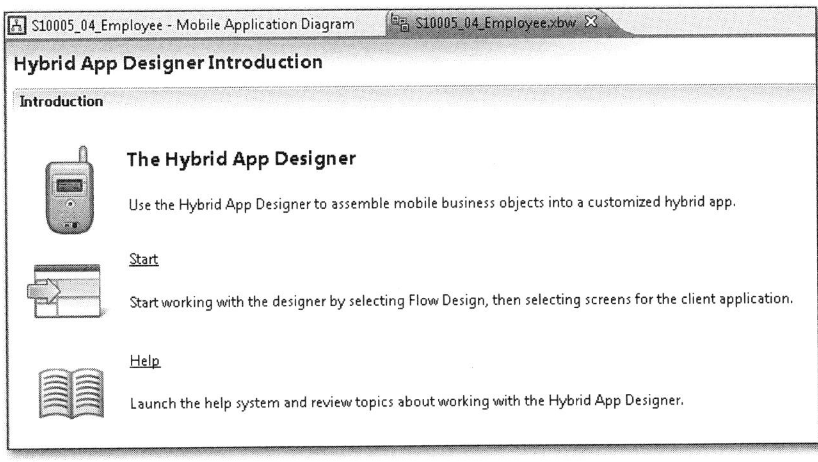

Figure 8.4 Introduction to Hybrid App Designer

8.1.2 Designing Screen Flow

The development process utilizes a graphical Flow Design Editor that displays all the hybrid app's screens as nodes. The logical flow between these screens is shown as connections between the nodes. This flow design has a single starting point, and each screen in the design has a unique name. The PALETTE view provides starting points, screens, and connections. An example of flow design is shown in Figure 8.5.

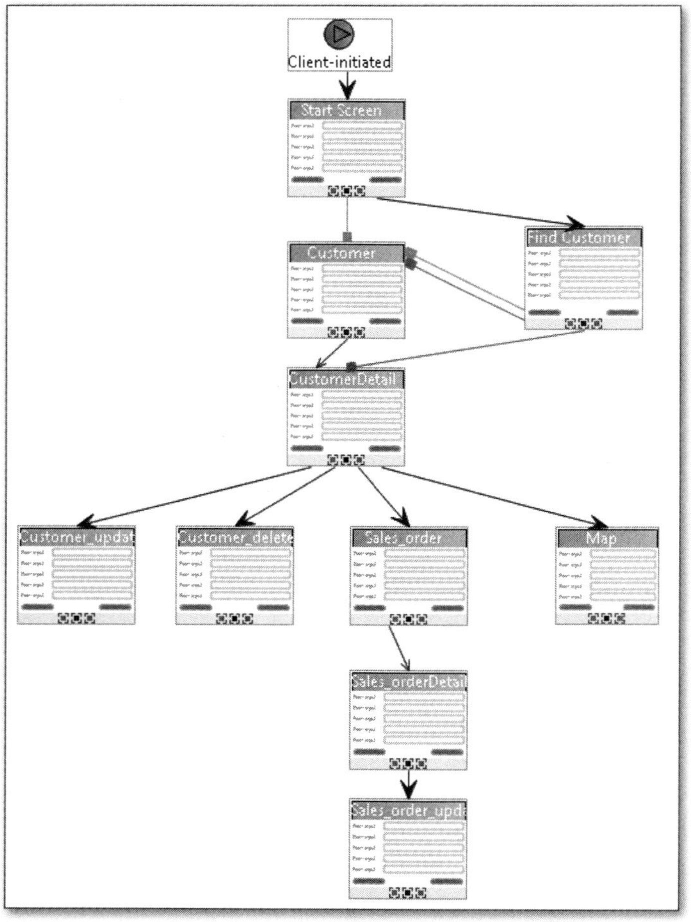

Figure 8.5 Flow Design

Clicking either START or the FLOW DESIGN tab reveals the initial flow design, as shown in Figure 8.6.

Figure 8.6 Initial Flow Design

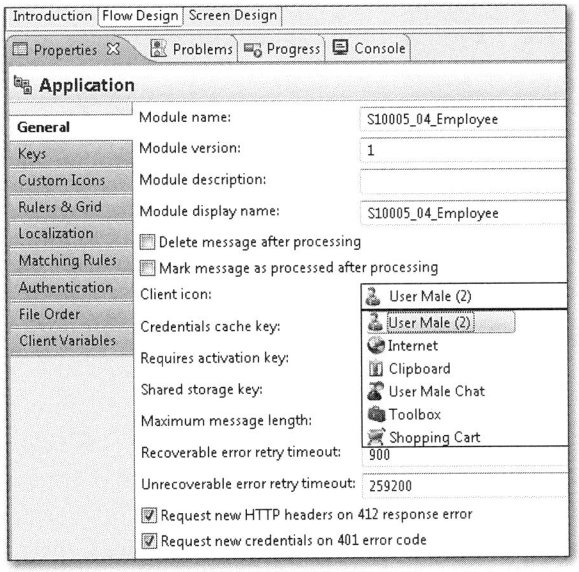

Figure 8.7 Setting Icon Property for the Hybrid App

This flow design consists of a client-initiated starting point and a screen with the name START SCREEN. When the flow design is being displayed, a series of tools exist under the PALETTE view that can be used to create screens, connections between screens, and any of the four starting points. (This is described in more detail in Section 8.2.) In the Flow Design Editor, the PROPERTIES view displays properties related to the hybrid app. Selecting the path FLOW DESIGN • PROPERTIES • GENERAL displays the GENERAL properties for this hybrid app. The property change, as shown in Figure 8.7, is used to select the icon displayed for this hybrid app on the end user's device. In this example, an icon of an individual is being selected for this app, which displays employee data.

8.1.3 Designing Screens

After the flow design is completed, you must design the content of the screens. Using the Screen Design Editor in the Hybrid App Designer, you can utilize screen controls that both display data and generate events (i.e., buttons and menus) controlling the flow between the screens, as described in the flow design. An example screen design is shown in Figure 8.8. During the development of the hybrid app, the designer flips back and forth between the Flow Design and Screen Design Editors.

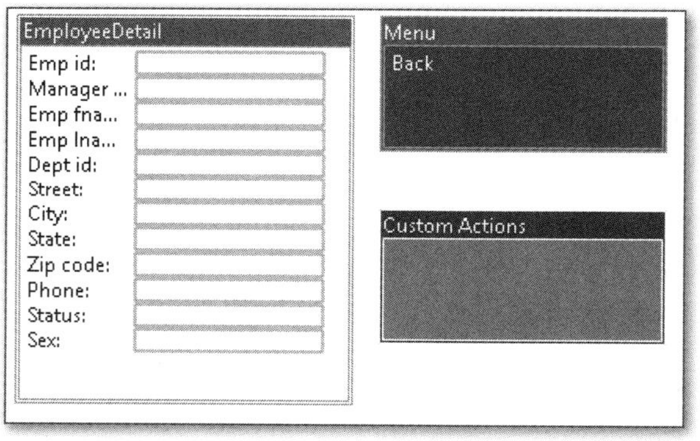

Figure 8.8 Screen Design

While the Flow Design Editor displays the entire hybrid app, the Screen Design Editor focuses on a single screen. Several approaches exist to have the Screen Design Editor focus on a particular screen. One approach is to select by clicking the desired screen and then select the Screen Design Editor. An alternative is to double-click the desired screen. As shown in Figure 8.9, the editor displays the screen, menu, and custom actions. The PROPERTIES view on the bottom displays properties for the selection, which in this case is the screen. The PALETTE view on the right can be used to add menu items, custom actions, and different controls that would make up the screen. When a specific control or menu item is added or selected, the PROPERTIES view focuses on the properties for that control or menu.

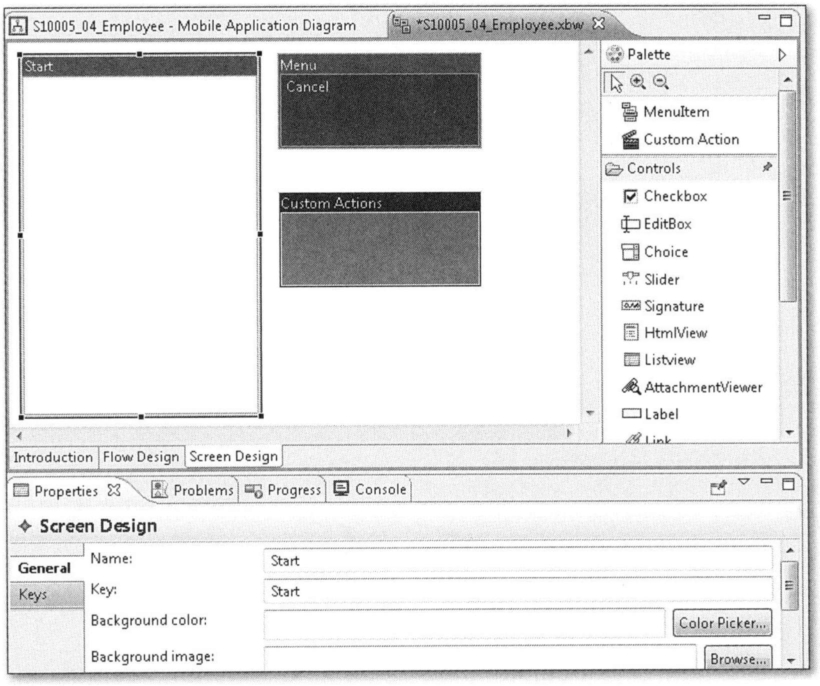

Figure 8.9 Screen Design Editor Focusing on the Start Screen

To add a control to this currently selected screen, click the control in the PALETTE view and then click the screen. As an example, click the HTMLVIEW control and then click the START SCREEN. The HTMLVIEW control allows you to enter a string of HTML as one of the properties of the control. The resulting output created by this HTML string is then displayed on the screen. After the HTMLVIEW control is added, the START SCREEN appears, as shown in Figure 8.10.

With this control selected, the next step is to enter the required properties. Every screen and control must be assigned a unique key. The user can enter a key value or can let the system suggest a value. For this example, we click the NEW KEY button and then OK to accept the suggestion, which is KEY1. For the default value, we enter an HTML string, such as this:

```
<h2>Hello World, from Dave</h2>
```

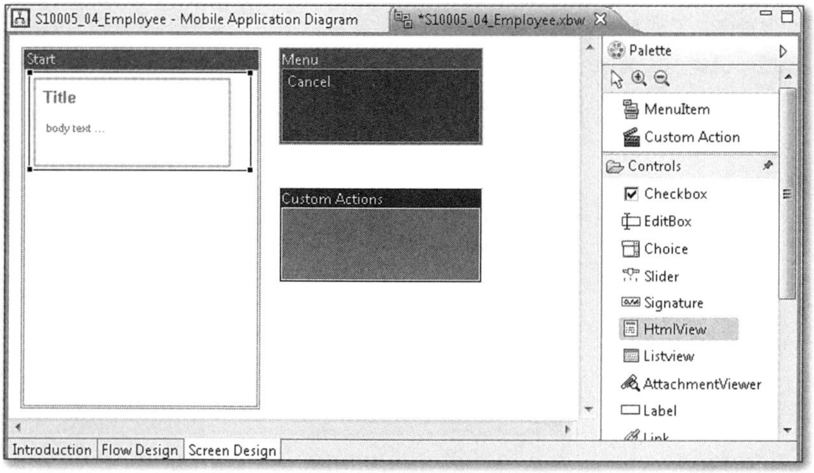

Figure 8.10 Adding the HTMLView to the Start Screen

The PROPERTIES view then appears, as shown in Figure 8.11. Note that even though the HTML text is entered, in the case of the HTMLVIEW control, the screen itself displays only the standard control. It does not replace the control with the results of the HTML string. To see this output, you have to wait until the hybrid app is run.

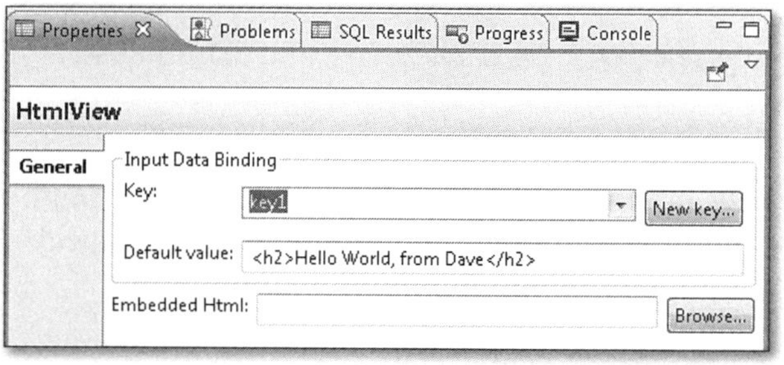

Figure 8.11 Properties of the HTMLView Control

At this point, the design is complete for a very simple hybrid app. The app simply displays the results of the HTML string, "Hello World, from Dave," on the first screen. This is all the app does!

8.1.4 Generating Mobile Packages

The next step is to generate the mobile package. You accomplish this by right-clicking a whitespace area of the editor and selecting the GENERATE HYBRID APP menu, as shown in Figure 8.12.

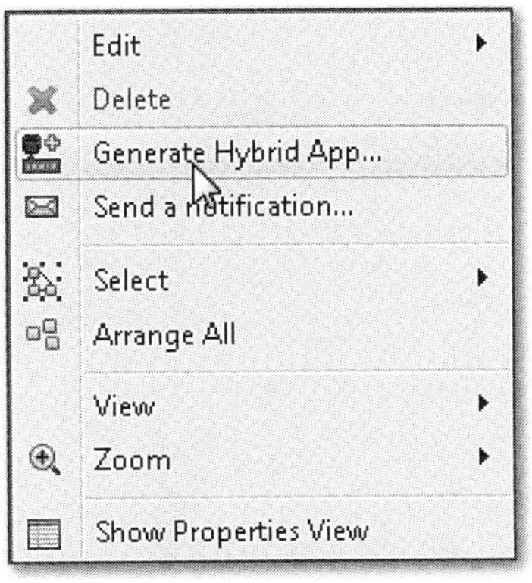

Figure 8.12 Generating the Hybrid App

The first dialog box requests that you save all changes. Click YES, as shown in Figure 8.13.

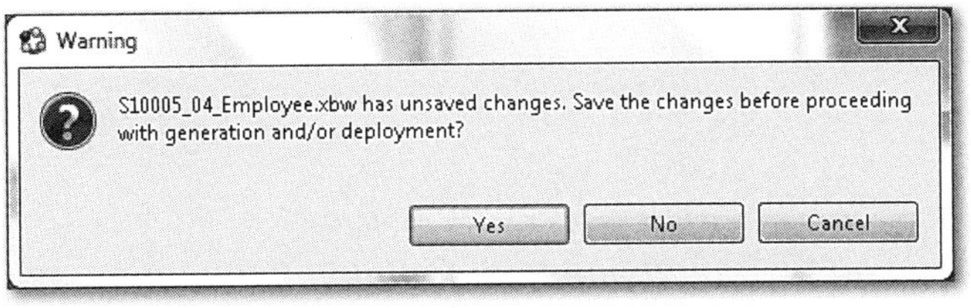

Figure 8.13 Save Changes Dialog Box

The second dialog box presents a wizard for generating the hybrid app: the Hybrid App Generation Wizard. This wizard provides a number of options that can be selected to impact the generation of the hybrid app. First we need to select the GENERATE PACKAGE option. The only change we need to make in our example is to enter the user's name (in this case, we use "S40000D") of the mobile device(s) to which we want to assign the hybrid app. (A discussion of device names was presented in Chapter 5; a discussion of the other parameters is presented later in this chapter.)

Once all the parameters are entered, as shown in Figure 8.14, the next step is to click the FINISH button. This wizard then does the following:

- Generates the code for the mobile app and save that code in the project folder
- Deploys that hybrid app (code) to the Unwired Server
- Assigns the package (from the Unwired Server) to one or more devices

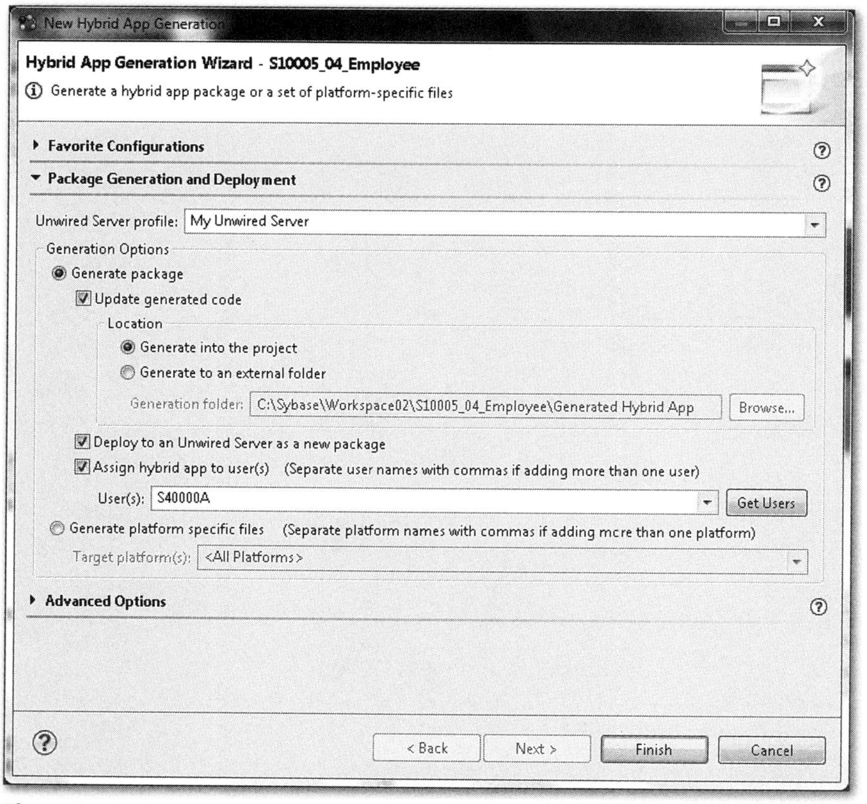

Figure 8.14 Generating Mobile Package

The status of this wizard can be viewed in the CONSOLE tab in the PROPERTIES view, as shown in Figure 8.15. If an error occurs at any stage, a message appears in this log.

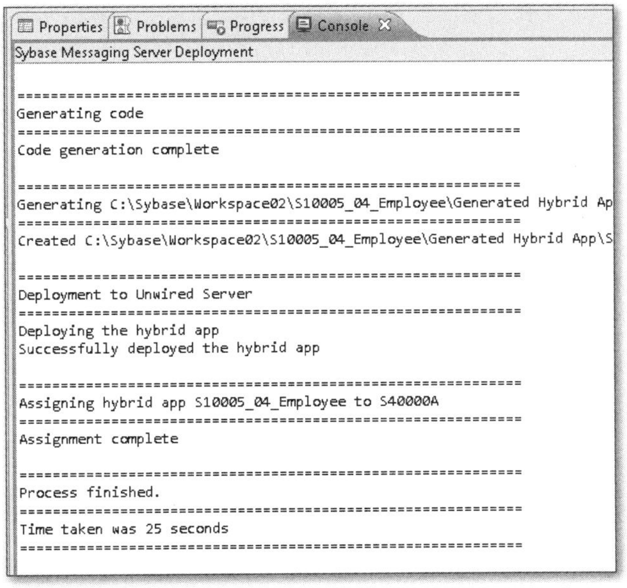

Figure 8.15 Log Showing the Generation of the Hybrid App

8.1.5 Testing Apps

Once the app has been assigned to the mobile device, the final step is to test the app, as shown in Figure 8.16. In this example, the app is being tested on an Apple iPad. On the left-hand side of the screen is the list of hybrid apps that have been deployed on this mobile device. The second app shows the icon of a person and has the name S10005_04_Employee, which is the sample app we have been developing. Clicking that menu item causes the hybrid app to run. In this example, the Start Screen appears and displays the text, "Hello World, from Dave."

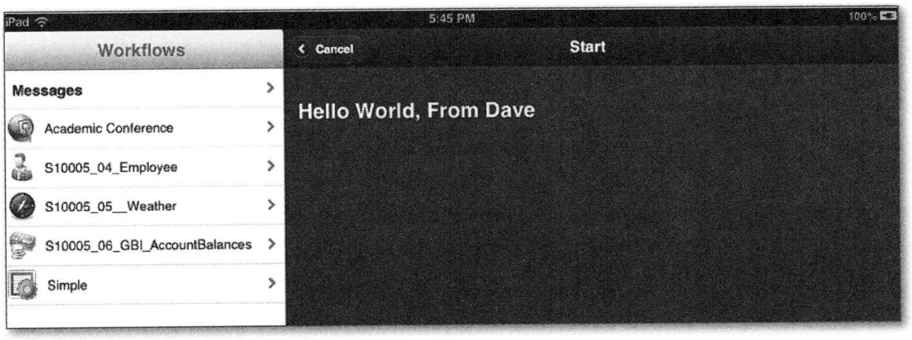

Figure 8.16 Sample App Running on an iPad

Now that we've given you a brief overview of the hybrid app development cycle, the following sections provide a more in-depth view of the flow design, screen design, and code generation processes.

8.2 Designing Screen Flow

The Flow Design Editor is used to create the screens (not the screen content) and the flow that takes place among those screens. This section focuses on the Flow Design Editor. We describe page properties, starting points for a hybrid app, and creating and connecting screens (using both drag and drop and the Palette).

8.2.1 Flow Design Page Properties

A number of properties exist for each of the flow design objects, including the flow design itself. To view and change the various properties, select the object and then the Properties view. If no object is selected, then the properties displayed are for the project itself, as shown in Figure 8.17. The first tab is used to show the app's General properties. A description of these properties is presented in Table 8.1.

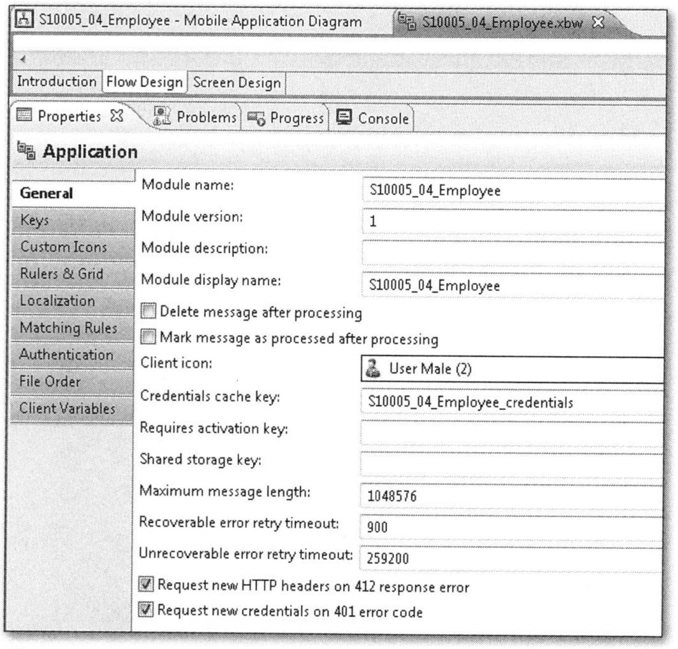

Figure 8.17 General Properties View for the Flow Design Page

Property	Description
MODULE NAME	The name of the module.
MODULE VERSION	The version of the module.
MODULE DESCRIPTION	Optional description.
MODULE DISPLAY NAME	Name that appears on the client's device.
DELETE MESSAGE AFTER PROCESSING	Messages are removed from In Box after being processed.
MARK MESSAGE AS PROCESSED	Messages are marked as processed and left in the inbox.
CLIENT ICON	Icon that appears on client device.
CREDENTIALS CACHE KEY	If present, credential sharing can take place among different apps (optional).
REQUIRES ACTIVATION KEY	If absent, each app must be activated separately; if present, multiple apps can have the same activation key—activate one, and all are activated (optional).
SHARED STORAGE KEY	If present, storage keys can be shared among apps (optional).
MAXIMUM MESSAGE LENGTH	Length in characters.
RECOVERABLE ERROR RETRY TIMEOUT	Time in seconds. Only 409 and 503 are recoverable.
UNRECOVERABLE ERROR RETRY TIMEOUT	Time in seconds.
REQUEST NEW HTTP HEADERS ON 412 RESPONSE ERROR	If selected, requests new HTTP headers when the backend throws a 412.
REQUEST NEW CREDENTIALS ON 401 ERROR CODE	If selected, requests credentials again when authentication fails (401 error code).

Table 8.1 General Properties View for the Flow Design Page

The KEYS tab, as shown in Figure 8.18, provides a list of keys that have been created for this hybrid app. The screen shows the list of keys, their data types, and whether they are bound to any object.

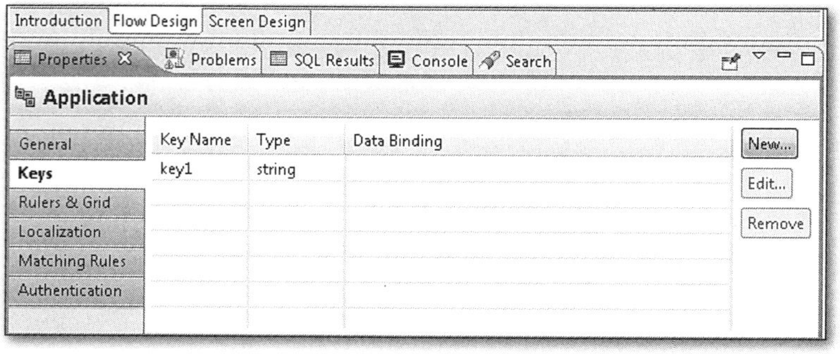

Figure 8.18 Key Properties View for Flow Design Page

The RULERS & GRID tab, as shown in Figure 8.19, does not impact the hybrid app. These properties impact only the appearance of the flow design. Options for including both a grid and a ruler provide the ability to create more organized or cleaner-looking flow design. This not only is helpful during the design phase, but also can produce better drawings for documentation.

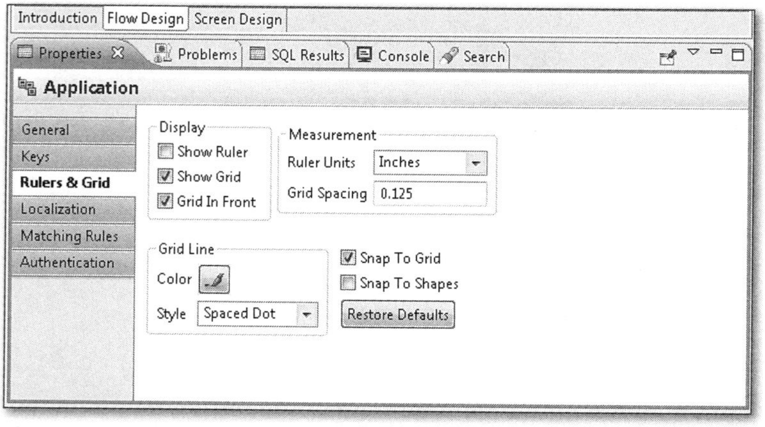

Figure 8.19 Rules & Grid Tab for the Flow Design Page

Right-clicking the background (either in the flow design or the screen design) produces a dialog box with several options. A description of some of these options is shown in Table 8.2, and the menu is shown in Figure 8.20.

Option	Description
EDIT/DELETE	Edit or delete the selected object.
GENERATE MOBILE HYBRID APP...	Generates the hybrid app package.
SEND A NOTIFICATION	Sends a notification to test server-initiated apps.
SELECT	Select (1) all, (2) all shapes, or (3) all connections on the page.
ARRANGE ALL	Arranges the shapes on the page.

Table 8.2 Descriptions of Context Options

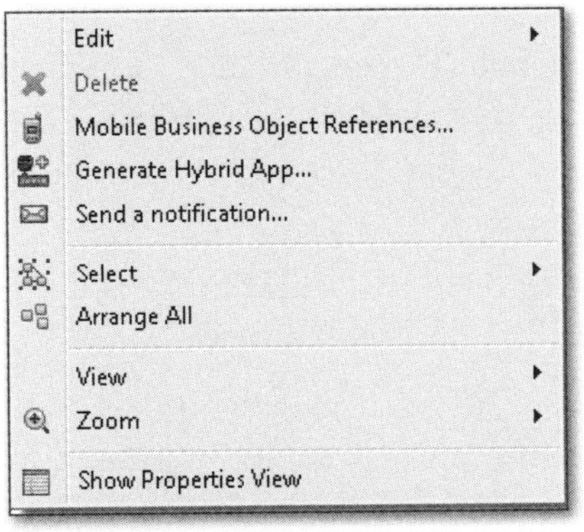

Figure 8.20 Context Options from Right-Clicking the Design Page

8.2.2 Starting Points

A starting point describes where and how execution begins when a hybrid app is opened. A list of the different types of starting points is shown in Table 8.3. Every app must have at least one starting point, and in some cases, it can have more than one starting point.

Option	Description
CLIENT-INITIATED	This starting point is activated when the user opens the hybrid app on the device.

Table 8.3 List of Starting Points

Option	Description
SERVER-INITIATED	This starting point is activated when a notification message is sent to the device that matches the matching rules specified in the hybrid app.
ACTIVATE	This starting point is activated the first time the user opens the hybrid app. It is not activated subsequent times the hybrid app is opened.
CREDENTIAL REQUEST	This starting point is activated when the hybrid app requires a user name and password and the app does not have a valid one in its cache.

Table 8.3 List of Starting Points (Cont.)

The CLIENT-INITIATED starting point is used for hybrid apps that are started by the user on the user's device and that do not require any additional security. The CREDENTIAL REQUEST serves the same function with the exception that the user must provide a user ID and a password before proceeding to run the app. This information is saved in a cache on the device for a specified period of time. As long as the information is current in the cache, this starting point appears to the user the same as CLIENT-INITIATED.

The SERVER-INITIATED starting point is used when the Unwired Server is used to notify the user of an event by sending the user a message. When the user opens the message, the user can then activate the hybrid app (which is stored on the Unwired Server) and view the app. A hybrid app must have one, and only one, of this type of starting point. The final starting point (ACTIVATE) can also be added to any hybrid app. This starting point is typically used to perform some initialization and is executed only the first time the app is run.

The starting point can be added either when the hybrid app is created (see Figure 8.3) or later, manually. You can add a starting point to the flow design manually as follows:

1. Click the type of starting point from the PALETTE view.
2. Click the blank workspace, and the starting point appears.

The flow design with an ACTIVATE starting point appears, as shown in Figure 8.21.

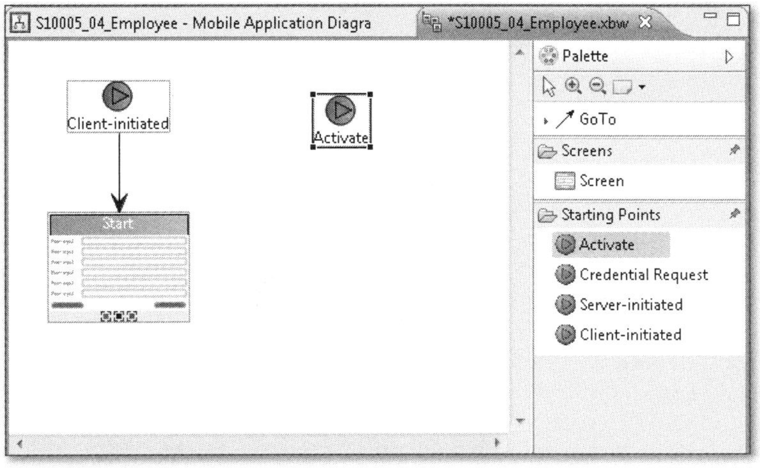

Figure 8.21 Palette with the Four Starting Point Options

8.2.3 Drag and Drop

The drag-and-drop approach can be used to add screens and connections that support both MBOs and their operations to the flow design. The MBOs created for each project are saved in a folder called the MOBILE BUSINESS OBJECT folder. Drilling down into that folder reveals both the list of attributes and the operations associated with each MBO.

Next, we'll look at two examples of the drag-and-drop approach in action.

Employee Example

In the employee example shown in Figure 8.22, the project has the name S10005_04_EMPLOYEE. This project contains a single MBO with the name EMPLOYEE. This MBO consists of 12 attributes and does not contain any operations.

XBW File

If we look at the bottom of the project folder, we see a file with the extension XBW. This is the file that contains the hybrid app design we are currently developing, including both the flow design and screen design. If the current design view is closed, it can be reopened by right-clicking on this XBW file and then selecting OPEN.

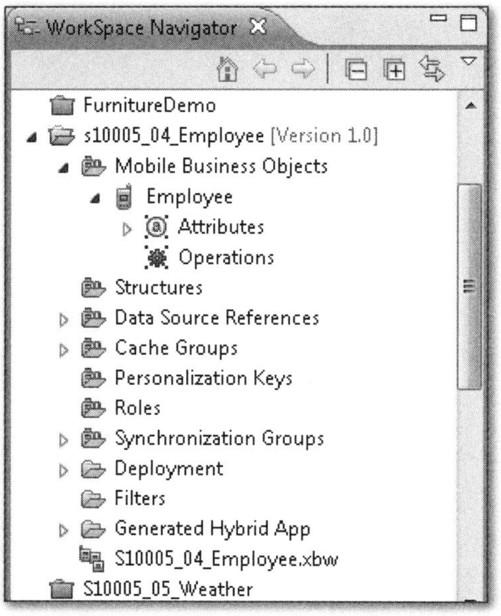

Figure 8.22 Employee MBO as Part of the Database Project

The screens and logic associated with this MBO can be created in the flow design using drag-and-drop as follows:

1. Left-click the MBO (EMPLOYEE).
2. Hold down the left mouse button, drag the MBO, and drop it onto the flow design.

In the case of a single MBO with no operations, this drag-and-drop process generates two new screens. The first screen, in this case with the name EMPLOYEE, is used to display the list of all employees. The second screen, in this case with the name EMPLOYEEDETAILS, is used to display all the attribute values for a single employee. These two screens are also connected with a link. This link includes the necessary logic (i.e., query) so that, when you select (i.e., click) an employee in the first screen, it is replaced by the second screen, which displays the attribute values of that employee. The resulting flow design, after this drag-and-drop process, appears as shown in Figure 8.23.

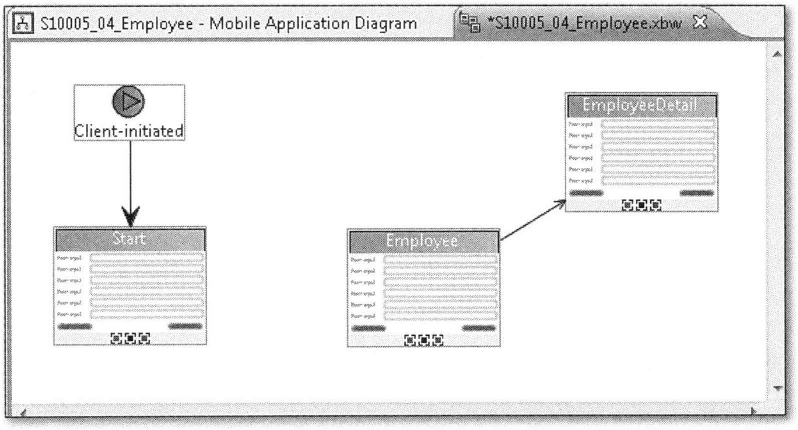

Figure 8.23 Screens Created by Dragging an MBO onto the Flow Design

To complete the flow design for this simple app, the final step involves connecting the START SCREEN to the EMPLOYEE screen. This connection or link can be created as follows:

1. Left-click GOTO in the Palette.
2. Left-click the START SCREEN and hold down the left mouse button.
3. Drag the link being created to the EMPLOYEE screen and release the mouse button.

This creates the flow design, as shown in Figure 8.24.

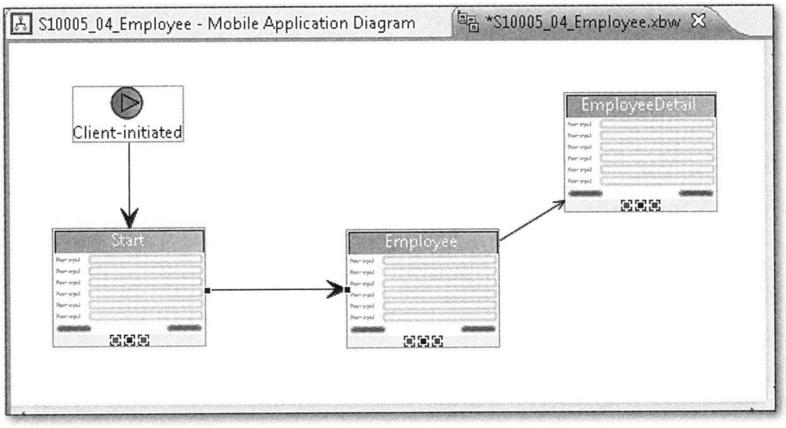

Figure 8.24 Connecting the Start Screen with MBO Creating Screens

This flow design is now complete and can be used to generate a hybrid app that would execute as follows:

1. The START SCREEN appears with a menu item that says OPEN SCREEN EMPLOYEE.
2. Clicking that menu item opens the EMPLOYEE screen.
3. The EMPLOYEE screen displays a list of all the employees.
4. Clicking a specific employee on that list opens the EMPLOYEE_DETAIL screen.
5. The detail fields of data for that employee are displayed.

Note

At this point, we are discussing only the flow design. For this hybrid app to function as described above, some changes will need to be made to the screen menus and their controls to make the hybrid app function and to improve appearance. These changes are discussed in Section 9.1.3 of Chapter 9.

Travel Request

The second example demonstrates the drag-and-drop approach to create screens associated with an MBO and includes all three of the operations. For this example, we are going to rename the previous TRAVEL REQUEST MBO (see Section 7.5 in Chapter 7) as simply T, as shown in Figure 8.25.

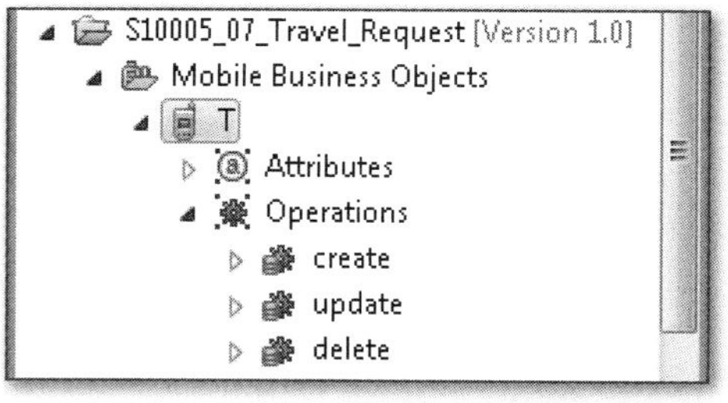

Figure 8.25 T MBO with Three Operations

When you then drag and drop the T MBO onto the flow design, you see that five new screens are created, as shown in Figure 8.26. The following is the function of each of these screens:

- The T and TDETAIL screens are the same as the ones we created in the previous example. The T screen provides a list of all the travel requests. The TDETAIL screen displays the contents of all the attributes associated with a specific travel request.
- The TUPDATEINSTANCE screen is used to update the values for the data attributes for a specific travel request.
- The TDELETEINSTANCE screen is used to delete a specific travel request.
- The TCREATE screen is used to enter attribute values and add a new travel request.

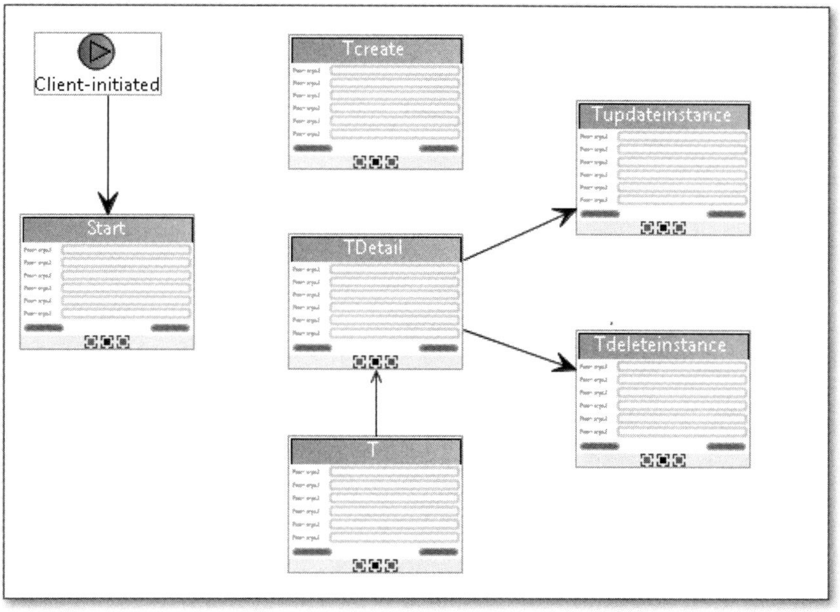

Figure 8.26 Screens Created by T MBO with All Three Operations

To complete this hybrid app, the START SCREEN needs to be connected to both the T screen and the TCREATE screen using GOTO under PALETTE. Once these changes are made, the flow design appears, as shown in Figure 8.27.

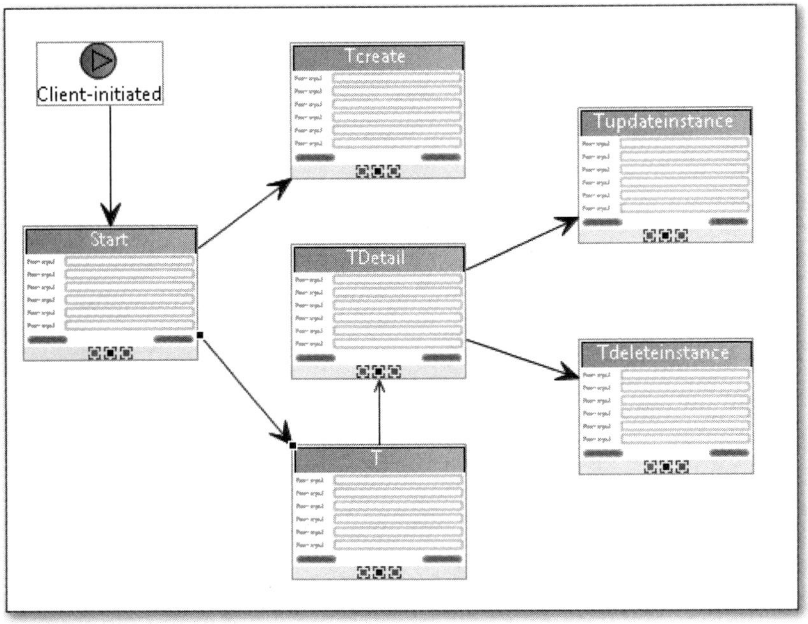

Figure 8.27 Connection of T and T_Create MBO

8.2.4 Palette Options

In the previous section, we have shown how many of the screens of a hybrid app can be generated by using the drag-and-drop approach utilizing the MBOs. However, in many hybrid apps, it may be necessary to add additional screens and linkages during the flow design process. If we look at the travel request example, the T screen provides a list of all travel requests. If this list is long, we might want to add, for example, the functionality to display only the travel requests that are for travel to a specific city. To add that functionality, we would then need to add an additional screen that offers the ability to request the name of the city and then only display travel requests to that city. This new screen could be added using the following procedure:

1. Left-click SCREEN in the Palette.
2. Left-click the flow design below START SCREEN. A new screen appears. Give the new screen the name FIND BY CITY.
3. Use the GOTO in the Palette to create a link between the START SCREEN and the FIND BY CITY screen.
4. Use the GOTO in the Palette to create a link between FIND BY CITY and T.

With this new flow design, the user can choose on the START SCREEN menu to either display all travel requests on the T screen or to go to the FIND BY CITY screen and then enter or select a city and display all travel requests for that city on the T screen, as shown in Figure 8.28.

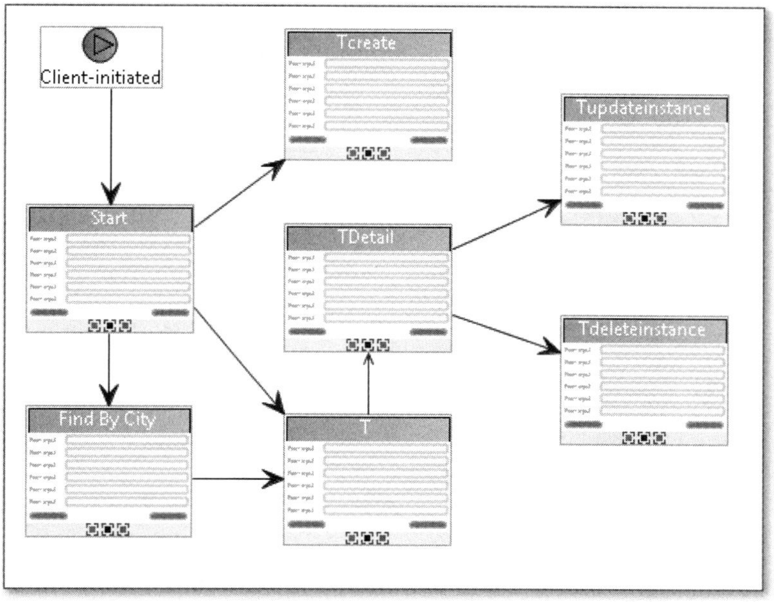

Figure 8.28 Adding the Find by City Screen

8.3 Designing Screens

Screen design is the process of configuring a single screen at a time. The easiest way to move from screen to screen is to first display the flow design, make one of the screens the current screen by clicking it, and then select the SCREEN DESIGN tab. Once the screen is selected and displayed, you are able to do any of the following:

- Define the GENERAL properties related to the screen.
- Add new controls to the screen.
- Define values for control properties.
- Add menu items and custom actions.
- Define values for menu item and custom action properties.

We describe each of these actions in more detail next.

8.3.1 Defining General Properties

The screen, like all objects, is defined by a series of properties. The GENERAL properties for the screen include a name, a key, and information about the background color or image. These properties are displayed by first highlighting the screen and then following the path SCREEN DESIGN • PROPERTIES • GENERAL, as shown in Figure 8.29. For this example, the following have been entered as the values for these GENERAL properties:

- NAME is START SCREEN.
- KEY is START_KEY.
- BACKGROUND COLOR is #80FFFF, which is light blue.
- BACKGROUND IMAGE is blank.

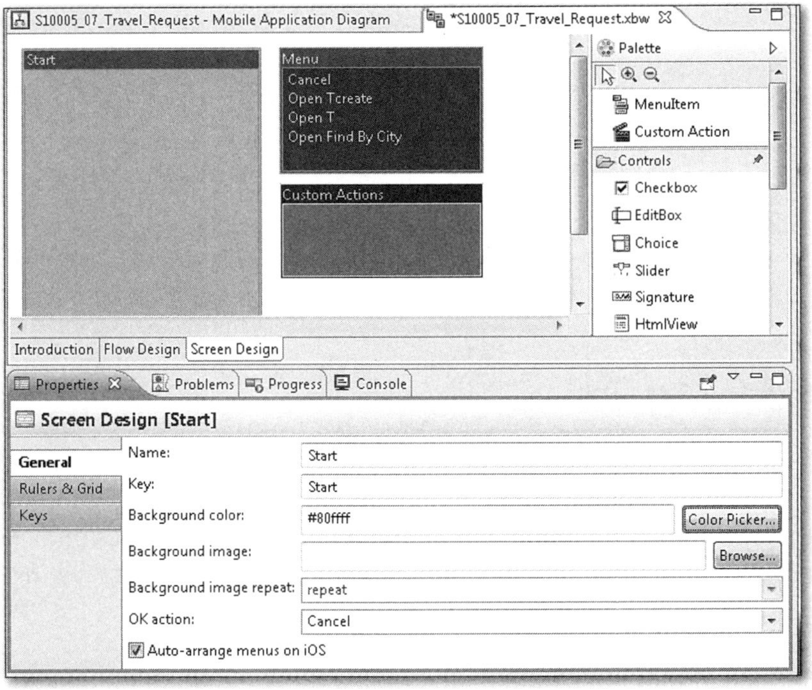

Figure 8.29 Screen Design for an Example Start Screen

Note that the properties do not have the concept of global or default settings, which are common among all screens. For example, if you would like every screen in your hybrid app to have the same background color, you need to change the background color for each screen one at a time. Many of the objects, including screens, have more than one grouping of properties. In the case of screens, we see two groups (i.e., tabs) of properties:

- General
  These are the most common properties. Some of these properties are common for all objects.
- Keys
  This tab displays the keys associated with this screen and their properties. (This topic is covered in more depth in Chapter 9.)

A list of all the properties for a screen is shown in Table 8.4.

Level	Property	Description
General	Name	The name that will appear on the top of the screen when this screen is displayed in the hybrid app.
	Key	Unique identifier. Click New Key to create a new key for the control unique identifier.
	Background Color	The Color Picker is used to select from a color palette. There is no global screen color. You must make the selection for each screen.
	Background Image	An image can be used in lieu of a background color for the screen.
	Background Image Repeat	Determines whether a background image is repeated or not.
	Ok Action	Determines whether the action taken upon clicking OK is canceling or saving.
	Auto Arrange Menus on iOS	Default is selected.
Keys	Key Name	Name.
	Type	Data type.
	Binding	Where bound.

Table 8.4 Properties for a Screen

8.3.2 Adding Controls

One or more controls can be added to a screen. These controls are used to request input from the users, display information to the user, control how information is displayed on the screen, and in some cases, even transfer control between screens. In the case of developing hybrid apps, the number of supported controls is limited, since it is possible to support only controls that exist and have a similar function in all four major mobile operating systems. A list of the controls currently supported is shown in Figure 8.30.

Figure 8.30 Controls Available for Designing Hybrid Apps

Some of these controls have some limitations when used on different mobile operating systems. Some of them also have a different appearance on different platforms. A description of each of these controls, along with some of their limitations, is presented in Table 8.5.

Control	Description	Limitation
Checkbox	Displays a true/false option on the screen.	N/A

Table 8.5 Controls Available for Designing Screens

Control	Description	Limitation
EditBox	EditBox has a logical type—used to enter text or assign values to strings, dates, decimals, integers, etc.	N/A
Choice	Displays a collection of items as a dropdown list from which the user can make a choice.	N/A
Slider	Represents numerical values within a range.	Not supported by Windows Mobile. When Optimize for Performance is selected, only supported on BlackBerry.
Signature	Adds a signature box to the screen—the value is the set of points that make up the signature.	Not supported by Windows Mobile. When Optimize for Performance is selected, only supported on BlackBerry.
HtmlView	Allows text with HTML markup to be displayed on the screen.	N/A
Listview	Displays lines that can be clicked to show further details for the selected line.	Can have only one list view per screen.
AttachmentViewer	Allows the user to download and view attachments on the device. The appropriate file viewer (for PDF, DOC, XLS, etc.) must be available on the device.	N/A
Label	Allows you to enter text that describes the control on the user interface.	N/A

Table 8.5 Controls Available for Designing Screens (Cont.)

Control	Description	Limitation
Link	Transfers control to another app, such as SEND MAIL.	Control cannot be returned back to app for iOS.
Button	Transfers to a different screen, similar to a menu item.	On Windows Mobile, only text is supported (no images).
Image	Displays an image on the screen.	Dynamic images not supported on Windows Mobile.
Collapsible	Creates an area on the screen that can be collapsible.	Supported only on iOS, Android, and BlackBerry.
Accordion	Allows for one or more collapsible controls.	Supported only on iOS, Android, and BlackBerry.
Switch	Users can drag back and forth between two exclusive choices.	Supported only on iOS, Android, and Windows Mobile Professional.
MAChartView	Displays a MAKit chart to provide graphical analytics capabilities.	Supported only on iOS and Android.
Checkbox Group	Groups one or more CHECKBOX controls.	N/A

Table 8.5 Controls Available for Designing Screens (Cont.)

Any control is added to a screen as follows:

1. Click the desired control in the PALETTE view.
2. Click on the screen where the control is required.

8.3.3 Defining Values for Control Properties

Once the control is placed on the screen, the next step is to define the values for properties associated with the control. For example, the properties associated with an EDITBOX control is shown in Figure 8.31.

Figure 8.31 Properties under the General Tab for the EditBox Control

Several properties are common to most controls, and some properties are unique to a specific control. A list of the properties that are common to many of the controls, along with their descriptions, is shown in Table 8.6. Many of the controls have several groups of properties. In the case of the EDITBOX control, we have a group called GENERAL properties and a group called ADVANCED properties. The common controls usually appear under the control's GENERAL group.

Property	Description
LABEL	Enter the text that describes the control on the user interface. Does not apply to HTMLVIEW or LISTVIEW.
LABEL POSITION	The position of the label in relation to the control it describes: LEFT, RIGHT, TOP, CENTER, CENTER_TOP or CENTER_BOTTOM.
INPUT DATA BINDING KEY	Unique identifier. Click NEW KEY to create a new key for the control.
DEFAULT VALUE	Default value for the control.
READ ONLY	Control is read only. Does not apply to HTMLVIEW or LISTVIEW. Not supported for Windows Mobile for CHECKBOX, SLIDER, or CHOICE controls.

Table 8.6 General Properties Used for Many of the Controls

In addition to the common properties, the EditBox has one control-specific property under the General tab, which is the Logical type. The EditBox control also has a number of control-specific properties under the Advanced group or tab, as shown in Figure 8.32.

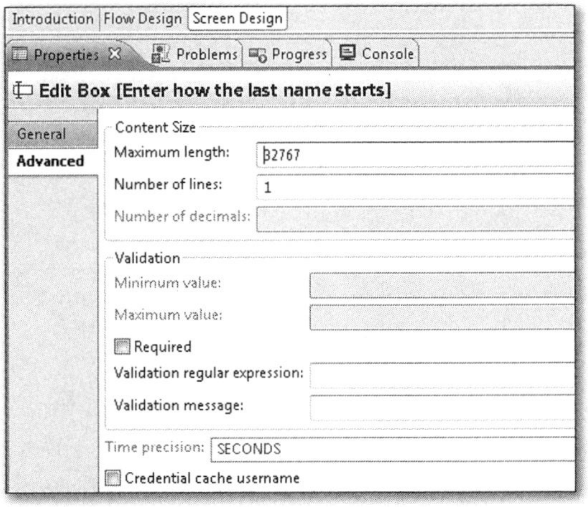

Figure 8.32 Properties under the Advanced Tab for EditBox Control

A list of each the controls and the properties that are specific to those controls is presented in Table 8.7.

Control	Level	Property	Description
Checkbox	Advanced	Required	Submission is possible only when the checkbox is selected.
		Validation Message	An error message appears if input validation is not matched.
EditBox	General	Password	If selected, text entered in this box is replaced with asterisks.
		Locale Specific	Enabled only for DATE, DATE-TIME, or TIME.
		Logical Type	TEXT, NUMERIC, DATE DATETIME, TIME.

Table 8.7 List of Each Control and Its Specfic Properties

Control	Level	Property	Description
EditBox (Cont.)	Advanced	Content Size	Maximum length, number of lines, number of decimals.
		Validation	Minimum value, Maximum value, required, validation expression, validation message.
		Time Precision	Precision of time portion displayed to user.
		Credential Cache Username	Treat EditBox as a username.
		Credential Cache Password	Treat EditBox as a password.
Choice	General	Logical Type	TEXT, NUMERIC.
		Options	Static, dynamic.
	Advanced	Content Size	Number of decimals.
		Validation Required	If required, message if missing.
Slider	General	Minimum Value	Specify a minimum.
		Maximum Value	Specify a maximum.
	Advanced	Validation Message	Message if input validation does not match.
Signature	Advanced	Validation Required	If required, message if missing.
HtmlView	General	Embedded HTML	Link to path.

Table 8.7 List of Each Control and Its Specfic Properties (Cont.)

Control	Level	Property	Description
Listview	General	Alternative Row Color	Use color picker to select color.
		On Empty List	This text appears when list is empty.
		Show Header	Column titles appear in list.
		Wrap Data in PRE and CODE Tags	Used when text contains HTML, HTML tags, HTML or XML escape characters, or <....>.
		Local Specific Display	Used for DATE, DATE-TIME, and TIME.
		Listview Details Screen	Screen that is opened when user clicks a row.
	Cells	Cell Lines	Lines that appear for each row.
		Fields for Cell Line	Fields that appear in each cell line.
	Advanced	Numbered List	Each item is numbered.
		Insert List	Insert block, rounded corners.
		Use Page Theme	Dividers use same theme.
		Filter Bar	Limit rows to those that match filter.
		Default placeholder	First filter value.
		Generate Dividers	Separate lists, no sorting.
		Case-sensitive	Determines whether dividers are case sensitive.
		Divider Position	Beginning or end of group.

Table 8.7 List of Each Control and Its Specfic Properties (Cont.)

Control	Level	Property	Description
Listview (Cont.)		Generate How	Default, custom, or key field.
		Generate Count Bubble	Count appears in each divider.
AttachmentViewer	General	Content	Object query, MIME type, file name and size.
Label		N/A	N/A
Link	General	HREF	Choose logical type.
Button	General	Has Image	Indicates that the button uses an image.
		Key	Used to identify the button image source.
		Default Value	Path of the image displayed.
		Image Dimension	Width and height in pixels.
Image	General	Alternative Text	Appears if image is invalid.
		Image Dimension	Width and height in pixels.
		Options	Static and dynamic.
Collapsible	General	Background Color	Color of control background.
		Background Image	Default no image.
		Background Image Repeat	How image repeats.
		Initial Collapse	Default.

Table 8.7 List of Each Control and Its Specfic Properties (Cont.)

Control	Level	Property	Description
Collapsible (Cont.)		Compact	If selected reduces size of control, default is unselected.
		Icon Position	Left, right, bottom, or top.
Accordion	General	Background Color	Color of control background.
		Background Image	Default no image.
		Background Image Repeat	How image repeats.
Switch	General	Toggle Options	On value, off value, on display, off display.
MAChartView	General	View Type	Full view (simple, zooming drilldown) or split view (simple).
		Name	Text identifying the chart, in HTML.
		Chart Type	Bar, line, pie, horizontal bar, bubble.
		Chart Title	Label on the screen.
		MAKit Query	Name of query in *Custom.js*.
	Category	Column	Column in chart that category is mapped.
		Format	Data format.
		Display Name	Lable on the column chart.
		Level (Zooming)	Identify category's position.
	Series	Column	Unique column to use for series.

Table 8.7 List of Each Control and Its Specfic Properties (Cont.)

Control	Level	Property	Description
MAChartView (Cont.)	Values	EXPRESSION	Combination of functions, variables, and column names.
	What-if Columns	TYPE	Float or integer.
		VALUE	Computed expression.
		MIN, MAX	Smallest and largest value.
Checkbox Group	General	LEGEND	Option text identify control.
		HORIZONTAL	Horizontal if selected, vertical is default.

Table 8.7 List of Each Control and Its Specfic Properties (Cont.)

MAChart View Control

The MACHARTVIEW control is only one option for including mobile analytics charts in your hybrid app. You can define a chart outside of the Hybrid App Designer by using the Mobile Analytics Kit (MAKit) component of the Sybase Mobile SDK and then you can add that chart to the HTMLVIEW control. This approach is presented in detail in Chapter 15. For full information about MAKit charts and metadata, we recommend the *Mobile Analytics Kit (MAKit) Developer Guides: HTML5*.

8.3.4 Adding Menu Items and Custom Actions

Menu items and custom actions are also defined as part of the screen design. Both of these actions result in a method being generated when the mobile app is built. If the action is associated with a menu, then a menu item is generated for the hybrid app. In the case of iOS, the menu is created using HTML; for the other platforms, a native menu item is generated. These appear in the screen, as shown in Figure 8.33.

If the menu item or action invokes an object query, it also creates an output key, which is a list that contains the keys that are bound to the attributes of the MBO. This output key points to the data that is passed to the next screen following the execution of the query.

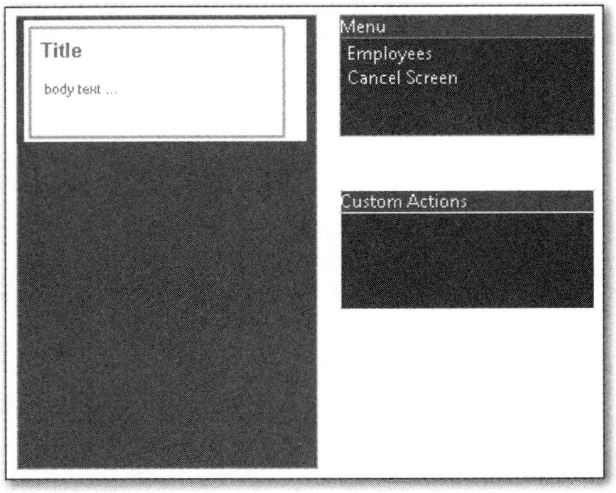

Figure 8.33 Menu and Custom Actions as Part of the Screen Design

A menu item or custom action can be added to the screen as follows:

1. Click either MENUITEM or CUSTOMACTION under PALETTE.
2. Click either the MENU or the CUSTOM ACTION box.

The PALETTE view is shown in Figure 8.34.

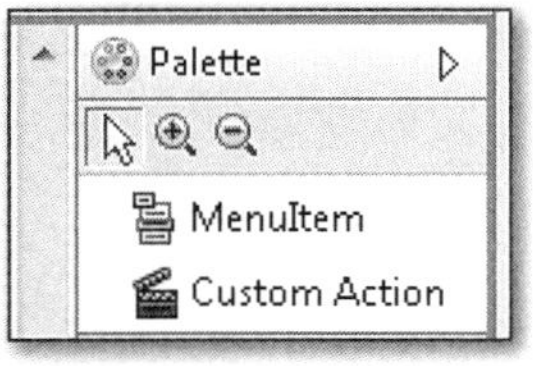

Figure 8.34 Selecting Either Menu Item or Custom Action from the Palette

8.3.5 Defining Values for Menu Item and Custom Action Properties

Each menu item or custom action is defined by specifying values for its various properties. For example, we look at a menu item that is used to find all the data for a specific MBO and then pass that data on to the next screen so that data can be displayed. For this menu item, a method that invokes the previously defined query `findAll`, which finds all data associated with the EMPLOYEE MBO, is generated. A list of the data is then passed to the EMPLOYEE screen for display using a

LISTVIEW control. The values of the GENERAL properties for this specific menu are shown in Figure 8.35 and the SUCCESS/ERROR properties are shown in Figure 8.36.

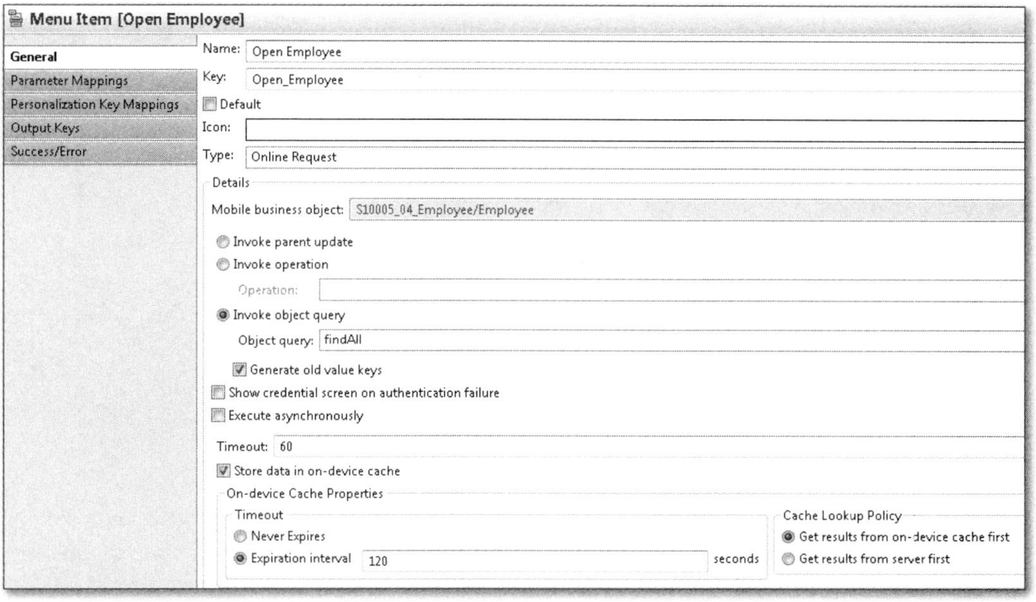

Figure 8.35 General Properties

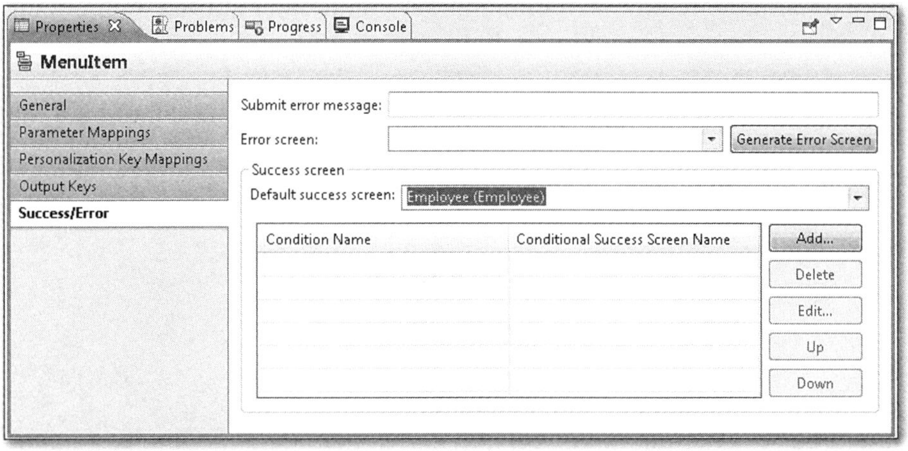

Figure 8.36 Success/Error Properties

As with the screen design, there are a number of properties that are in common for creating all menu items and custom actions. These properties and descriptions are shown in Table 8.8.

Property	Description
NAME	A valid name for menu or custom action (10 or fewer characters).
KEY	Unique key to bind the item.
DEFAULT	For menu item, select to display menu item prominently on the screen.
TYPE	SAVE, OPEN, CLOSE HYBRID APP, ONLINE REQUEST, SUBMIT HYBRID APP, SELECT CERTIFICATE, RESET CREDENTIALS, RESET ACTIVATION, CUSTOM, ADD/UPDATE/DELETE KEY COLLECTION.
DETAILS	Configure actions, setting properties based upon type.

Table 8.8 General Properties for Menu Item and Custom Action Properties

As you can see, the TYPE property defines the type of method that is generated for the menu item. The best method of describing the menu- or action-specific properties is to group them together by function based on similar types.

The first group includes four types related only to which screen (or lack of screen) is displayed next. Selecting the SAVE, CLOSE, or CANCEL menu types results in the current screen (or app) being closed. Every app should have a CLOSE or CANCEL menu item so the user can exit that app. The OPEN type is used to simply go to (or display) another screen. The properties for these menu types are shown in Table 8.9.

Property		Description
TYPE	SAVE	Close the current screen and save the input made to the collection of response values.
	CLOSE	Leave the client-side component.
	CANCEL	Close the current open screen without doing any data validation.
	OPEN	Screen to be opened with GoTo connection.
SCREEN		Only used for open; this is the screen to be opened.

Table 8.9 Properties for Type and Screen

The second group is the SUBMIT HYBRID APP menu item, which allows the app to update, delete, or create data even when the app is not connected to the network. The method associated with this menu item supports the *sometimes connected* scenario. When this menu item is executed, the method saves and queues up the necessary transaction and then closes the app. The next time the user is connected, the transaction is executed. With this type, the communication between the app and the backend is referred to as *asynchronous communication*. The method associated with this menu item requires an MBO for which one or more operations have been defined. It is one of these operations that is performed by the method. A message related to the success or failure is then sent to the specified device. The specific properties associated with the SUBMIT HYBRID APP menu item is shown in Table 8.10.

Specific Properties	Description
MOBILE BUSINESS OBJECT	MBO from which the menu item is associated.
INVOKE PARENT UPDATE	No changes made to the parent MBO, but CUD operations are performed on descendents.
INVOKE OPERATION	Invoke the selected MBO operation with which the menu item is associated.
SUBMIT CONFIRMATION MESSAGE	Enter text to show in message box when finished successfully (optional).
RESUBMIT CONFIRMATION MESSAGE	Text to show when trying to submit hybrid app a second time.
ERROR SCREEN	Screen to navigate to in case of error.
SUBJECT	If error occurs, message is sent to this device.
FROM	If error occurs, text appears in the FROM field.

Table 8.10 Specfic Properties for Submit Hybrid App Menu Item

The third group is the ONLINE REQUEST menu item, which creates a synchronous call to the Unwired Server and then waits for a response. To use this menu item, the method assumes the user is connected to the Unwired Server. While this menu item can be associated with the MBO operators, it can also be used with any of the MBO queries for retrieving or updating information. The method created with this menu item requires that the MBO has been defined with either a query or an operation. The most frequent use of this menu type is the retrieval of a KEY LIST of data to be passed to another screen for display. The specific properties associated with the ONLINE REQUEST menu is shown in Table 8.11.

Specific Properties	Description
MOBILE BUSINESS OBJECT	The MBO with which the menu item is associated.
INVOKE PARENT UPDATE	No changes made to the parent MBO, but CUD operations are performed on descendents.
INVOKE OPERATION	Invoke the selected MBO operation with which the menu item is associated.
INVOKE PARENT QUERY	Invoke the object query from the menu item.
GENERATE OLD VALUE KEYS	Used for update or parent update (default).
SHOW CREDENTIAL SCREEN ON AUTHENTICATION FAILURE	If selected, credential screen is shown on failure.
SUBMIT ERROR MESSAGE	Message user sees if an error occurs.
TIMEOUT	Enter time, in seconds, to wait before timing out.
ON DEVICE CACHE TIMEOUT	The time, in seconds, during which query results are cached on the device.
ERROR SCREEN	Screen to go to when error occurs when submitting data to the server.
SUCCESS SCREEN	Screen you go to upon successful communication with the server.

Table 8.11 Specific Properties of Online Request Menu Item

The final group is the CREATING A KEY COLLECTION menu item, which can be used only for a screen that contains a LISTVIEW control. The specific properties associated with this menu item are shown in Table 8.12.

Specific Properties		Description
TYPE	ADD KEY COLLECTION	Add.
	UPDATE KEY COLLECTION	Update.
	DELETE KEY COLLECTION	Delete.
MOBILE BUSINESS OBJECT		MBO with which the menu item is associated.
INVOKE PARENT UPDATE		No changes made to the parent MBO, but CUD operations are performed on descendents.
INVOKE OPERATION		Invoke the selected MBO operation with which the menu item is associated.
INVOKE OBJECT QUERY		Invoke the object query from the menu item.
GENERATE OLD VALUE KEYS		Required for update operations (default).
LISTVIEW		Select the list key.

Table 8.12 Specific Properties for Creating a Key Collection Menu Item

8.4 Generating a Hybrid App

Once the flow design is complete for the app and the screen design is completed for each screen, the final step is to generate the package for the hybrid app. The package is generated as follows:

1. Right-click on some white space on either the flow design or screen design.
2. Select GENERATE HYBRID APP, as shown in Figure 8.37.

This brings up the wizard for creating the app, which is shown in Figure 8.38. The wizard consists of a section for PACKAGE GENERATION AND DEPLOYMENT:

- UPDATE GENERATED CODE
  This section specifies whether code should be generated for this project and saved locally. The UPDATE GENERATED CODE option is usually selected, and the code that is generated is normally saved in the workspace folders. There are

instances when you may not want to update the code, such as when the code was previously generated and then the developer modified certain portions of that code. Code customization is the focus of Chapter 10 and Chapter 11.

- Deploy to Unwired Server as a New Package
  This section specifies whether you want to deploy this package to the Unwired Server as a first time deployment. If selected and an existing package is present, the existing package will be erased.
- Assign Hybrid App to User(s)
  This section is used to assign and deploy the hybrid app to the specific mobile devices and emulators. If names are entered, they must match exactly the device names that have been previously defined. If any names in the list do not match, the deployment will fail. The result of this section is that the mobile app is downloaded to any and all end users' devices that are provided.
- Generate Platform-Specific Files
  You can also specify whether you want to generate platform specific files.
- Advanced Options
  The wizard also has a section for advanced options. In this section, you might also want to change the jQuery mobile theme; for example, Theme B generates text with a white shadow under it (which is sometimes hard to read on screens with a light background); personal preference might dictate changing this to Theme A.

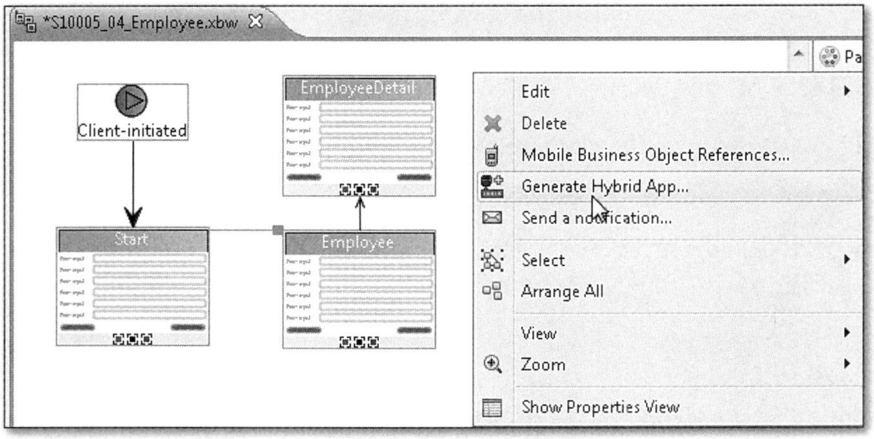

Figure 8.37 Generating the Hybrid App Package

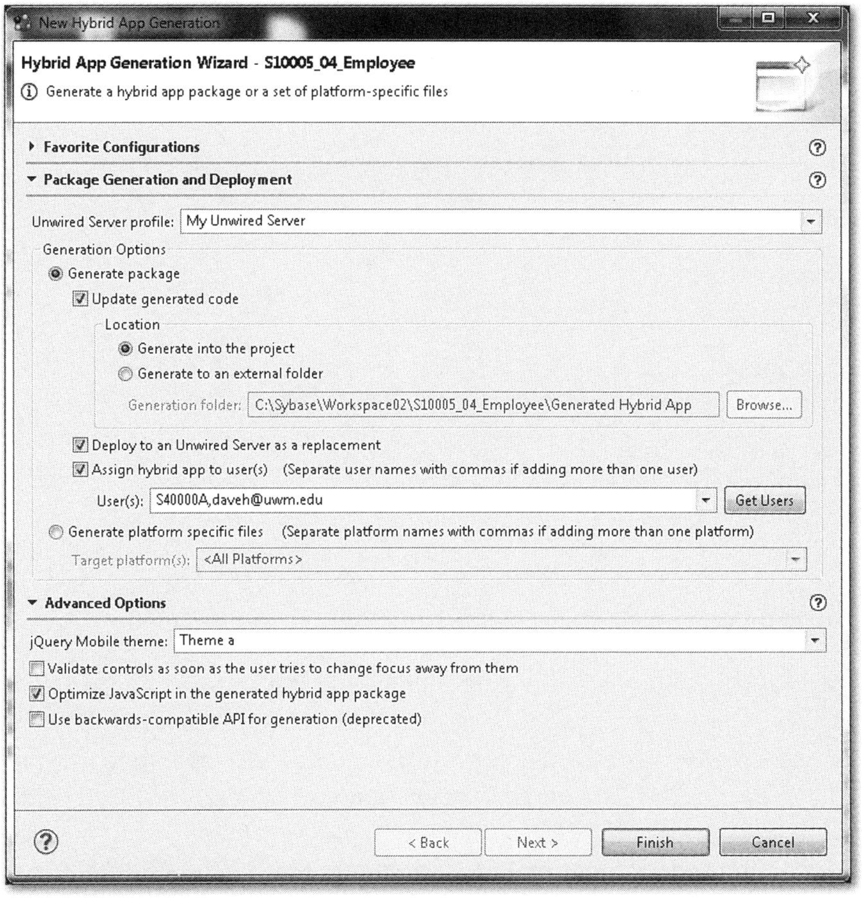

Figure 8.38 Wizard for Generating the Mobile Package

When the deployment finishes, a log describing the success or failure of each step is displayed under the CONSOLE tab, as shown in Figure 8.39. This log describes the success and failure of each of the steps:

- Generating code and saving code in the workspace
- Deploying the app to the Unwired Server
- Assigning the app to individual users

```
========================================================
Generating code
========================================================
Code generation complete

========================================================
Generating C:\Sybase\Workspace02\S10005_04_Employee\Generated Hybrid
========================================================
Created C:\Sybase\Workspace02\S10005_04_Employee\Generated Hybrid Ap

========================================================
Deployment to Unwired Server
========================================================
Deploying the hybrid app
Successfully deployed the hybrid app

========================================================
Assigning hybrid app S10005_04_Employee to S40000A
========================================================
Assignment complete

========================================================
Process finished.
========================================================
Time taken was 1 minutes and 15 seconds
========================================================
```

Figure 8.39 Console Log Showing the Status of the Mobile App Generation

8.5 Summary

This chapter provided a description of the individual steps required to create a hybrid app. There are two major activities that are performed in this iterative design process. The flow design activity involves defining the screens that make up the app and the menus that provide connections among those screens. The screen design activity involves placing the desired controls on each of the screens. A major effort in both of these design processes involves specifying values for each of the properties associated with each object. Once the design is complete, the code is generated for the app, saved in the user's workspace, deployed to the Unwired Server, and assigned to mobile devices.

This chapter shows some examples of hybrid app development in action.

9 Examples of Hybrid App Development

This chapter will build upon the concepts introduced in Chapter 7 and Chapter 8 to walk you through practical examples of building hybrid apps. The chapter will present several examples that demonstrate a number of different design techniques. The examples that have been selected are directed more toward illustrating concepts than designing commercial apps. (In the last chapter of this book, we will present a complete case study that demonstrates many of these concepts in a realistic app.)

When considering the role of information retrieval with respect to mobile apps, there are two major types of apps based on how they are initiated: client-initiated apps and server-initiated apps. A client-initiated app, which is much more common, has a role to retrieve information from the various EIS sources. A server-initiated app's role is to notify or send a message to the end user. In this chapter, we explain the process of designing both types of apps. Section 9.1, Section 9.2, and Section 9.3 show examples of client-initiated apps; Section 9.4 shows an example of a server-initiated app. All sections walk you through each of the steps involved in designing the app in question.

9.1 Employee Database Example

The first client-initiated example is a hybrid app that allows the user to view a list of employees, select a specific employee, and then view the detail data for the selected employee.

9.1.1 MBO

This hybrid app uses the EMPLOYEE MBO (which was presented in Section 7.3.1 of Chapter 7). This MBO is used to map to a database table. The MBO, as shown in Figure 9.1, contains most of the attributes in the `Employee` table but was designed without using any of the operators. The MBO uses the two standard queries: `findAll` and `findByPrimaryKey`.

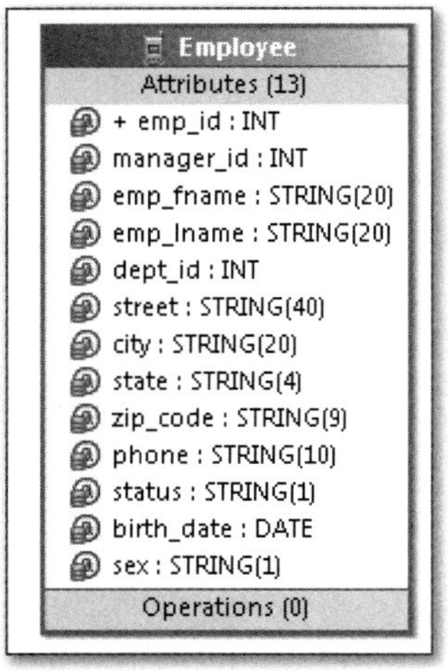

Figure 9.1 Employee MBO Built with findAll and findByPrimaryKey Queries

9.1.2 Flow Design

The EMPLOYEE flow design, as shown in Figure 9.2, was created as follows:

1. Create a client-initiated app, which creates the START SCREEN.
2. Drag the MBO onto the workspace, which creates the EMPLOYEE and the EMPLOYEEDETAIL screens.
3. Using the GoTo control, add a link between the START SCREEN and the EMPLOYEE screen. The control will also add a menu item to the START SCREEN.

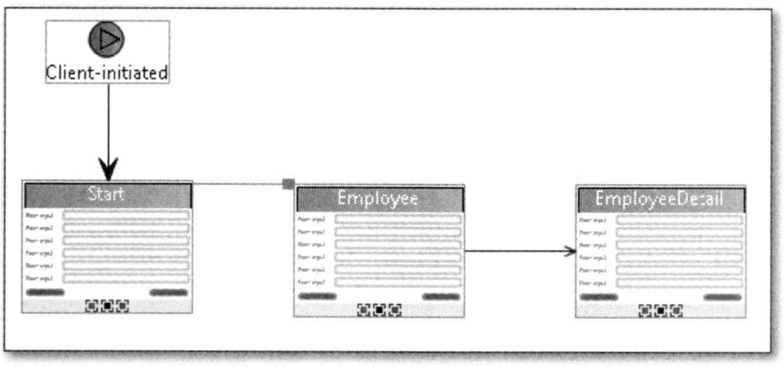

Figure 9.2 Employee App Flow Design

9.1.3 Screen Design

The following is a description of the changes made to the various screens.

Start Screen

An HTMLVIEW control is first added to the START SCREEN, as shown in Figure 9.3, to provide a description for this hybrid app. The two required properties required for this control are the `key` and the HTML string that is entered into the default value.

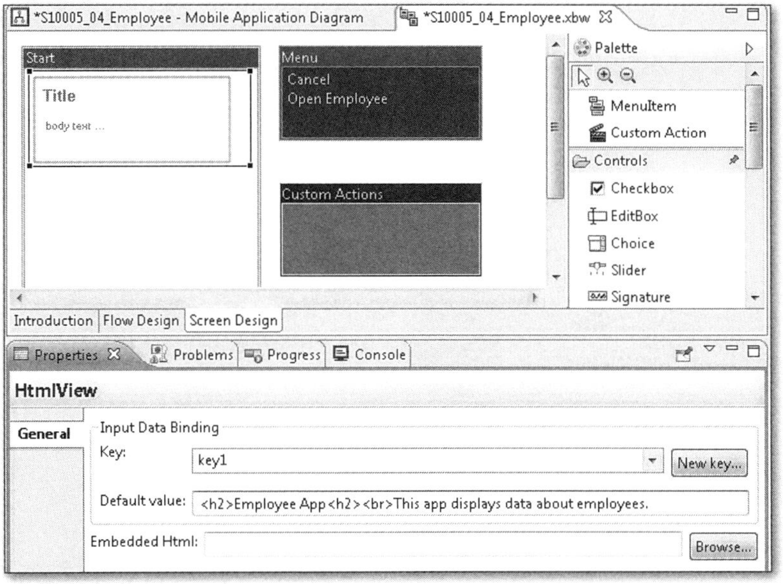

Figure 9.3 Property Values for the HTMLView Control on the Start Screen

The menu item (OPEN EMPLOYEE) is added to the START SCREEN when the GOTO link is created. The default method type assigned to the menu type is OPEN. This method simply opens the EMPLOYEE screen when the menu item is clicked, which is not sufficient for this app. For this app, it is also necessary to retrieve the data that will be displayed in the EMPLOYEE screen.

This is accomplished as follows:

1. The method type needs to be changed to ONLINE REQUEST.
2. The ONLINE REQUEST method needs to invoke the `findAll` query associated with the EMPLOYEE MBO. This retrieves a list of data that is saved as the output key.

3. Upon the successful completion of the query, the app displays the list of data on the EMPLOYEE screen. The name of the output key that contains the list of data is EMPLOYEE.

The values for the GENERAL properties for this menu item are shown in Figure 9.4, the values of the OUTPUT KEYS properties are shown in Figure 9.5, and the values of the SUCCESS/ERROR properties are shown in Figure 9.6.

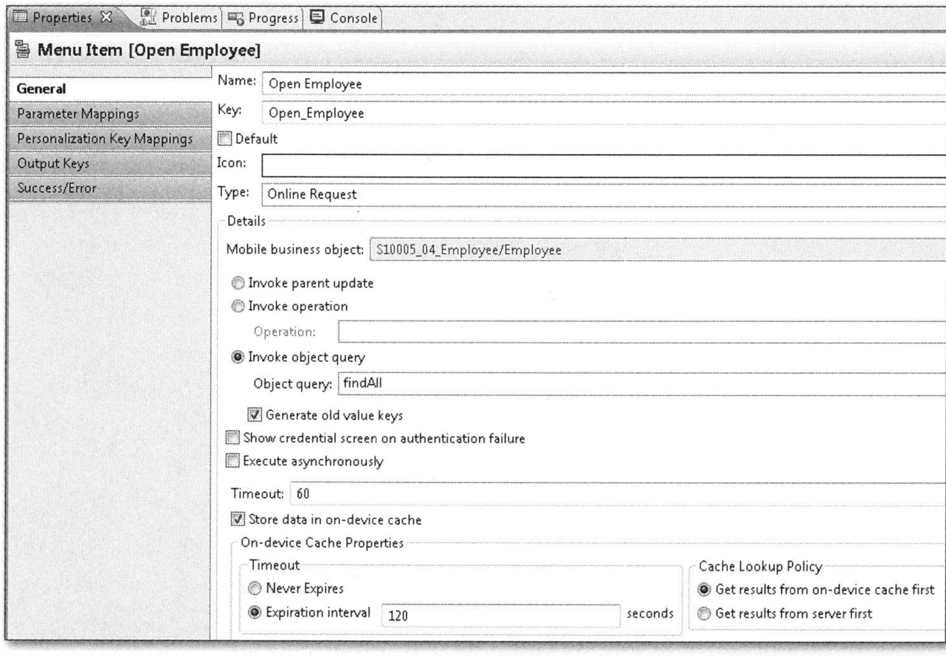

Figure 9.4 General Properties

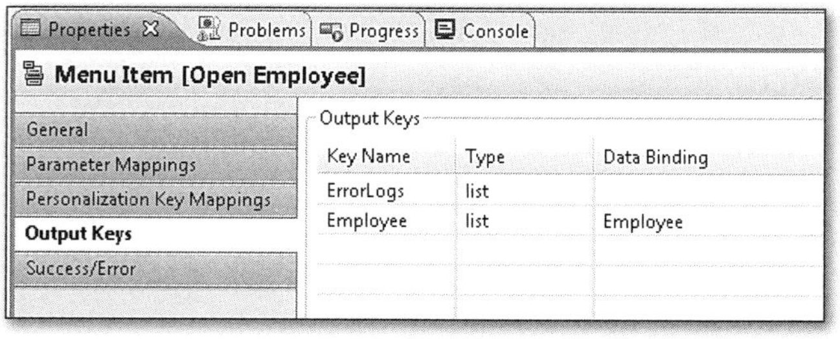

Figure 9.5 Output Keys Properties

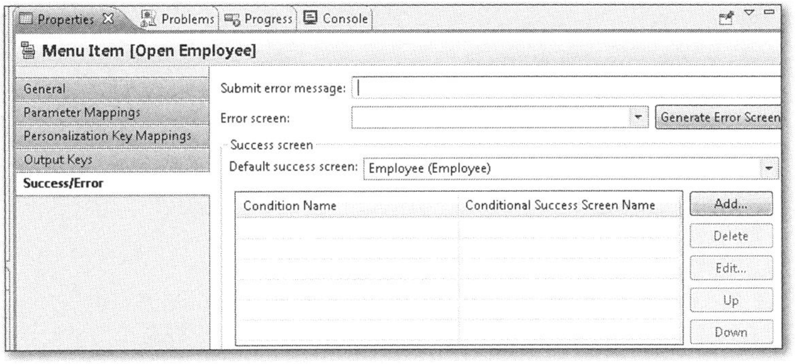

Figure 9.6 Success/Error Properties

Employee Screen

The Employee screen was created by dragging and dropping the Employee MBO onto the flow design. This screen contains a single Listview control that lists all the Employees in the list generated by the Start Screen. The default Listview, as shown in Figure 9.7, shows a cell for each employee. Clicking that cell displays the Employees Detail screen. The default properties for the Listview have each attribute being displayed on its own line in the cell, as shown in Figure 9.8. As you can see, each cell consists of a number of lines with one attribute per line.

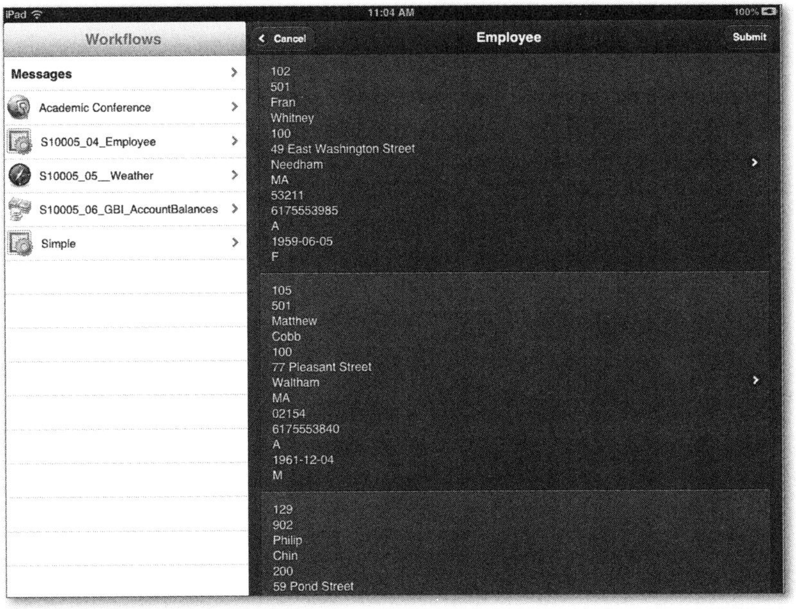

Figure 9.7 Employee Screen with the Default Listview Properties

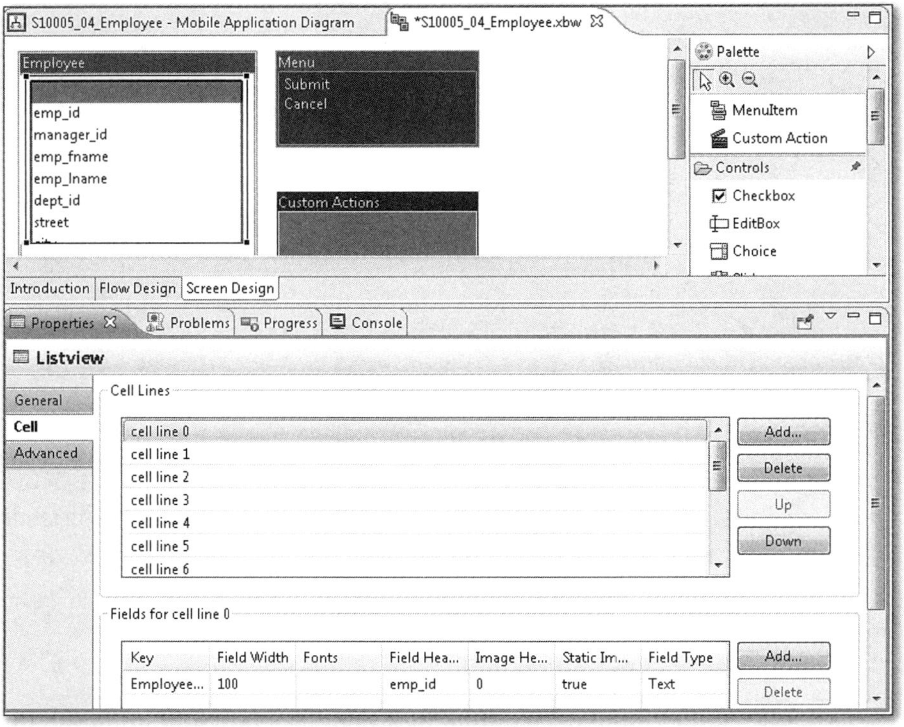

Figure 9.8 Cell Properties for the Default Listview

This default version of the Listview uses a lot of screen real estate for each Employee cell. There is no need to show all the attributes in each cell, since their values are provided on the EmployeeDetail screen. A more effective approach would be to display only one line per cell with two or three key attributes displayed on that line. The following are the changes required for the Cell properties to accomplish this:

1. Select cell line1 to cell line 20 and delete them.
2. Select cell line 0, select the first field (ID-ATTRIBUTE-KEY), edit it by changing the width to 20% and field header to ID, and then save.
3. With cell line 0 still selected, add a second field (FNAME-ATTRIBUTE-KEY) with a width of 30% and header of First, and save.
4. With cell line 0 still selected, add a third field (LNAME-ATTRIBUTE-KEY) with a width of 40% and header of Last, and save.

The cell properties now appear as shown in Figure 9.9.

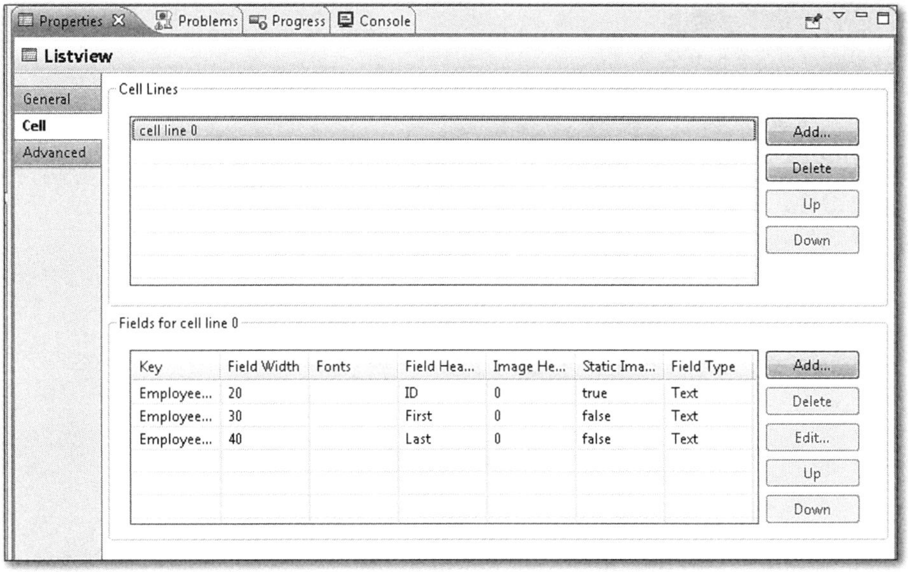

Figure 9.9 Cell Properties for Revised Listview

9.1.4 Review App

With this change, the hybrid app is now ready to be deployed and assigned to the end user's device. The screens that this app creates, as shown in Figure 9.10, are described as follows:

- The START SCREEN provides a description of the hybrid app and contains a menu item called EMPLOYEES.
- When the EMPLOYEES menu item is clicked, the `findAll` query method is executed, and then a list of all the resulting data is downloaded from the Unwired Server to the mobile device. The EMPLOYEE screen is then displayed.
- The EMPLOYEE screen contains a LISTVIEW control that displays a cell of data for each employee. Each cell is a single-line cell that displays the employee's ID and FIRST and LAST NAME.
- Clicking any of these cells displays the EMPLOYEEDETAIL screen, which consists of EDITBOX controls that display the value for each of the attributes for that employee.

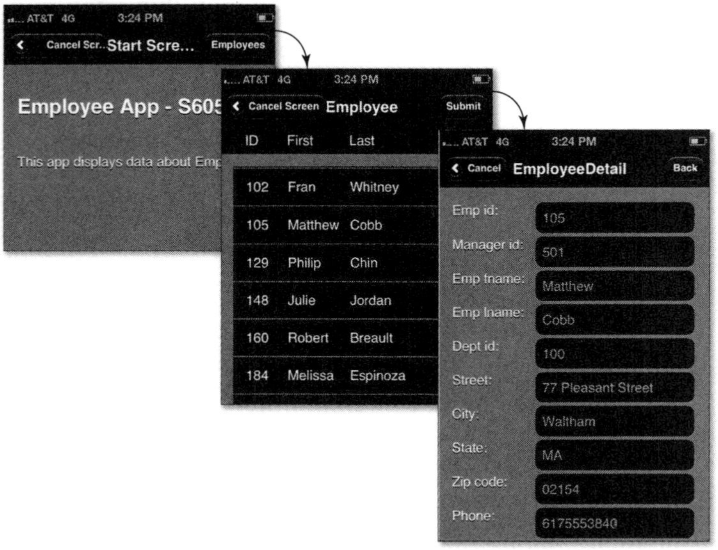

Figure 9.10 Screens for the Employee Hybrid App

9.2 Customer/Sales Order Database Example

The second example illustrates the power of using relationships between MBOs. The example consists of a CUSTOMER and a SALES ORDER MBO that are related via a one-to-many relationship (as presented previously in Section 7.7 of Chapter 7). The relationship allows the Unwired Server to manage many of the details required when working with multiple MBOs. The example also introduces the concept of operations. Both the CUSTOMER and SALES ORDER MBOs have all three operations defined: update, delete, and create. For this example, the end user can retrieve a list of sales orders related to a specific customer.

9.2.1 MBOs

The first MBO is created using the drag-and-drop approach within a database table with the name of `Customer`. All three operations are selected for this MBO. The second MBO is created in a similar manner, using the `Sales_Order` table. A relationship is created using the ID attribute in the CUSTOMER MBO and the `Cust_Id` in the SALES ORDER MBO, as shown in Figure 9.11. As you can see in Figure 9.12, the relationship is one-to-many, composite, and bi-directional.

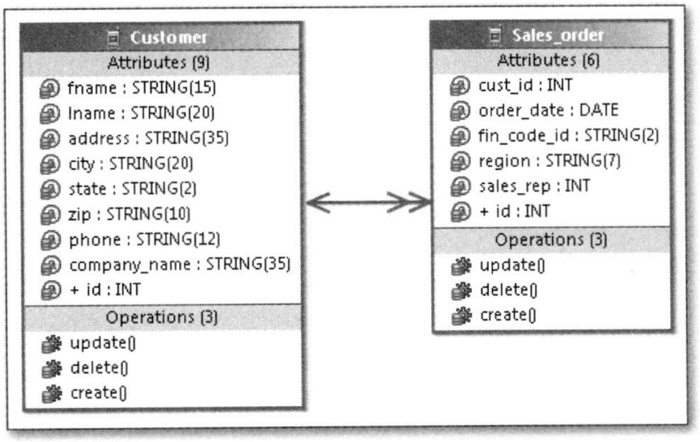

Figure 9.11 MBOs for the Customer-Sales Order Example

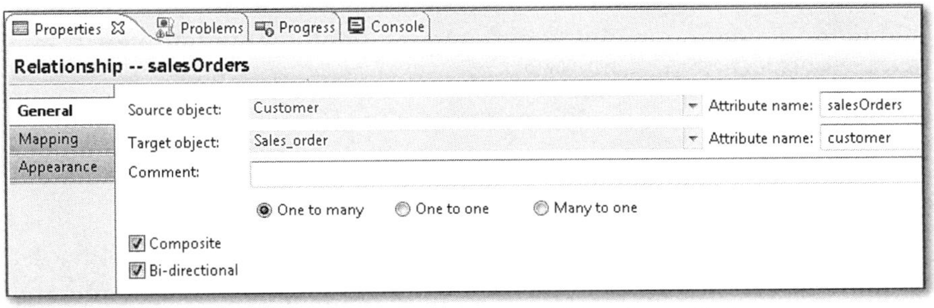

Figure 9.12 Relationship Properties for Customer and Sales Order MBO

There are four queries, as shown in Figure 9.13, that are defined for the CUSTOMER MBO:

- `findAll`
- `findByPrimaryKey`
- `findAllSortedByLname`
- `findByLname` (looks for sub-strings)

There are also two queries, as shown in Figure 9.14, that are defined for the SALES ORDER MBO:

- `findAll`
- `findByPrimaryKey`

Once the two MBOs are created, the queries are defined, and the relationship is created, the project is then deployed to the Unwired Server.

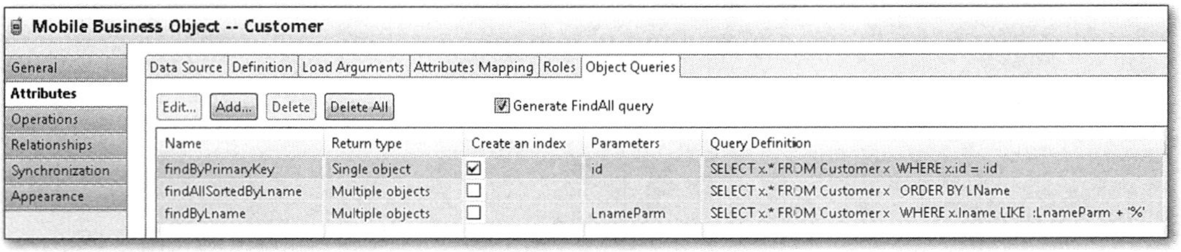

Figure 9.13 Queries for the Customer MBO

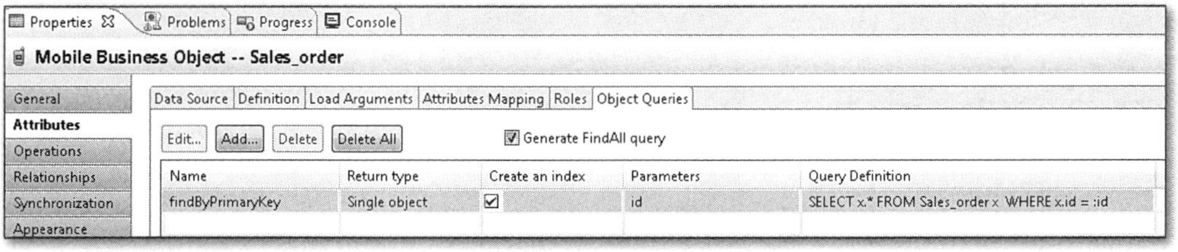

Figure 9.14 Queries for the Sales Order MBO

9.2.2 Flow Design

The hybrid app for this example is initially created with a client-initiated starting point and a Start Screen. Using the drag-and-drop approach, dragging the Customer MBO onto the flow design, as shown in Figure 9.15, creates nine additional screens. These screens are as follows:

- Customer MBO
  - Customer
    Provides a list of all customers.
  - Customer Detail
    Provides a list of all attribute values for a specific customer.
  - Customer Update
    Provides the ability to update attribute values for a specific customer.
  - Customer Delete
    Allows the deletion of the specific customer.
  - Customer Create
    Provides the creation of a new customer (initially not connected to any screen).

- SALES ORDER MBO
  - SALES ORDER
    Provides a list of all sales orders for a specific customer.
  - SALES ORDER DETAIL
    Provides a list of all attribute values for a specific sales order.
  - SALES ORDER UPDATE
    Provides the ability to update attribute values for a specific sales order.
  - SALES ORDER ADD
    Provides the creation of a new sales order for a specific customer.

Most of the connections among the screens are already provided with the necessary menus and queries.

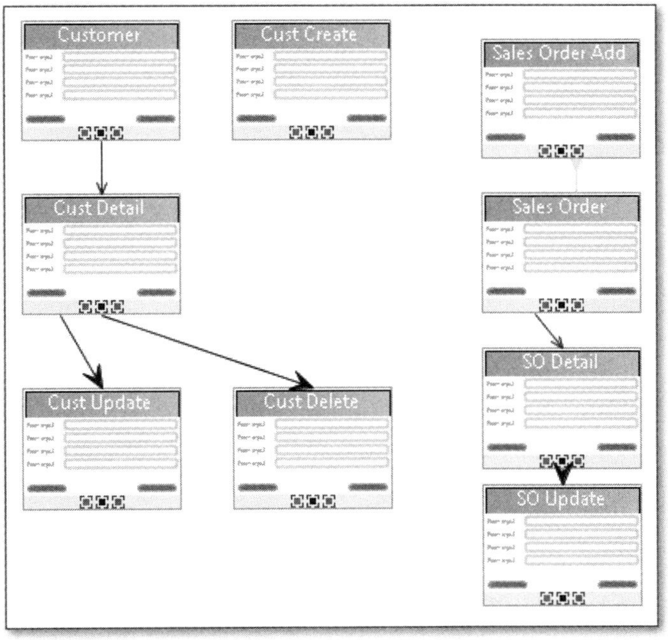

Figure 9.15 Created Screens

There are some areas where the designer needs to make additional connections and add the necessary queries. One area involves connecting the START SCREEN to EMPLOYEES, modifying EMPLOYEES so that there is one line per cell and adding the ability to search for a specific employee. The revised flow design for these changes is shown in Figure 9.16. The following steps are required to create this completed structure:

1. Create a new screen called FIND CUST.
2. Add a GoTo between START SCREEN and FIND CUST.
3. Add a GoTo between FIND CUST and CUSTOMER.
4. Add a GoTo between START SCREEN and CUSTOMER.
5. Add a GOTO between START SCREEN and CUST CREATE.

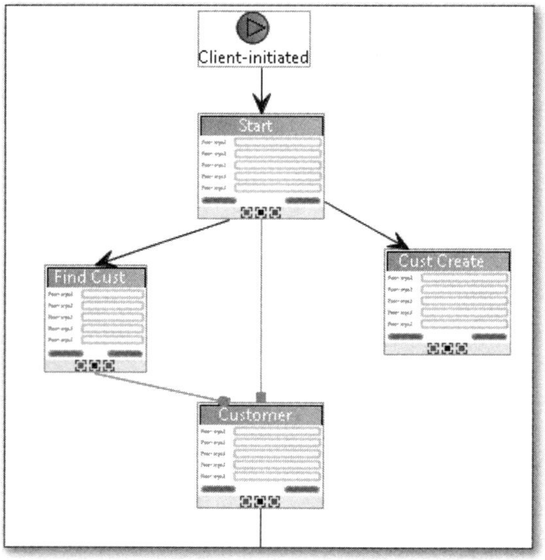

Figure 9.16 Flow Design for the Customer Portion of the Hybrid App

9.2.3 Screen Design

Several of the generated screens require modifications.

Start Screen

The START SCREEN now has four menu items, as shown in Figure 9.17. The menu items are configured as follows:

- The CUSTOMER LIST menu item is configured as an online request method that calls the `FindAllSortedByLname` query. This query retrieves all of the employees and then orders them alphabetically by last name for display on the CUSTOMER screen. The GENERAL properties for this menu item are shown in Figure 9.18 and the SUCCESS/ERROR properties are shown in Figure 9.19.

- The FIND CUSTOMER menu item, whose properties are shown in Figure 9.20, is of the type OPEN, which simply displays the new FIND CUST screen.
- The CREATE CUSTOMER menu item, whose properties are shown in Figure 9.21, is also of the type OPEN and displays the CUST CREATE screen.

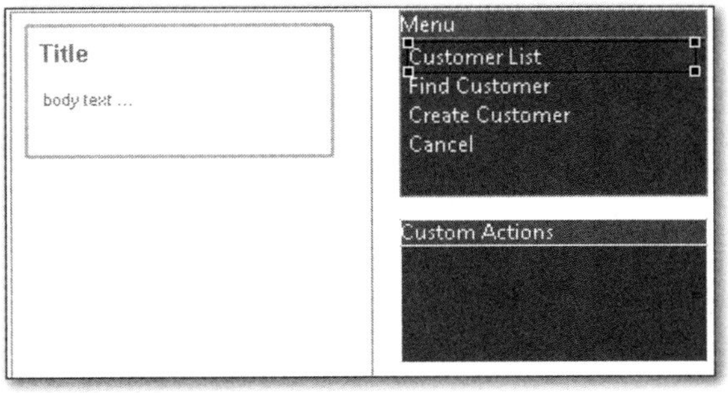

Figure 9.17 Menu Items on the Start Screen

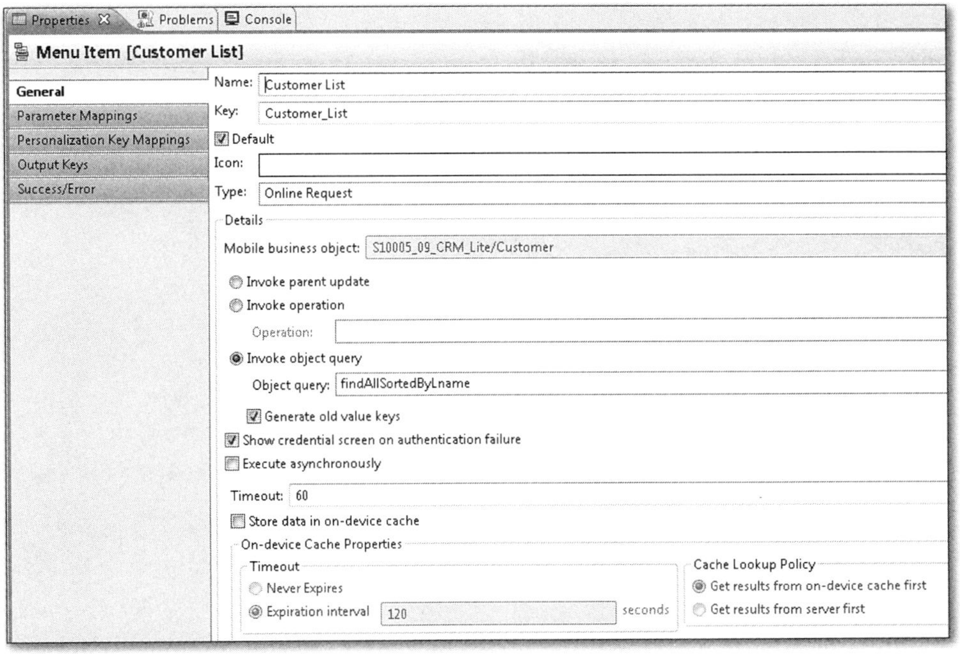

Figure 9.18 General Properties for the Customer List Menu Item

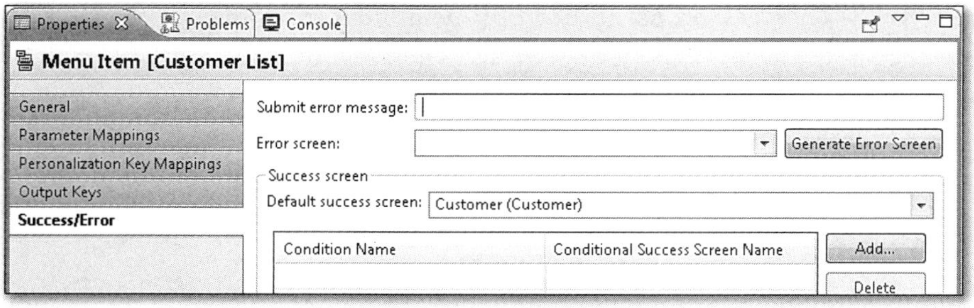

Figure 9.19 Success/Error Properties for the Customer List Menu Item

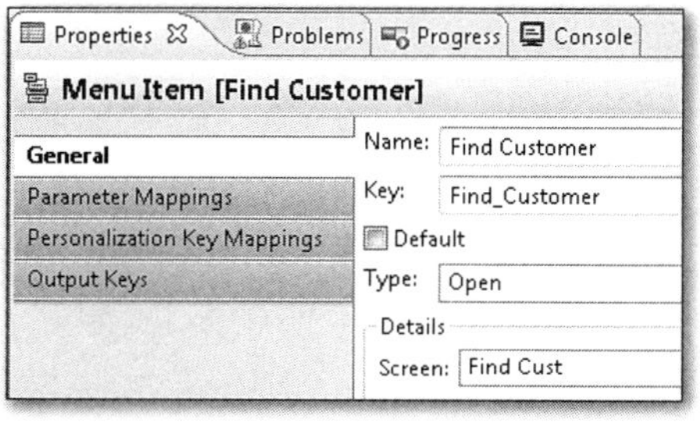

Figure 9.20 Properties for Find Customer Menu Item

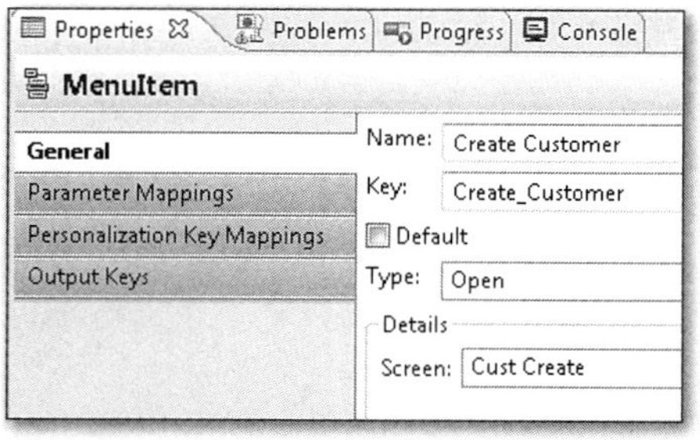

Figure 9.21 Properties for Create Customer Menu Item

Find Customer

The new FIND CUST screen, as shown in Figure 9.22, is constructed as follows:

1. Add an EDITBOX control. The user enters a text string in this box. The properties for this control are shown in Figure 9.23. This string is used by the `FindByLastName` query.
2. Add a BUTTON control. When the user clicks this button, an online request is made using the `FindByLastName` query. The GENERAL properties for this button

are shown in Figure 9.24 and Figure 9.25, and the SUCCESS/ERROR properties are shown in Figure 9.26. The PARAMETER MAPPINGS properties, as shown in Figure 9.27, map the sub-string entered in the EDITBOX (LAST NAME) screen to the parameter (`LnameParm`) used in the query.

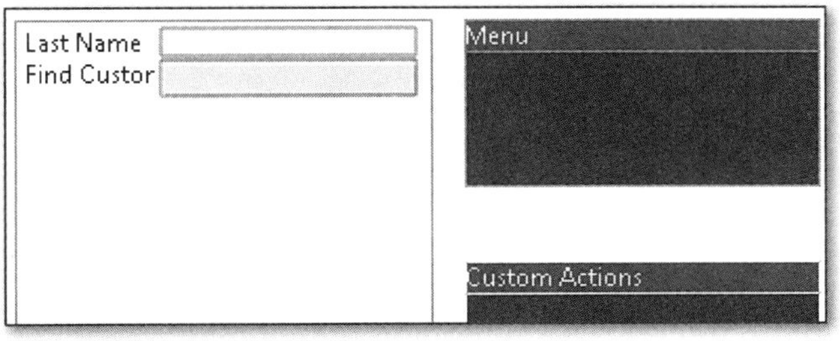

Figure 9.22 The New Find Customer Screen

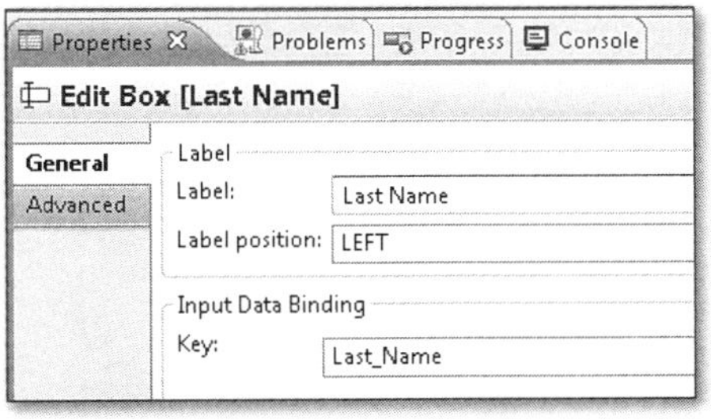

Figure 9.23 Properties for EditBox on Find Customer Screen

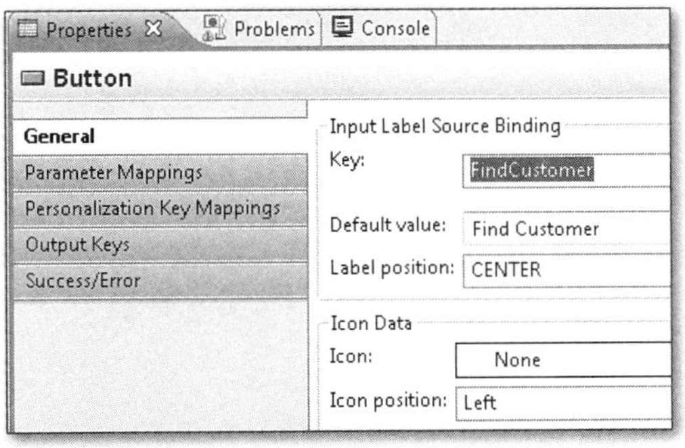

Figure 9.24 General Properties for Button on Find Customer Screen (Part 1)

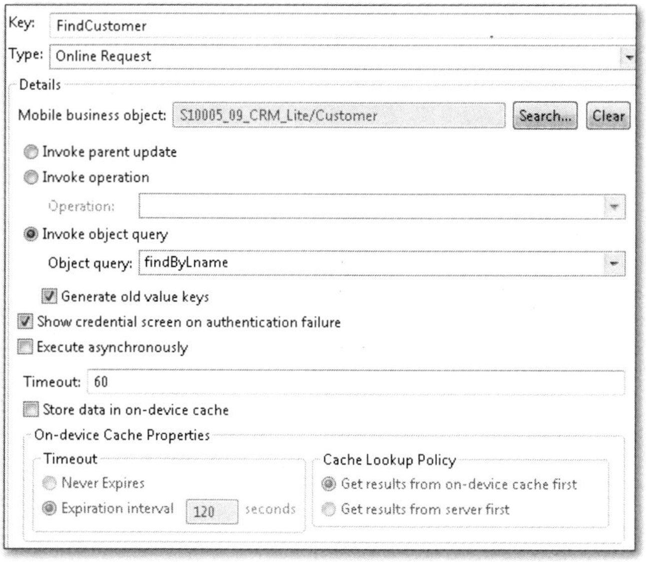

Figure 9.25 General Properties for Button on Find Customer Screen (Part 2)

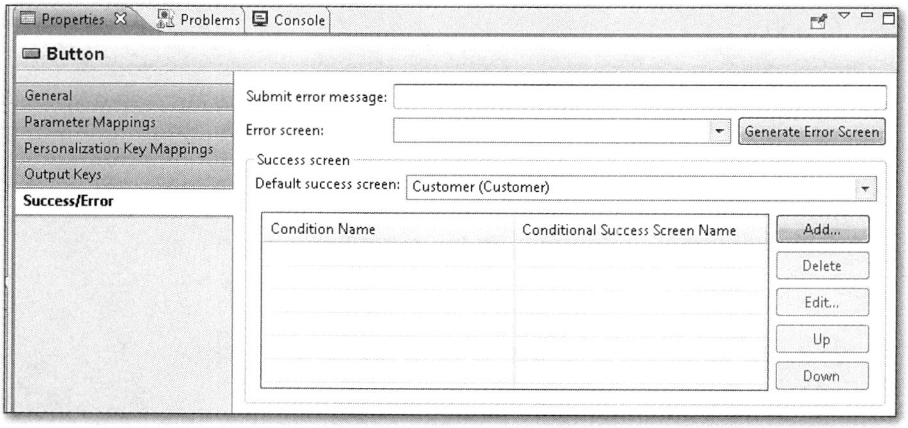

Figure 9.26 Success/Error Properties for Button on Find Customer Screen

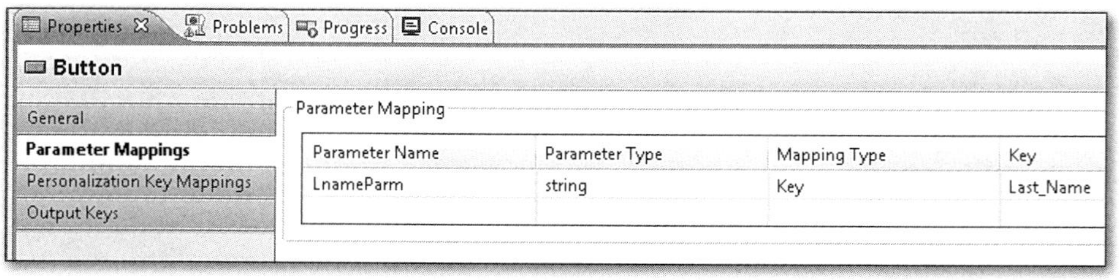

Figure 9.27 Parameter Mapping for Button on Find Customer Screen

9.2.4 Review App

At this point, it would be good to take a look at this portion of the app. For this example, we look at screenshots taken from an iPhone 5. The screens for the first two customer functions, as shown in Figure 9.28, are described as follows:

- Customer List
  - The Customer List menu item is selected from the Start Screen menu.
  - The Customer screen appears with a Listview control that lists all the customers' names, in alphabetical order.
  - The user scrolls down the list and clicks on a specific customer.
  - The Customer Detail screen, which displays the values of all the attributes for that specific customer, appears.
- Customer Create
  - The Customer Create menu item is selected from the Start Screen menu.
  - The Customer Create screen appears with blank values in the EditBoxes for each of the Customer attributes.
  - The user enters the values into the EditBox and then clicks the Create menu item, which creates a new customer.

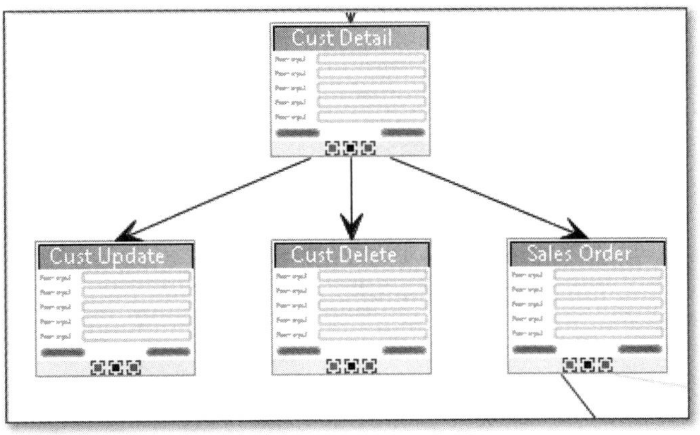

Figure 9.28 Customer App with Customer Create, Customer List, and Customer Detail

The screens for the third customer function, as shown Figure 9.29, are described as follows:

- Find Cust
  - The Find Customer menu item is selected from the Start screen menu.
  - The Find Cust screen appears with a blank EditBox control and a Find Customer button.
  - The user enters one or more letters into the EditBox and then clicks the button.
  - The Customer screen appears, and all customers whose last name begins with the characters entered are displayed in the Listview control.
  - The user scrolls down the list and clicks a specific customer.
  - The Cust Detail screen, which displays the values of all the attributes for that specific customer, appears.

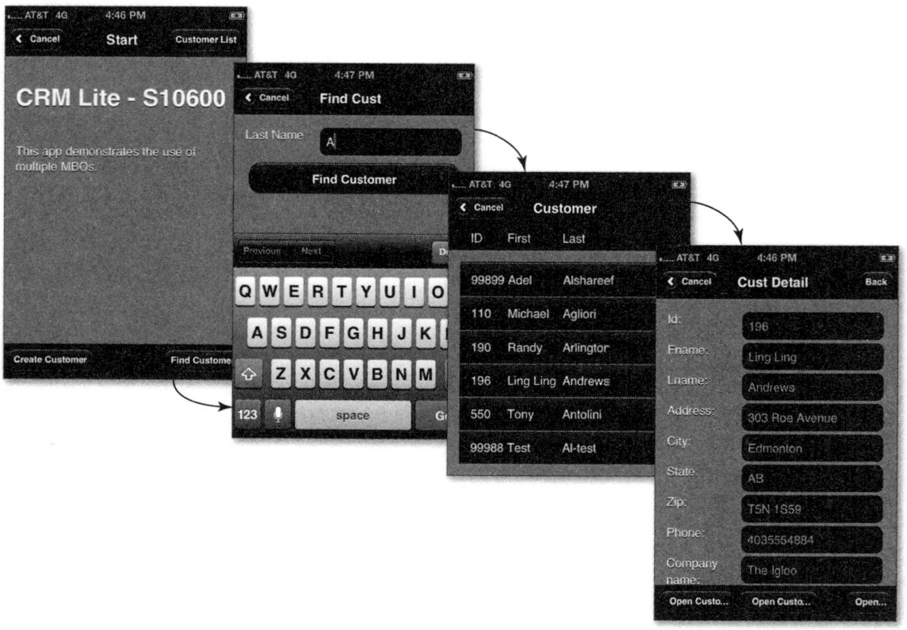

Figure 9.29 Customer App with Find Customer, Customer, and Customer Detail

Once a specific customer is located and displayed using the Cust Detail screen, as shown in Figure 9.30, menu items are provided for the following three functions:

- Cust Update
  - The current values for each attribute of the current customer are initially displayed.
  - The end user can modify any of those values.
  - The end user can then click the Update menu item, and these values are saved.
- Cust Delete
  - The current values for each attribute of the current customer are initially displayed.
  - The end user can click the Delete menu item, and this customer is deleted.
- Sales Order
  - A list of all the sales orders for the given customer is displayed.

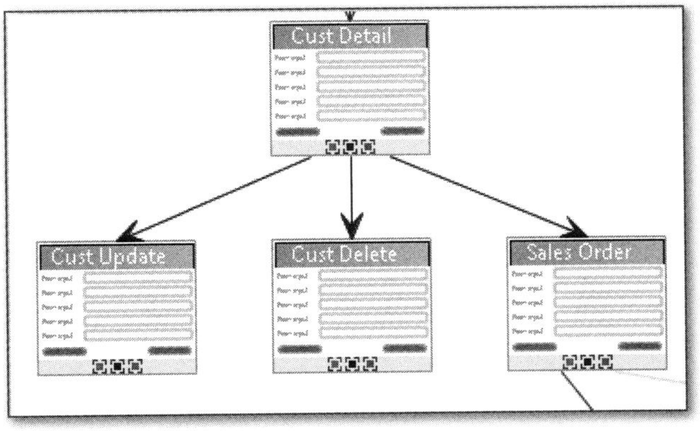

Figure 9.30 Customer App with Customer Update, Customer Delete, and Sales Order

The screens for these three options are shown in Figure 9.31.

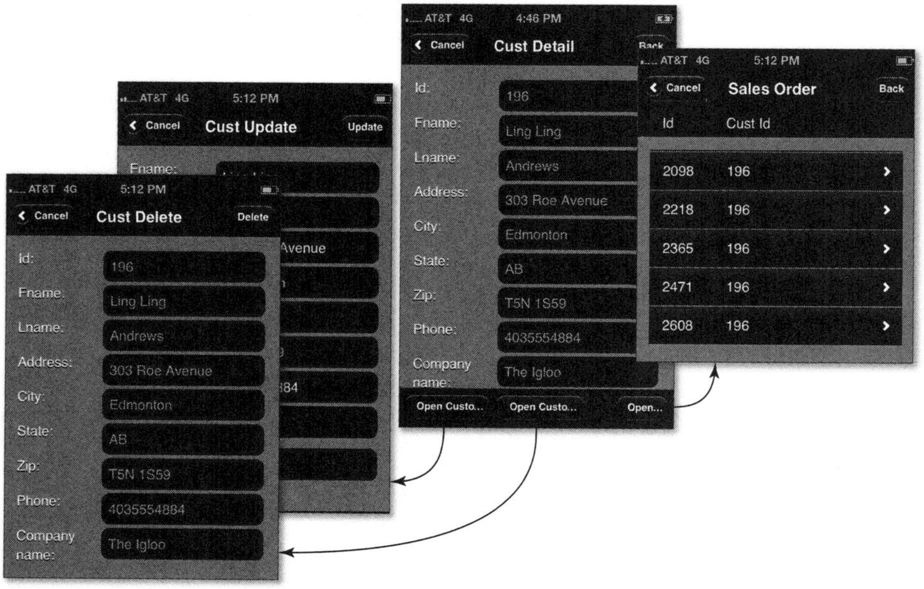

Figure 9.31 Customer App with Customer Update, Customer Delete, and Sales Order Screens

The last group of screens is related to the SALES ORDER screen, as shown in Figure 9.32, and includes the following functionality:

- SALES ORDER ADD
  - The SALES ORDER ADD menu item is selected from the SALES ORDER screen menu.
  - The SALES ORDER ADD screen appears with blank values in the EDITBOXES for each of the SALES ORDER attributes.
  - The user enters the values into the EDITBOXES and then clicks the CREATE menu item, which creates a new SALES ORDER.
- SALES ORDER DETAIL
  - The SALES ORDER DETAIL screen appears with values in EDITBOXES for each attribute.

The following two menu items provide the functionality described:

- Sales Order Update
  - The current values for each attribute of the current sales order are initially displayed.
  - The end user can modify any of those values.
  - The end user can then click the Update menu item, and these values are saved.
- Sales Order Delete
  - The current values for each attribute of the current sales order is initially displayed.
  - The end user can click the Delete menu item, and this sales order is deleted.

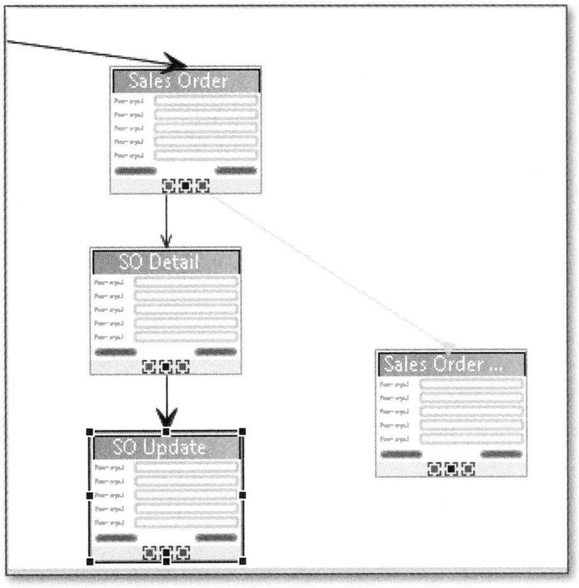

Figure 9.32 Customer App with Sales Order Detail, Sales Order Update, and Sales Order Create

The screens for this functionality are shown in Figure 9.33.

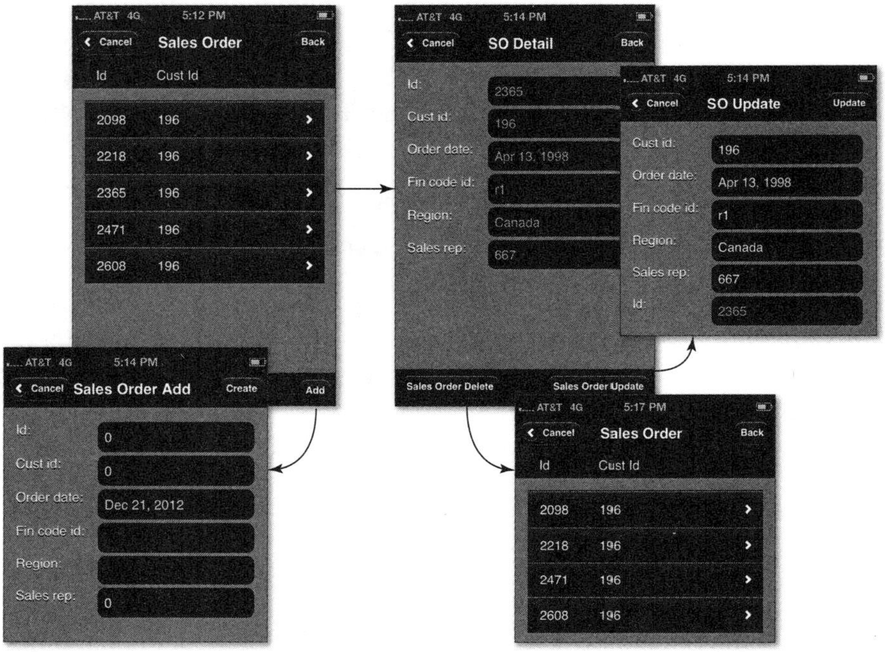

Figure 9.33 Customer App with Sales Order Detail, Sales Order Add, Sales Order Update, and Sales Order Delete Screens

The overall flow design for this CUSTOMER—SALES ORDER app is shown in Figure 9.34.

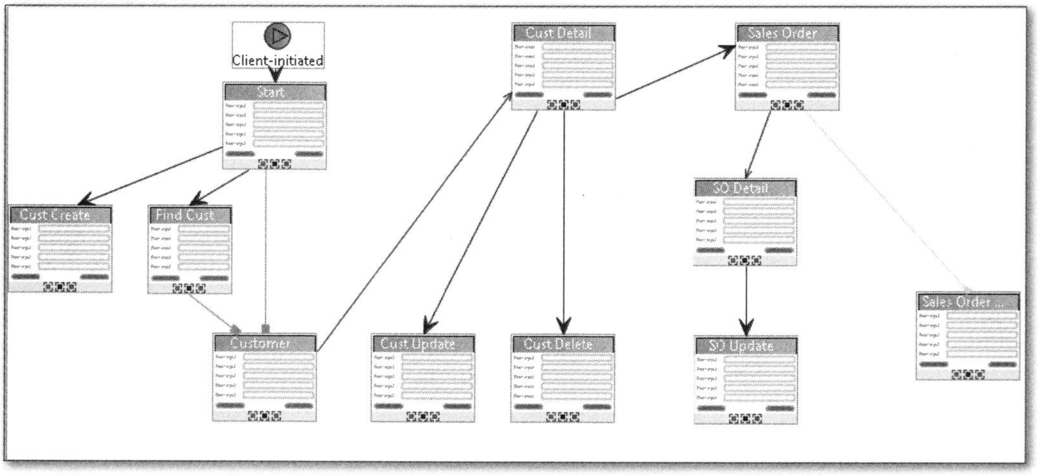

Figure 9.34 Complete Flow Design for Customer-Sales Order App

9.3 General Ledger Account Balances from SAP Example

The third client-initiated example (introduced previously in Section 7.3.3 of Chapter 7) is a hybrid app that allows the end user to access general ledger account balances from an SAP system.

9.3.1 MBO

The MBO for this app is built using CREATE MOBILE BUSINESS OBJECT under the PALETTE view, with the designer searching for the BAPI that provides the required data. In this example, the selected BAPI was `BAPI_GL_GETACCPERIODBALANCES`, with the details shown in Figure 9.35. Unlike the two previous examples, this example requires that the end user provide values for input arguments before being able to view the results. This example requires specifying the following four input arguments:

- `COMPANYCODE`
  The top-level identifier in an SAP system is the company code. In our example, we are using a system that has a company code of *US00*.
- `CURRENCYTYPE`
  This attribute is used to specify which currency should be used when reporting the data. For this example, we are using *10* for US dollars.
- `FISCALYEAR`
  This attribute is used to identify which year to use when selecting the data. For this example, we are using *2012*.
- `GLACCT`
  This attribute specifies the general-ledger account number, which is a 10-digit string. For this example, we select the bank account, which is coded as *0000100000* for our case.

The created BAPI, as shown in Figure 9.36, includes three of the input parameters and a number of output parameters. The data returned is a list, with each row in the list representing a fiscal period within the fiscal year. In this case, there are 16 rows of data, with 12 rows for the months and four rows representing the four quarters. The input parameter that is missing is CURRENCY TYPE (a two-digit string). This issue occurs any time an attribute is required as an input, yet the attribute is not returned as a result. To resolve this issue, it is necessary to add an additional attribute to the MBO.

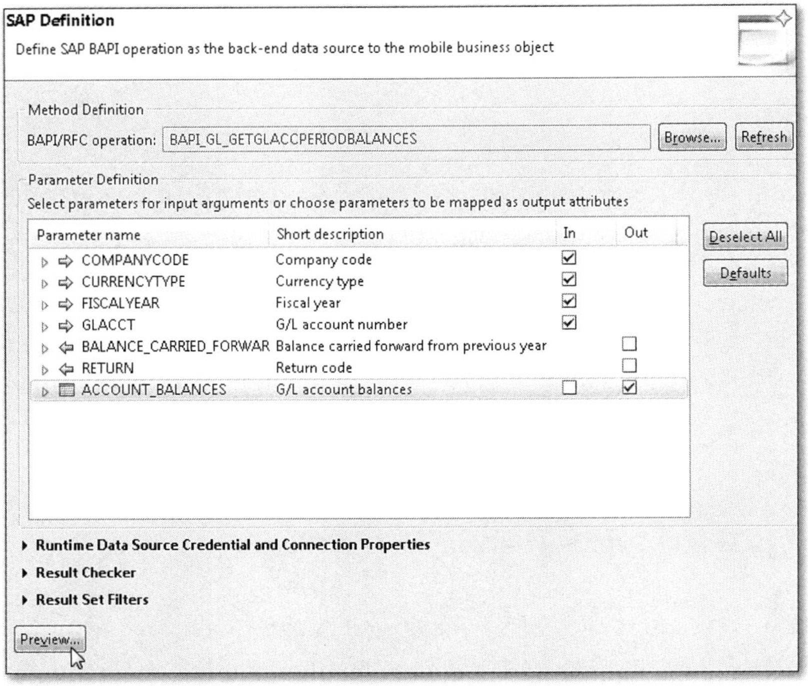

Figure 9.35 Defining the SAP BAPI for Accessing General Account Balances

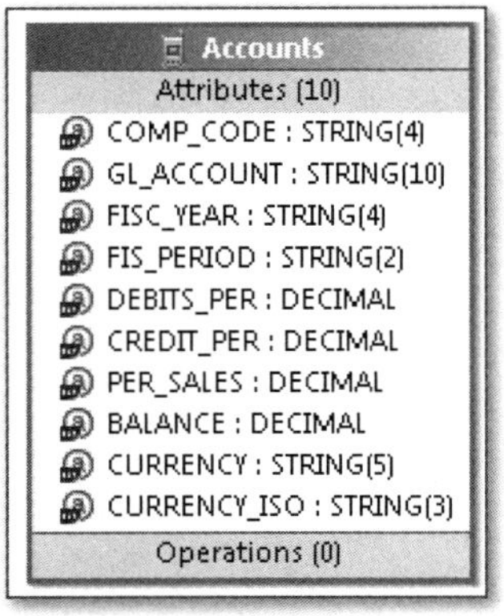

Figure 9.36 MBO Created by for BAPI_GL_GETACCPERIODBALANCES

To add an additional attribute to the MBO:

1. Highlight the MBO.
2. Follow the path PROPERTIES • ATTRIBUTES • ATTRIBUTES MAPPING.
3. Click ADD.
4. Rename the new argument `CUR_TYPE` with a data type of `String(2)`, and deselect NULLABLE.

The attribute list then appears, as shown in Figure 9.37.

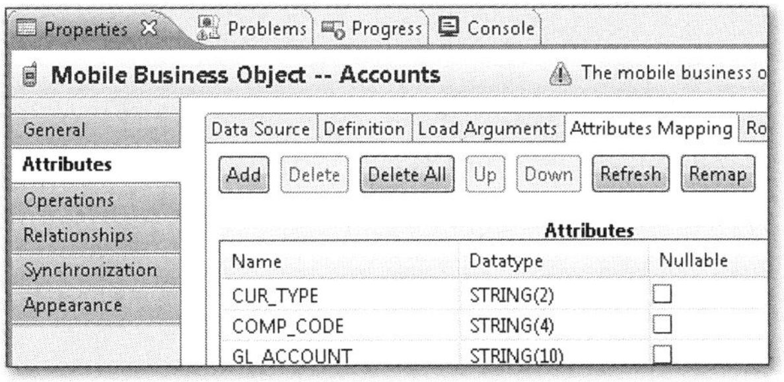

Figure 9.37 MBO Attributes after Adding CUR_TYPE

The next step in building the MBO involves creating personalization keys for each of the four input arguments. The keys are created by right-clicking the PERSONALIZATION folder under the project and then selecting new. Once the keys are added, the folder appears, as shown in Figure 9.38.

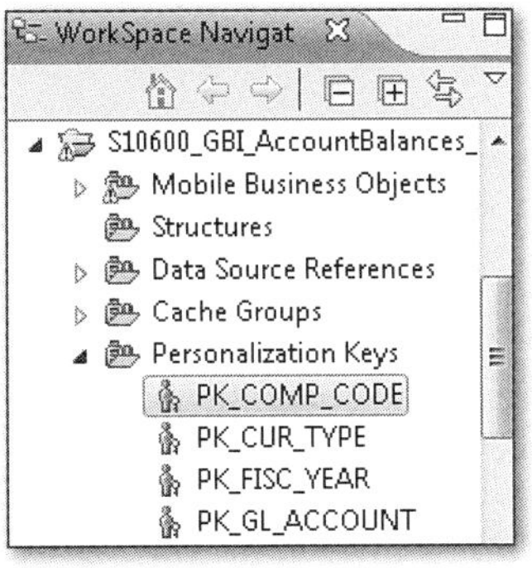

Figure 9.38 New Personalization Keys

The properties required to create each of the personalization keys are shown in Table 9.1.

Name	Type	Default Value	Storage
PK_COMP_CODE	STRING(4)	US00	TRANSIENT
PK_CUR_TYPE	STRING(2)	10	TRANSIENT
PK_FISC_YEAR	STRNG(4)	2012	TRANSIENT
PK_GL_ACCOUNT	STRING(10)	0000100000	TRANSIENT

Table 9.1 Properties Used for Creating Personalization Keys

The next step involves mapping the attributes and the personalization keys to the input arguments. This mapping takes places as follows:

1. Highlight the ACCOUNTS MBO.
2. Follow the path PROPERTIES • ATTRIBUTES • LOAD ARGUMENTS.
3. The four BAPI arguments appear.
4. Using the dropdowns, match these four input arguments first to their corresponding MBO attribute, and then to their personalization key.

These mappings should appear as shown in Figure 9.39.

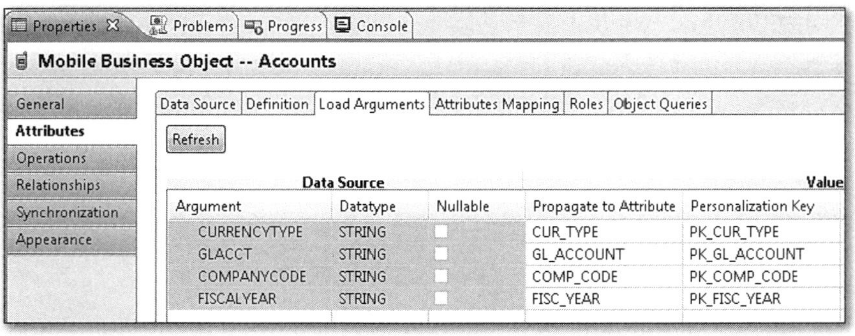

Figure 9.39 Mapping of the BAPI Input Arguments to Attributes and Personalization Keys

The final step involves adding a new query used to retrieve an account from the SAP system using the BAPI. This query is created as follows:

1. Highlight the Accounts MBO.
2. Follow the path Properties • Attributes • Object Queries.
3. Click Add.
4. Enter `findAccount` for the query name.
5. Create four parameters and map them to their respective MBO attributes.
6. Click Generate to create the query, as shown in Figure 9.40, and then click OK.

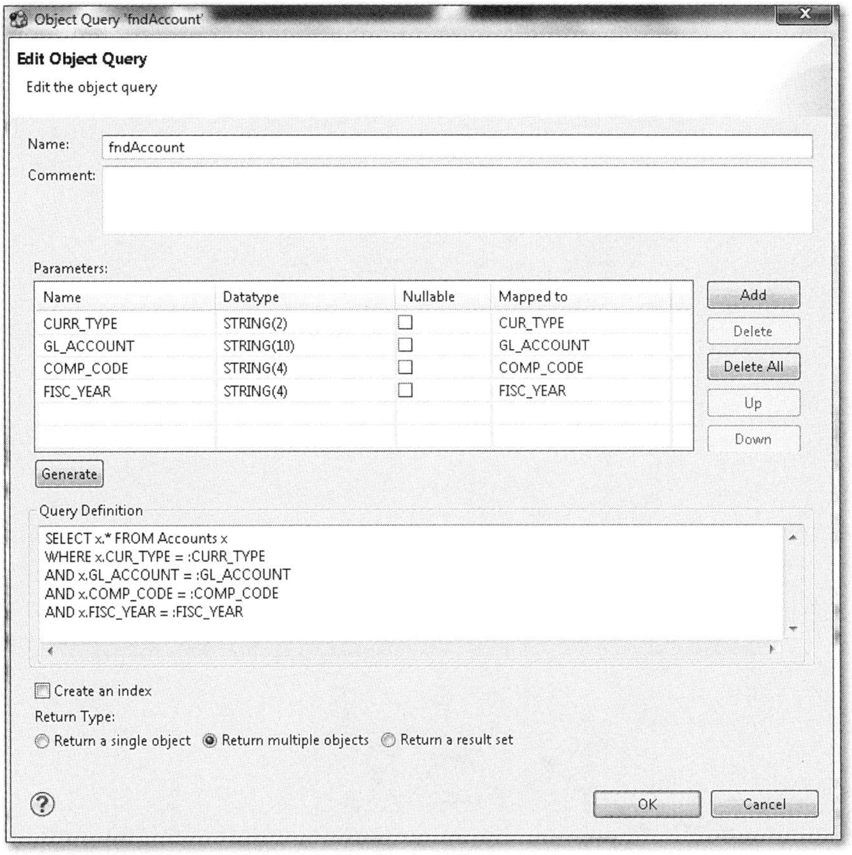

Figure 9.40 Creating the Query Used to Retrieve an Account Using the BAPI

The final step in the MBO development is to deploy this project.

9.3.2 Flow Design

The creation of the hybrid app begins with a drag and drop of the MBO onto the workspace. This creates the BALANCES and DETAILS screens, as shown in Figure 9.41.

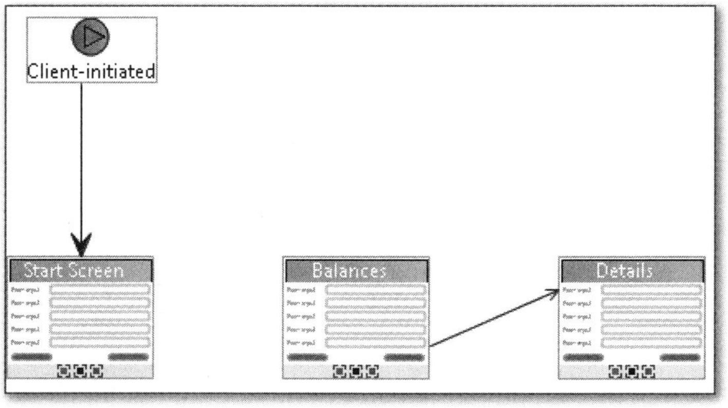

Figure 9.41 Resulting Screens after the Drag and Drop of the MBO

9.3.3 Screen Design

In this example, changes need to be made to the BALANCES screen and START SCREEN.

Balances Screen

The BALANCES screen lists the account balance for each of the periods. This screen, as shown in Figure 9.42, requires the standard modification whereby the number of lines per cell is reduced to one, and three key attributes are placed on that same line.

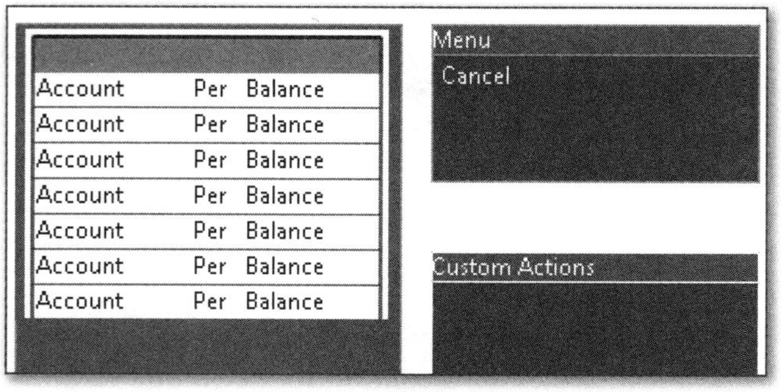

Figure 9.42 Revised Screen Design for the Balances Screen

Start Screen

What is different about this hybrid app is the need to request values for the four parameters prior to calling the method that retrieves the data for the BALANCES screen. For this app, it was decided to request these input parameters on the START SCREEN, as shown in Figure 9.43

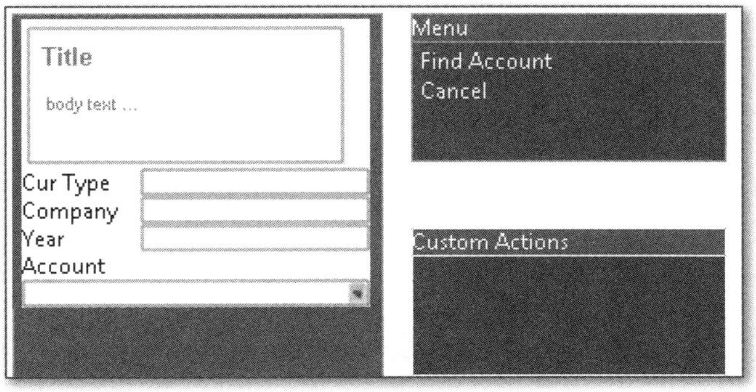

Figure 9.43 Start Screen Used to Input Parameters

To begin with, three EDITBOXES are used so that the user can enter the required values. These values are normally the same for each request. In order to simplify the app, a default value is entered and saved for each of these EDITBOXES, as shown in Figure 9.44.

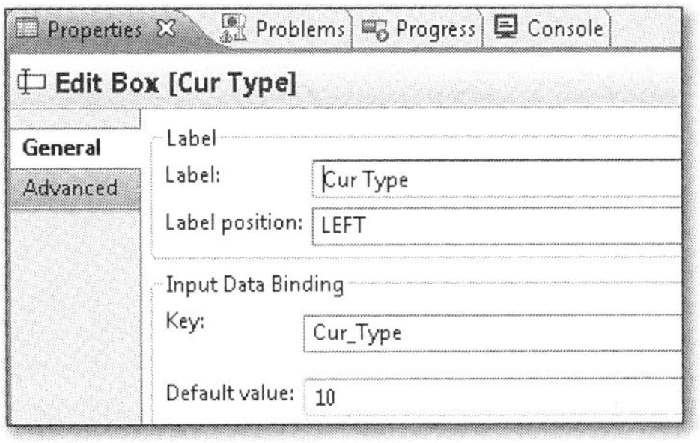

Figure 9.44 Properties for EditBox Used for Currency Type with a Default Value of 10

The fourth input argument requires that the user enter a unique 10-digit account number in order to view a specific account. In order to simplify this entry, a CHOICE control is used. The CHOICE control allows the end user to look up the account name, such as BANK or SALES REVENUE, from a dropdown list, and it is mapped to the appropriate 10-digit code. The properties for this choice control, as shown in

Figure 9.45, provide for the creation of a lookup table. As you can see, the dropdown box displays Bank as a choice for the end user. If Bank is chosen, the account is then assigned the value of *0000100000*. For this example, the table being built uses the static option. This means that the values are manually entered into this table by the designer. The second option, called *dynamic*, allows the designer to dynamically build the table by retrieving data from a backend source. This requires the use of an MBO and the necessary query to retrieve the data required to dynamically build the table. This technique is presented in the last chapter.

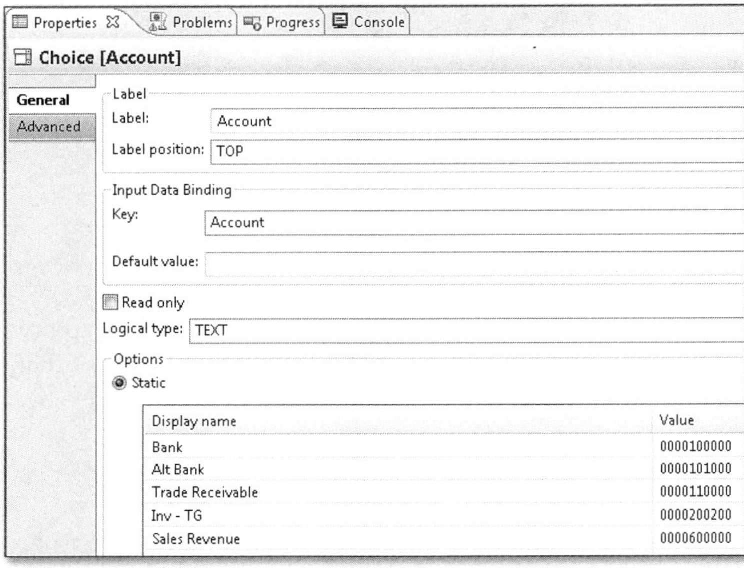

Figure 9.45 Properties for Creating the Choice Control for Account

The final step for this example is to build a menu item that links between the Start Screen and the Balances screen. We create a menu item with the name of Find Account. This menu item consists of an Online Request method that calls the `findAccount` query, with the General properties as shown in Figure 9.46 and the Success/Error properties as shown in Figure 9.47. The Parameter Mapping properties, as shown in Figure 9.48, are created automatically. The designer must add the four personalization key mappings, as shown in Figure 9.49.

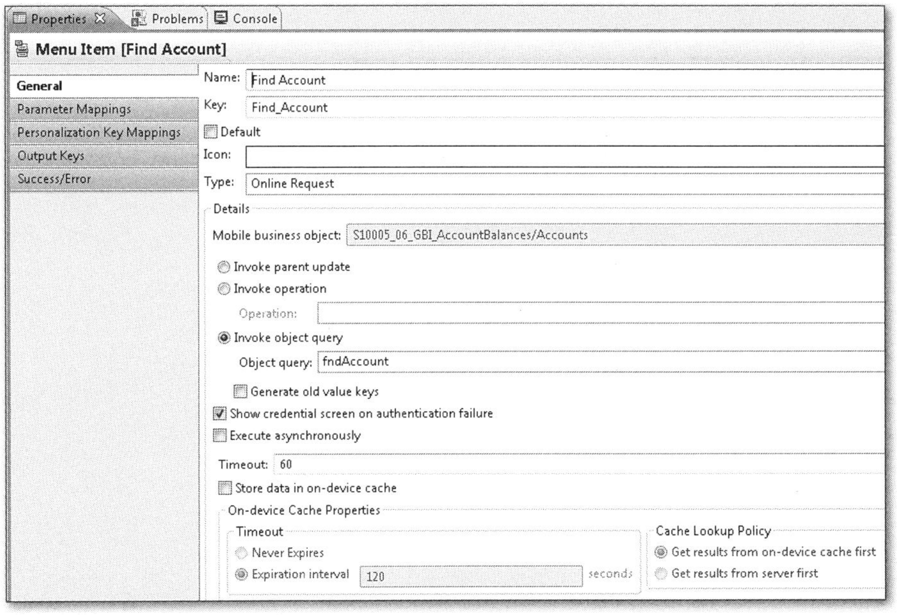

Figure 9.46 General Properties for the Find Account Menu Item

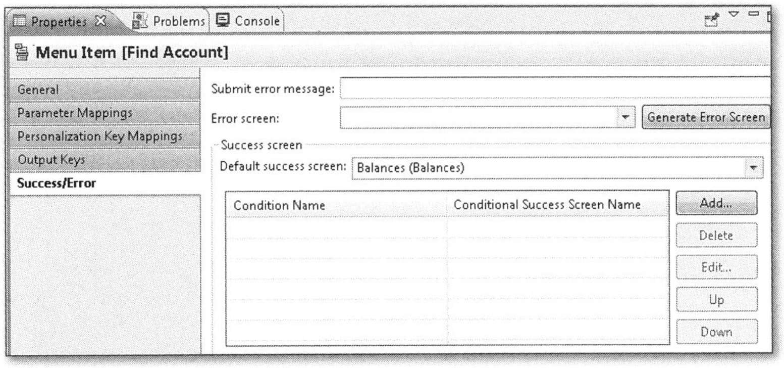

Figure 9.47 Success/Error Properties for the Find Account Menu Item

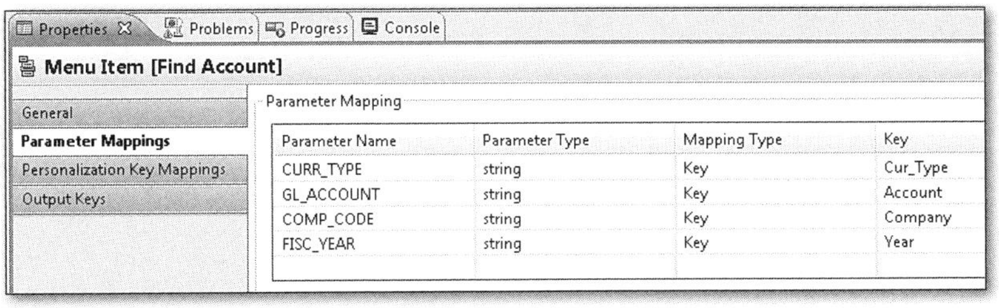

Figure 9.48 Parameter Mappings Properties for the Find Account Menu Item

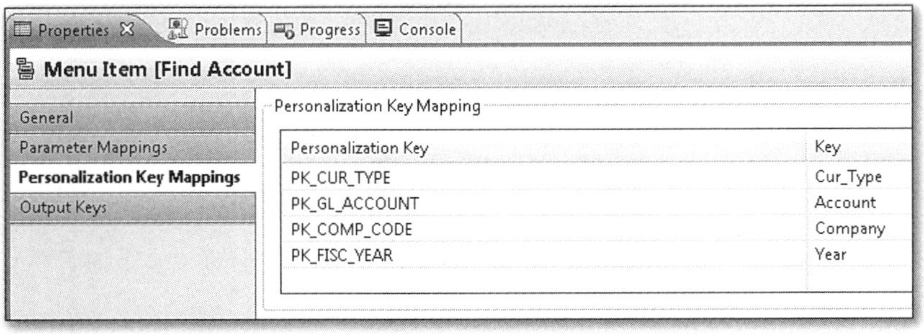

Figure 9.49 Personalization Key Mappings Properties for the Find Account Menu Item

9.3.4 Review App

The screens for the completed hybrid app are shown in Figure 9.50.

Figure 9.50 Find General Ledger Account Hybrid App

9.4 Sales Order Notification Database Example

The previous hybrid apps described in this chapter have one characteristic in common—they are all client-initiated apps. They all have client-initiated starting points and were assigned to users' devices. The hybrid app we are going to describe in this section has a server-initiated starting point and is not assigned to a user's device. These types of apps are initiated by a notification sent to the Unwired Server from either a backend system or possibly another app.

Server notifications take three forms:

- **Message and hybrid app sent by another mobile app**
  For example, a user may require an approval from a manager. This scenario requires an email account to route messages.
- **Data change notification (DCN) sent by an EIS**
  For example, a sales order is created in the system, and a program on the EIS sends a DCN to the Unwired Server, which sends a notification to an app used by a customer service representative. This scenario requires development of the program to create and transmit the DCN on the EIS.
- **Scheduled data load between the EIS and the Unwired Server occurs**
  This requires a cache group on the Unwired Server with a scheduled update policy.

This example simulates the DCN scenario. The specific scenario is that a customer service representative is notified when a new sales order is created. A program in the EIS determines that the customer is assigned to the app user and sends a notification to the Unwired Server with the new sales order ID. The Unwired Server sends a notification to the user's device. The user can open the message to retrieve the sales order details. For this exercise, a test notification is sent from the SDK to simulate the DCN from the EIS.

9.4.1 MBO

The SALES ORDER MBO is created by using the drag-and-drop approach from the `Sales_Order` table, with the three operations not being selected. The created SALES ORDER MBO is shown in Figure 9.51.

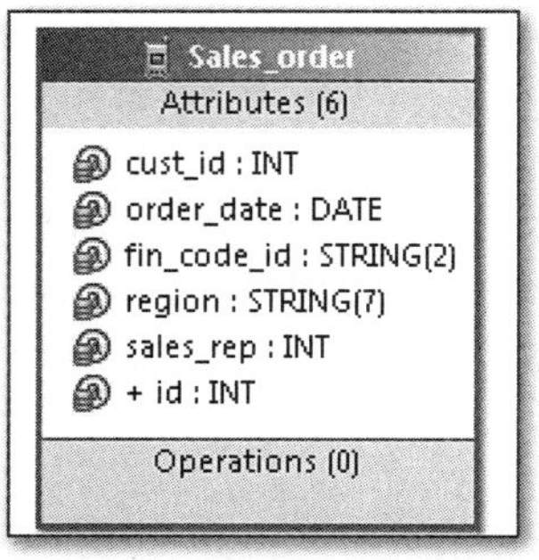

Figure 9.51 Sales Order MBO

The notification generated for this MBO wants the user to focus specifically on the new sales order. We need to modify this MBO so that it returns a single sales order, as opposed to all the sales orders in the data table. This is accomplished as follows:

1. Select the SALES ORDER MBO.
2. Follow the path PROPERTIES • ATTRIBUTES • DEFINITION.
3. Edit the SQL statement to include the following statement:
   `WHERE id =:order_id`

The definition then appears, as shown in Figure 9.52. Note that this query is modifying the definition of the MBO so that it contains only one sales order. Adding this condition to the definition also creates the load argument shown in Figure 9.52. The properties need to be changed as follows:

1. Add the `order_id` parameter.
2. Change the data type to `INT`.

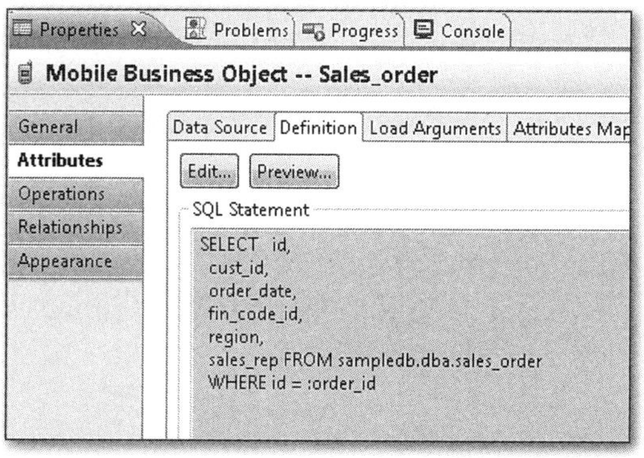

Figure 9.52 Modified SQL Statement that Defines the Sales Order MBO

3. Select ID for PROPAGATE TO ATTRIBUTE.
4. Enter "0" for DEFAULT VALUE.

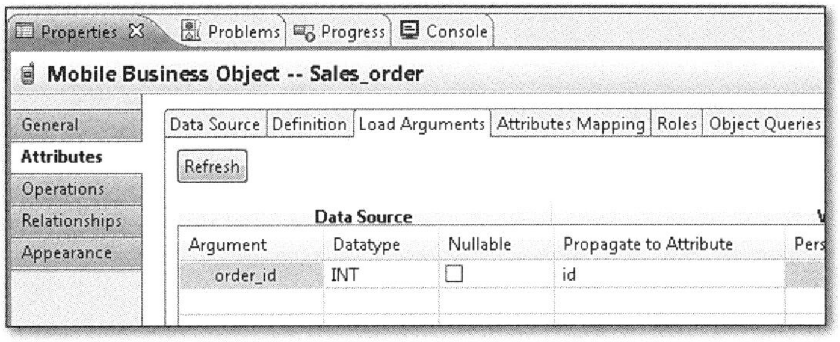

Figure 9.53 Load Arguments Created for the Sales Order MBO

The `findByParameter` object query is constructed by the system, as shown in Figure 9.54, based on the load argument values.

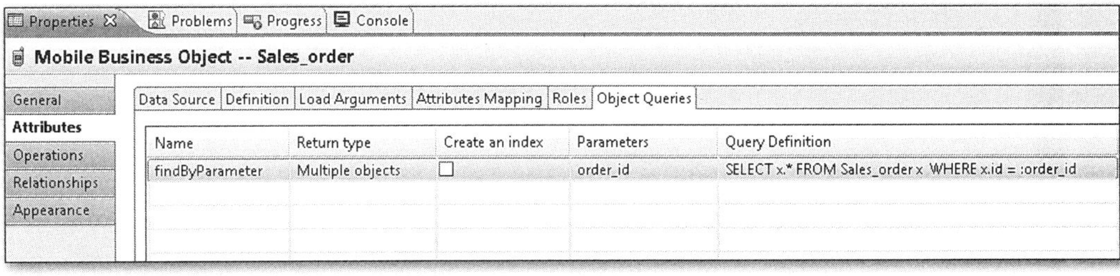

Figure 9.54 Object Query Created for the Sales Order MBO

The caching policy for the SALES ORDER MBO, as shown in Figure 9.55, needs to be modified as follows:

1. Expand the project structure in the Workspace Navigator and right-click the DEFAULT CACHE group.
2. Select PROPERTIES.
3. Select POLICIES and set the policy to ONLINE. This causes the app to load data from the server each time the app retrieves data and prevents the Unwired Server from caching the data.

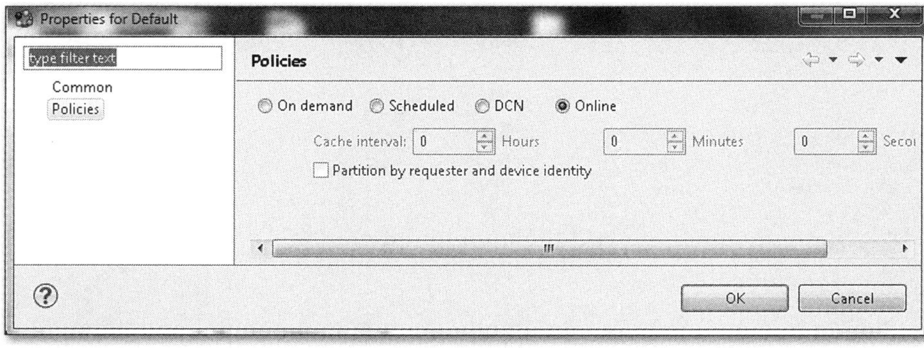

Figure 9.55 Cache Policy for the Sales Order MBO

The SALES ORDER MBO is added as an entry to this default cache group, as shown in Figure 9.56.

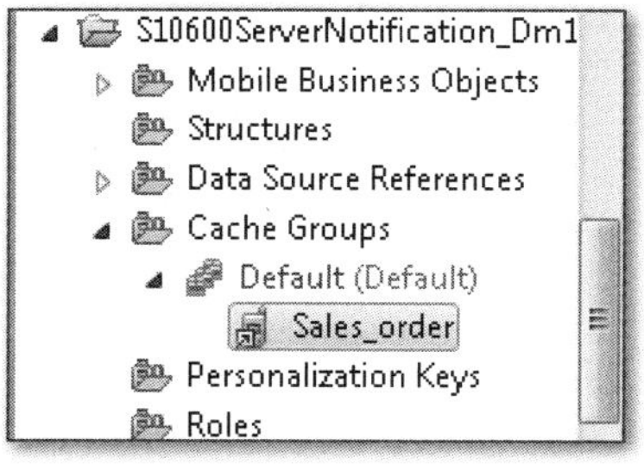

Figure 9.56 Sales Order Added to the Default Cache Policy

The MBO is complete, so the project needs to be deployed.

9.4.2 Create Hybrid App and Define Server Notification

Create the hybrid app as follows:

1. Right-click the project name in the Workspace Navigator.
2. Select NEW • HYBRID APP DESIGNER.
3. Enter the app name and click NEXT.
4. As shown in Figure 9.57, select RESPONDS TO SERVER-DRIVEN NOTIFICATIONS and click NEXT.

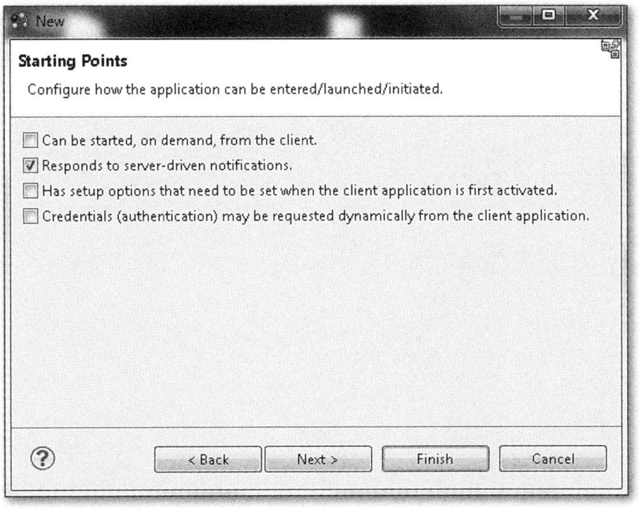

Figure 9.57 Selecting the Server-Driven Notification Starting Point

Select the MBO and object query as follows:

1. Click the SEARCH button to find and select the SALES ORDER MBO.
2. Select the `findByParameter` object query.
3. The parameter is filled in by the system, as shown in Figure 9.58.
4. Click NEXT.

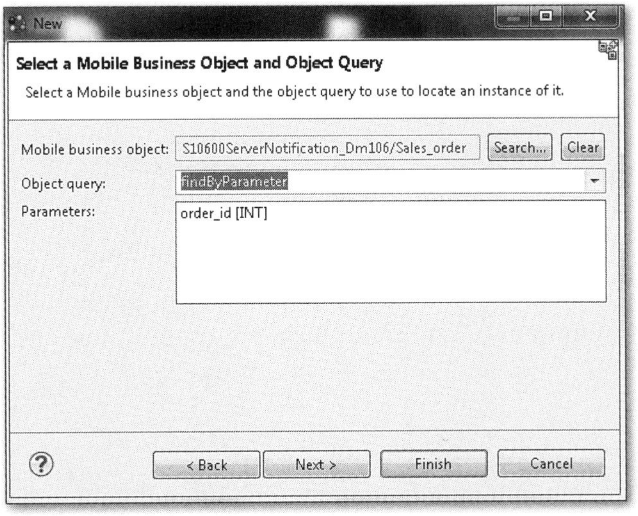

Figure 9.58 Selecting the MBO and Object Query

To specify the sample notification, do the following:

1. Enter "Order (2001) created" in the SUBJECT field.
2. Click NEXT.

The sample notification is shown in Figure 9.59.

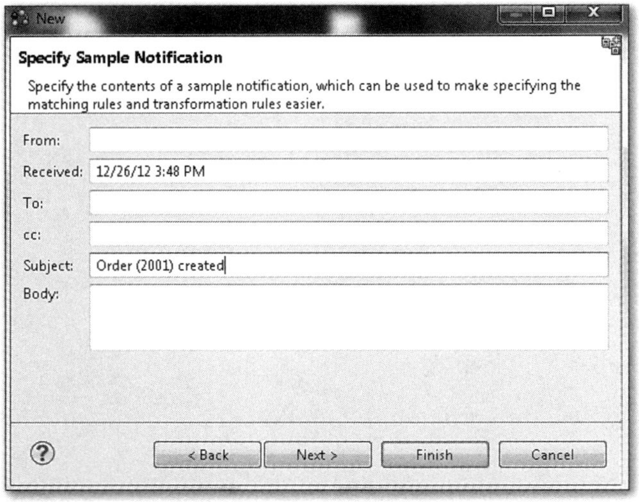

Figure 9.59 Defining the Sample Notification

To create the matching rule, proceed with the following steps:

1. Highlight the word ORDER.
2. Right-click the highlighted portion and select SELECT AS MATCHING RULE. Click NEXT.
3. On the next screen, select ORDER_ID for the EXTRACTION RULES and select SUBJECT for the field.
4. Click the button with the ellipses (...) next to the START tag, highlight ORDER, and click OK.
5. Click the button with the ellipses next to END tag, highlight CREATED, and click OK.

This creates the matching rule, as shown in Figure 9.60, which allows the Unwired Server to recognize a notification as belonging to this app. The subject line must be formatted like this string, with the number between the parentheses replaced with the relevant sales order ID.

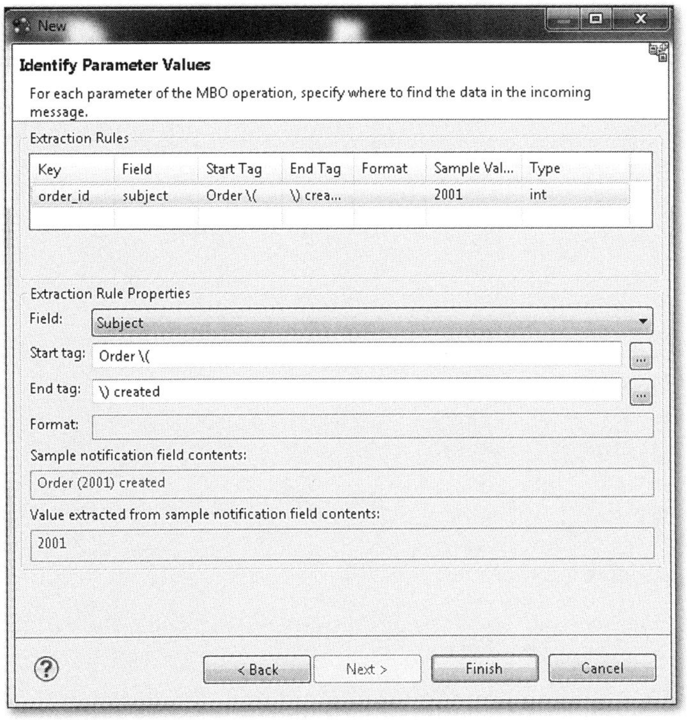

Figure 9.60 Matching Rule

To complete:

1. Click FINISH.
2. Click START or the FLOW DESIGN tab.
3. The SDK creates the basic app, as shown in Figure 9.61.

Figure 9.61 Created Server-Initiated Hybrid App

9.4.3 Flow Design

The following final modification is required:

1. Select the SALES_ORDER screen and delete it.
2. Click GOTO in the PALETTE on the right side of the screen, click the server-initiated starting point, and drag it to the SALES_ORDERDETAILS screen. The final flow design is shown in Figure 9.62.

Figure 9.62 Completed Hybrid App

9.4.4 Test App

Right-click the background of the Flow Design window and select Generate Hybrid App.

The final step involves Send as Notification to test the hybrid app as follows:

1. Start your emulator or device.
2. Right-click the Flow Design background in the SDK and select Send a notification, as shown in Figure 9.63.

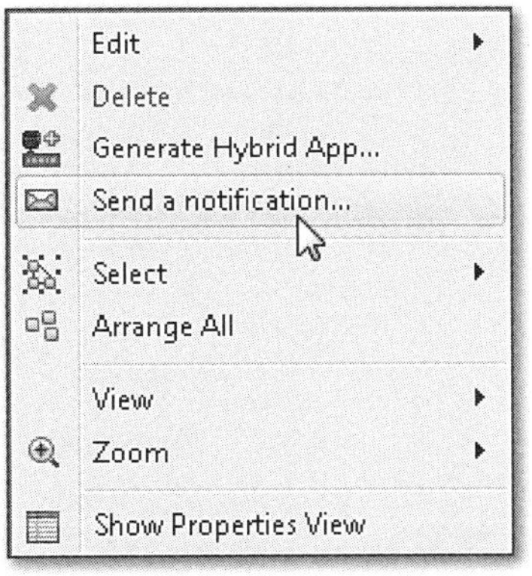

Figure 9.63 Testing the Hybrid App by Sending a Notification

3. Select My Unwired Server in the Unwired Server Profile.
4. Click Get Device Users.
5. Select your user ID for the To field.
6. Enter "Order (2003) created" in the Subject field, as shown in Figure 9.64.
7. Click Send.

The notification should be processed on your device as follows:

1. A notification should then appear on the Message screen, as shown in Figure 9.65.
2. The user clicks the message to open the message detail.
3. The user clicks the detailed message to open the app. (If you get a message to wait for the app to be loaded, click OK, wait a few minutes, and try again.)
4. The app is loaded using the sales order number in the subject line of the notification.

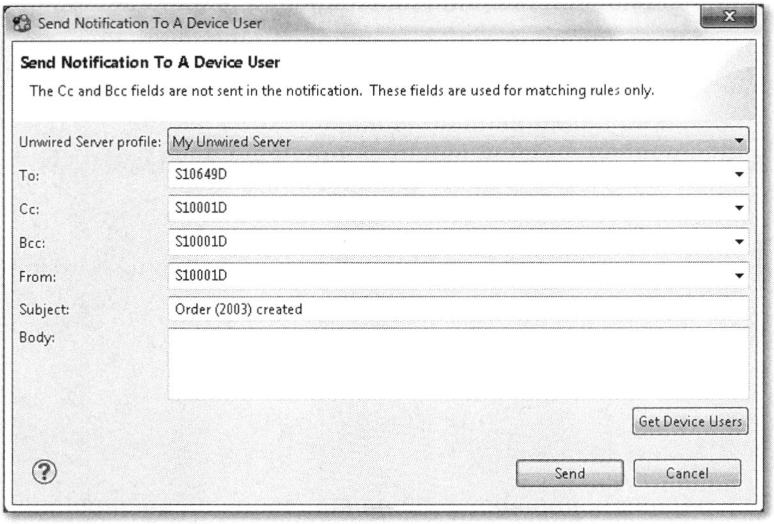

Figure 9.64 Sending a Notification.

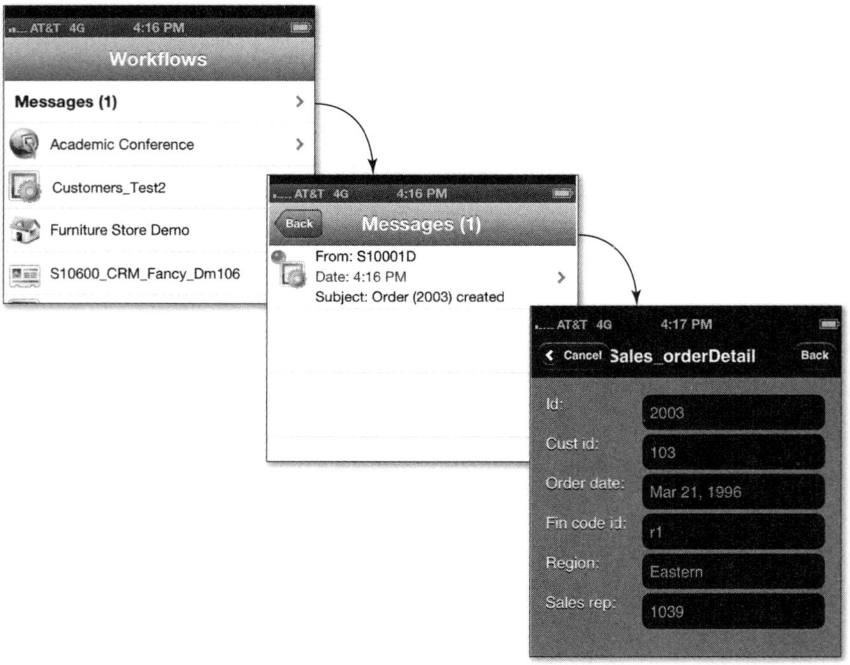

Figure 9.65 Server-Initiated Sales Order Hybrid App

9.5 Summary

This chapter presented the design of several complete hybrid apps. These examples brought together the material presented in Chapter 7 and Chapter 8. The examples were based on both database and SAP EISs and covered both client-initiated and server-initiated apps.

PART III
Customizing Hybrid Apps

This chapter introduces methods that developers can use to go beyond the capabilities provided by the Hybrid App Designer and customize the user interface of hybrid apps.

10 Customizing Hybrid App User Interfaces

The Hybrid App Designer (formerly called the Mobile Workflow Forms Editor) described in Chapter 8 provides the tools to develop full-featured apps. However, there are times when a developer may want more control over the user interface or the ability to implement a feature that is not available in standard hybrid apps. This chapter describes techniques developers can use to customize an app's user interface. The first section explains how to generate the project code files, which are required for customizing. The remainder of the chapter then describes four examples of customizing the interface, most of which involve customizing the HTML files that implement the user interface:

- Adding a jQuery button group
- Creating a custom list that improves on the standard LISTVIEW control
- Displaying a Google map using a jQuery plug-in
- Implementing conditional navigation so that destination screens can be determined at runtime

Because the Sybase Mobile SDK overwrites HTML files, editing them directly is problematic. As a result, most of these customizations are accomplished by adding custom JavaScript that manipulates the HTML at runtime to the *Custom.js* file.

The flow design for the examples in this chapter is shown in Figure 10.1. The app can be used to view the products in inventory, look up supplier details for products, and view a map showing the supplier's location. This app can be implemented without customization. Even the map can be shown if the data in the database are in the correct format (see Chapter 11 for details).

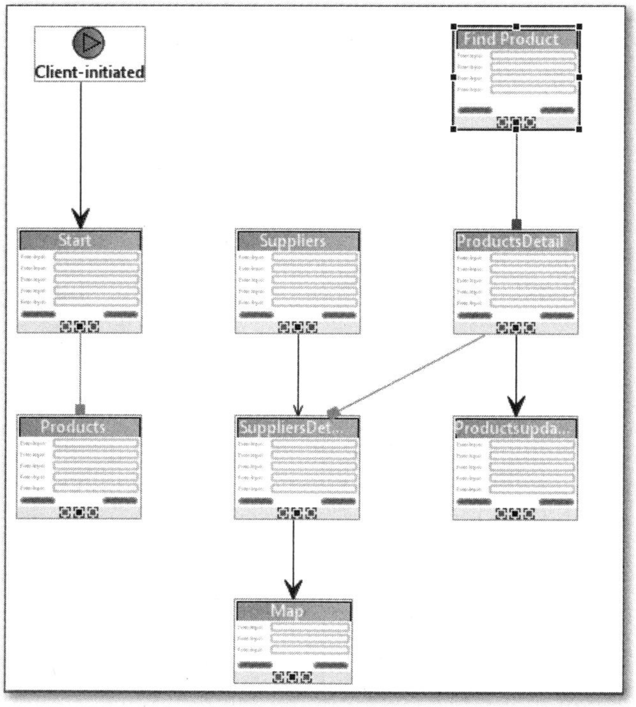

Figure 10.1 Sample App Flow Design

10.1 Generating the Project Code

Customization involves modifying the project code that is generated by the Hybrid App Generation Wizard. As described in Chapter 8, this wizard performs three functions: generates the project code, deploys the app to the Unwired Server, and assigns the app to one or more users. You can enable each of these functions independently, using checkboxes in the dialog. To customize an app, the project code must first be generated, but not necessarily deployed to the server. This is accomplished by deselecting the checkboxes that enable the deployment and assignment functions.

Figure 10.2 shows the section of the dialog used to specify code generation settings.

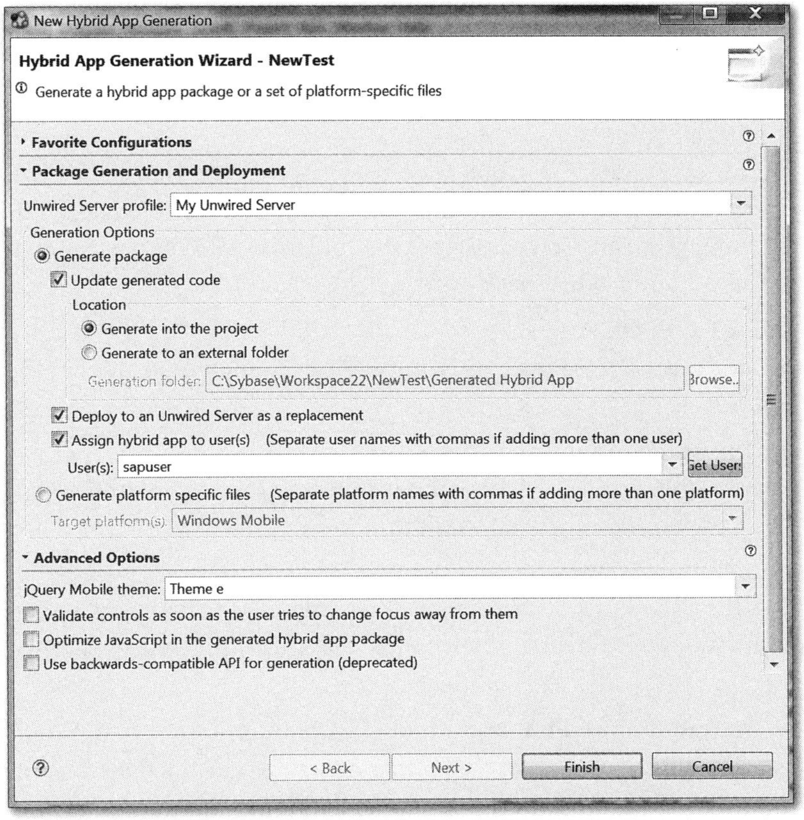

Figure 10.2 Code Generation Settings in the Hybrid App Generation Wizard

The following fields can be configured:

- UNWIRED SERVER PROFILE
  You can select the Unwired Server connection profile to which you wish to deploy the app. Even if you choose not to deploy the app, this selection determines which credentials to use if you are using static credentials.
- GENERATE PACKAGE
  If you choose this option, the code is generated for all supported platforms.
- UPDATE GENERATED CODE
  If you select UPDATE GENERATED CODE, the wizard generates the various files that contain the project's code. Most of the project files are overwritten each time the wizard is invoked. However, there are exceptions, which are noted in Section 10.1.1. If you wish to deploy the app to the server without overwriting

the existing code, deselect this option. Note that if you do this, any changes to the project made in the Hybrid App Designer since the last time the code was generated will not be reflected in the project code.

- Generate into the Project
  By default, the files are saved into a folder called Generated Hybrid App, located in the project directory structure in the SDK workspace. The wizard creates this folder the first time it is run. You can access the folder in the Workspace Navigator within the Unwired Workspace.
- Generate to an External Folder
  You can instruct the wizard to create a ZIP file containing the project files in the specified folder.
- Deploy to an Unwired Server as a Replacement
  This option causes the wizard to deploy the hybrid app to the Unwired Server and replace any previously deployed versions.
- Assign Hybrid App to Users
  You can assign the app to one or more users. You only have to assign an app to a user once. Subsequent deployments will update the app for previously assigned users.
- Generate Platform Specific Files
  If you choose this option, the wizard will only generate files for the platform you choose. The resulting files appear under a folder named Platform Specific Files, which is created under the Generated Hybrid App folder in the project workspace.

In the Advanced Options section, you can configure the following fields:

- jQuery Mobile Theme
  You can choose a jQuery Mobile theme, which determines the look and feel of the app on most devices. More information about the jQuery Mobile themes can be found at *http://jquerymobile.com/demos/1.1.1/*.
- Validate Controls as Soon as the User Tries to Change Focus From Them
  This will cause the app to perform validation for a control as soon as it loses focus.
- Optimize JavaScript in the Generated Hybrid App Package
  If this option is selected, the hybrid app's JavaScript files are concatenated into

a single file (e.g., *SUP0.js* or *SUP1.js*). If this option is not selected, the JavaScript files are handled separately.

- Use Backward-Compatible API for Generation (Deprecated)
  If this option is selected, the JavaScript files will conform to the conventions prior to version 2.2 service pack 2 of the Sybase Mobile SDK.

The Hybrid App Generation Wizard generates app files based on the properties defined within the Hybrid App Designer. The visual interface is defined by HTML5 and CSS, while the data access and navigation logic is exposed as a JavaScript API. The following sections describe the contents of the generated files.

10.1.1 Project Files

Figure 10.3 shows a portion of the Generated Hybrid App folder in the Workspace Navigator. The files shown define the project structure. The following files should not normally be edited:

- *<hybrid app>.zip*
  Contains all of the app's files. This file is always overwritten by the Hybrid App Generation Wizard, even if you deselect the option to update the generated code. This file is deployed to the Unwired Server.
- *Manifest.xml*
  Specifies how the contents of the *<hybrid app>.zip* file are organized. This file is always overwritten by the Hybrid App Generation Wizard, even if you deselect the option to update the generated code.
- *Datajs-version.js*
  This contains a JavaScript library of functions for OData and native device services. These functions allow you to incorporate third-party JavaScript SDKs in your app. This file is always overwritten.
- *WorkflowClient.xml*
  Specifies how the data sent by the server maps to MBO operations and object queries.
- Look-and-feel XML files
  Link mobile platforms (Windows Mobile Professional, BlackBerry, BlackBerry6, iOS, and Android) to specific HTML files. Platforms can share the same HTML files or use different ones.

Figure 10.3 XML and XSD Files

10.1.2 HTML Files

The generated HTML files can be edited directly; however, they are overwritten each time the project code is generated. There are three ways to customize the generated HTML:

- Add an HTMLVIEW control to a form. The contents of the default value field for the control are inserted into the HTML files.
- Use jQuery or JavaScript to insert or manipulate HTML at runtime.
- Edit the HTML directly and then elect not to generate the project code when you deploy it to the server.

Figure 10.4 shows the generated HTML files in the Workspace Navigator. The HTML file used by a device depends, in part, on whether the developer wishes to optimize for appearance or performance. Set this option in the Unwired Workspace by selecting WINDOW • PREFERENCES • SYBASE, INC. • MOBILE DEVELOPMENT • HYBRID APP DESIGNER and choosing the preferred option on the right-hand side of the dialog.

The three HTML files are as follows:

- *hybridapp.html*
  This file contains basic HTML formatting. It is used on the BlackBerry, Android, and iOS platforms when OPTIMIZE FOR PERFORMANCE is selected. It is always used on Windows Mobile devices.

- *hybridapp_Custom.html*
  This file contains HTML with CSS styles. It is used with BlackBerry 5.0 when OPTIMIZE FOR APPEARANCE is selected.
- *hybridapp_JQM.html*
  This file contains HTML styled with jQuery Mobile CSS styles. It is used for BlackBerry 6.0, Android, and iOS devices when OPTIMIZE FOR APPEARANCE is selected.

Figure 10.4 Three Generated HTML Files

All the screens in an app are contained within a single file within separate <div> elements. The detailed contents of the HTML files vary, but they all have the following general structure, as shown in Listing 10.1:

```
<body [...]>
      <div id=Screen1KeyScreenDiv" [...] >
        [...]
        <form id="Screen1KeyForm" [...] >
        [...]
        </form>
        [...]
      </div>
      <div id=Screen2KeyScreenDiv" [...] >
        [...]
        <form id="Screen2KeyForm" [...] >
        [...]
        </form>
        [...]
      </div>
</body>
```

Listing 10.1 Structure of HTML Files

The IDs for all HTML elements are derived from the keys created in the Hybrid App Designer. For example, screen IDs have the format `ScreenKeyScreenDiv`, where `ScreenKey` is the key assigned to the screen. In general, the keys assigned to interface elements are used to create the IDs assigned to them in HTML. No two elements can have the same ID because JavaScript uses IDs to reference HTML elements. In practice, this means that no two objects of the same type (i.e., two EDITBOX controls) can have the same key.

Figure 10.5 Project JavaScript Files

10.1.3 JavaScript Files

The generated JavaScript API provides the primary means to customize navigation and data manipulation. The JS folder shown in Figure 10.5 contains the JavaScript files used by the app. Most of these files contain functions that can be used by developers, and the *Custom.js* file provides functions in which the developer can insert custom code.In addition, any JavaScript file copied into this folder is automatically included in the project when the code is generated. *The Custom.js* file is the primary entry point for adding custom code to a project. If the code you are adding is extensive, you may find it more convenient to place your code in separate files; however, you have to call this code from the *Custom.js* file. Some of the files are generated by the Hybrid App Generation Wizard if they do not exist. A comment at the top of each file indicates whether it is generated or not.

Some of the files of interest are the following:

- *API.js* and *Utils.js*
  Contain a variety of useful methods, including methods to manipulate dates, manipulate HTML, and communicate with the Hybrid Web Container.
- *Callbacks.js*
  Contains methods to handle callbacks from the container used for event handlers that are asynchronous.
- *Camera.js*
  Contains methods used to access the device's camera or photo library.
- *Certificate.js*
  Contains methods for handling X.509 certificates.
- *Custom.js*
  Contains methods in which the developer can enter custom code. Many of the methods are invoked automatically at different points during the app lifecycle. The relevant lifecycle events are `WorkflowLoad`, `Submit`, `NavigateForward`, `NavigateBackward`, `ShowScreen`, `MenuItemClick`, and `Save`. `Before` and `After` methods exist for each event. If the `Before` methods return `false`, the corresponding event is cancelled. Details about the methods can be found in the Sybase documentation, and the *Custom.js* file contains examples of how to use many of them. This file is generated by the Hybrid App Generation Wizard only if it does not already exist.
- *ExternalResource.js*
  Contains methods used to access resources on HTTP servers.
- *hwc-comms.js, hwc-utils.js and hwc-api.js*
  Native container functions.

- *Json2.js*
  Contains methods to process JSON.
- *data.PlatformIdentification.js*
  Contains functions that can be used to identify the device on which an app is running.
- *Resources.js*
  Contains methods used to access localized string resources.
- *SUPStorage.js*
  Contains methods used to store name value pairs in a local cache.
- *Timezone.js*
  Contains methods to extract and format dates and times.
- *HybridApp.js*
  Contains methods for menu, screen, and database operations. A method is created in this file for each menu item created in the Hybrid App Designer.
- *WorkflowMessage.js*
  Contains methods to manipulate the data contained in workflow messages.

Contained within the JS folder are seven additional folders. There are four folders that contain the platform specific files for the Cordova (formerly PhoneGap) API libraries. Cordova allows a hybrid app to access device resources, and is described in Chapter 17. The JQUERY folder contains the jQuery and jQuery Mobile libraries. jQuery is a JavaScript library that simplifies HTML manipulation, event handling, animation, and AJAX interactions. jQuery Mobile is designed to build lightweight, cross-platform apps for mobile devices. (You can find more information about jQuery Mobile at *http://jquerymobile.com/*.) The MAKIT folder contains files uses with the MAKit graphics library. Finally, the WIDGETS folder contains JavaScript files that implement the screen controls.

10.1.4 CSS Files

jQuery Mobile CSS defines the default look and feel. While jQuery Mobile provides a number of themes, you are free to edit the CSS files or add your own to customize the app's appearance. The CSS files are generated by the Hybrid App Generation Wizard only if they do not already exist. You could, for example, create a branded look and feel based on your organization's web standards. All files in this folder are automatically included in the project, so you can copy your own CSS files to this folder.

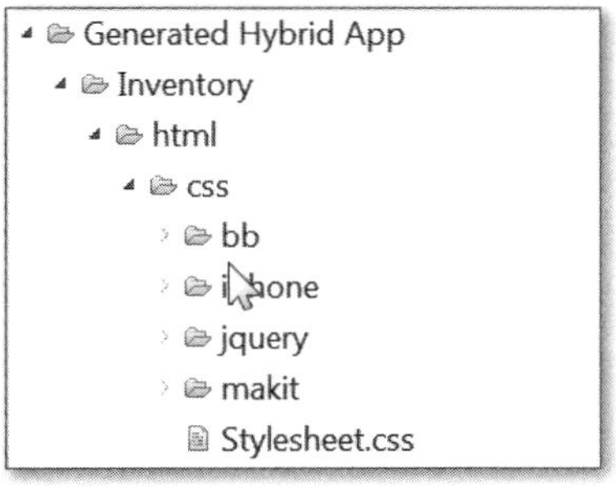

Figure 10.6 CSS Files

10.2 Adding a jQuery Button Group

jQuery provides powerful techniques to create and manipulate dynamic interfaces. Many of these techniques are illustrated at *http://jquerymobile.com/test/.* One technique that is especially useful for mobile apps is a button group. A button group can be used to implement a toolbar-like grouping of buttons, which minimizes the screen real estate they occupy. In addition, the buttons leverage the JavaScript generated by the Unwired Workspace, so they require minimal code.

Next, we'll walk you through the steps involved in adding a jQuery button group.

10.2.1 Modifying the Hybrid App

The first step is to create menu items that implement the required actions and configure their properties. When a menu item is added to a project, the code generator creates a function in the *Workflow.js* file with the following signature:

```
menuItemCallback<Screen Key><MenuItem Key>()
```

`<Screen Key>` is the key created for the screen on which the MENUITEM appears, and `<MenuItem Key>` is the key created for the menu item.

10.2.2 Adding the HTML

Once the menu items are working, add an HtmlView control to the screen on which you want the button group to appear. An HtmlView control is used to insert custom HTML into the generated HTML files. Enter code similar to that shown in Listing 10.2 in the Default value field. The default value field initially shows only one line, making it difficult to see the code you have entered. To see multiple lines, click off the control and then return to it. The result should look similar to Figure 10.7.

```
<div data-role="controlgroup" data-type="horizontal" align="center">
  <a href="javascript:menuItemCallbackStartProducts();" data-
role="button"  style="width:125px;">Products</a>
  <a href="javascript:menuItemCallbackStartLow_Inventory();" data-
role="button" style="width:125px;">Reorder</a>
  <a href="javascript:navigateForward('Find_Product');" data-
role="button" style="width:125px;">Search</a>
  <a href="javascript:menuItemCallbackStartExit();" data-role="button"
style="width:125px;">Exit</a>
</div>
```

Listing 10.2 HTML for jQuery Button Group

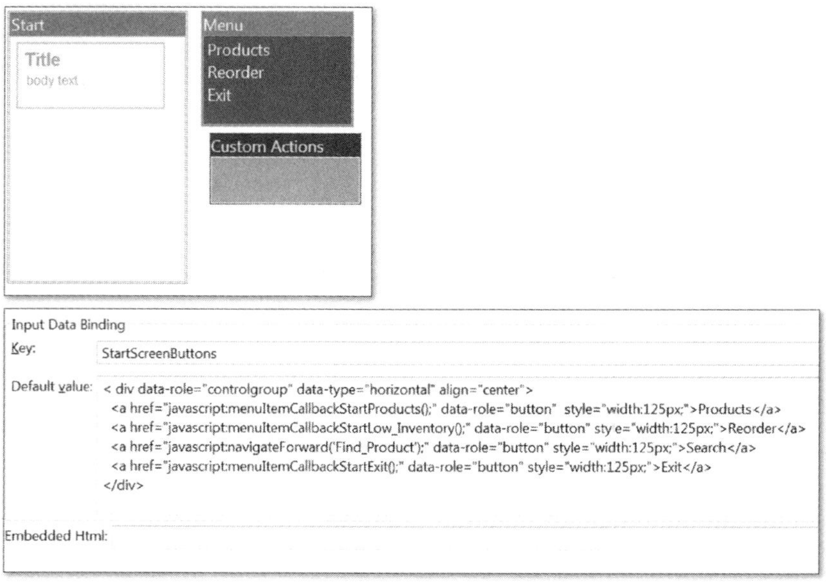

Figure 10.7 Creating a jQuery Button Group

The data-role="controlgroup" option in the <div> element creates the jQuery button group. Each <a> creates a link, and the data-role="button" tag indicates to jQuery that these should be displayed as buttons.

The first, second, and fourth buttons execute the functions created for the menu items on the screen. The third button uses the utility function navigateForward() found in the *API.js* file to open the FIND_PRODUCT screen. The resulting screen is shown in Figure 10.8. The figure shows a horizontal and vertical button group. For a smaller screen change, the <a> tags to something like this:

```
<a href="javascript:menuItemCallbackStartProducts();" data-
role="button"  data-mini="true" style="width:70px;font-size:x-
small;">Products</a>
```

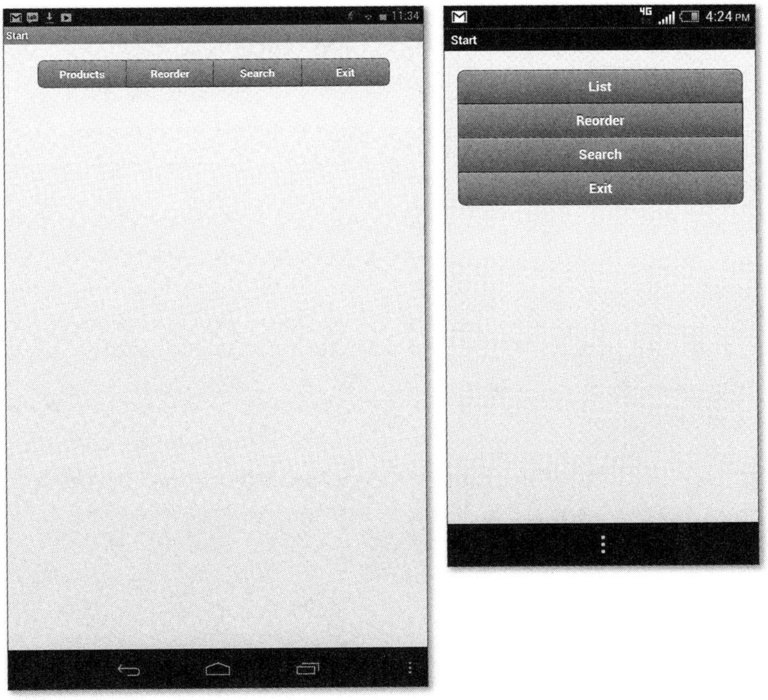

Figure 10.8 Orienting a jQuery Button Group Vertically or Horizontally

10.2.3 Hiding the Menu Items

One problem you may encounter with the button group on iOS devices is that menu items are displayed as buttons in the corners of the screen. This may be somewhat confusing for the user because there are multiple buttons to perform the same task on the screen. This is remedied by using jQuery to hide the menu items on iOS devices. This technique does not work on the BlackBerry or Android platforms. However, on these platforms, the menus are hidden from the user unless the user activates the menu, so this is not a significant drawback.

As described in Section 10.1.3, custom JavaScript/jQuery is normally added to the *Custom.js* file. Since the menu items should be hidden before any screen is shown, the code should be entered into the `customBeforeWorkflowLoad()` function, which is invoked when the app loads but before any data is loaded or screens displayed.

To hide the menu items, you must reference their HTML elements. The menu items are implemented in the HTML files as `<a>` elements like the following:

```
<a href="javascript:void(0)" id="StartScreenDivProducts"
name="Products" onclick="menuItemCallbackStartProducts();"> Products</a>
```

Note that the `onclick` method executes the same menu item callbacks used in the button group.

One way to obtain a reference to the menu item is to use a jQuery ID selector. As you can see in the code snippet above, IDs for menu items have the format `<Screen key>ScreenDiv<MenuItem key>`. The screen key in the code above is START, and the menu item key is `Products`.

To hide the menus shown in Figure 10.7, open the *Custom.js* file and insert the code in Listing 10.3 into the `customBeforeWorkflowLoad()` method:

```
$('#StartScreenDivExit').hide();
$('#StartScreenDivProducts').hide();
$('#StartScreenDivSearch').hide();
```

Listing 10.3 Hiding the Menus

The `$` invokes jQuery, and the # indicates that the selection is based on the HTML element's ID. The hide() method renders the element invisible. The `customBeforeWorkflowLoad()` function is shown in Figure 10.9.

jQuery Selectors

This simple example of jQuery selectors barely scratches the surface of their sophistication. A complete reference for the jQuery API can be found at *http://api.jquery.com/*.

```
//Use this method to add custom html to the top or bottom of a form
function customBeforeWorkflowLoad() {
    // Please note some scenarios that should not be handled here:
    // 1) There is no screen available before workflow load finishes
    // 2) You cannot called any SharedStorage functions here because the url
    // parameters
    // have not yet been parsed and workflow setting are not yet initialized.

    $('#StartScreenDivExit').hide();
    $('#StartScreenDivProducts').hide();
    $('#StartScreenDivSearch').hide();

    return true;
}
```

Figure 10.9 Code to Hide the Menus

10.3 Creating a Custom List

The LISTVIEW control provided by the Unwired Workspace is a simple, flexible tool, but it could be improved. This example replaces the standard LISTVIEW control with a custom list using JavaScript and jQuery. It also illustrates techniques for accessing the data in the workflow message returned by the Unwired Server. (These techniques are covered in more detail in Chapter 11.)

Next, we walk you through the steps involved in creating a custom list.

10.3.1 Modifying the Hybrid App

Create a screen with a LISTVIEW control using the Hybrid App Designer as described in Section 8.2.3 of Chapter 8. This ensures that the detail screen is created correctly. Next, delete the LISTVIEW control, and then add an HTMLVIEW control. In the default value property of the HTMLVIEW control, enter the following HTML: `<div id="CustomList"></div>` (Figure 10.10).

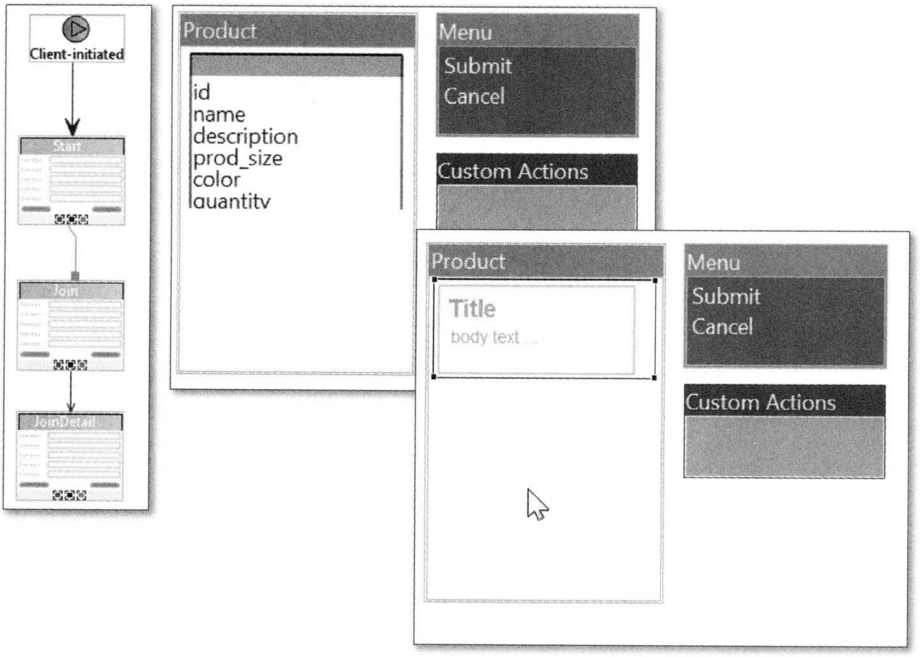

Figure 10.10 Creating a Custom List

10.3.2 Adding Custom JavaScript

The custom list must be constructed after the workflow message is returned by the Unwired Server but before the destination screen is displayed. A method that satisfies these requirements is the `customAfterNavigateForward(screenKey, destScreenKey)` method in the *Custom.js* file.

The code shown in Listing 10.4 creates a list of products and inserts a list divider for different product categories. The object query that retrieves the data from the MBO must sort the data on the category name field. Open the *Custom.js* file and insert code similar to that shown in Listing 10.4.

```
if (destScreenKey == "Products"){
  //Retrieve the message value collection
  var mvc = getCurrentMessageValueCollection();
  //Retrieve the data in ProductsKey
  var itemList = mvc.getData("Products");
```

```
  //Retrieve the actual values
  var items = itemList.getValue();
  //Retrieve the number of records
  var numOfItems = items.length;
  var i = 0;
  //Start a string variable to hold the HTML and
  // start the list
  var htmlOutput = '<div id="ProductList">
<ul data-role ="listview" data-
autodividers="true"  data-divider-theme="b"
data-filter="true">';
  var currCat = '';
  // iterate through the results and build our list
  while (i < numOfItems) {
    //Retrieve the current record
    var currItem = items[i];
    //Retrieve the values in the record
    var name = currItem.getData("Products
   _productName_attribKey").getValue();
    var inStock = currItem.getData("Products
   _unitsInStock_attribKey").getValue();
    var onOrder = currItem.getData("Products
   _unitsOnOrder_attribKey").getValue();
    var reOrder = currItem.getData("Products
   _reorderLevel_attribKey").getValue();
    var cat = currItem.getData("Products
   _categoryName_attribKey").getValue();
    //If the category changes create a new list
    // divider
    if (cat != currCat){
   currCat = cat;
   htmlOutput += '<li class = "group" data-
      role="list-divider">' + cat + '</li>';
    }
    //Create a list item using the record key as
    //the ID
    htmlOutput += '<li><a class="listClick" data-
   icon="arrow-l" id ="'
    htmlOutput += currItem.getKey() + '">';
    htmlOutput += '<h2>' + name + '</h2>';
    htmlOutput += 'In stock: ' + inStock + ' On
```

```
 Order: ' + onOrder + ' Reorder: ' + reOrder ;
    htmlOutput += '</a></li>';
    i++;
   }// while
   htmlOutput += '</ul></div>';
   //Insert the generated html into the products
   //form
   $('#CustomList')
.html(htmlOutput);
     //Tell JQuery to reformat the list
   $.mobile.listview.prototype.enhanceWithin(
$.mobile.activePage);

   //Attach a handler for the click event of the
//list items
   $(".listClick").click(function(){
     $(this).addClass("ui-btn-active");
     //Use the navigateForward utility function to
  //navigate to the detail screen and pass the
  //ID from the list item
     navigateForward("ProductsDetail", this.id );
     // special case for bb
     if (isBlackBerry()) {
       return;
     }
   });
   }
```

Listing 10.4 Custom List

The code shown in Listing 10.5 determines whether the destination screen is the screen on which the list is displayed. Next, the data is extracted from the workflow message value collection using the `getCurrentMessageValueCollection(listKey)` method, which is included in the *API.js* file. The message value collection is an in-memory object that contains data used to update the controls on the screen. (Interacting with the message value collection is described in more detail in Chapter 11.) `Products` is the list key that contains the results of the object query. The `itemList.getValue()` returns an array of the items in the Products list key, as shown in Listing 10.5.

```
var message = getCurrentMessageValueCollection();
//Retrieve the data in ProductsKey
var itemList = message.getData("Products");
//Retrieve the actual values
var items = itemList.getValue();
//Retrieve the number of records
var numOfItems = items.length;
```

Listing 10.5 Retrieving an Array of Product Records

The code loops through the items in the products list, extracting the values of individual attributes, as shown in Listing 10.6.

```
var currItem = items[i];
//Retrieve the values in the record
var name = currItem.getData("ProductsKey
   _productName_attribKey").getValue();
var inStock = currItem.getData("ProductsKey
   _unitsInStock_attribKey").getValue();
var onOrder = currItem.getData("ProductsKey
   _unitsOnOrder_attribKey").getValue();
var reOrder = currItem.getData("ProductsKey
   _reorderLevel_attribKey").getValue();
var cat = currItem.getData("ProductsKey
   _categoryName_attribKey").getValue();
```

Listing 10.6 Retrieving Values of Individual Attributes

An HTML list element is created for each item using the surrogate key from the message value collection as the element's ID, as shown in Listing 10.7.

```
//Create a list item using the record key as the ID
htmlOutput += '<li><a class="listClick" data-
   icon="arrow-l" id ="'
htmlOutput += currItem.getKey() + '">';
htmlOutput += '<h2>' + name + '</h2>';
htmlOutput += 'In stock: ' + inStock + ' On Order: '
 + onOrder + ' Reorder: ' + reOrder ;
htmlOutput += '</a></li>';
```

Listing 10.7 Creating HTML for List Elements

When the category name changes, the code creates a divider list element, as shown in Listing 10.8.

```
if (cat != currCat){
  currCat = cat;
  htmlOutput += '<li class = "group" data-role="list-
   divider">' + cat + '</li>';
}
```

Listing 10.8 Adding a List Divider for Each Product Category

Once the loop completes, the list is inserted into the app's HTML.

This statement inserts the list code into the `<div>` that was inserted into the HTMLVIEW control:

```
$('#CustomList'). html(htmlOutput);
```

The first time the page list is created, jQuery formats it properly. However, after the first time, jQuery must be instructed by this statement to format it:

```
$.mobile.listview.prototype.enhanceWithin($.mobile.activePage);
```

Finally, a click event handler is added to the list items so that they respond to user interaction. The jQuery in this code uses a selector based on the class of the element, which is indicated by the period, as shown in Listing 10.9.

```
$(".listClick").click(function(){
  $(this).addClass("ui-btn-active");
  //Use the navigateForward utility function to navigate to the detail
screen and pass the ID from the list item
  navigateForward("JoinDetail", this.id );
  // special case for bb
  if (isBlackBerry()) {
    return;
  }
});
```

Listing 10.9 Adding a Click Handler to the List Items

The resulting list is shown in Figure 10.11. The list dividers provide some organization to the list, and the FILTER ITEMS field allows the user to find items quickly on a long list.

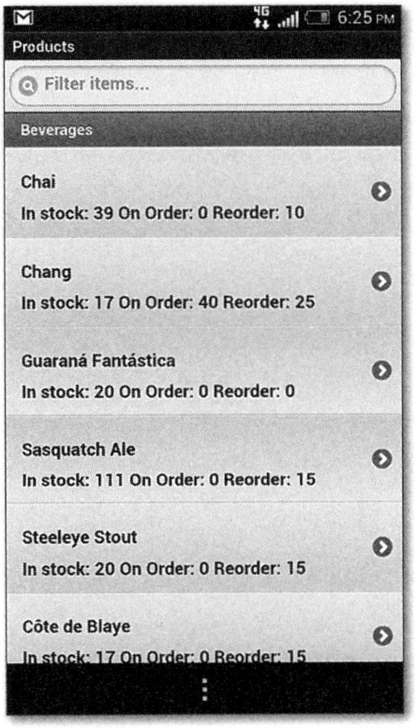

Figure 10.11 The Custom List

10.4 Displaying a Map Using a jQuery Plug-in

Displaying a map using Google's map API is easy using the jQuery map plug-in. jQuery plug-ins are JavaScript libraries that provide specific functionality based on jQuery. There are many jQuery plug-ins available, some of which may be useful for mobile apps. You can find a description of many plug-ins at *http://archive.plugins.jquery.com/*.

Next, we walk you through the steps of displaying a map using a jQuery plug-in.

10.4.1 Copying the jQuery Map Plug-in

To use a plug-in in a project, copy the plug-in file into the JQUERY folder in the Workspace Navigator. References to the files in this folder are automatically included in the project when the code is generated.

Google's Terms of Service

Use of the Google Map API is subject to certain restrictions that you should understand before using it in a production app. You are free to use the API if the app is freely accessible to the public. The app cannot operate behind a firewall or on an internal network unless you have entered into an agreement with Google. You are, however, free to use the API in an app that is sold on Google Play. For more information on Google's terms of service, visit *https://developers.google.com/maps/terms#section_9_1*.

The jQuery Google map plug-in can be found at *http://code.google.com/p/jquery-ui-map/*. The archive contains multiple plug-in files, along with a number of demos. The example in this book uses the *jquery.ui.map.js* plug-in. Download the plug-in archive, extract the files, and then copy the correct file into the JQUERY folder in the Workspace Navigator.

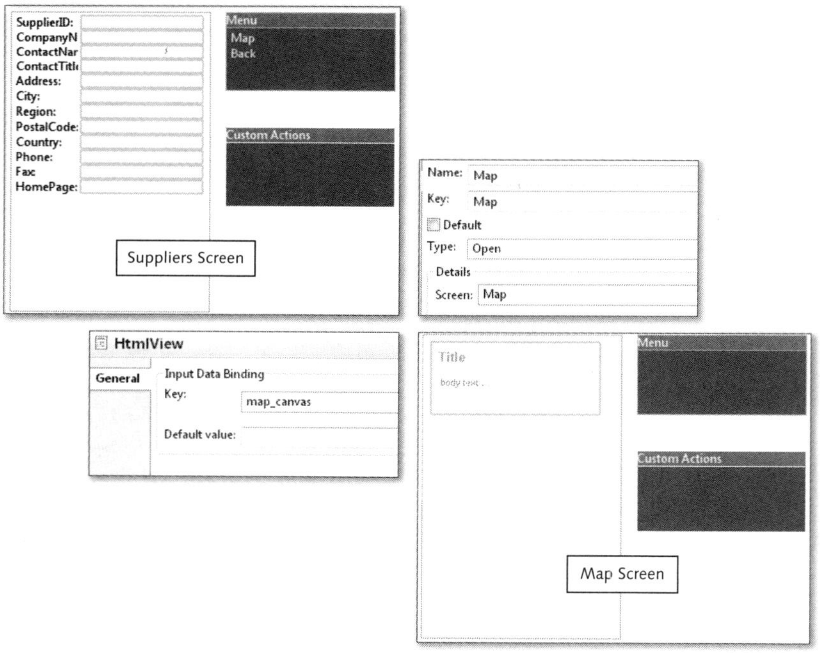

Figure 10.12 The Supplier Details Screen and Map Screen

10.4.2 Modifying the Hybrid App

Figure 10.12 shows the setup for this example. The first screen shows the details for a supplier, including the address. The supplier detail screen includes a MENUITEM control that opens a screen called MAP. The MAP screen has an HTMLVIEW control that has a key of map_canvas. This creates a <div> element in the project's HTML with ID map_canvas, in which the map is displayed.

10.4.3 Adding Custom JavaScript

The code that displays the map is added to the customAfterNavigateForward function in the *Custom.js* file. This function is invoked after a forward screen transition. The code is shown in Listing 10.10.

```
//make sure we're navigating to the map screen
if(destScreenKey == "Map"){
  //Delete a map if one was created before
  $('#map_canvas').gmap('destroy');
  //Get a reference to the form on the screen that called the Map
screen (SupplierDetail)
  var form = document.forms[screenKey + "Form"];
  if(form){
    //Construct an address string to pass to the geocoding service
    //Retrieve the data to construct the address

     var address = form.Suppliers_address_attribKey.value + ", ";
     address += form.Suppliers_city_attribKey.value + ", ";
     address += form.Suppliers_region_attribKey.value + ", ";
     address += form.Suppliers_postalCode_attribKey.value;
} else {
  //If the address isn't available, notify the user
  //and cancel the map retrieval
  return false;
}
//Use JQuery to get a reference to the window and get the height of the
window less 100 px
//The height must be a fixed value or the map will not be displayed
var ht = parseInt($(window).height())-100;

//Use JQuery to set the style of the div that will display the map.
$('#map_canvas').css({
  'height': ht  + 'px',
  'width': '100%',
  'border-style': 'groove'
});
```

```
//Get a reference to the geocoder service.
var geocoder = new google.maps.Geocoder();
//Call the geocoder service and pass a callback
//function that creates the map
geocoder.geocode( { 'address': address}, function(results, status) {
  //Make sure the address was coded successfully
  if (status == google.maps.GeocoderStatus.OK) {
   //Use Jquery to get a reference to the
   //map_canvas div and use the gmap function to
   //attach the map
$('#map_canvas').gmap({'center':
results[0].geometry.location,
'zoom': 10,
'disableDefaultUI':true,
'callback': function() {
  //callback function places a marker on the map at
  //the address location
  var self = this;
  self.addMarker({
'position': this.get('map').getCenter()
   //Attach a click event handler to the marker
   //which displays the address
}).click(function() {
self.openInfoWindow({ 'content': address }, this);
});
}});
} else {
  //Notify the user if the geocode is not successful
 //alert("Geocode was not successful for the
  // following reason: " + status);
}
});
}
```

Listing 10.10 Displaying a Google Map with the jQuery Map Plug-in

The first step is to determine whether the MAP screen is the screen being shown:

```
if(destScreenKey == "Map"){
```

Next, any previously created map is deleted:

```
$('#map_canvas').gmap('destroy');
```

The next step is to construct the address using the values in the fields on the SUPPLIER DETAILS screen. The code is shown in Listing 10.11. The address is used later in the code to pass to Google's geocoder service.

```
var form = document.forms[screenKey + "Form"];
if(form){
  var address = form.Suppliers_address_attribKey.value + ", ";
   address += form.Suppliers_city_attribKey
.value + ", ";
   address += form.Suppliers_region_attribKey.value
+ ", ";
   address += form.Suppliers_postalCode_
attribKey.value;
```

Listing 10.11 Constructing the Address

Next, the `<div>` element that will display the map is styled. The height property of the `<div>` element must be a fixed value, or else the map does not display. The code retrieves the height of the window using the jQuery height function. jQuery is then used to add CSS properties to the `map_canvas <div>` element.

```
var ht = parseInt($(window).height())-100;

$('#map_canvas').css({
  'height': ht  + 'px',
  'width': '100%',
  'border-style': 'groove'
});
```

A reference to the Google geocoder service is obtained:

```
var geocoder = new google.maps.Geocoder();
```

Finally, the geocoder service is called. The geocoder service returns the longitude and latitude of the address. Geocoding is processor intensive, so in a production app, it would be better to use this service when the supplier record is created and have the latitude and longitude stored with the supplier data rather than calling it each time the map is loaded. The arguments passed to the geocode method are the address string and a callback function that is executed when the geocoder service returns:

```
geocoder.geocode( { 'address': address}, function(results, status) {
```

If the service returns successfully, the gmap method from the jQuery map plug-in is called, and the map parameters are passed to it:

```
  if (status == google.maps.GeocoderStatus.OK) {
  $('#map_canvas').gmap({'center':
   results[0].geometry.location,
'zoom': 10,
'disableDefaultUI':true,
```

A callback is passed to `gmap`, which places a marker on the map at the address location:

```
'callback': function() {
  var self = this;
  self.addMarker({
'position': this.get('map').getCenter()
```

Finally, an event handler is added to the marker so that the address is displayed when the marker is clicked:

```
}).click(function() {
self.openInfoWindow({
'content': address }, this);
});
}});
```

The last step is to insert a `<script>` element into the project HTML that will initialize the Google Map service. Using jQuery to insert this line into the HTML at runtime does not work, so it must be added at design time. Open the HTML file located in the HTML folder of the hybrid app and insert the following code shown below in the `<head>` section of the file:

```
<script type="text/javascript" src="http://maps.google.com/maps/api/js?
sensor=false&libraries=places"></script>
```

When you deploy the app, make sure you deselect the option that causes the project code to be generated because this will cause the HTML files to be overwritten. The resulting app is shown in Figure 10.13.

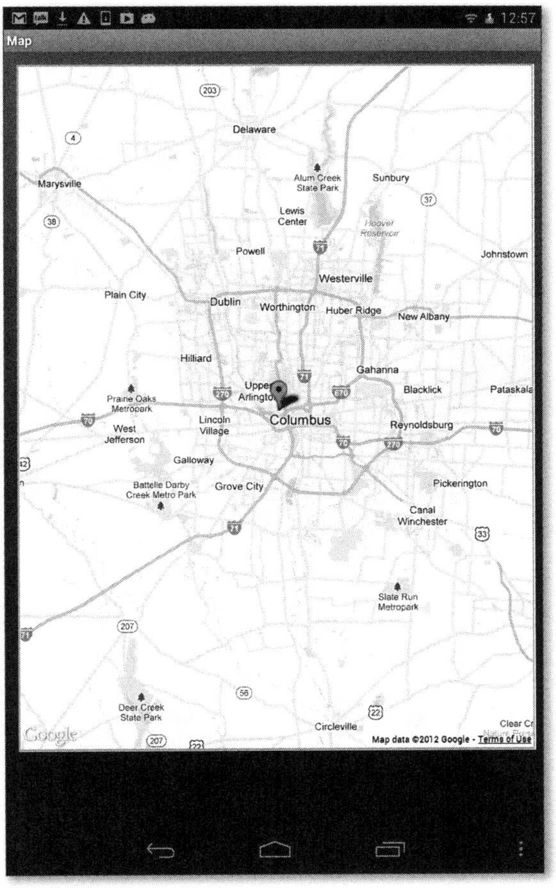

Figure 10.13 The Map Displayed on an Android Device

The jQuery map plug-ins can be used to implement many other map features. Refer to the documentation and demos provided to see how this is done. In addition, Chapter 17 shows how to use the Google JavaScript API to show a map.

10.5 Implementing Conditional Navigation

There are times when the target of a navigation should be determined at runtime. Consider a search that may return one or more records. If multiple records are returned, a list should be shown. The user can then select an item to see the details. However, if only one record is returned from the search, the app should navigate directly to the details screen. This section demonstrates how to implement conditional navigation.

10.5.1 Modifying the Hybrid App

The first step is to assign conditions to the menu item or button that initiates the navigation. Select the menu item or button and then select the PROPERTIES view. Conditions are created at the bottom of the GENERAL PROPERTIES sheet. As shown in Figure 10.14, the DEFAULT SUCCESS screen represents the default destination screen, and a condition is created for each potential target screen. When creating the condition, a name is created to identify the condition, and the target screen associated with it is selected. In this example, the condition `ONE_RECORD` is assigned to the PRODUCTDETAILS screen, and the condition `MULTIPLE_RECORDS` is assigned to the PRODUCTS1 screen.

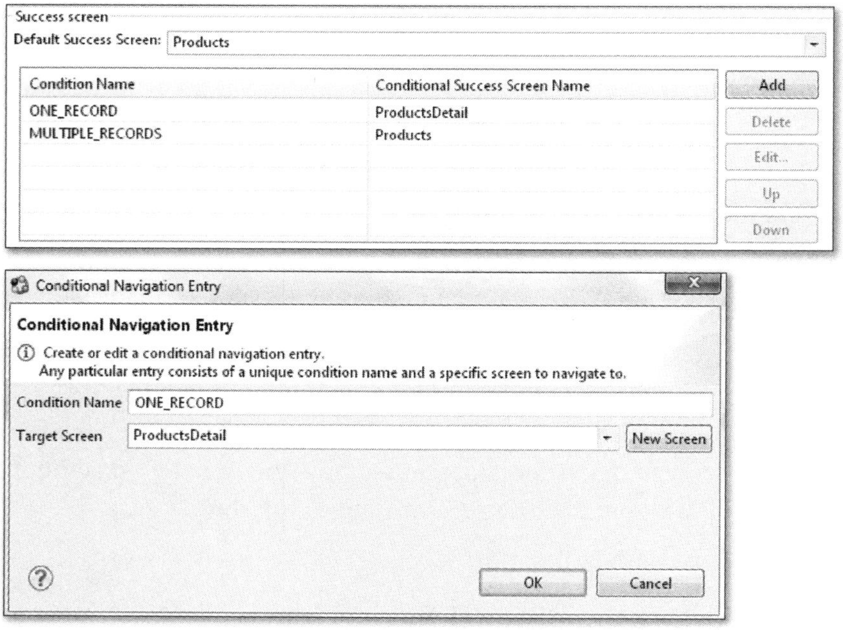

Figure 10.14 A Condition Created for Each Potential Destination Screen

10.5.2 Adding Custom JavaScript

The JavaScript that implements the conditional navigation is inserted into the `customConditionalNavigation(currentScreenKey, actionName, defaultNextScreen, conditionName, workflowMessage)` function in the *Custom.js* file. The parameters for this function are as follows:

- **`currentScreenKey`**
  The screen from which the navigation was initiated.
- **`actionName`**
  The key for the menu item or button that initiated the navigation.
- **`defaultNextScreen`**
  The screen specified in the Default Success Screen property.
- **`conditionName`**
  The name of the condition being tested.
- **`workflowMessage`**
  The workflow message returned by the server.

Enter the code shown in Listing 10.12.

```
//Check which screen and action initiated the
// navigation
if ((currentScreenKey === 'Find_Product') &&
   (actionKeyName ==='Search')){
  //Check for the result of only one record returned
  if (conditionName === 'ONE_RECORD'){
    //Get the workflow message which contains
    //the data passed to the target screen
    var values=workflowMessage.getValues();
    //get the data in the customer list key
    var products = values.getData("Products");
    //Get the list
    var pList = products.getValue();
    //Get the number of items in the list
    var count = pList.length;
    //Check to see if there is only one record
    if (count == 1){
      //Get the data row from the list
      var row1 = pList[0];
      //update the workflow message with the
```

```
//single row so that the detail keys
      //will be available to the product detail
//text fields
      workflowMessage.updateValues(row1);
      //returning true causes the application to navigate to the screen
associated
      //with the condition
      return true;
    }
  }
  else if (conditionName == 'MULTIPLE_RECORDS'){
    //returning false causes the application to navigate to the
defaultNextScreen
    return false;
  }
}
```

Listing 10.12 The Code Checking for Each Condition

The code first checks for the initiating screen and action. Next, each of the conditions is checked to determine whether it applies. The function is called for each condition that is defined. If the function returns `false` in every case, the default navigation is performed. In this example, the destination screen is dependent on the number of records returned in the products list key by the server. For the `ONE_RECORD` condition, the code first extracts the data from the workflow message. The statement `products.getValue()` returns an array:

```
var values=workflowMessage.getValues();
var products = values.getData("Products");
var pList = products.getValue();
```

Next, the array size is checked to see whether only one record is returned. If there is only one record, the workflow message is updated as a single record. Returning `true` causes the app to navigate to the screen associated with this condition.

```
var count = pList.length;
if (count == 1){
  var row1 = pList[0];
  workflowMessage.updateValues(row1);
  return true;
}
```

The final `else if conditional` is not necessary since the method returns `false` by default, but it is included for completeness.

10.6 Summary

This chapter introduced some options for configuring the user interface. More examples are shown in Chapter 17. In the next chapter, techniques for accessing and manipulating the data sent by the server are described.

This chapter illustrates techniques to customize the functionality of hybrid apps.

11 Customizing Hybrid App Functionality

This chapter builds upon Chapter 10 by describing additional ways that customization can extend hybrid apps. While the previous chapter concentrated on methods to enhance the user interface, this chapter describes ways to enhance the functionality of apps. Among the more important techniques is the ability to access and manipulate the data messages transmitted to and from the Unwired Server. Section 11.1 describes how this is done. Section 11.2 describes custom actions, which are similar to menu items but are invoked programmatically rather than by the user. Finally, Section 11.3 describes the SUPStorage API, which is used to store data on the device.

11.1 Working with Data

Customization often involves programmatically accessing the hybrid app's data to modify it or take actions based on the values of attributes. Within hybrid apps, data elements are represented as keys. The data elements that derive from an MBO are mapped to attributes of the MBO (see Figure 11.1). The names of keys mapped to MBO attributes take the form of `<MBO name>_<Attribute name>_attribKey`. You can also create additional keys to hold data elements that are not mapped to MBO attributes.

Most keys contain single data elements, but a list type key can contain multiple items and consist of multiple individual keys. Figure 11.2 shows a list key called EMPLOYEE. The key dialog indicates that the data in the key is sent by the server as a result of an object query on the EMPLOYEE MBO. The figure also shows the individual keys assigned to the list.

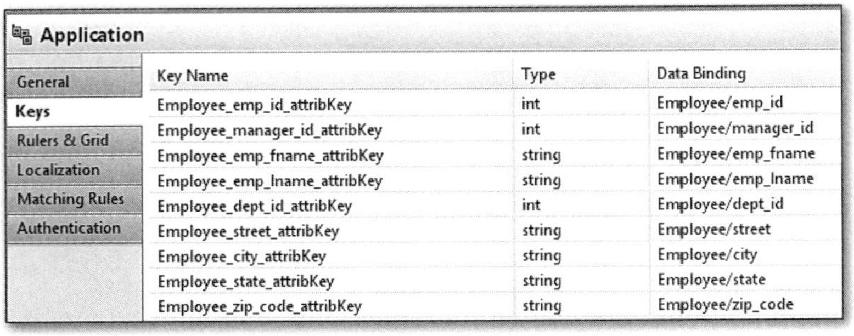

Figure 11.1 Data in a Container App Represented as Keys

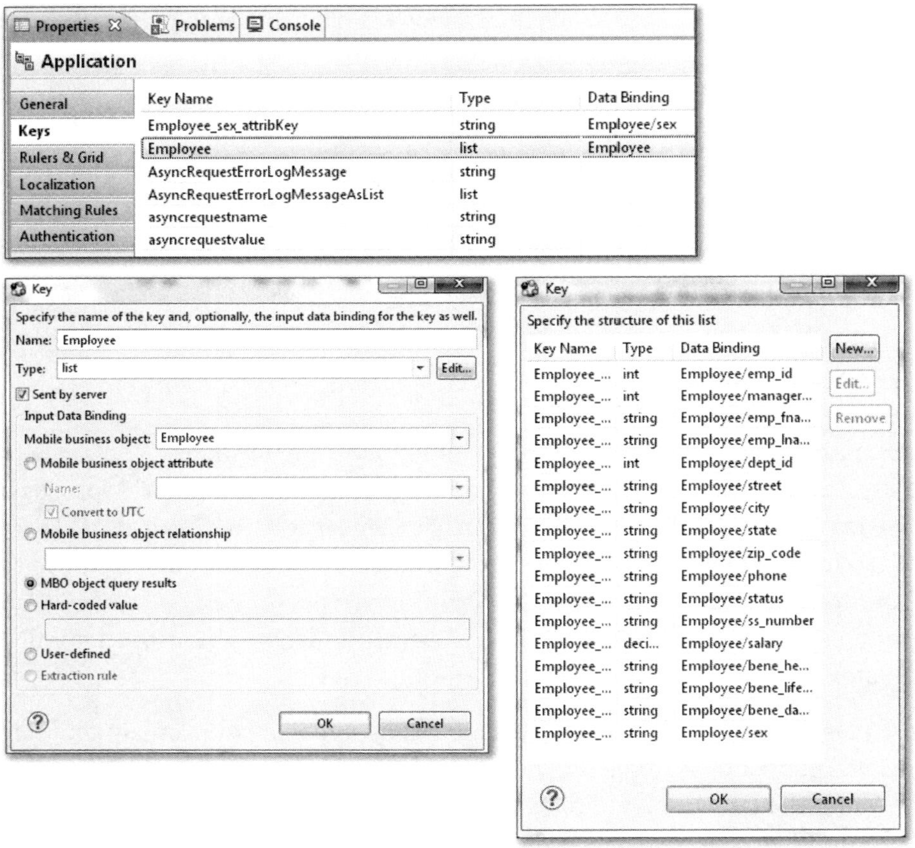

Figure 11.2 List Key Containing Multiple Items Made Up of Multiple Keys

Keys can be assigned to controls in the Hybrid App Designer (formerly called the Mobile Workflow Forms Editor), allowing the app to manage updates between the control and the app's data structures. Figure 11.3 shows an EDITBOX control mapped to the key `Employee_emp_id_attribKey`. Most controls are mapped to single item keys, but LISTVIEW controls are mapped to list keys.

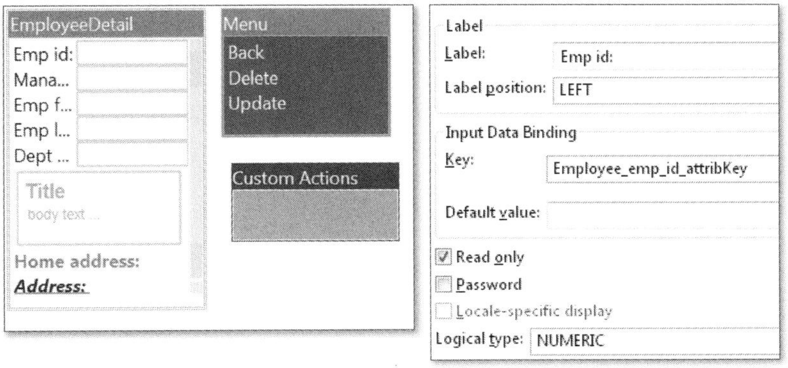

Figure 11.3 Keys Mapped to Controls in the Hybrid App Designer

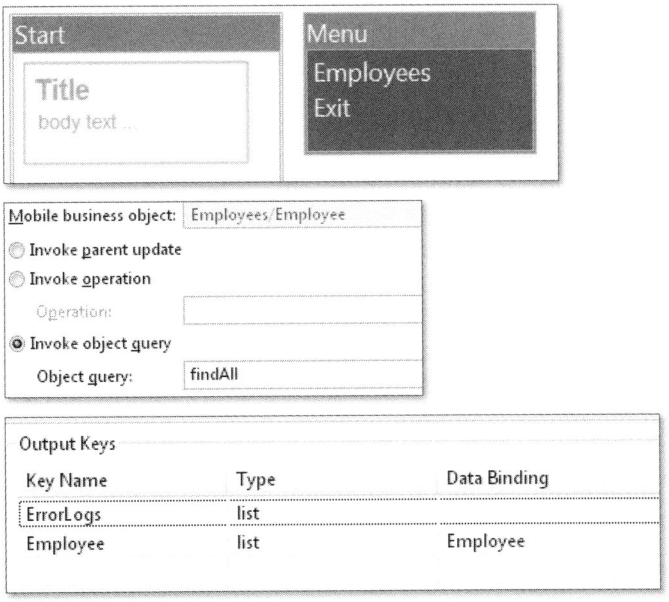

Figure 11.4 Output Keys Determining the Data Available after a Screen Navigation

MENUITEM and BUTTON controls that cause navigations to other screens are assigned output keys designating the data that is available on the destination screen. Figure

11.4 shows a menu item called EMPLOYEES that invokes the `findAll` object query in an MBO called EMPLOYEE. The results of the object query fill the EMPLOYEE list key. Because this key is designated as an output key for the menu item, the data is available on the destination screen. The SDK is able to determine that, in this case, a list key is required because the object query returns multiple items. The output keys for an object query that returns a single item are the individual keys that correspond to the return values from the object query. If you encounter a situation in which a screen appears with no data, a likely cause is a problem with output keys.

Section 11.1.1 describes how app data is structured, and Section 11.1.2 describes the JavaScript API that is used to manipulate the data.

11.1.1 Understanding the WorkflowMessage Structure

A hybrid app communicates with the Unwired Server using an XML message called `WorkflowMessage`. Figure 11.5 shows a portion of a workflow message in XML format. Near the top of the message, the XML identifies the destination app screen (`Employee`), the menu item or button that initiated the data request (`Employees`), and the list key that contains the data (`Employee`).

```
<?xml version="1.0"?>
<XmlWidgetMessage>
    <Header>
    </Header>
    <WidgetScreen>Employee</WidgetScreen>
    <RequestAction>Employees</RequestAction>
    <Values>
        <Value key="Employee" type="LIST" >
            <Values key="100001" s="" >
                <Value key="_surrogateKey" type="TEXT" >100001</V>
                <Value key="Employee_emp_id_attribKey" type="NUMBER" >102</V>
                <Value key="Employee_manager_id_attribKey" type="NUMBER" >501</V>
                <Value key="Employee_emp_fname_attribKey" type="TEXT" >Fran</V>

            </Values>
            <ValueS key="100002" s="" >
                <Value key="_surrogateKey" type="TEXT" >100002</V>
                <Value key="Employee_emp_id_attribKey" type="NUMBER" >105</V>
                <Value key="Employee_manager_id_attribKey" type="NUMBER" >501</V>
                <Value key="Employee_emp_fname_attribKey" type="TEXT" >Matthew</V>

            </Values>
```

Figure 11.5 Portion of a Workflow Message in XML Format

List keys contain a number of `<Values>` sections that correspond to data records. Each item in the list includes a surrogate key assigned by the Unwired Server and

<Value> elements that correspond to the attribute keys assigned to the list in the Unwired Workspace. Each value element consists of a key, a value, and a type. The possible data types are `TEXT`, `NUMBER`, `DATETIME`, `BOOLEAN`, and `LIST`.

The data is accessible to the hybrid app via three objects defined in *WorkflowMessage.js*. The objects are `WorkflowMessage`, `MessageValue`, and `MessageValueCollection`. The `MessageValue` object consists of a single key value pair and corresponds to the <value> elements in Figure 11.5.

`MessageValueCollection` contains the data used to update the controls on a screen. Using the example in Figure 11.5 on a screen with a LISTVIEW, `MessageValueCollection` consists of the EMPLOYEE value list.

Multiple `MessageValueCollection` objects can exist in memory. The currently displayed screen is updated from the current `MessageValueCollection`. For example, assume a LISTVIEW control is mapped to the EMPLOYEE list key from Figure 11.5. When a user selects an item on the LISTVIEW control, a detail screen, which shows the values from the selected item, is displayed. The `MessageValueCollection` object used to update the LISTVIEW control is called the parent of the collection used to update the detail screen.

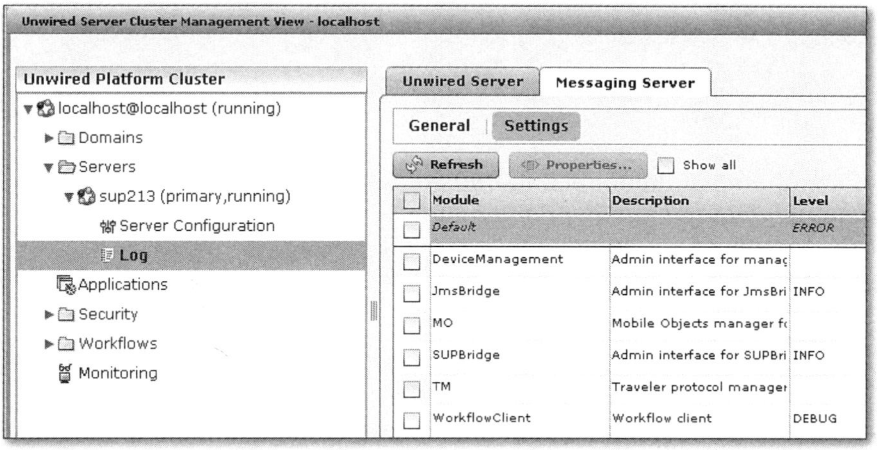

Figure 11.6 Setting the Debug Level to DEBUG to Create a Log

The workflow message is written to the workflow client log on the Unwired Server if the log level is set to DEBUG. The log level is set in the SCC, as shown in Figure 11.6. The log file can be found at *C:\Sybase\UnwiredPlatform\Servers\UnwiredServer\logs\WorkflowClient* on the server machine. The log contains all messages transmitted between the app and the Unwired Server, including XML messages. Files with "OBMO" in the name contain synchronous requests resulting from online requests. Files with "ampservice" in the name contain asynchronous requests resulting from a server notification or submit workflow.

Another way to view the workflow message is to insert the following line in the *Custom.js* file:

```
alert(hwc.getDataMessage().serializeToString());
```

This line retrieves the workflow message, converts it to a string, and displays it in a dialog box.

11.1.2 Accessing Data using the JavaScript API

The functions described in this section are defined in either the *API.js* or *WorkflowMessage.js* file. Studying these files is helpful, and there is a complete function list in the documentation. The list of functions in this section is not exhaustive, but it does include many of the most useful ones.

Table 11.1 shows functions used to reference the `WorkflowMessage` and `MessageValueCollection` objects. These functions are defined in *API.js*.

Method	Description
hwc.getDataMessage()	Returns a reference to the `WorkflowMessage` object.
`getCurrentMessageValueCollection(listKey)`	Returns the `MessageValueCollection` of the current screen. `listKey` is an optional argument that is used to narrow the data returned.
`getMessageValueCollectionForOnlineRequest( screenKey, requestAction, keys, keyTypes)`	Returns a reference to the `MessageValueCollection` to be sent for an online request. This allows manipulating the data before it is sent to the server.

Table 11.1 Functions Used to Access Data in the Workflow Message

The functions in Table 11.2 can be used to access the `WorkflowMessage` object. These functions are defined in *WorkflowMessage.js*.

Method	Description
`WorkflowMessage.getValues()`	Returns the `MessageValueCollection`.
`WorkflowMessage.getRequestAction()`	Returns the request action of the `WorkflowMessage`.
`WorkflowMessage.getWorkflowScreen()`	Returns the screen of the `WorkflowMessage`.
`WorkflowMessage.serializeToString()`	Returns a string version of the `WorkflowMessage`. A useful debugging tool.
`WorkflowMessage.updateValues(sourceValues, listViewValuesKey)`	Updates the values of a `WorkflowMessage`.

Table 11.2 Functions to Access the WorkflowMessage Object

Table 11.3 shows functions used to manipulate `MessageValue` objects. These functions are defined in *WorkflowMessage.js*.

Method	Description
`MessageValue.getKey()`	Returns the key of the `MessageValue`.
`MessageValue.setKey(key)`	Sets the key of the `MessageValue`.
`MessageValue.getValue()`	Returns the value of the `MessageValue`.
`MessageValue.setValue(value)`	Sets the value of the `MessageValue`.
`MessageValue.getType()`	Returns the type of the `MessageValue`.
`MessageValue.setType(type)`	Sets the type of the `MessageValue`.

Table 11.3 Functions to Access the MessageValue Object

Table 11.3 shows functions to use with the `MessageValueCollection` object. These functions are defined in *WorkflowMessage.js*.

Method	Description
`MessageValueCollection.getData(key)`	Returns the data specified by key.
`MessageValueCollection.getKey()`	Returns the key of the `MessageValueCollection`.
`MessageValueCollection.setKey(key)`	Sets the key of the `MessageValueCollection`.
`MessageValueCollection.add(key,value)`	Adds a new value to the `MessageValueCollection`.
`MessageValueCollection.clear()`	Clears the `MessageValueCollection`.
`MessageValueCollection.remove(key)`	Removes the value with the specified key.
`MessageValueCollection.getCount()`	Returns the number of values in the `MessageValueCollection`.
`MessageValueCollection.getValues()`	Returns an array of the values in the `MessageValueCollection`.
`MessageValueCollection.getKeys()`	Returns an array of keys in the `MessageValueCollection`.

Table 11.4 Functions to Access the MessageValueCollection

Data is transferred between the controls on a screen and the current `MessageValueCollection` using the following two functions defined in *API.js*:

```
updateUIFromMessageValueCollection(screenKeyName, values)
```

and

```
updateMessageValueCollectionFromUI(values, screenName, keys, keyTypes,
updateModifiedValue)
```

The `Values` parameter is a `MessageValueCollection` object; the `screenName` and `screenKeyName` are the open screen; `Keys` is an optional parameter that is an array of keys used to limit the controls updated on the screen; and `keyTypes` is an optional array of the key types. The last parameter is an optional Boolean argument indicating to update the modified value and not the actual message value. Next we describe examples of using these methods.

Accessing Lists and Individual Values

The first example is from the custom list example described in the previous chapter. Now that you have some understanding of the data access methods, the example may be clearer. This example shows how to process a list key.

The portion of the code of interest is shown in Listing 11.1.

```
var message = getCurrentMessageValueCollection();
var itemList = message.getData("Products");
var items = itemList.getValue();
[…]
var numOfItems = items.length;
var currItem = items[i];
[…]
var name = currItem.getData("Products _productName_attribKey").
getValue();
```

Listing 11.1 Retrieving a List Key

The first statement obtains a reference to the current message value collection, which contains the list key called `Products`. The `Products` key is included because it was an output key for the menu item that initiated the navigation.

```
var message = getCurrentMessageValueCollection();
```

The second statement retrieves an object that contains the data stored in the `Products` key.

```
var itemList = message.getData("Products");
```

The method `getValue()` returns an array of the items in the `Products` key.

```
var items = itemList.getValue();
```

The previous three statements could be combined into the following:

```
var items = getCurrentMessageValueCollection().getData("Products")
.getValue();
```

Next, the length of the array is retrieved:

```
var numOfItems = items.length;
```

An individual item is retrieved using array notation:

```
var currItem = items[i];
```

Finally, within the while loop, the values of individual key value pairs are retrieved:

```
var name = currItem.getData("Products_productName_attribKey").
getValue();
```

The previous example illustrates how to retrieve individual key values from a list in a `MessageValueCollection`. If the `MessageValueCollection` does not contain a list key, the code looks like Listing 11.2:

```
var message = getCurrentMessageValueCollection();
//Retrieve the MessageValue object
var nameObj = message.getData("Products
 _productName_attribKey");
//Retrieve the value
var name = nameObj.getValue();
```

Listing 11.2 Retrieve an individual key.

This could be combined into the following:

```
var name = getCurrentMessageValueCollection()
.getData("Products _productName_attribKey").getValue();
```

Adding Values to the MessageValueCollection

This section demonstrates how to add `MessageValue` objects to the `MessageValueCollection` and use them to update controls on a screen. The scenario involves using a link control to display a map using the Google Map API in the device's browser. The map example shown in Section 10.4 of Chapter 10 used a jQuery plug-in, which, while more complex, has some advantages. For example, the map in Chapter 10 is displayed on a screen within the app, so it is possible to include other controls on the same screen as the map. Also, the example in Chapter 10 only hinted at the capabilities of the various jQuery map plug-ins. However, under the right circumstances, the method shown here provides a viable alternative.

Figure 11.7 shows a screen with a LINK control in the Hybrid App Designer. The contents of the link control's PREFIX and SUFFIX properties are appended before and after the control's value and are then executed. In this case, the PREFIX is configured to execute the `showUrlInBrowser()` method, which opens the device's browser. This method is defined in the *API.js* file. The URL in this case invokes the Google Map API, which expects a parameter that is an address in the form `<street>, <city>, <state>, <zip code>`.

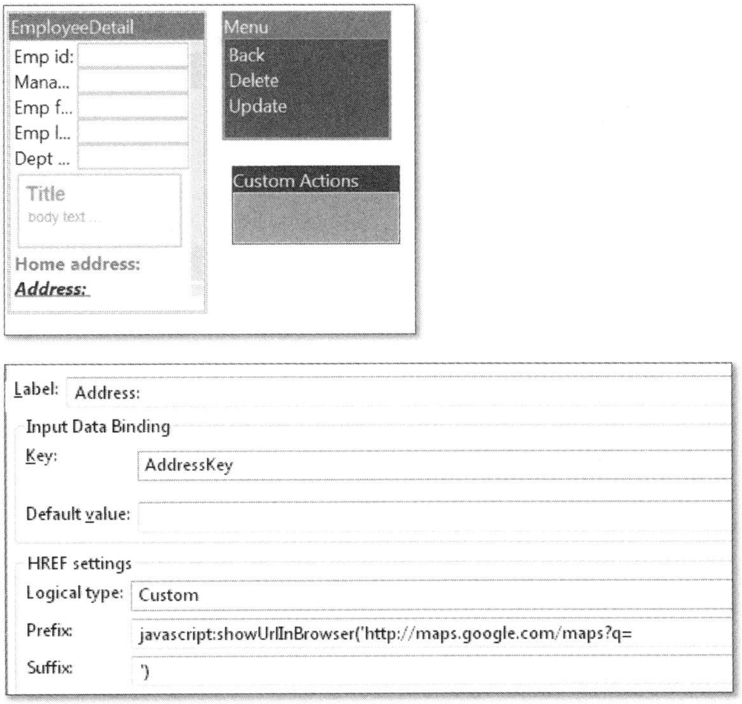

Figure 11.7 Link Control Used to Open the Device's Browser

The value of the Link control provides the address. If the address were stored in the database in the correct format, it would be a simple matter to map the address attribute key to the link control. However, in this case, the address is stored as four separate fields that are mapped to separate attribute keys (see Figure 11.8). The procedure used to update the Link control with the correct value is as follows:

1. Retrieve the address values from the `MessageValueCollection` and build the address string.
2. Create a new `MessageValue` object.
3. Set the `MessageValue` object's value and type properties.
4. Set the `MessageValue` object's key equal to the key assigned to the link control.
5. Add the `MessageValue` object to the `MessageValueCollection`.
6. Call `updateUIFromMessageValueCollection()` to update the screen controls.

Employee_street_attribKey	string	Employee/street
Employee_city_attribKey	string	Employee/city
Employee_state_attribKey	string	Employee/state
Employee_zip_code_attribKey	string	Employee/zip_code

Figure 11.8 Elements of the Employee Address Stored in Separate Keys

The code shown in Listing 11.3 is added to the `customAfterNavigateForward` function in the *Custom.js* file. This function is invoked after a new screen is opened. This ensures that the message value collection contains the data for the controls on the screen and that the controls are available for update.

```
if(destScreenKey == 'EmployeeDetail'){
  //Retrieve the MessageValueCollection
  var message = getCurrentMessageValueCollection();
  //If the MessageValueCollection exists
  if (message){
    //Build a string with the address elements
    var add =             message.getData(
  "Employee_street_attribKey").getValue();
    add += ", " +
  message.getData("Employee_city_attribKey")
  .getValue();
    add += ", " + message.getData(
  "Employee_state_attribKey").getValue();
    add += ", " + message.getData(
  "Employee_zip_code_attribKey").getValue();
    //Create a new MessageValue Object
    var MessageVal = new MessageValue();
    //Set the object's key equal to the key mapped to
    //the Link control
    MessageVal.setKey("AddressKey");
    //Set the value equal to the address string
    MessageVal.setValue(add);
    //Set the value type
    MessageVal.setType(MessageValueType.TEXT);
    //Add the MessageValue object to the
    //MessageValueCollection
    message.add(MessageVal.getKey(), MessageVal);
    //Update the screen controls from the
    //MessageValueCollection
```

```
updateUIFromMessageValueCollection
  (destScreenKey, message);
  }
}
```

Listing 11.3 Create the Address String

The result is shown in Figure 11.9.

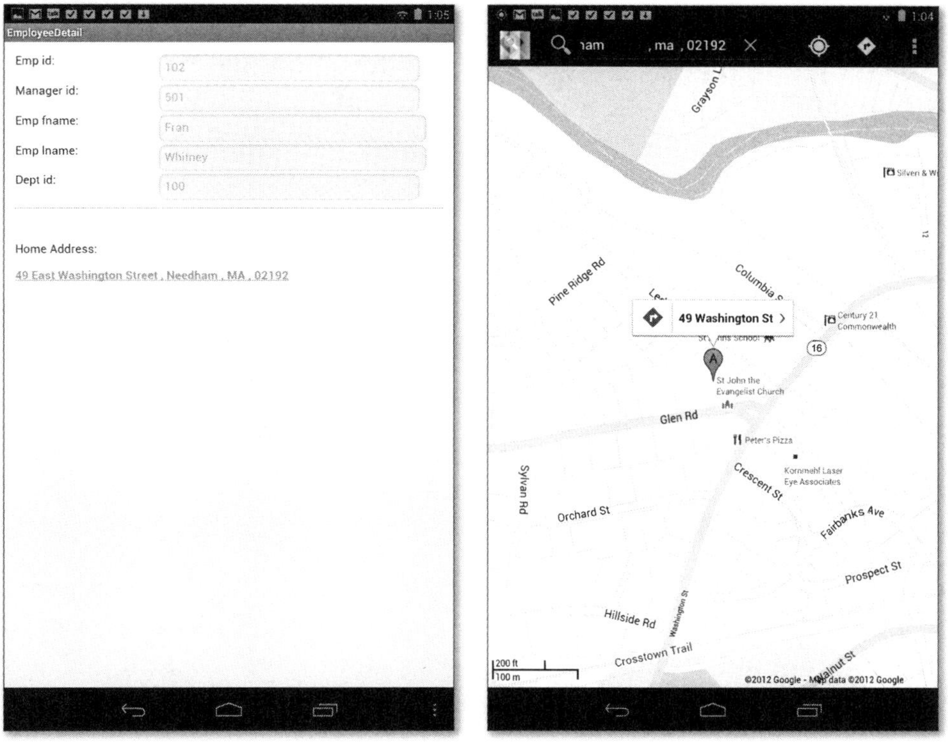

Figure 11.9 Using a Link Control to Display a Google Map

Contrast this code with that shown in Listing 11.4, which is used to create the address string for the jQuery map example in Chapter 10.

```
var form = document.forms[screenKey + "Form"];
  if(form){
    //Construct an address string to pass to the geocoding service
    //Retrieve the data to construct the address
```

```
   var address = form
.Suppliers_address_attribKey.value + ", ";
   address += form.Suppliers_city_attribKey.value
+ ", ";
   address += form.Suppliers_region_attribKey.value
+ ", ";
   address += form
.Suppliers_postalCode_attribKey.value;
```

Listing 11.4 Access Form Elements

This code retrieves the address values from the controls on the SUPPLIER DETAILS screen instead of from the `MessageValueCollection`. This is necessary because the menu item that invokes the map screen is an open type that has no output keys, so there is no `MessageValueCollection` for the map screen.

11.2 Using Custom Actions

Custom actions are similar to menu items. Their properties are configured the same way, and the Hybrid App Generation Wizard creates a function identical to MENU-ITEM functions in the *Workflow.js* file. The difference is that custom actions do not have a visual component, so they must be executed programmatically. Custom actions are used when the developer wishes to initiate a request for data from the server without user intervention.

Consider the example in Figure 11.10. On this screen, the user can select a postage type using a CHOICE control. When the CHOICE control value changes, the app retrieves the postal rate, adds the postage amount to the order amount, and then updates the interface.

This section describes two ways to use custom actions to implement this scenario. In the first scenario, the options in the CHOICE control and the postage rates are retrieved from two different MBOs. This requires two custom actions. In the second scenario, the options and rates are retrieved from the same MBO, requiring only one custom action.

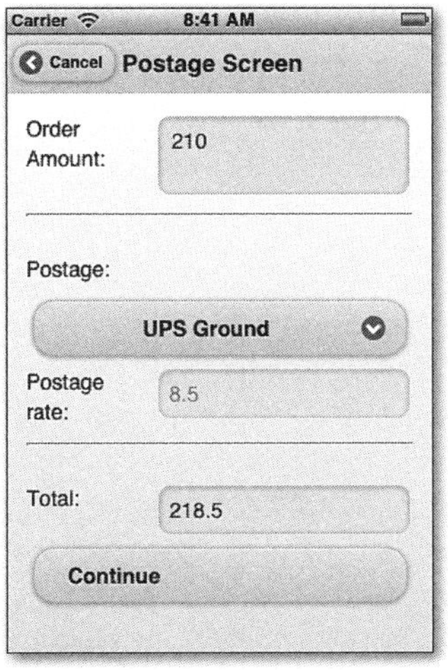

Figure 11.10 The Postage App

11.2.1 Two Custom Actions

Configuring the two custom actions for the postage app can be divided into two steps: creating the custom actions, and then creating the JavaScript. Next, we describe both.

Creating the Custom Actions

The first custom action retrieves the data that provides the options for the CHOICE control. The app must retrieve this data when the POSTAGE screen loads. This is accomplished with the custom action `findPostage`. The second custom action is used to retrieve the postage rate associated with the option selected in the CHOICE control. This custom action, called `findPostageAmount`, is executed when the user changes the CHOICE control selection.

findPostage

Figure 11.11 shows the configuration of `findPostage`. Notice the properties are exactly the same as a MENUITEM control that invokes the `findAll` object query.

The default success screen for `findPostage` is the same screen on which the CHOICE control is located. It may seem unnecessary to specify the default success screen if the screen does not change, but leaving this field blank results in an error, indicating that the values required for the CHOICE control are unavailable. Figure 11.12 shows the configuration of the CHOICE control. The value key and display name key are mapped to MBO attributes available from the output key associated with `findPostage`.

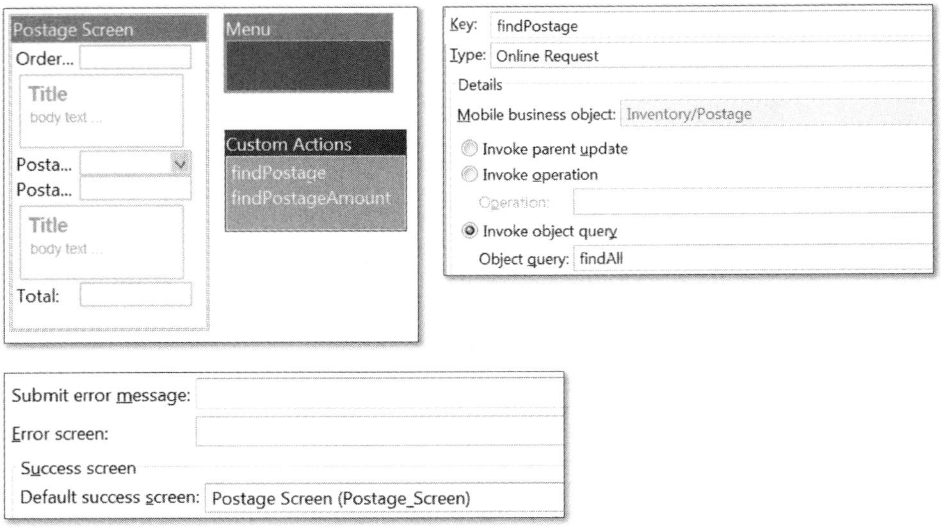

Figure 11.11 Configuring Custom Actions the Same Way as Menu Items

findPostageAmount

The second custom action uses the `findByPrimaryKey` object query, and the query attribute is mapped to the Choice control key. The properties of the Choice control and the custom action are shown in Figure 11.12.

Creating the JavaScript

Because custom actions do not have a visual component, you must add code to *Custom.js* to execute them. The names of the functions follow the same format as those created for menu items. The format is `menuItemCallback<screen key><custom action name>`. For example, the function for `findPostage` on the Postage_Screen screen is named `menuItemCallbackPostage_ScreenfindPostage`.

Implementing this scenario requires the following steps:

1. Invoke the `findPostage` function to retrieve the data for the Choice control when the screen is first shown.

2. Add an event handler to the CHOICE control, which calls the `findPostageAmount` function when the CHOICE control changes.
3. Calculate the total when a postage type is selected.

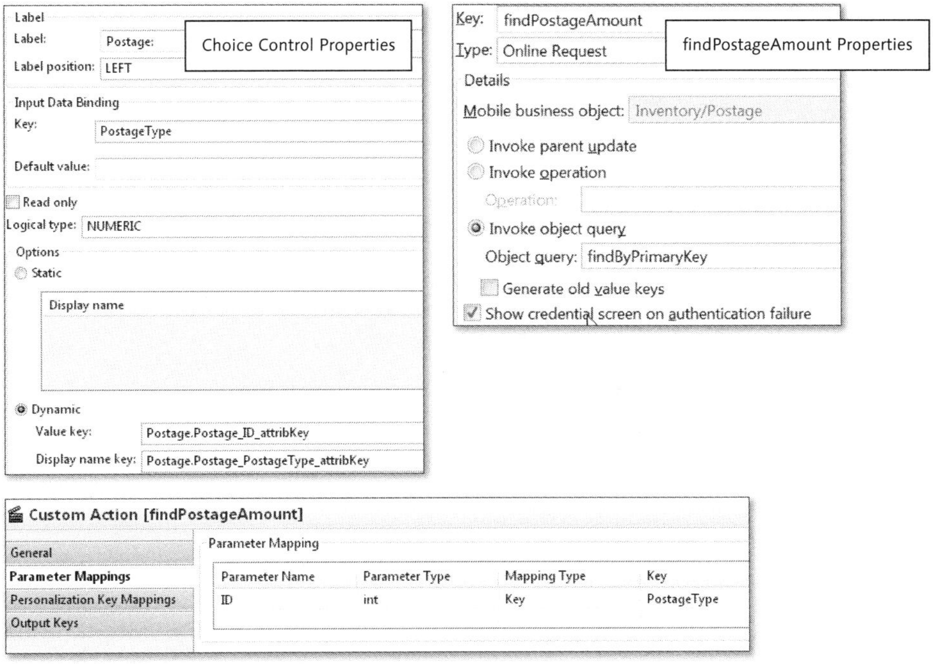

Figure 11.12 Configuration of the findPostageAmount Custom Action

Retrieving Postage Types with findPostage

The code shown in Listing 11.5 invokes the function created for the `findPostage` custom action. The code is added to `customAfterNavigateForward`, which is called after a navigation to a new screen occurs. The code executes when the app navigates to the POSTAGE_SCREEN.

```
function customAfterNavigateForward(screenKey, destScreenKey) {
  //If the app is navigating to this
  //screen from another screen
  if (destScreenKey === "Postage_Screen") {
   //Call the method created for the custom action
   menuItemCallbackPostage_ScreenfindPostage();
```

Listing 11.5 Invoking the findPostage Custom Action

Adding an Event Handler to the Choice Control

Next, a change event handler is added to the CHOICE control. The event handler calls the `findPostageAmount` function retrieving the postage rate. The code shown in Listing 11.6 is added immediately after the previous code in the `customAfter-NavigateForward` function.

```
    //Add a change event handler to the choice control
    $('#PostageType').change(function(){
     //The event handler calls the findPostageAmount
     //custom action to retrieve the postal rate
     menuItemCallbackPostage_
ScreenfindPostageAmount();
     });
    }
}
```

Listing 11.6 Adding an Event Handler to the Choice Control

Calculating the Total When a Postage Type is Selected

When the user changes the selected postage type, the change event handler invokes the `findPostageAmount` custom action. When the results from `findPostageAmount` return, they are processed in the `customAfterDataReceived` function by the code in Listing 11.7.

```
function customAfterDataReceived(incomingWorkflowMessage) {
if (incomingWorkflowMessage.getRequestAction()==="findPostageAmount"){
   //Retrieve the MessageValueCollection
     //from the incoming WorkFlowMessage
   var mvc = incomingWorkflowMessage.getValues();
   //Retrieve the postage rate
   var rateVal = mvc.getData
          ("Postage_rate_attribKey").getValue();
   //Retrieve the order amount
   var orderAmt =
          mvc.getData("Order_Amount").getValue();
   //Retrieve the total MessageValue
   var totObj = mvc.getData("Total")
   //Add the order amount and rate
   var total = parseFloat(orderAmt) +
          parseFloat(rateVal) ;
   //set the value of the total value in
     //the total MessageValue
```

```
    totObj.setValue(total);
    //Update the interface    updateUIFromMessageValueCollection(
            "Postage_Screen", mvc);
}
```

Listing 11.7 Calculating the Total

This code, shown in Listing 11.8, executes only if the incoming workflow message was created by the `findPostageAmount` object query. It then retrieves the current `MessageValueCollection` and extracts the postage rate and the order amount.

```
      //Retrieve the MessageValueCollection
      //from the incoming WorkFlowMessage
    var mvc = incomingWorkflowMessage.getValues();
    //Retrieve the postage rate
    var rateVal = mvc.getData
           ("Postage_rate_attribKey").getValue();
    //Retrieve the order amount
    var orderAmt =
            mvc.getData("Order_Amount").getValue();
```

Listing 11.8 Retrieving the Postage Rate and Order Amount

Next, a `MessageValue` for the total key is retrieved, and the value of this `MessageValue` object is set equal to the sum of the order amount and rate (see Listing 11.9).

```
      //Retrieve the total MessageValue
    var totObj = mvc.getData("Total")
    //Add the order amount and rate
    var total = parseFloat(orderAmt) +
            parseFloat(rateVal) ;
    //set the value of the total value in
      //the total MessageValue
      totObj.setValue(total);
```

Listing 11.9 Calculating the Total and Updating the MessageValue Object

Finally, the user interface is updated from the `MessageValueCollection`:

```
      //Update the interface    updateUIFromMessageValueCollection(
            "Postage_Screen", mvc);
```

The last step is to calculate the total when the screen first loads. If you do not do this, the only way the user could select this first option is to select another option and then reselect the default option. The change event handler fires only if the selected postage rate changes. The CHOICE control defaults to the first option, so you must retrieve the rate for the first option when the screen loads. The code shown in Listing 11.10, added to the customAfterDataReceived function, retrieves the rate for the first option.

```
function customAfterDataReceived
      (incomingWorkflowMessage) {
  if (incomingWorkflowMessage.getRequestAction() ===
      "findPostage") {
  //The setTimeout() function is  necessary to
  //ensure the updateUIFromMessageValueCollection()
  //has completed
  setTimeout(function() {
    //Retrieve the postage rate
   menuItemCallbackPostage_
        ScreenfindPostageAmount();
  },250);
 }
}
```

Listing 11.10 Retrieving the Postage Rates

The code runs only if the request action from the incoming workflow message is findPostage, which happens only when the POSTAGE screen first loads. The setTimeout function introduces a delay to ensure that the updateUIFromMessageValueCollection function has completed before calling menuItemCallbackPostage_ScreenfindPostageAmount. The callback function retrieves the postage rate associated with the selected option, which is the first option in this case.

The situation is different if the postage type and rate are both retrieved from the same MBO. In this case, a single object query can retrieve both, so only one custom action is required. The next section illustrates this situation.

11.2.2 One Custom Action

If the postage rates and postage options are stored in one MBO, they can be retrieved by one object query. The first thing to do is change the VALUE KEY property of the CHOICE control so that it is mapped to Postage_rate_attribKey, as shown in Figure 11.13. This allows the code to access the rate directly from the control.

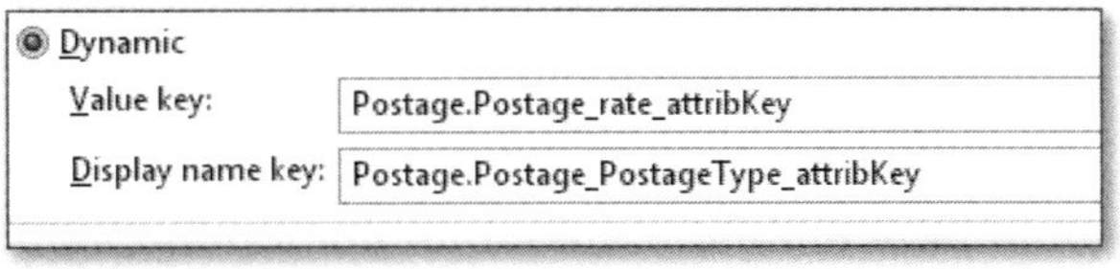

Figure 11.13 Mapping the Value Key Field to the Postage Rate Key

Next, change the code in the `customAfterNavigateForward` function, as shown in Listing 11.11.

```
if (destScreenKey === "Postage_Screen") {
  // Call the method created for the custom action
  menuItemCallbackPostage_ScreenfindPostage();
  $('#PostageType').change(
    function() {
        var rate = this.value;
        updatePostageScreen(rate);
  });
}
```

Listing 11.11 Updating the Change Event Handler

The change event handler now retrieves the postage rate directly from the Choice control, rather than calling a custom action.

```
var rate = this.value;
```

The `updatePostageScreen` function, shown in Listing 11.12, is added to the *Custom.js* file.

```
function updatePostageScreen(rate){
  //Retrieve the current MessageValueCollection
  var mvc = getCurrentMessageValueCollection();
  //Retrieve the rate  MessageValue
  var rateObj = mvc.getData("Postage_rate");
  //Update the value in the MessageValue
  rateObj.setValue(rate);
  //Retrieve order amount
  var orderAmt = mvc.getData("Order_Amount")
      .getValue();
  //Calculate the total
  var total = parseFloat(orderAmt) +
      parseFloat(rate);
  //Retrieve the total MessageValue
  var totalObj = mvc.getData("Total");
```

```
    //Update the value
    totalObj.setValue(total);
    //Update the interfae
    updateUIFromMessageValueCollection
        ("Postage_Screen",mvc);
}
```

Listing 11.12 Updating the Postage Screen

Another difference in this version is that the `Postage_rate` EDITBOX control cannot be mapped to `Postage_rate_attribKey` since the `MessageValueCollection` contains a list and not an individual item. The EDITBOX control is mapped to a new key called `Postage_rate`. The code shown in Listing 11.13 retrieves a reference to the `MessageValue` object of this key and sets the value equal to the selected rate.

```
 //Retrieve the rate  MessageValue
var rateObj = mvc.getData("Postage_rate");
//Update the value in the MessageValue
rateObj.setValue(rate);
```

Listing 11.13 Updating the Postage_rate Key

Similar code, shown in Listing 11.14, is used to set the value of the `Total` key mapped to the `Total` EDITBOX control.

```
 //Retrieve the total MessageValue

var totalObj = mvc.getData("Total");

//Update the value

totalObj.setValue(total);
```

Listing 11.14 Updating the Total Key

Updating the `MessageValue` objects does not directly update the EDITBOX controls on the interface. To accomplish this, the last line of code calls `updateUIFromMessageValueCollection`.

The last step is to update the code in `customAfterDataReceived` that initially updates the screen using the default when the screen first loads, as shown in Listing 11.15.

```
if (incomingWorkflowMessage.getRequestAction() === "findPostage") {
  setTimeout(function() {
    //Retrieve the currently selected rate
    //from the PostageType select element
```

```
    var rate = $("select#PostageType").val();
    //Update the screen
    updatePostageScreen(rate);
  },250);
```

Listing 11.15 Initializing the Screen When It Loads

This code retrieves the rate from the CHOICE control using jQuery and then calls the `updatePostageScreen` function.

11.3 Using the SUPStorage API

Very often, you may find it useful to be able to store data on the device rather than having to retrieve it from the server every time it is needed. For example, you may have some setup parameters, such an employee ID, that a user can enter to personalize an app. The SUPStorage API provides functions that allow you to cache key value pairs on the device. The cached data is encrypted and persisted between uses of the app. This means the data is not only accessible from multiple screens in the app, but also available each time the app is invoked, until the data is removed. The amount of data that can be stored is limited only by the memory available on the device. The functions used to access SUPStorage are shown in Table 11.5.

Method	Description
`SUPStorage(store)`	Creates a cache named "store." Store is optional.
`SharedStorage()`	Returns a reference to a cache shared among container applications.
`getSharedStorageKey()`	Returns the key to shared storage.
`isSharedStorageEnabled()`	Determines whether shared storage is enabled for the current app.
`SUPStorage.prototype.length()`	Returns the number of key value pairs in the cache.
`SUPStorage.prototype.key(index)`	Returns the key at index. Keys remain at the index unless modified.

Table 11.5 Functions Used to Access SUPStorage

Method	Description
SUPStorage.prototype.getItem(key)	Returns a copy of the value associated with the key. If the key does not exist, null is returned.
SUPStorage.prototype.setItem(key,value)	Stores the key value pair in the cache. If the key exists, the value is replaced with the new value. Both keys and values are stored as strings.
SUPStorage.prototype.clear()	Clears the cache.
SUPStorageException(code,message)	Creates an SUPStorageException object.

Table 11.5 Functions Used to Access SUPStorage (Cont.)

To illustrate the use of SUPStorage, the postage rate example from Section 11.2 will be extended in three ways:

- The selected postage type will be saved on the device, so when the user returns to the POSTAGE screen, the previously selected value will be the default value.
- The postage types will be saved on the device to avoid a round trip to the server.
- The app will be modified to use dynamic credentials.

11.3.1 Saving the Chosen Postage Type as a Default Value

First, modify the `customBeforeNavigateForward` as shown in Listing 11.16. This code saves the selected value from the CHOICE control when the user navigates away from the `Postage_Screen`.

```
function customBeforeNavigateForward(screenKey, destScreenKey) {

  if (screenKey === "Postage_Screen" ){
      //Initialize the data store
    var store = new SUPStorage();
      //Retrieve the selected postage type from
      //the postage type select element
    var PostageID = $("select#PostageType").val();
      //Save the postage type in the data store
```

```
        store.setItem("DefaultPostageID", PostageID);
    }
    return true;
}
```

Listing 11.16 Saving the Select Value from the Choice Control

The code first checks to see whether the screen being navigated from is the `Postage_Screen`:

```
if (screenKey === "Postage_Screen" ){
```

It then creates an `SUPStorage` object. The cache is not named, assuming only one will be used (Listing 11.17).

```
//Initialize the data store
var store = new SUPStorage();
```

Listing 11.17 Creating an SUPStorage Object

Next, the value of the CHOICE control is retrieved using jQuery, and the value is stored with the key `DefaultPostageID` as shown in Listing 11.18.

```
//Retrieve the selected postage type from
//the postage type select element
var PostageID = $("select#PostageType").val();
//Save the postage type in the data store
store.setItem("DefaultPostageID", PostageID);
```

Listing 11.18 Save the Selected Value

Finally, the `customAfterDataReceived` function is modified as shown in Listing 11.19 to retrieve the saved value.

```
function customAfterDataReceived(incomingWorkflowMessage) {
  if (incomingWorkflowMessage.getRequestAction()
        === "findPostage") {
    setTimeout(function() {
    //Create the data store
    var store = new SUPStorage();
    //Retrieve the store postage id using the
    //key used to store it
    var rate = store.getItem("DefaultPostageID");
    //Check to see if the value exists
    if (rate) {
```

```
        //Update the postage type select element
        $("select#PostageType").val(store
           .getItem("DefaultPostageID"));
      } else {
         //If no stored value is saved, retrieve
         //the selected rate from the control
       var rate = $("select#PostageType").val();
      }
      //Update the interface
      updatePostageScreen(rate);
       },250);
    }
}
```

Listing 11.19 Setting the Choice Control

The code in Listing 11.20 initializes the cache and retrieves the value associated with the `DefaultPostageID` key.

```
//Create the data store
var store = new SUPStorage();
//Retrieve the store postage id using the
//key used to store it
var rate = store.getItem("DefaultPostageID");
```

Listing 11.20 Initializing the Cache and Retrieve the Key

The first time the app runs, there is no saved value, so the code checks to be sure the value exists.

```
//Check to see if the value exists
if (rate) {
```

Listing 11.21 Checking that the Rate Has Been Saved Previously

If the saved rate exists, it is used to set the selected option of the CHOICE control using jQuery (Listing 11.22).

```
//Update the postage type select element
$("select#PostageType").val(store
      .getItem("DefaultPostageID"));
```

Listing 11.22 Setting the Choice Control to the Saved Rate

If the saved rate does not exist, the rate is retrieved from the CHOICE control (Listing 11.23).

```
//If no stored value is saved, retrieve
//the selected rate from the control
var rate = $("select#PostageType").val();
```

Listing 11.23 Retrieving the Saved Rate from the Choice Control

Finally, the screen is updated:

```
updatePostageScreen(rate);
```

The first time the `Postage_Screen` is opened, the CHOICE control defaults to the first option. When the user navigates away from the screen, the selected postage type is saved, and the next time the POSTAGE screen appears, the CHOICE control defaults to the saved option.

11.3.2 Saving the Postage Types on the Device

In this example, the postage types and rates are saved on the device rather than being retrieved each time the POSTAGE screen is loaded. This example illustrates how to process a local cache with multiple key value pairs.

First, you need to make some changes to the interface. Since you will be loading the options into the CUSTOM control, you must change the CHOICE control to a static type. You have to add a dummy item or you receive an error. The configuration is shown in Figure 11.14.

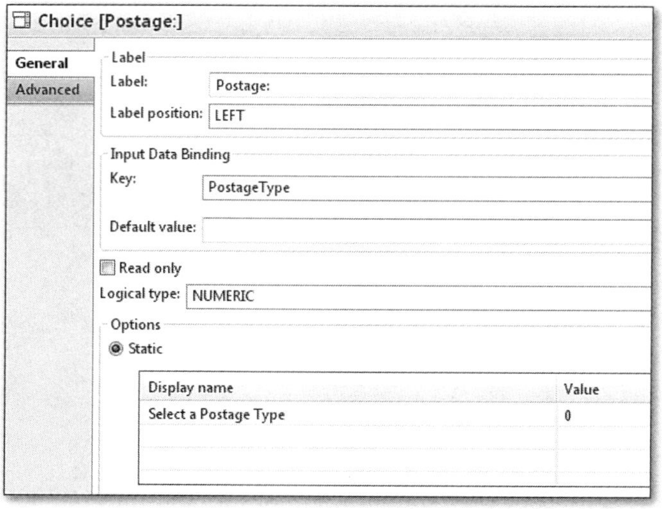

Figure 11.14 Changing the Custom Control Options to Static and Adding a Dummy Option

Next, add a menu item that the user can use to refresh the postage rates. This is necessary since, if the rates on the backend change, the user must have some way to refresh the local cache. The configuration is shown in Figure 11.15. The type of the menu is CUSTOM.

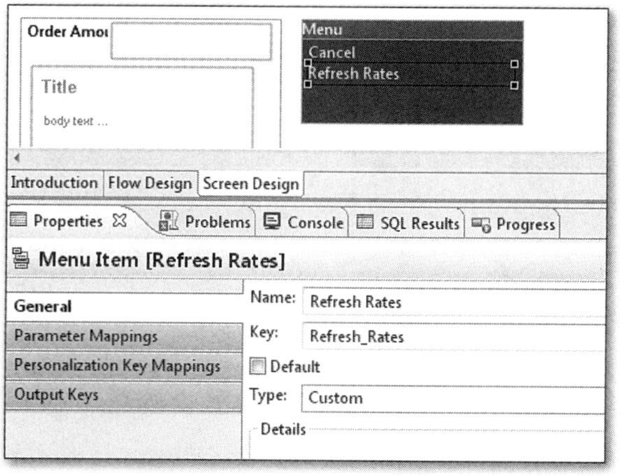

Figure 11.15 Creating a Menu Item of Type Custom

Add the code in Listing 11.24 to the `customBeforeMenuItemClick` in the *Custom.js* file. This code checks for the `Refresh_Rates` menu item and then calls the custom action to retrieve the postage rates.

```
function customBeforeMenuItemClick(screen, menuItem){
     if(menuItem == "Refresh_Rates"){
     menuItemCallbackPostage_ScreenfindPostage();
   }

   return true;
}
```

Listing 11.24 Retrieving the Postage Rates

Create a global variable in the *Custom.js* file to reference the `SUPStorage` object. Add the following line to the top of the *Custom.js* file.

```
var store = null;
```

Add the function shown in Listing 11.25 to the *Custom.js file*. This function implements the `getLength` method of the SUPStorage prototype and then instantiates the cache and assigns it to the global variable. The cache is named using the cache argument passed to the function.

```
function initSUPStorage(cache) {
  //implement the getLength prototype method
  SUPStorage.prototype.getLength = function() {
    return store.length();
  }
  //Instantiate the cache
  store = new SUPStorage(cache);
}
```

Listing 11.25 Initialize the SUPStorage Cache

Next, update the `customAfterNavigateForward` function as shown in Listing 11.26. The function initializes the `PostalRates` cache and then checks for existing values using the `getLength` method. If there are existing values, the fillPostageList function is called to configure the CHOICE control options. If there are no existing values, the custom action's callback is called to retrieve the values from the server. Finally, the change event handler is added to the CHOICE control.

```
function customAfterNavigateForward(screenKey, destScreenKey) {
  //Check for the postage screen
  if (destScreenKey === "Postage_Screen") {
    //Initialize the cache
    initSUPStorage("PostalRates");
    //Determine if the cache has values
    if (store.getLength()>0) {
```

```
        //If it does fill the choice control
        fillPostageList();
      } else {
        //else call the custom action
menuItemCallbackPostage_ScreenfindPostage();
      }
      //Assign the change handler to the list
      $('#PostageType').change(
       function() {
        var rate = this.value;
updatePostageScreen(rate);
     });
   }
}
```

Listing 11.26 Updating the Choice Control Items

The `fillPostageList` function is shown in Listing 11.27. This function uses jQuery to retrieve a reference to the <select> HTML element the SDK generates for the CHOICE control. It then empties the list before looping through the items in the local cache, adding an option element for each one to the list.

```
function fillPostageList(){
  var type,rate,option;
  //Retrieve a reference to the Choice control
  //select HTML element
  var list = $("select#PostageType");
  //Delete existing options
  list.empty();
  //For each item in the cache
  for (var i = 0; i < store.getLength(); i++) {
    //Create an option element
    option = document.createElement("option");
    //Retrieve the postage type and rate
    type = store.key(i);
    rate = store.getItem(store.key(i));
    //Add the option to the control
    list.prepend('<option value="' + rate + '">'
+ type + '</option>');
  }
  //Refresh the list
  list.selectmenu('refresh');
}
```

Listing 11.27 Adding Choice Control Options

The customAfterDataReceived function is modified as shown in Listing 11.28, so the retrieved postal rates are added to the local cache before calling fillPostage-List. The difference between this version and the last version is that the postage rates and types are added to the local cache before the fillPostageList function is called to add the options to the CHOICE control.

```
function customAfterDataReceived
(incomingWorkflowMessage) {

  //Check for the custom action
  if (incomingWorkflowMessage.getRequestAction()
 === "findPostage") {
    setTimeout(function() {
      //Retrieve the data from the WorkflowMessage
      var mvc = getCurrentMessageValueCollection();
      var rates = mvc.getData("Postage").getValue();
      var rate,type;

      for(var i = 0; i<rates.length; i++){
        //Retrieve the rate and type
        rate = rates[i].getData(
"Postage_rate_attribKey").getValue();
     type = rates[i].getData(
"Postage_PostageType_attribKey")
.getValue();
       //Add the key value pair to the cache
     store.setItem(type, rate);
      }
      //Add the options to the Choice control
      fillPostageList();
      //Calculate the initial total
      var rate = $("select#PostageType").val();
      updatePostageScreen(rate);

    },250);
  }
}
```

Listing 11.28 Saving the Postage Rates on the Device

The last bit of code is the updatePostageScreen function shown in Listing 11.29. If the CHOICE control is filled from the local cache, it is possible that a workflow message was not processed, so the Postage_rate and TotalMessageValue objects do not exist in the current MessageValueCollecticn. One option is to call the updateMessageValueCollectionFromUI function, which creates the objects and then proceeds as in the example described in Section 11.2.2. Another possibility is to access the HTML elements directly, as illustrated in Listing 11.29. The code retrieves a reference to the Postage_Screen Form, which is a <form> element that contains the controls for the postage screen. Next, the value of the rate field is updated, the order amount is retrieved, and the total field is updated with the calculated value. Finally, updateMessageValueCollectionFromUI is called to update the current MessageValueCollection. This is necessary if the rate or total values will be passed on to subsequent screens.

```
function updatePostageScreen(rate){
  //Retrieve a reference to the Postage screen form
  var form = document.forms["Postage_ScreenForm"];
  if(form){
    //set the postage rate control to the
    //selected rate
    form.Postage_rate.value = rate;
    //Retrieve the order amount
    orderAmt = form.Order_Amount.value;
    //Calculate the total
    var total = parseFloat(orderAmt) +
 parseFloat(rate);
    //Update the total field
    form.Total.value = total;
  }
  //Update the current MessageValueCollection
updateMessageValueCollectionFromUI(
getCurrentMessageValueCollection(),
getCurrentScreen());
}
```

Listing 11.29 Calculating the Total and Updating the Interface

11.3.3 Using Dynamic Credentials

Credentials in hybrid apps were described in Section 8.2.1 of Chapter 8. This example illustrates the use of dynamic credentials. Although no code is required, dynamic credentials are stored in the local cache. This means that multiple hybrid apps can share the cached credentials if they specify the same credentials cache key.

The first step is to create a new screen on which the user can enter a username and password. Add a CREDENTIAL REQUEST starting point and connect it to the new screen. The configuration is shown in Figure 11.16.

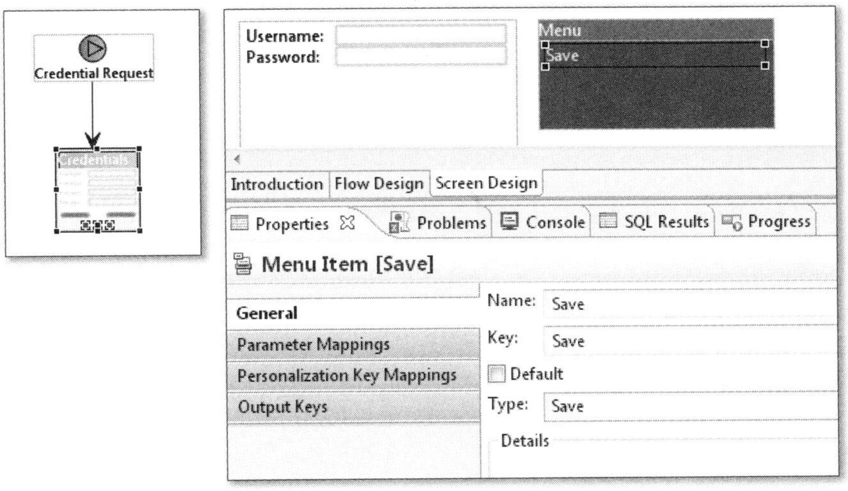

Figure 11.16 Adding a Credential Request Starting Point and a Credential Screen with Two EditBox Controls

Add EDITBOX controls to the new screen—one for the username and one for the password—and assign keys. On the GENERAL PROPERTIES tab, select PASSWORD for the password EDITBOX control and select CREDENTIAL CACHE PASSWORD on the ADVANCED PROPERTIES tab. For the username EDITBOX control, select CREDENTIAL CACHE USERNAME on the ADVANCED PROPERTIES tab. Add a menu item and set its type to SAVE. Figure 11.17 shows the configuration for the password EDITBOX control. Finally, deselect USE STATIC CREDENTIALS on the AUTHENTICATION PROPERTIES tab of the flow design.

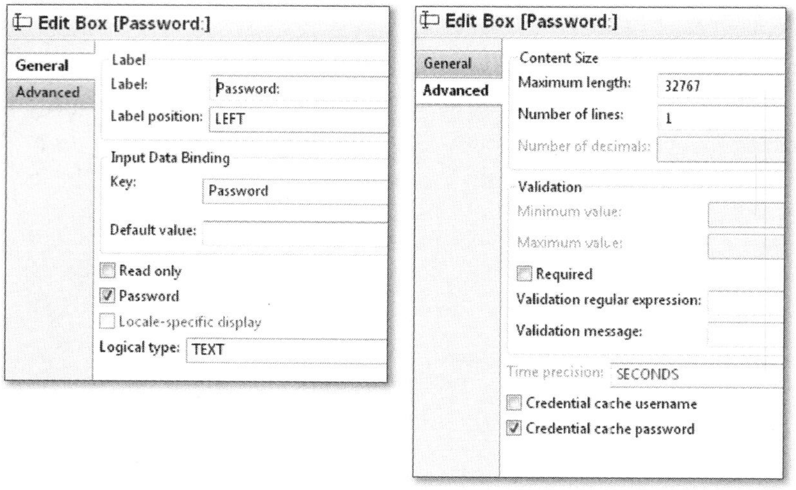

Figure 11.17 Configuration of a Password EditBox Control

When the user runs the app, the credentials screen appears if the cached credentials do not exist or if authentication with the saved credentials fails (see Figure 11.18).

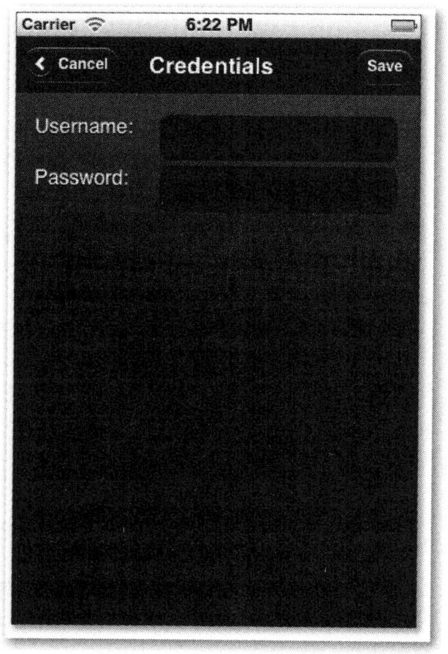

Figure 11.18 The Credentials Screen

11.4 Summary

This chapter and the previous chapter described various techniques you can use to extend the capabilities of hybrid apps. These techniques only scratch the surface of what is possible. The next three chapters introduce native apps, which provide ultimate control over the interface and the device.

PART IV
Developing Native Apps

Native apps using the Object API take advantage of mobile devices' capabilities and provide full offline operation.

12 Introduction to Developing Object API Native Apps

There are two types of native apps in SAP Mobile Platform: Object API apps and OData SDK apps. Figure 12.1 shows how these apps are supported by SAP Mobile Platform. The boxes with dashed lines indicate the portion of the development that is performed using the Unwired Workspace. For both types of native apps, data access elements are developed in the Unwired Workspace. The created artifacts are deployed to the Unwired Server, and then the Unwired Workspace is used to generate APIs in a platform's native language. These APIs implement the Application Specialization and Core Services layers. The generated APIs can be imported into whichever IDE is used to develop the app.

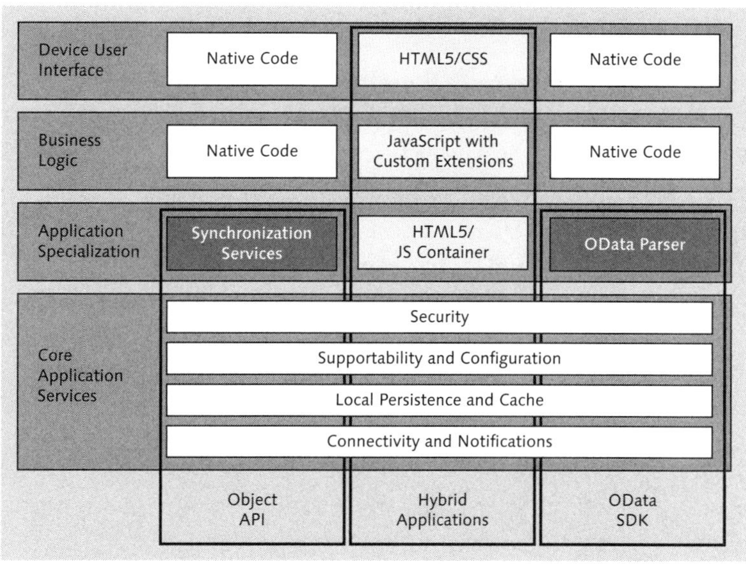

Figure 12.1 Structure of Apps Supported by SAP Mobile Platform

The core app services provide a common set of services across all supported platforms, facilitating the development of apps for multiple platforms. The application specialization layer in Figure 12.1 differentiates between two types of native apps.

Object API apps, also called *synchronization apps*, are developed using the native language of a device: Java for Android and BlackBerry, Objective-C for iOS, and C# for Windows Mobile. Like hybrid apps, apps based on the Object API rely on MBOs or the Data Orchestration Engine to access data. One major difference between hybrid and Object API apps is data synchronization and the ability to store data on the device. This allows Object API apps to be fully functional even without connectivity to the Unwired Server.

Object API apps consist of both server-side and client-side components:

- Server-side components consist of MBO or Data Orchestration Engine (DOE) packages. These components control the interaction between the enterprise system and the Cache Database (CDB) on the Unwired Server.
- Client-side components control the interaction between the CDB and the device data store and provide offline data capabilities.

With MBOs, the app accesses data cached on the Unwired Server, and the enterprise system is updated from the Unwired Server. With the DOE, data is cached on the DOE, and the Unwired Server forwards messages between the device and the DOE. Data on the enterprise system is updated from the DOE.

OData SDK apps rely on the Open Data (OData) protocol. OData was originally released by Microsoft under the Open Specification Promise. The purpose of OData is to provide a standard, cross-platform way to access and update data. It relies on web technologies such as RESTful services, Uniform Resource Identifiers (URI) for resource location, HTTP verbs (`GET`, `PUT`, `POST`, and `DELETE`) for operations and JavaScript Object Notation (JSON) for data representation. Development of OData is now guided by the OData Technical Committee in OASIS and is a widely used protocol.

OData SDK apps use the Unwired Server's Online Data Proxy runtime option, which allows RESTful web services to connect through the Unwired Server to service documents published on SAP NetWeaver Gateway. Unlike with Object API apps, data is not cached on the Unwired Server, so OData SDK apps rely on a request/response model without synchronization.

The focus of our book is mobile app development with SUP. Because OData apps do not cache data on the Unwired Server, their development process is quite different and beyond the scope of this book. As such, this chapter introduces you to the development process for Object API apps. We explain the synchronization process, walk you through the various functionalities an Object API app must perform, and provide a description of the APIs included in the Object API. The subsequent two chapters go into the details surrounding the actual development for the Android and iOS platforms.

12.1 Synchronization

Key features of Object API apps are synchronization and transactions that support reliable updates and offline capabilities. The goal of synchronization is to maintain data consistency among the client, CDB, and enterprise system. When changes occur on one tier, synchronization mechanisms ensure that the changes are replicated on the other tiers. For example, a sales representative can create sales orders on an Object API app while offline at a remote customer location. Synchronization ensures that the data is replicated to the other data tiers when the app regains connectivity. Synchronization-based apps rely on replication-based synchronization (RBS), which uses bulk transfers and transactions to ensure efficient and reliable data transfers. Figure 12.2 shows the components involved in synchronization.

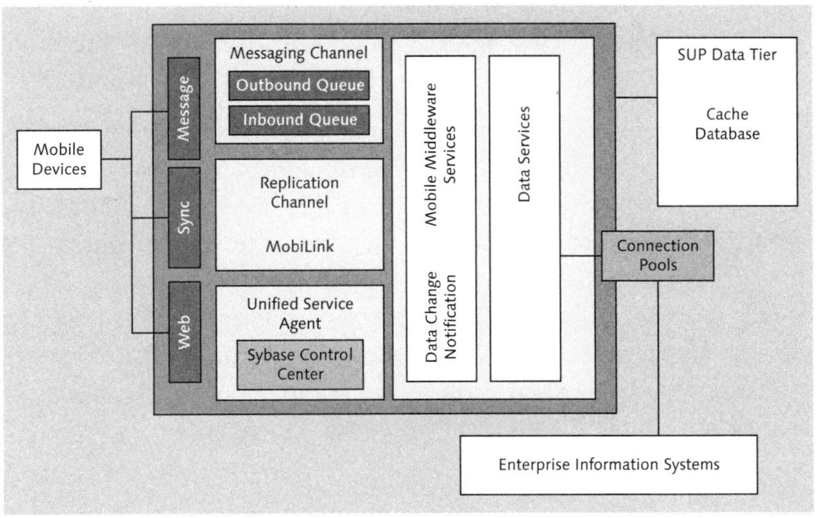

Figure 12.2 Unwired Server Components Involved in Synchronization

Mobile channel interfaces provide the channels through which data is transferred to and from mobile devices. Channels can be synchronous or asynchronous and can vary in service quality (proprietary or standards based). There are two main channel interface types:

- **Messaging channel**
  The messaging channel is used for data notifications. There are two types of messages sent via the messaging channel:
  - Data change notifications (for example, BlackBerry Enterprise Service and Apple Push Notification Service) can be sent by the Unwired Server when it detects data changes in the enterprise system. By default, this feature is disabled, but it can be enabled for each synchronization group. You can define an onSynchronize callback in the app to respond to change notifications.
  - When changes are sent from the device to the Unwired Server, the messaging channel is used to send a `replayFinished` message to the device. The device must initiate another synchronization to retrieve the results.
- **Replication channel**
  The replication—or synchronization—channel is used to synchronize data between the device and the Unwired Server.

Mobile middleware services (MMS) manage communications between the mobile channel interfaces and device requests.

Data services provide the channel used to access data in the enterprise system. Together with MMS, data services manage the Cache Database as it is synchronized with the device and enterprise system.

Replication-based synchronization consists of an upload and a download phase. During the upload phase, data change replay records on the device are transferred and replayed on the CDB by the replay logic within the MMS. The changes are replayed against the data service interfaces to the enterprise system. During the download phase, changes in the cached data on the Unwired Server are transferred to the device. Each phase is conducted within a transaction to ensure data integrity. Changes to data in the enterprise system are replicated to the device in accordance with policies created in the data package. These changes are replicated on the device the next time the client app invokes synchronization.

12.2 Object API App Functionalities

Regardless of the platform, the basic development flow for an Object API app is the same, requiring the following steps. The data package is developed in the Unwired Workspace and deployed to the Unwired Server. Next, the APIs in the native language are generated within the Unwired Workspace. Finally, these APIs are then imported into the development tool used to develop the native app. The APIs include all the functionality required to develop an Object API app, including interacting with the Unwired Server and manipulating data. The common features of Object API apps are described in this section, and sample code for the Android and iOS platforms is provided.

12.2.1 Registering an App on the Unwired Server

The first time an app connects to the Unwired Server, it registers itself creating an app connection. After the first time, the app can connect to the existing app connection. The processes involved with interacting with the Unwired Server are described next.

Setting Up App Properties

The `Application` object contains authentication credentials and connection information required to connect to the Unwired Server. It is also registered with the database class providing access to local database objects. The code shown in Listing 12.1 (for Android) and Listing 12.2 (for iOS) instantiates the `Application` object, associates it with the database, and then configures the `ConnectionProperties`, which contain the information used to connect to the Unwired Server.

```
//Create an instance of the Application object
Application app = App.getInstance();
//The identifier of the app deployed to the Unwired
//Server
app.setAppIdentifier("CustomersMBO");
//Set the Android context
app.setAppContext(getAppContext());
//Retrieve the ConnectionProperties
ConnectionProperties connProps =
   app.getConnectionProperties();
// Set connection properties
ConnectionProperties cp =
```

```
   app.getConnectionProperties();
cp.setServerName("smp.demo.com");
cp.setPortNumber(5001);
cp.setLoginCredentials(new
   LoginCredentials("username",""));
cp.setActivationCode("123");
```

Listing 12.1 Android

```
//Create an instance of the Application object
SUPApplication *app = [SUPApplication        getInstance];
//The identifier of the application deployed to the Unwired
//Server
[app setApplicationIdentifier:@"CustomersMBO"];
//Retrieve the ConnectionProperties
SUPConnectionProperties *properties = [app
        connectionProperties];
//Set the connection properties
[properties setServerName:@"smp.demo.com"];
[properties setPortNumber:5001];
[properties setNetworkProtocol:@"http"];

SUPLoginCredentials *loginCred =
         [SUPLoginCredentials getInstance];
[loginCred setUsername:@"username"];
[loginCred setPassword:@""];
[properties setLoginCredentials:loginCred];
[properties setActivationCode:@"123"];
```

Listing 12.2 iOS

Registering an App

An app registers with the Unwired Server using the `registerApplication` method of the `Application` object. After the first time, the app can use the `startConnection` method instead. Before registering, the `Application` object is associated with the app ID assigned to the MBO package deployed to the Unwired Server. Examples for Android and iOS are shown in Listing 12.3 and Listing 12.4.

```
CustomersMBODB.setApplication(app);
if (app.getRegistrationStatus() !=
  RegistrationStatus.REGISTERED) {
  app.registerApp();
} else {
```

```
  app.startConnection();
}
```

Listing 12.3 Android

```
 [CustomersMBOCustomersMBODB setApplication:app];
if([SUPApplication registrationStatus] ==
      SUPRegistrationStatus_REGISTERED){
      [app startConnection:0];
} else {
      [app registerApplication:0];
}
```

Listing 12.4 iOS

Setting Up the Connection Profile

The connection profile contains information about the local database. You must configure it before accessing the database. Information that can be included in the profile includes the database location, page size, and encryption key. Examples for Android and iOS are shown in Listing 12.5 and Listing 12.6.

```
// Initialize the device database connection profile (if
// needed)
ConnectionProfile cp = CustomersDB.getConnectionProfile();
// encrypt the database
cp.setEncryptionKey("an encryption key");
```

Listing 12.5 Android

```
SUPConnectionProfile* cp = [CustomerMBOCustomerMBODB
getConnectionProfile];
[cp setEncryptionKey:@"an encryption key"];
```

Listing 12.6 iOS

Setting Up the Synchronization Profile

The synchronization profile contains information about the Unwired Server's data synchronization channel. This information can include the server host, port, domain name, certificate, and public key. The settings are automatically pushed over the Unwired Server's messaging channel to the app and inherited from the app connection template configured on the Unwired Server used to register the app. Alternatively, the app can specify this information. Examples for Android and iOS are shown in Listing 12.7 and Listing 12.8.

```
ConnectionProfile cp = CustomersMBODB.getSynchronizationProfile();
cp.setServerName("smp.demo.com");
cp.setPortNumber(2480);
cp.setNetworkProtocol("http");
cp.setDomainName("default");
cp.save();
```

Listing 12.7 Android

```
SUPConnectionProfile *connProfile =
          [CustomersMBOCustomersMBODB
          getSynchronizationProfile];
[connProfile setNetworkProtocol:@"http"];
[connProfile setPortNumber:2480];
[connProfile setServerName:@"smp.demo.com"];
[connProfile setDomainName:@"default"];
```

Listing 12.8 iOS

Creating and Deleting the Local Database

The local database is created automatically by the Object API when needed, if it does not exist. The Object API also includes methods to create, delete, and configure the database. Examples for Android and iOS are shown in Listing 12.9 and Listing 12.10.

```
if (!SUP101DB.databaseExists()){
        SUP101DB.createDatabase();
    }
```

Listing 12.9 Android

```
if (![SUP101SUP101DB databaseExists]) [SUP101SUP101DB createDatabase];
```

Listing 12.10 iOS

Setting Up Callbacks

You can register callback handlers that respond to app, database, and synchronization events:

- The `application` callback handler monitors changes in app settings, messaging connection status, and app registration status.
- The `persistence` callback handler monitors changes related to the database as a whole or for individual MBOs.
- The `synchronization status listener` monitors the synchronization process. You can use this to debug and tune the synchronization process and provide feedback on the user interface.

Synchronizing

You can synchronize a specified subset or all synchronization groups at once. The methods in the generated database allow you to enable SSL synchronization, initiate synchronization, and enable change notifications for synchronization groups. You can also specify personalization and synchronization parameters. Examples for Android and iOS are shown in Listing 12.11 and Listing 12.12.

```
CustomersMBODB.synchronize();
```

Listing 12.11 Android

```
[CustomersMBOCustomersMBODB synchronize];
```

Listing 12.12 iOS

12.2.2 Accessing MBO Data

An Object API app must also be able to access the local MBO data using object queries or dynamic queries.

Object Queries

The generated code includes methods for each object query defined in the MBOs. For example, assume an MBO includes an object query called `findByCity` that filters the CUSTOMER MBO on the state field. The `Customer MBO` class includes the method `public static Customer findByCity(String city)` in the Android Object API and `+ (SUPObjectList*) findByCity:(NSString*)city` in the iOS Object API. Examples for Android and iOS are shown in Listing 12.13 and Listing 12.14.

```
GenericList<Customer> customers = Customer.findByCity("Milwaukee");
```

Listing 12.13 Android

```
SUPObjectList *customers =v[Customer findByCity:@"Milwaukee"];
```

Listing 12.14 iOS

Dynamic Queries

Queries can also be created dynamically using the `Query` object. Dynamic queries can include complex selection criteria and sort criteria, as well as MBO joins. Examples for Android and iOS are shown in Listing 12.15 and Listing 12.16.

```
//Create the query object
Query query = new Query();
//Create a sort criteria object
SortCriteria sort = new SortCriteria();
//Configure the sort criteria
sort.add("company_name", SortOrder.ASCENDING);
//Add the sort criteria to the query object
query.setSortCriteria(sort);
//Create an attribute test object
AttributeTest aTest = new AttributeTest();
//Set the test attribute
aTest.setAttribute("state");
//Set the test value
aTest.setTestValue("NY");
//Set the test condition
aTest.setTestType(AttributeTest.EQUAL);
//Add the attribute test to the query object
query.setTestCriteria(aTest);
//Execute the query
GenericList<Customer> customerIds =
   Customer.findWithQuery(query);
```

Listing 12.15 Android

```
//Create the query object
SUPQuery *query = [SUPQuery getInstance];
//Configure the SQL
[query select:@"*"];
[query from:@"Customer":@"c"];
//Create a sort criteria object
SUPSortCriteria *sort = [SUPSortCriteria
         getInstance];
//Set the sort attribute
[sort add:[SUPAttributeSort ascending:
```

```
            @"company_name"]];
//Add the sort critera to the query object
Query.sortCriteria = sort;
//Add an attribute test
query.testCriteria = [SUPAttributeTest match:@"state" :@"NY"];
//Execute the query
SUPObjectList* customers = [CustomersMBOCustomer
            findWithQuery:query];
The Query object can be used to create dynamic queries.
```

Listing 12.16 iOS

12.2.3 Manipulating Data

Object API apps modify MBO data using the operation methods that are created in the MBO. When an operation such as `update` is invoked, the updated data enters a pending state. The `submitPending` method creates operation replay records that are run on the Unwired Server during the next synchronization. Examples for Android and iOS are shown in Listing 12.17 and Listing 12.18.

```
//Create a customer instance
Customer customer = new Customer();
//Set the required fields for the customer
...
//Call the create operation
customer.create();
//Create the replay records
customer.submitPending();
//Sync the changes to the Unwired Server
CustomerMBODB.synchronize();

//Retrieve an existing customer instance
Customer customer = Customer.findByPrimaryKey(101);
//Update an attribute
customer.setCity("Dublin");
//Call the update operation
customer.update();
//Create the replay records
customer.submitPending();
//Sync the changes to the Unwired Server
CustomerMBODB.synchronize();

//Retrieve an existing customer instance
```

```
Customer customer = Customer.findByPrimaryKey(101);
//Call the delete operation
customer.delete();
//Create the replay records
customer.submitPending();
//Sync the changes with the Unwired Server
CustomerMBODB.synchronize();
```

Listing 12.17 Android

```
//Create a new customer instance
SUP101Customer *customer = [[SUP101Customer alloc] init];
//Set the required fields for the customer
...
//Call the create opoeration
[newcustomer create];
//Create the replay records
[newcustomer submitPending];
//Sync the changes to the Unwired Server
[CustomerMBOCustomerMBO synchronize];

//Find an existing customer instance
Customer *customer = [Customer findByPrimaryKey:101];
//Update an attribute
customer.city = @"Dublin";
//Call the update operation
[customer update];
//Create the replay records
[newcustomer submitPending];
//Sync the changes to the Unwired Server
[CustomerMBOCustomerMBO synchronize];

//Find an existing customer instance

Customer *customer = [Customer findByPrimaryKey:101];
//Call the delete operation
[customer delete];
//Create the replay records
[newcustomer submitPending];
//Sync the changes to the Unwired Server
[CustomerMBOCustomerMBO synchronize];
The generated code includes methods for the CUD operations.
```

Listing 12.18 iOS

12.3 APIs Included in the Object API

This section provides partial descriptions of some of the APIs included in the Object API. For a complete list, see the developer guide Object API documentation, which can be found in the folders shown in Table 12.1.

Platform	Folder
Android	*<UnwiredPlatform_InstallDir>\MobileSDK213\ObjectAPI\apidoc\android*
BlackBerry	*<UnwiredPlatform_InstallDir>\MobileSDK213\ObjectAPI\apidoc\rim*
iOS	*<UnwiredPlatform_InstallDir>\MobileSDK213\ObjectAPI\iOS*
Windows Mobile	*<UnwiredPlatform_InstallDir>\MobileSDK213\ObjectAPI\apidoc\cs*

Table 12.1 Documentation for the Object APIs Installed with the Unwired Workspace

> **Note**
>
> The Object API is extensive, so a complete description is not possible here. In addition, many of the details, such as method signatures, property names, etc., vary among the different platforms. This section can provide only general descriptions. For more detailed descriptions, use the documentation at *http://infocenter.sybase.com* or the documentation included with the Unwired Workspace.

12.3.1 Application API

The Application API manages mobile app registrations, connections, and context. It includes four classes: `Application`, `ConnectionProperties`, `ApplicationSettings`, and `ConnectionPropertyType`.

Application

The `Application` class is used to manage the `Application` object and to register and connect to the Unwired Server. Table 12.2 shows some of the properties and methods included in the `Application` class.

Property or Method	Description
getInstance	Returns an instance of the Application object.
setApplicationIdentifier	Sets the identifier of the Application object. This must match the name of the MBO package deployed to the Unwired Server.
registrationStatus (iOS) getRegistrationStatus (Android)	Determines whether the app has been registered on the Unwired Server.
registerApplication	Registers the app on the Unwired Server. You must set the app identifier and connection properties before calling this method. It should be called only if the app is not already registered.
setApplicationCallback	Sets a callback to monitor changes in registration and connection status as well as app settings. This method is optional.
setApplicationContext (Android)	Sets the Android app context. This is required before calling registerApplication or startConnection.
startConnection	Connects to the Unwired Server if the app is already registered.
ConnectionStatus property (iOS) getConnectionStatus (Android)	Returns the status of the connection.
getConnectionProperties	Returns the connection properties. Connection properties must be set before registering or connecting to the Unwired Server.
ApplicationSettings property (iOS) getApplicationSettings (Android)	Returns the app settings returned by the Unwired Server.
stopConnection	Closes the connection to the Unwired Server.
unregisterApplication	Deletes the app's registration on the Unwired Server.

Table 12.2 Properties and Methods in the Application API

ConnectionProperties

The `ConnectionProperties` class is used to configure properties required to register and connect to the Unwired Server. It includes the methods and properperites shown in Table 12.3. The properties are part of the iOS API, while the `getter` and `setter` methods are part of the Android API.

Property or Method	Description
`activationCode` property (iOS) `getActivationCode` (Android) `setActivationCode` (Android)	Retrieves or sets the activation code used to register the app if one is required.
`networkProtocol` property (iOS) `getNetworkProtocol` (Android) `setNetworkProtocol` (Android)	Retrieves or sets the network protocol. Defaults to HTTP.
`loginCertificate` property (iOS) `getLoginCertificate` (Android) `setLoginCertificate` (Android)	Retrieves or sets the login certificate used to authenticate if digital certificates are used.
`loginCredentials` property (iOS) `getLoginCredentials` (Android) `setLoginCredentials` (Android)	Retrieves or sets the username and password used to authenticate with the Unwired Server.
`portNumber` property (iOS) `getPortNumber` (Android) `setPortNumber` (Android)	Retrieves or sets the port number for the channel used to register the app or connect to the server.
`serverName` property (iOS) `getServerName` (Android) `setServerName` (Android)	Retrieves or sets the server name.
`securityConfigurationProfile` property (iOS) `getSecurityConfigurationProfile` (Android) `setSecurityConfigurationProfile` (Android)	Retrieves or sets the security configuration. If not specified, the server selects the security profile by matching the app identifier with the app connection template defined on the server.

Table 12.3 ConnectionProperties Methods and Properties

Property or Method	Description
urlSuffix property (iOS) getUrlSuffix (Android) setUrlSuffix (Android)	Retrieves or sets the URL suffix for the server URL. This is only used if connecting through a Relay Server or proxy server. If it is not set, the app attempts to determine it from the default value on the server.
farmId property (iOS) getFarmId (Android) setFarmId (Android)	Retrieves or sets the farm ID. This is used only if connecting through a Relay Server or proxy server.
httpHeaders property (iOS) getHttpHeaders (Android) setHttpHeaders (Android)	Retrieves or sets any custom HTTP headers if connecting through a Relay Server or proxy server.
httpCookies property (iOS) getHttpCookies (Android) setHttpCookies (Android)	Retrieves or sets any custom cookies if connecting through a Relay Server or proxy server.
httpCredentials property (iOS) getHttpCredentials (Android) setHttpCredentials (Android)	Retrieves or sets credentials for basic HTTP authentication if connecting through a Relay Server or proxy server.

Table 12.3 ConnectionProperties Methods and Properties (Cont.)

ApplicationSettings

Table 12.4 contains properties and methods of the ApplicationSettings class used to configure the Application object.

Property or Method	Description
isApplicationSettingsAvailable	Determines whether app settings are available from the Unwired Server.
getStringProperty	Returns a string value of the specified ApplicationSettings property.
getIntegerProperty	Returns an integer value of the specified ApplicationsSettings property.

Table 12.4 ApplicationSetting Properties and Methods

Property or Method	Description
getBooleanProperty	Returns a Boolean value of the specified ApplicationsSettings property.
Custom 1 - 4 properties (iOS) getCustom1 - 4 (Android)	Creates up to four custom properties for the app to use.
connectionId property (iOS) getConnectionId (Android)	Returns the connection ID.
domainName property (iOS) getDomainName (Android)	Returns the domain name.

Table 12.4 ApplicationSetting Properties and Methods (Cont.)

ConnectionPropertyType

The ConnectionPropertyType class contains properties to configure password policies. The available properties are shown in Table 12.5. The properties are accessed using the ApplicationSettings methods getStringProperty, getIntegerProperty, and getBooleanProperty shown in Table 12.4.

Property	Description
PwdPolicy_Enabled	Determines whether a password policy is enabled.
PwdPolicty_Default_Password_Allowed	Determines whether the app is allowed to use the default password for the data vault.
PwdPolicy_Length	Determines the minimum password length.
PwdPolicy_Has_Digits	Determines whether the password must have digits.
PwdPolicy_Has_Upper	Determines whether the password must have uppercase letters.
PwdPolicy_Has_Lower	Determines whether the password must have lowercase letters.
PwdPolicy_Has_Special	Determines whether the password must have special characters.

Table 12.5 Properties of the ConnectionPropertyType Class

Property	Description
PwdPolicy_Expires_In_N_Days	Determines the number of days a password is valid.
PwdPolicy_Min_Unique_Char	Determines the minimum number of unique characters in a password.
PwdPolicy_Lock_Timeout	Determines the timeout value (in seconds) the data vault is locked. A value of 0 indicates no lockout time.
PwdPolicy_Retry_Limit	Determines the number of unsuccessful login attempts before the data vault is locked.

Table 12.5 Properties of the ConnectionPropertyType Class (Cont.)

12.3.2 Afaria APIs

The Afaria APIs allow the app to retrieve configuration data from an Afaria server. The app must be installed on the device from an Afaria app policy that includes the connection information. Afaria is beyond the scope of this book, so this API is not described here.

12.3.3 Connection API

The Connection API includes methods and properties used to manage the local database, establishing a connection and authenticating with the Unwired Server. The Connection API includes the `ConnectionProfile` and `SynchronizationProfile` classes.

ConnectionProfile

The `ConnectionProfile` class manages local database properties. An instance of the `ConnectionProfile` is retrieved using the getConnectionProfile method of the database class.

An example for Android is:

```
ConnectionProfile cp = CustomerMBODB.getConnectionProfile();
```

An example for iOS is:

```
SUPConnectionProfile* cp = [CustomerMBOCustomerMBODB
getConnectionProfile];
```

Table 12.6 shows the properties and methods used to configure the `ConnectionProfile`.

Property or Method	Description
`openConnection`	Opens a connection to the database. Any database interaction opens or creates the database if it does not exist, so you do not need to call this method in most cases. You must set all `ConnectionProfile` properties before opening the database, or else they do not take effect.
`closeConnection`	Closes a connection to the database.
`generateEncryptionKey`	Generates an encryption key, stores it in a data vault, and sets the encryption key in the connection profile.
`cacheSize`	Manages the cache size of the database.
`maxDbConnections`	Allows the database to be accessed from multiple threads, which improves performance. For example, the app could display the first page of a long list, while the remainder of the list data is loaded in a background thread. The default value is 2.
`setProperty`	Sets the `ConnectionProfile` properties. For example, the database filename property can be set in Android using: `ConnectionProfile cp = SUP101DB.getConnectionProfile();` `cp.setProperty("databaseFile", "SUP101.ulj");` `cp.save();`

Table 12.6 ConnectionProfile Methods and Properties

SynchronizationProfile

The `SynchronizationProfile` class contains information used to establish connections with the Unwired Server's data synchronization channel. The information, which includes the server address, domain, protocol, port, and certificate, is automatically provisioned from the server and inherited from the app connection template used to register the app. This allows the administrator to determine the synchronization parameters based on the specific deployment details (e.g., whether a Relay Server or a certificate is used).

By default, synchronization is performed asynchronously. During the upload phase, operation replay records from the device are uploaded to the server, where they are replayed against the Cache Database. The synchronize method returns without waiting for the results. The download phase downloads any changes from the Cache Database but does not download the results of the operation replays that have not completed. This behavior, called *asynchronous operation replay*, can be changed so that the synchronization process waits until all changes on the Cache Database have completed before returning. A callback method can be created to respond to events in the synchronization process.

An instance of the `SyncrhonizationProfile` object is retrieved using the `getSynchronizationProfile` method of the database class. Examples for Android and iOS are shown in Listing 12.19 and Listing 12.20.

```
ConnectionProfile sp = SUP101DB.getSynchronizationProfile();
```

Listing 12.19 Android

```
SUPConnectionProfile* sp = [CustomerMBOCustomerMBODB
getSynchronizationProfile];
```

Listing 12.20 iOS

Property or Method	Description
`setServerName`	Sets the name of the server.
`setUser`	Sets the username.
`setPassword`	Sets the password.
`setNetworkProtocol`	Sets the network protocol (HTTP or HTTPS).
`setPortNumber`	Sets the port number.
`setNetworkStreamParams` (iOS) `getStreamParams` (Android) `setStreamParams` (Android)	Sets parameters, such as compression of the synchronization channel.

Table 12.7 Properties and Method used to Connect to the Server

Property or Method	Description
`setDomainName`	Sets the domain name.
`setAsyncReplay`	Enables or disables asynchronous operation replay. If disabled, the synchronization operation waits after the upload phase of the synchronization until the operation replay records are completed before beginning the download phase.
`beginSynchronize`	Synchronizes but only completes the upload phase.
`save`	Must be called for the synchronization profile parameters to take effect.

Table 12.7 Properties and Method used to Connect to the Server (Cont.)

12.3.4 Authentication API

The Authentication API includes methods for logging on to the server using username and password or X.509 certificates. The Authentication API supports the use of data vaults for certificates and single sign-on.

12.3.5 Personalization API

The Personalization API is used to manage personalization keys. Personalization keys can be used to save authorization credentials for enterprise systems or mapped to synchronization parameters to filter the data downloaded from the Cache Database.

An instance of the `PersonalizationParameter` object is retrieved using the `getPersonalizationParameters` method of the database class. Examples for Android and iOS are shown in Listing 12.21 and Listing 12.22.

```
PersonalizationParameters pp = CustomerMBODB.
getPersonalizationParameters();
```

Listing 12.21 Android

```
SUP101PersonalizationParameters *pp = [CustomerMBOCustomerMBODB
getPersonalizationparameters];
```

Listing 12.22 iOS

Table 12.8 shows properties and methods of the `PersonalizationParameter` class.

Property or Method	Description
`<personalization key name>` (iOS) `set<personalization key name>` (Android)	Sets the value of the personalization key. A set method is created for each personalization key.
`save`	Must be called for the personalization key values to take effect.

Table 12.8 Properties and Methods of the PersonalizationParameter Class

12.3.6 Synchronization API

The Synchronization API is used to manage the synchronization process. Methods are included that allow setting and saving synchronization parameters and performing synchronization.

Setting Synchronization Parameters

If synchronization parameters are defined for an MBO, their value must be set using the `SynchronizationParameter` class before synchronization is performed. An instance of the `SynchronizationParameter` object can be retrieved using the `getSynchronizationParameters` method of the MBO class. Examples for Android and iOS are shown in Listing 12.23 and Listing 12.24.

```
CustomerSynchronizationParameters sp = Customer.
getSynchronizationParameters();
```

Listing 12.23 Android

```
CustomerSynchronizationParameters *sp = [Customer
getSynchronizationParameters];
```

Listing 12.24 iOS

Property or Method	Description
`<Sync parameter name>` (iOS) `set<Sync parameter name>` (Android)	Sets the value of the synchronization parameter.
`save`	Saves the synchronization parameter values. Synchronization parameters do not take effect unless you call `save`, even if default values are configured.

Table 12.9 Properties and Methods of the SynchronizationParameter Class

Performing Synchronization

Synchronization can be performed individually for each synchronization group or for all synchronization groups. Table 12.10 shows the properties and methods that control the synchronization process.

Method	Description
`synchronize`	Synchronizes either a specific synchronization group if the name is passed as parameter or all synchronization groups if no parameter is passed.
`isReplayQueueEmpty`	Determines whether there are unsent operation replay records.

Table 12.10 Synchronization Properties or Methods of the Database Class

12.3.7 Message-Based Synchronization API

The Message-Based Synchronization API is used to allow an app to subscribe to a data package. Apps receive notifications when changes occur in any of the MBOs in synchronization groups to which they have subscribed. The Unwired Server uses a messaging channel to send change notifications to a device. The `onSynchronize` callback handler can be created to respond to the notifications. Table 12.11 shows the properties and methods involved in configuring data change notification.

Property or Method	Description
`enabledSIS` property (iOS) `getEnabledSIS` (Android)	Returns whether change notification is enabled for a synchronization group.
`setEnabledSIS`	Sets the change notification status for a synchronization group.
`setInterval`	Sets the interval at which notifications are sent for a synchronization group.

Table 12.11 Properties and Methods of the SynchronizationGroup Class Used to Configure Data Change Notification

12.3.8 Log Record APIs

The Log Record APIs allow you to customize logging and retrieve server-based and client-based logs.

12.3.9 Change Log APIs

The Change Log APIs allow you to retrieve entity changes from the backend.

12.3.10 Security APIs

The Security APIs allow you to configure many aspects of security. These include the use of certificates for authentication, creating and managing secure data vaults, implementing end-to-end encryption, and encrypting the database.

12.3.11 Callback and Listener APIs

The Callback and Listener APIs allow you to create callback handlers that respond to device, database, application, and synchronization events. There are three separate components of this API: Callback Handler API, ApplicationCallback API, and SyncStatusListener API.

Callback Handler API

The `CallbackHandler` API interface is invoked for any database event for any database or MBO for which you have registered a callback handler. You can create a handler by extending the `DefaultCallbackHandler` class or by implementing the com.sybase.persistence.CallbackHandler interface. Table 12.13 shows some of the callbacks you can override.

Callback	Description
onImport	Called when an import message is applied to the database. It does not imply that the import was committed because there may be multiple import messages.
onReplaySuccess onReplayFailure	Called when a server reports the failure or success of operation replays.
onLoginFailure onLoginSuccess	Called when a login failure or success message is received from the Unwired Server.
onSubscribeSuccess onSubscribeFailure	Called when a subscribe success or failure message is received from the Unwired Server.
onSearchSuccess onSearchFailure	Called when a backend search succeeds or fails.
onSynchronize	Called at different stages of the synchronization process.
onTransactionCommit onTransactionRollback	Called when a transaction is committed or rolled back.
onConnectionStatusChange	Called when the connection status changes.

Table 12.12 Callbacks in the CallbackHandler Class

Examples for Android and iOS are shown in Listing 12.25 and Listing 12.26.

```
//Callback implementation
private class MyCallbackHandler extends DefaultCallbackHandler
{
        @Override
      public int onSynchronize(GenericList
<SynchronizationGroup> groups,
SynchronizationContext context)
      {
        switch (context.getStatus())
        {
          case SynchronizationStatus.STARTING:
```

```
            Log.d("Customers","Syncing");
            break;
           case SynchronizationStatus.FINISHING:
             Log.d("Customers","sync done");
             break;
             }
      }
}

//Register the callback
CallbackHandler handler = new MyCallbackHandler();
CustomerMBODB.registerCallbackHandler(handler);
```

Listing 12.25 Android

```
// Interface file
#import "SUPDefaultCallbackHandler.h"

@interface CallbackHandler : SUPDefaultCallbackHandler
{
}

@end

//Implementation file
#import "CallbackHandler.h"
#import "SUPConnectionUtil.h"

@implementation CallbackHandler
+ (CallbackHandler*)getInstance
{
    CallbackHandler* _me_1 = [[CallbackHandler alloc] init];
    [_me_1 autorelease];
    return _me_1;
}

- (SUPSynchronizationActionType)onSynchronize:(SUPObjectList *)
syncGroupList withContext:(SUPSynchronizationContext *)context
{
  switch ([context status]) {
      case SUPSynchronizationStatus_STARTING:
          // perform necessary action
          break;
      case SUPSynchronizationStatus_ERROR:
```

```
            // perform necessary action
            break;
        }
    return SUPSynchronizationAction_CONTINUE;
}

@end

//Register the callback handler
CallbackHandler* handler = [CallbackHandler newHandler];
[CustomerMBODB registerCallbackHandler:handler];
[handler release];
```

Listing 12.26 iOS

ApplicationCallback API

The ApplicationCallback API responds to application events. You register an `ApplicationCallback` implementation with the `Application` object instance to receive the callbacks shown in Table 12.13.

Callback	Description
`onApplicationSettingsChanged`	Called when a server reports changes in app settings.
`onConnectionStatusChanged`	Called when the connection status changes.
`onDeviceConditionChanged`	Called when certain conditions, defined in the `DeviceCondition` class, change.
`onRegistrationStatusChanged`	Called when the app's registration status changes. The possible registration status values are defined in the `RegistrationStatus` class.
`onHttpCommunicationError`	Called when a server returns an HTTP communication error.

Table 12.13 ApplicationCallback Handlers

Examples for Android and iOS are shown in Listing 12.27 and Listing 12.28.

```
//Callback implementation
private class MyApplicationCallbackHandler extends
    DefaultApplicationCallback
    {
```

```
        @Override
        public void onConnectionStatusChanged(int
        connectionStatus, int errorCode, String
        errorMessage){
           switch (connectionStatus)
           {
           case ConnectionStatus.CONNECTED:
              Log.d("Customers","Connected");
              break;
           case ConnectionStatus.CONNECTION_ERROR:
              Log.d("Customers","Connection error");
              Log.d("Customes", errorMessage);
              break;
        ...
           }
        }
      }

//Register callback with Application object
Application app = Application.getInstance();
app.setApplicationCallback(new
   MyApplicationCallbackHandler());
```

Listing 12.27 Android

```
//Implementation file
#import "CallbackHandler.h"
#import "SUPConnectionUtil.h"

@implementation CallbackHandler
+ (CallbackHandler*)getInstance
{
    CallbackHandler* _me_1 = [[CallbackHandler alloc] init];
    [_me_1 autorelease];
    return _me_1;
}

- (void)sendNotification:(NSNotification *)notification
{
    [[NSNotificationCenter defaultCenter]
postNotification:notification];
    [notification release];
}
```

```
- (void)postNotification:(NSString *)notification withObject:(id)obj;
{
    NSNotification *n = [NSNotification notificationWithName:notificati
on object:obj];
    [n retain];
    [self performSelectorOnMainThread:@selector(sendNotification:)
withObject:n waitUntilDone:NO];
}
- (void)onConnectionStatusChanged:(SUPConnectionStatusType)
connectionStatus :(int32_t)errorCode :(NSString*)errorMessage
{
    NSLog(@"onConnectionStatusChanged: status = %d, code = %d, message
= %@",connectionStatus,errorCode,errorMessage);
        NSString *notification = nil;
    switch(connectionStatus)
    {
      case SUPConnectionStatus_CONNECTED:
         notification = ON_CONNECT_SUCCESS;
      break;
      case SUPConnectionStatus_DISCONNECTED:
          notification = ON_CONNECT_FAILURE;
          break;
      default:
      // Ignore all other status changes for this example.
      break;
    }

    if (notification != nil) [self postNotification:notification
withObject:nil];
}
@end



...

@end


//Create an instance of the callback handler in the class
//in which you manipulate data
...
```

```
CallbackHandler *ch = [CallbackHandler getInstance];
[self setCallbackHandler:ch];
[CustomerCustomerDB registerCallbackHandler:[self callbackHandler]];
...
```

Listing 12.28 iOS

SyncStatusListener API

The SyncStatusListener API provides a means by which the app can respond to events in the synchronization process. Examples for Android and iOS are shown in Listing 12.29 and Listing 12.30.

```
//Callback implementation
public class MySyncListener implements SyncStatusListener
{
  @Override
  public boolean objectSyncStatus(ObjectSyncStatusData arg0) {
    switch (data.getSyncStatusState()) {
      case SyncStatusState.SYNC_DONE:
     Log.d("Customers","Sync Done");
        break;
      case SyncStatusState.SYNC_STARTING:
     Log.d("Customers","Sync Starting");
        break;
     }
     return false;
   }
}

//Register the callback
SyncStatusListener listener = new MySyncListener();
CustomerMBODB.synchronize(listener);
```

Listing 12.29 Android

```
// The interface file
#import "SUPSyncStatusListener.h"
#import "SUPSyncStatusInfo.h"

@interface MySyncStatusListner : NSObject <SUPSyncStatusListener>
@end
```

```
// The implementation file
#import "MySyncStatusListner.h"
@implementation MySyncStatusListner

-(void) onGetSyncStatusChange:(SUPSyncStatusInfo*)info
{
  switch(info.state)
  {
      case SYNC_STATE_NONE:
          MBOLogDebug(@"SYNC_STATE_NONE");
          break;
      case SYNC_STATE_STARTING:
          MBOLogDebug(@"SYNC_STATE_STARTING");
          break;
   }
}
@end
```

Listing 12.30 iOS

12.3.12 Query APIs

The Query APIs provide the ability to query MBOs. Data can be retrieved using object queries defined in the MBO or through dynamic queries. Methods are created in the MBO classes for each object query. An instance of the query object is created using the query class' constructor. Examples for Android and iOS are shown in Listing 12.31 and Listing 12.32.

```
Query query = new Query();
```

Listing 12.31 Android

```
SUPQuery *query = [SUPQuery getInstance];
```

Listing 12.32 iOS

The methods and properties used to create dynamic queries are shown in Table 12.14.

Method	Description
`select`	Sets the list of attributes to retrieve to the `Query` object.
`where`	Adds the `attributeTest` object to the `Query` object.
`setTake`	Sets the number of rows to retrieve.
`setSortCriteria`	Adds the `sort` object to the `Query` object.
`setSkip`	Sets the number of rows to skip.
`orderBy`	Adds an order by condition to the `Query` object

Table 12.14 Properties and Methods of the Query Class

You can add one or more `SortCriteria` or `AttributeTest` objects to the `Query` object. Examples of configuring these objects were shown in Section 12.2.2.

12.3.13 Persistence API

The Persistence API includes methods to perform `create`, `update`, and `delete` operations on data in the device database and to submit the changes to cached MBOs. The Persistence API also includes methods that can be used to retrieve information about the state of app entities, such as database entities and MBOs. We discuss both of these APIs in the following section.

Operations API

The Operations API implements MBO operations. Code samples illustrating operations were shown in Section 12.2.3. Next we illustrate additional scenarios.

Deleting a Record in an MBO in a Relationship

Examples of deleting a record in an MBO in a relationship are shown in Listing 12.33 and Listing 12.34.

```
Customer cust = Customer.findById(101);
GenericList<Sales_order> orders = cust.getSalesOrders();
Sales_order order = orders.item(0);
order.delete();
//Call submitPending on the parent - customer
```

```
cust.submitPending();
CustomerMBODB.synchronize();
```

Listing 12.33 Android

```
CustomerMBOCustomer *cust = [ CustomerMBOCustomer find: 101]
SUPObjectList* orderlist = customer.orders;
SUP101Sales_Order* order = [orderlist item:0];
[order delete];
//Call submitPending on the parent - customer
[order.cust submitPending];
[CustomerMBOCustomerMBODB synchronize];
```

Listing 12.34 iOS

The save Operation

The `save` method can be used whether you are creating a new record or updating an existing record. In addition, `save` saves changes to children when called on a parent. Examples for Android and iOS are shown in Listing 12.35 and Listing 12.36.

```
Customer cust = Customer.findById(101);
GenericList<Sales_order> orders = cust.getSalesOrders();
Sales_order order = orders.item(0);
order.sales_rep = "Smith";
cust.state = "WI" ;
//Call submitPending on the parent - customer
cust.submitPending();
CustomerMBODB.synchronize();
```

Listing 12.35 Android

```
CustomerMBOCustomer *customer = [ CustomerMBOCustomer find: 101]
SUPObjectList* orderlist = customer.orders;
SUP101Sales_Order* order = [orderlist item:0];
order.sales_rep = @"Smith";
customer.state = @"WI" ;
[customer save];//Call submitPending on the parent - customer
[order.customer submitPending];
[CustomerMBOCustomerMBODB synchronize];
```

Listing 12.36 iOS

You can manage the pending state using the MBO's methods in Table 12.15.

Method	Description
cancelPending	Cancels pending CUD operations for a specific record.
submitPending	Submits operations for a specific record so they can be replayed on the Unwired Server.
submitPendingOperations	Submits all pending operations on the MBO.
cancelPendningOperations	Cancels all pending operations on the MBO.

Table 12.15 Methods to Manage the Pending State of MBOs

To show you how these methods could be used, Listing 12.37 and Listing 12.38 show examples for Android and iOS.

```
Customer customer = Customer.findById(101);
//Process record
...
if (errorHappened) {
    customer.cancelPending();
}
else {
    customer.submitPending();
}
```

Listing 12.37 Android

```
CustomerMBO *customer = [CustomerMBOCustomer find:101];
//Process record
...
If (errorHappened) {
  [customer cancelPending];
} else {
  [customer submitPending];
}
```

Listing 12.38 iOS

Object State API

The Object State API is used to determine the state of MBO entities. It includes the methods in Table 12.16.

Method	Description
isNew	Returns true if the entity is new but has not been created in the local database.
isCreated	Returns true if the entity is new and has been created in the local database, and the replay record has been successfully run on the server.
isDirty	Returns true if the entity has been changed but not saved to the local database.
isDeleted	Returns true if the entity was loaded into memory and deleted.
isUpdated	Returns true if the entity has been updated but the replay record has not been successfully run on the server.
pending	Returns true for any row that represents a pending CUD operation or a row that has cascading children with a pending operation.
pendingChange	If pending is true for an entity, this returns the type of change.
replayCounter	Returns a value that indicates how many times the entity has been changed on the client.
replayPending	When a pending row is submitted to the server, the replayCounter is copied to replayPending. This allows the app to determine whether the entity has changed since it was submitted to the server.
replayFailure	When the server returns a replayFailure message, the value of replayCounter is copied to replayFailure, and replayPending is set to 0.

Table 12.16 Methods of the Object State API

12.4 Summary

Now that you have a basic understanding of the Object API and the synchronization process, you are ready to develop some simple Object API apps. The next chapter presents a native Android app, and the following chapter presents an iOS app.

This chapter demonstrates the use of the Object API for Android apps. The basics of Android apps are outlined, and a complete example app that illustrates Object API fundamentals is presented.

13 Developing Android Apps

In recent years, Android has become the dominant mobile platform, making the Android Object API essential knowledge for mobile developers who use SAP Mobile Platform. This chapter demonstrates the fundamentals of developing Android Object API apps. We start by outlining some basics of Android development, including a description of the example app we'll be programming. Then, we dive straight into the steps of Android app development, which are the following:

1. Create the MBOs and deploy them to the Unwired Server.
2. Generate the Java classes that implement the Object API.
3. Create an Android project.
4. Import libraries and copy the generated Object API files to the new project.
5. Implement the app.
6. Run the app.

Background Knowledge

While the code in this chapter produces a working Android app, the focus is on the techniques for using the Object API. Many of the essential elements of a production app, such as validation, callback handlers, and error trapping, have been left out in order to simplify the code. A familiarity with Java and Android development will be helpful since explanations of the code focus only on the Object API.

13.1 Getting Started

This section gets you started with the basics of Android app development: the components of an Android app and how to prepare the development environment. We conclude with a description of the example app we will be coding in this chapter.

13.1.1 Components of an Android App

Here, we describe the basic components of Android apps. This description is very brief, so if you are new to Android development, consider this only a starting point.

Activities

The basic component of an Android app is the `activity`. Each screen in an app corresponds to a separate activity, although not every activity has a visual component on the user interface. The `activity` class file contains the Java code, while the interface description is contained in a separate layout file, which is described later.

Navigation between activities is accomplished using *intents*. Intents can pass data to the invoked activity by including a `Bundle`. Intents are either explicit, in which case they specify which activity they invoke, or implicit, in which case they specify a requirement and allow the system to determine which activity in which app can fulfill the requirement. Android apps can register themselves with the system to make their capabilities available to other apps. For example, the contacts app can be used by other apps, eliminating the need to reproduce that app's capability.

Permissions

Android includes a permissions system with a number of predefined permissions that a developer can assign to an app. For example, if an app needs to access the Internet, the permission must be explicitly included. In many cases, the user is presented with the permissions when the app is installed. If they refuse any permission, then the app cannot be installed.

Layouts

The user interface is defined in XML files called *layout files*. The user interface is built using `View` and `ViewGroup` objects. `View` objects are UI widgets, such as buttons and text fields. `ViewGroup` objects are containers used to organize and group other objects. `ViewGroup` objects have no visual component. Figure 13.1 shows how `ViewGroup` and `View` objects are organized hierarchically.

AndroidManifest.xml

The *AndroidManifest* file is an XML file that describes the various components and settings of the app. Every activity and permission must be included in the

AndroidManifest file. The *AndroidManifest* file also defines which activity is loaded first when the app runs.

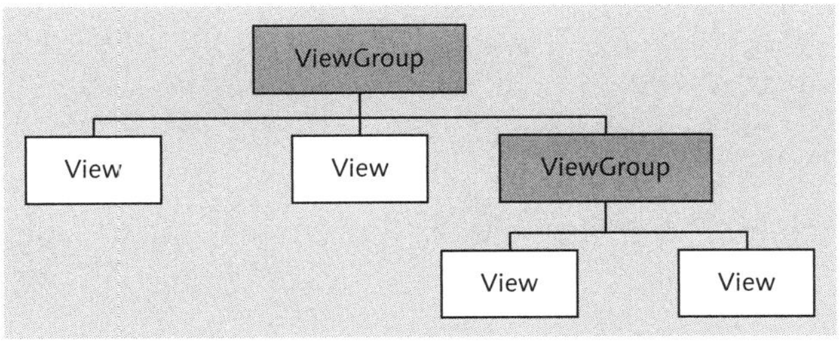

Figure 13.1 Hierarchy of View and ViewGroup Objects that Make Up the User Interface

13.1.2 Preparing the Development Environment

This chapter assumes that the development environment used is Eclipse. This is convenient because, with some additional configuration, the Unwired Workspace can be used for development of the MBOs as well as the Android app. If you already have the Unwired Workspace installed, there are three steps required to prepare it for developing Android apps:

1. Install the Android SDK.
2. Install the latest tools and platforms in the Android SDK Manager.
3. Install the Android Development Tools (ADT) plug-in for Eclipse.

Eclipse Versions

The Unwired Workspace does not use the latest version of Eclipse. For the most part, this should not pose a problem, but if you want to use the latest version for your Android development, you can find it at *http://www.eclipse.org*.

Installing the Android SDK and Tools in the Android SDK Manager

The process for completing the first two steps in the list above was described in Section 5.1.1 of Chapter 5. If you did not complete that process, do so now. If you did complete that process, open the Android SDK Manager and allow it to check for updates. The version of the SDK files must match those of the ADT plug-in that is installed in the next section. If there are any updates, install them.

Installing the ADT Plug-in for Eclipse

The Android Development Tools (ADT) plug-in for Eclipse includes all the tools required to develop Android apps. You can install multiple copies of Eclipse, which enables you to keep the Unwired Workspace separate from the Eclipse you use to develop Android apps. A simpler solution, if you want to keep them separate, is to use different workspace folders for each type of project. However you choose to proceed, keep in mind the warning from Chapter 4 about updating Eclipse components in the Sybase Mobile SDK. You should not update components other than the ADT in the Sybase Mobile SDK.

To install the ADT plug-in, use the HELP • INSTALL NEW SOFTWARE menu option in Eclipse. In the ADD REPOSITORY dialog, click Add and enter *https://dl-ssl.google.com/android/eclipse* in the LOCATION field and any value in the NAME field. When you click OK, Eclipse checks the location for updates. There may be a number of updates available unrelated to Android development that you can ignore. The ADT is labeled DEVELOPER TOOLS. Select it and click NEXT to continue the installation (Figure 13.2).

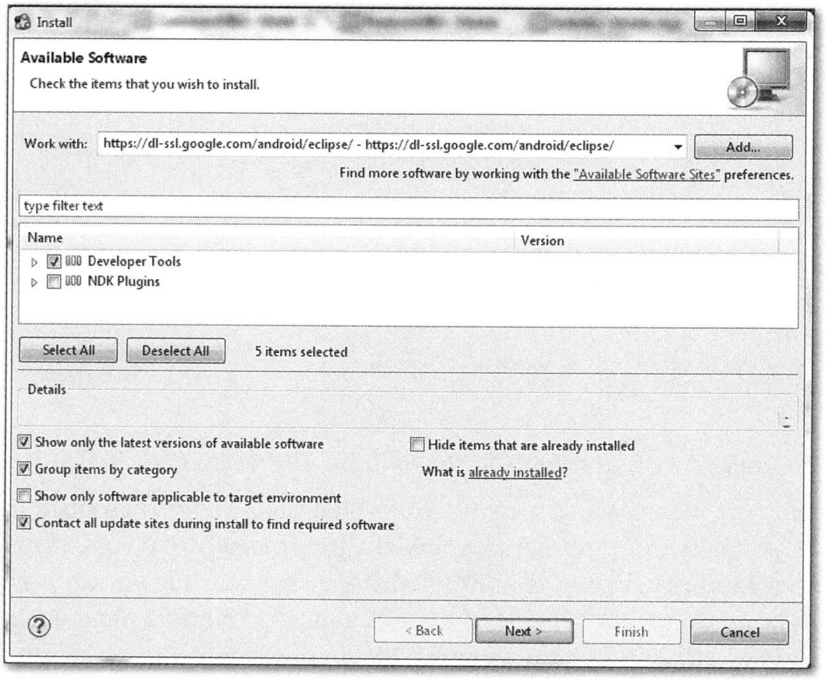

Figure 13.2 Tools Required to Develop Android Apps in Eclipse

When the installation is complete, you must tell Eclipse where the Android SDK files are located. Navigate to WINDOW • PREFERENCES in Eclipse and select ANDROID in the left pane. If the path is not already entered, use the BROWSE button to enter the path to the ANDROID-SDK folder. Once you have completed this step, Eclipse is ready to develop Android apps.

13.1.3 Introducing the Example App

The example app, called CustomerApp, allows a customer service representative to retrieve, edit, and create customer data while in the field. Once the data is synchronized, the app can be used normally in an offline mode. The simplicity of the app is consistent with the design philosophy that, for mobile apps, simpler is better. Unlike desktop applications, which tend to be complex and multipurpose, users in a mobile environment have come to expect simpler, special purpose apps. In the Android environment, the capability of apps to easily share functionality using implicit intents encourages the simple-is-better philosophy. For example, it would be relatively easy to integrate the sample app with existing device apps, such as the contacts, email, dialer, and camera apps.

Although the sample app is simple, it is sufficiently complex to demonstrate the fundamental concepts of the Android Object API. Once the app registers with and connects to the Unwired Server, the user can manually synchronize with the CDB by tapping a button on the first screen. The user can retrieve a complete list of customers or filter the list based on the company name. The user can also edit a customer's data, create a new customer, or delete the customer.

The app consists of a single MBO based on the `Customers` table that is included in the sample database. The MBO includes an object query and a synchronization parameter that is mapped to the state attribute.

The app consists of five activities:

- The `MainActivity` provides the home screen for the app. It also includes code that connects to the Unwired Server and synchronizes the customer data.
- The `CustomerListActivity` provides a screen showing a complete list of customers. The user can tap an item on the list to open the `CustomerDetailsActivity`.
- The `CustomerSearchActivity` allows the user to filter the customer list based on the company name. The results of a search are shown in a list. The user can open the `CustomerDetailsActivity` by tapping an item on the list.

- The `CustomerListAdapterActivity` has no user interface component of its own but creates the interface view that depicts the rows in the lists. It also manages the data shown in the lists.
- The `CustomerDetailsActivity` shows the customer details. The user can update data or delete the customer from this screen.

Screenshots of the four app screens and the flow of control between them are shown in Figure 13.3.

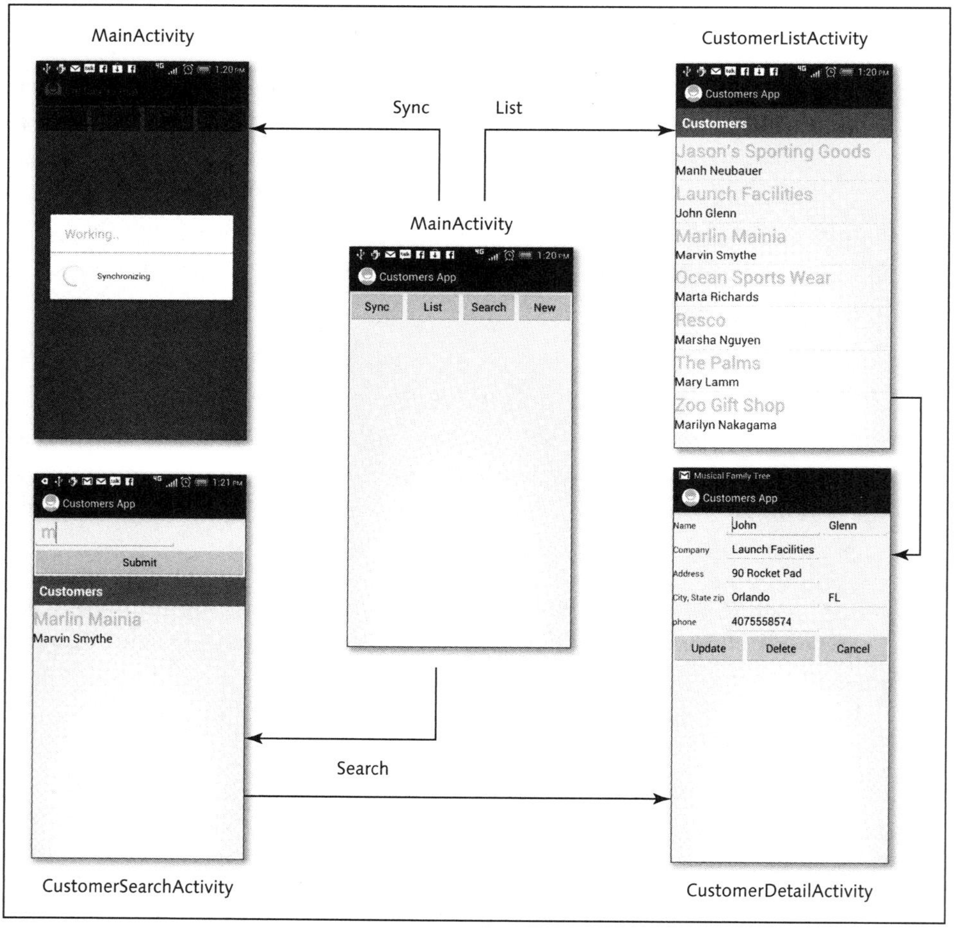

Figure 13.3 Four Screens of the Example App

13.2 Creating the MBOs

Developing MBOs for an Object API app is similar to developing hybrid apps, which we explained in great detail in Chapter 7; thus, we don't spend a lot of time on this discussion here. However, it's important to note that, because they are synchronizing with the CDB, Object API apps must have at least one non-online MBO. Data is not cached in the CDB for online MBOs. Figure 13.4 shows the definition of the MBO and the synchronization parameter `state_SP`. The synchronization parameter is used to filter the customer list downloaded to the device based on the state field of the customers table.

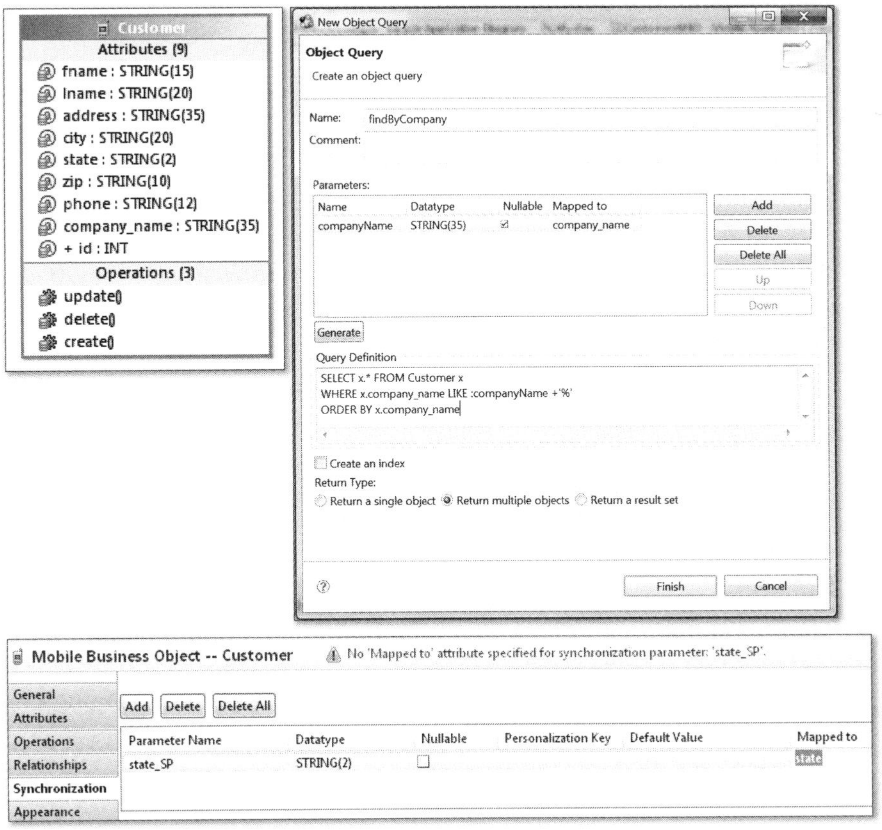

Figure 13.4 Single MBO Based on the Customer Table of the Sample Database

Once the MBOs are created, deploy them to the Unwired Server in the usual way.

13.3 Generating the Object API Classes

By default, the Object API classes are created in a folder called GENERATED CODE in the project workspace. If you intend to create apps for multiple platforms, it is a good idea to create subfolders in the GENERATED CODE folder for each platform. Simply right-click the GENERATED CODE folder and select NEW • FOLDER to create the folder.

Before starting the Hybrid App Generation Wizard, you must have an active connection to the MBO data sources. The *Android.jar* file must also be available.

To start the wizard, right-click the project name in the WORKSPACE NAVIGATOR or on the background of the Mobile Application Diagram Editor, and select GENERATE CODE. The first screen allows you to create a configuration you can use to recall the settings in the future.

On the next screen, select the MBOs you wish to include. The sample app has only one MBO, but an app could have a large number of MBOs, which would entail a long deployment time. If you are deploying the MBOs after making a change to a single MBO, you may choose to deploy only the MBO that was changed. The wizard calculates the page size after you select the MBOs based on the total attribute size. If the total attribute size is larger than the page size setting on the next screen, a warning is displayed.

The last screen of the wizard is shown in Figure 13.5. The following fields can be configured for Android:

- LANGUAGE
  Select JAVA FOR ANDROID.
- PLATFORM
  Select ANDROID.
- UNWIRED SERVER
  Select the Unwired Server connection profile to which you deployed the MBOs.
- DOMAIN
  Select the domain on the Unwired Server to which you deployed the MBOs.
- PAGE SIZE
  The value entered automatically in this field is based on the total size of the MBO attributes. Whatever size is entered here must be larger than the

accumulated size of the attributes of any of the MBOs selected and also compatible with the database. If you leave the field blank, the page size is set to 4 KB.

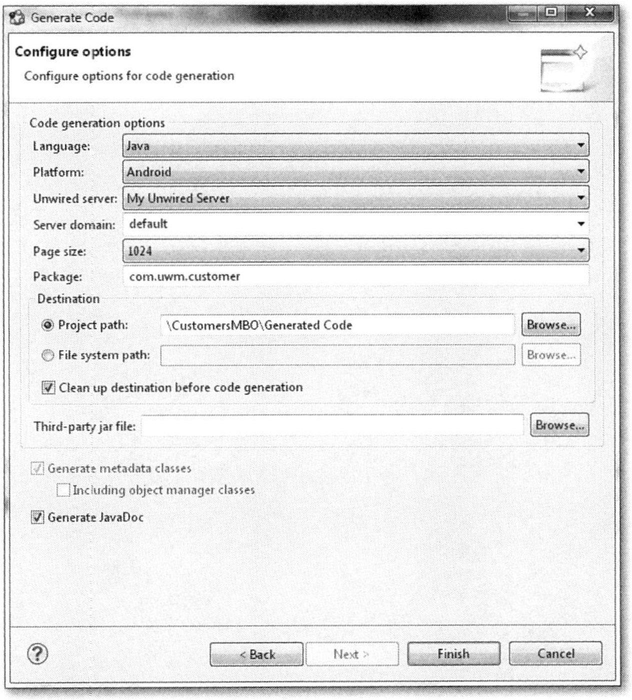

Figure 13.5 Object API Classes Generated by the Generated Code Wizard

- Package
  This field is optional. The value entered must follow Java package naming conventions.
- Destination
  You can choose to create the files within the project workspace or an external folder. The default folder is the Generated Code folder in the project workspace. You can also check the box to Clean up destination before code generation to delete all the files in the destination folder before generating new files.
- Third-party JAR File
  Enter the path to the *Android.jar* file. If you installed the ADT and specified the location of this file in the Unwired Workspace, this field is optional. If neither of these options is configured, you receive a warning.

- GENERATE JAVADOC
  If you select this option, the documentation for the generated classes is created in the destination folder.

Once you click FINISH, the code is generated. Figure 13.6 shows the results in the Workspace Navigator, assuming GENERATE JAVADOC was selected. The specific classes generated depend on the design of the data model.

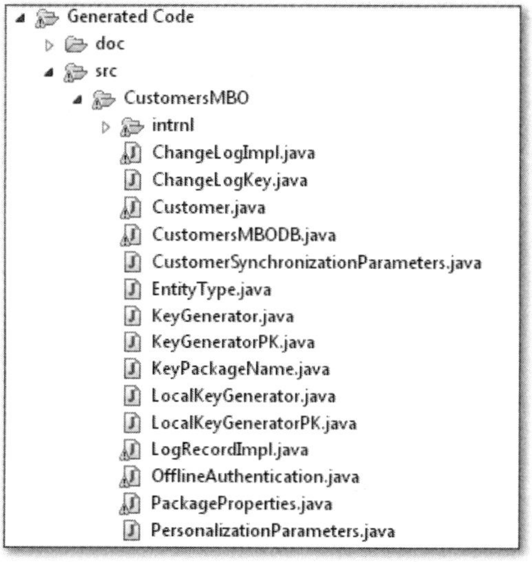

Figure 13.6 Generated Java Classes

Among the generated files, the following are referenced in the example app:

- `CustomersMBODB`
  Implements the methods required to access the device database.
- `Customer`
  Implements the attributes, operations, object queries, etc., of the CUSTOMER MBO. A similar class would be created for every MBO in the package.
- `CustomerSynchronizationParameters`
  Implements the synchronization parameter. This is created only because there is a synchornization parameter in this example.
- `KeyGenerator`
  Implements the generation of surrogate keys used to access MBO instances.

13.4 Creating the Project

Create a new Android appplication project in the Eclipse you wish to use for Android development. Choose ANDROID APPLICATION PROJECT, as shown in Figure 13.7.

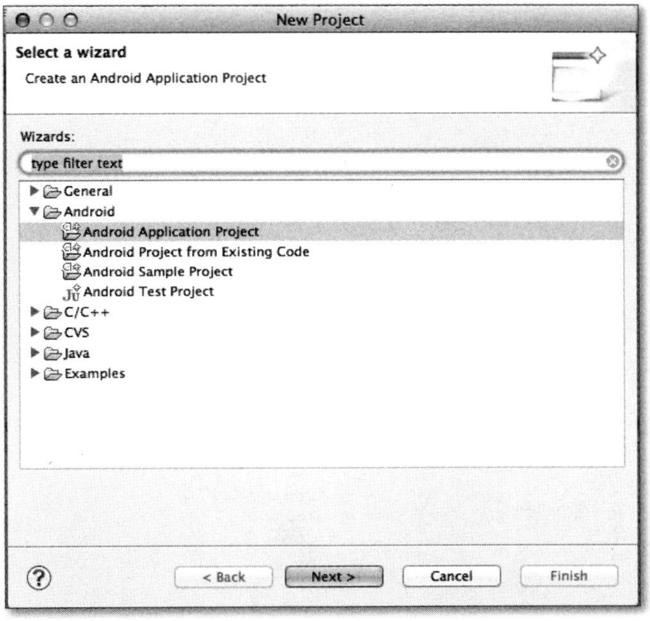

Figure 13.7 Creating an Android Project

When you choose the SDK requirements on the next screen, shown in Figure 13.8, keep in mind the API level of the device or emulator on which you are running the app. The app must be compiled with an SDK that matches the emulator or device. Also, you should make a note of the minimum and target SDK levels because these must be entered into the code for the *AndroidManifest.xml* file (which is provided in Section 13.6.1).

On the next dialog, choose to create an activity. You can choose to create a custom icon if you wish. If you check CREATE PROJECT IN WORKSPACE, the project files are created in the workspace you selected when Eclipse started. The defaults are shown in Figure 13.9.

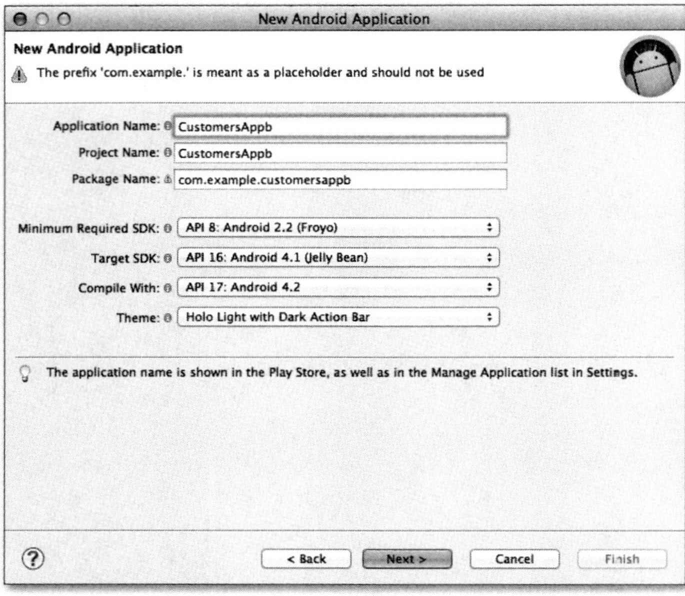

Figure 13.8 Setting a Target SDK Level that Matches the Device or Emulator You Use

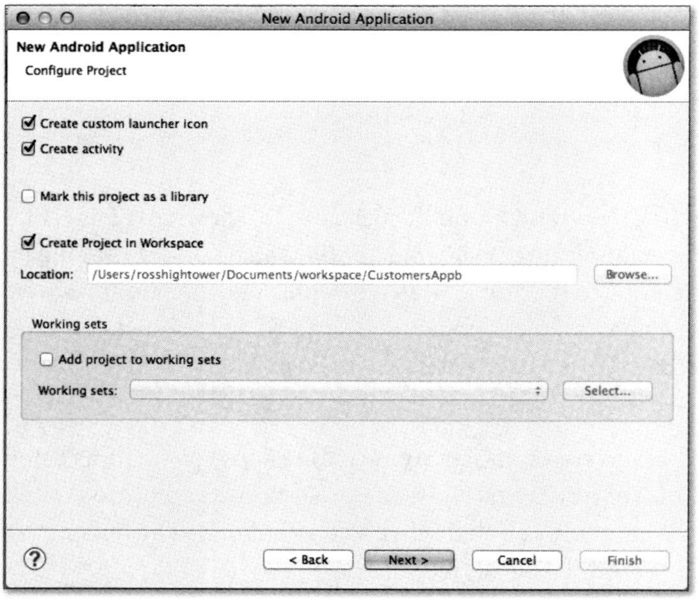

Figure 13.9 Choosing to Create an Activity

If you chose to create a custom icon, the next screen allows you to customize the icon. You can choose any options here. On the subsequent dialog, shown in Figure 13.10, choose to create a `BlankActivity`.

Figure 13.10 Creating a BlankActivity

On the last dialog, leave the defaults as shown in Figure 13.11. The name of the activity created should be `MainActivity`, and the layout should be `activity_main`. Click FINISH, and the project is created.

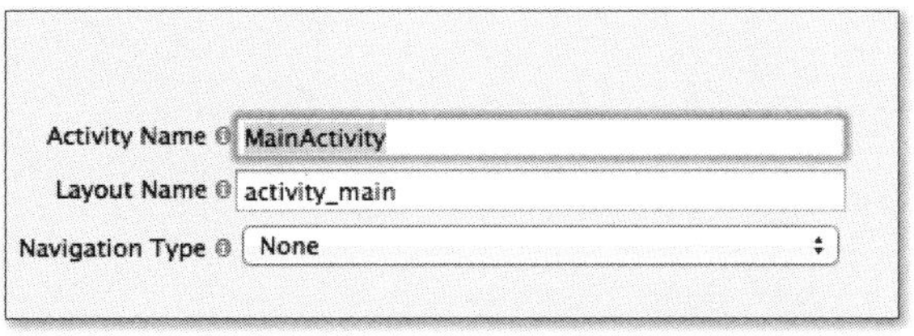

Figure 13.11 Leaving the Defaults on the Last Dialog

You now have the skeleton of the app's project. Before you can begin writing code, you must import some libraries and copy the Object API classes.

13.5 Importing Libraries

Copy the following files from *<Unwired Workspace Installation folder>\MobleSDK213\ObjectAPI\Android*:

- *clientLib.jar*
- *sup-client.jar*
- *UltraLiteJNI12.jar*
- ARMEABI folder

Paste them into the LIBS folder of your project. You can paste them directly in the Package Explorer within Eclipse.

Navigate to PROJECT • PROPERTIES in Eclipse, and select JAVA BUILD PATH in the left pane. Select LIBRARIES in the right pane and click ADD JARS. Navigate to the LIBS folder of your project and select the three JAR files you copied previously (Figure 13.12). Click OK until you exit PROPERTIES.

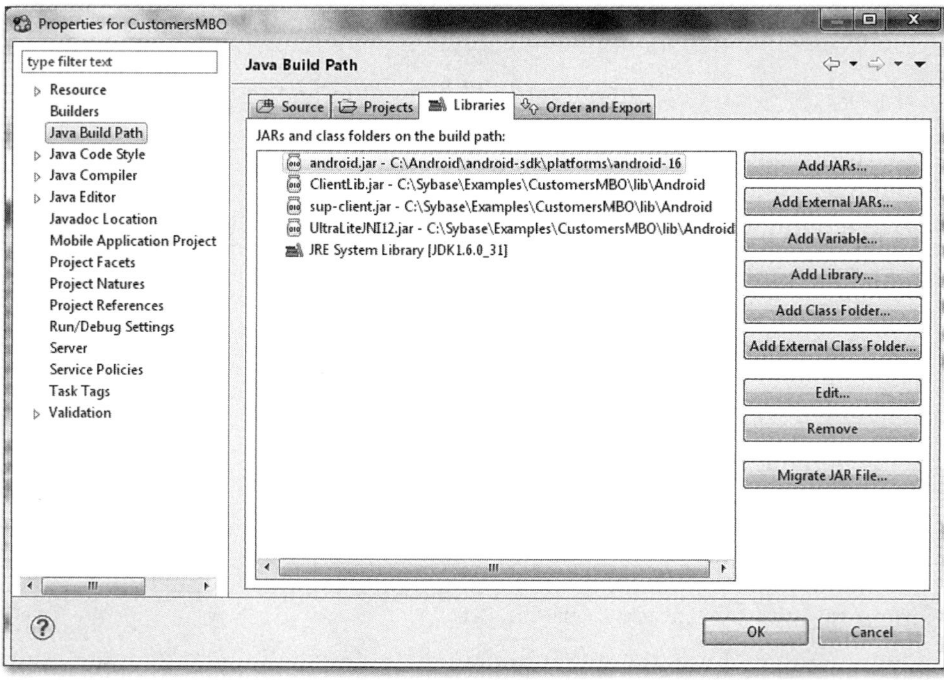

Figure 13.12 Adding the JAR Files to the Java Build Path

To copy the Object API files, return to your MBO project and copy the contents of the GENERATED CODE/SRC folder (do not include the SRC folder itself, just the files in the folder) and paste them into the SRC folder of your Android project. You can do this within Eclipse. The result is shown in Figure 13.13.

Figure 13.13 Copying the Object API Classes into the Project's src Folder

This concludes the preparation process. You can now begin to develop the app.

13.6 Implementing the App

In this section, we provide all the code required to implement our example app. Each section begins with the code listing and then includes explanations of those portions of the code that involve the Object API. The focus is on minimum functionality. There are many more options available within the Object API, which are described in the documentation.

The following sections describe the implementation of the components of the app. The first step in developing the app is to define the app structure in the *AndroidManifest.xml* file. Next, the structure of the user interface is defined in the layout files. Finally, the Java code that implements the five activities is shown.

13.6.1 Defining the App Structure in AndroidManifest.xml

The *AndroidManifest* file defines the structure of the project. For the sample app, there is an entry for each of the five activities. Another project characteristic defined in the *AndroidManifest* file is permissions. The sample app must be able to access the Internet to communicate with the Unwired Server. It must also be able to obtain information about the device and the device's network connectivity. The permissions related to WiFi connectivity are necessary if the device communicates using WiFi.

The *AndroidManifest* file is created automatically when you create the project. The contents reflect the choices you made when you created the project and define only the `MainActivity` that is created when you create an Android project. Navigate to the project folder in the Package Explorer and double-click the *AndroidManifest.xml* file to open it. The editor has a number of tabs with tools that allow you to edit the file indirectly. Since you want to replace the existing code with the code shown in Listing 13.1, click the tab labeled ANDROIDMANIFEST.XML. Replace the existing code with the code shown in Listing 13.1. This code reflects standard Android practice and requires little explanation. You should edit the contents to include the package name and minimum and target SDK levels you used when you created the project. Also, the string values are hardcoded in the code below (as they are in all the listings in this chapter), and you may wish to move those into a resource file.

```
<manifest xmlns:android="http://schemas.android.com/apk/res/android"
    package="com.example.customersapp"
    android:versionCode="1"
    android:versionName="1.0" >

    <uses-sdk
        android:minSdkVersion="8"
        android:targetSdkVersion="16" />

    <uses-permission android:name="android
.permission.INTERNET"/>
    <uses-permission android:name="android
.permission.READ_PHONE_STATE"/>
    <uses-permission android:name="android
.permission.CHANGE_WIFI_STATE"/>
    <uses-permission android:name="android
.permission.ACCESS_WIFI_STATE"/>
    <uses-permission android:name="android
.permission.ACCESS_NETWORK_STATE"/>


    <app
        android:icon="@drawable/ic_launcher"
        android:label="@string/app_name"
        android:theme="@style/AppTheme" >

    <activity
       android:name=".MainActivity"
       android:label="Customer App" >
       <intent-filter>
           <action android:name=
```

```
"android.intent.action.MAIN" />
          <category android:name=
"android.intent.category.LAUNCHER" />
       </intent-filter>
    </activity>

    <activity
       android:name=".CustomerListActivity"
            android:label="Customers App" >
     </activity>

     <activity
        android:name=".CustomerListAdapter"
        android:label="Customers App" >
     </activity>

     <activity
        android:name=".CustomerDetailActivity"
        android:label="Customers App" >
     </activity>

     <activity
        android:name=".CustomerSearchActivity"
        android:label="Customers App" >
     </activity>

</manifest>
```

Listing 13.1 AndroidManifest File

13.6.2 Creating the Layout Files

The next step is to create the layout files that define the user interface. Four of the activities have visual components, so each requires a layout file. In addition, two layout files are required to define the rows and header of the LISTVIEW control. As with the *AndroidManifest* file, these reflect standard Android programming practice, so we won't go into further explanation here.

To create a layout file, right-click the RES/LAYOUT folder in the PACKAGE EXPLORER and select NEW • OTHER. Select ANDROID XML LAYOUT FILE and click NEXT. Use the headings of the following sections as the names of the layout files.

activity_main.xml

The *activity_main* layout file defines the user interface for the `MainActivity` activity. The code in Listing 13.2 defines the four buttons on the home screen of the app. The resulting layout is shown in Figure 13.14.

Figure 13.14 MainActivity Screen

```
<?xml version="1.0" encoding="utf-8"?>
<LinearLayout xmlns:android="http://schemas.android.com/apk/res/
android"
    xmlns:tools="http://schemas.android.com/tools"
    android:layout_width="match_parent"
    android:layout_height="match_parent" >

    <Button
        android:id="@+id/button_sync"
        android:layout_width="0px"
        android:layout_height="wrap_content"
        android:layout_weight="1"
        android:text="Sync"/>

    <Button
        android:id="@+id/button_list"
        android:layout_width="0px"
        android:layout_height="wrap_content"
```

```
        android:layout_weight="1"
        android:text="List"/>

    <Button
        android:id="@+id/button_search"
        android:layout_width="0px"
        android:layout_height="wrap_content"
        android:layout_weight="1"
        android:text="Search"/>

    <Button
        android:id="@+id/button_create"
        android:layout_width="0px"
        android:layout_height="wrap_content"
        android:layout_weight="1"
        android:text="Create"/>

</LinearLayout>
```

Listing 13.2 MainActivity Layout File

customerlist.xml

The *customerlist* layout file defines the interface for the `CustomerListActivity` activity. The only component on this screen is a LISTVIEW control (see Listing 13.3). The resulting screen is shown in Figure 13.15.

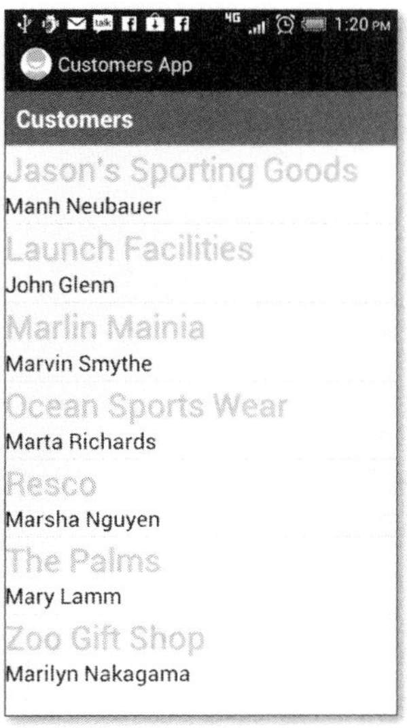

Figure 13.15 CustomerListActivity Screen

```
<?xml version="1.0" encoding="utf-8"?>
<LinearLayout xmlns:android="http://schemas.android.com/apk/res/
android"
    android:layout_width="match_parent"
    android:layout_height="match_parent"
    android:orientation="vertical" >

    <ListView android:layout_width="fill_parent"
          android:id="@+id/listView1"
          android:layout_height="wrap_content"></ListView>

</LinearLayout>
```

Listing 13.3 CustomerListActivity Layout File

customersearch.xml

The *customersearch* layout file defines the user interface for the `CustomerSearchActivity` activity. It defines an EditText control that the user can use to enter a search term, a Button control used to initiate the search, and a ListView control to display the results (see Listing 13.4). The resulting screen is shown in Figure 13.16.

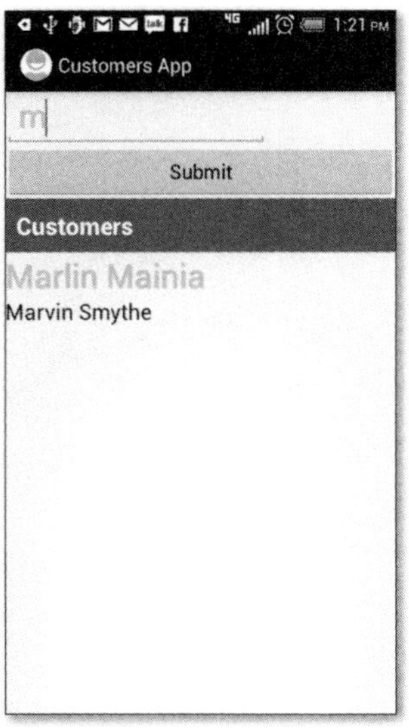

Figure 13.16 The CustomerSearchAtivity Screen

```
<?xml version="1.0" encoding="utf-8"?>
<LinearLayout xmlns:android="http://schemas.android.com/apk/res/
android"
    android:layout_width="match_parent"
    android:layout_height="match_parent"
    android:orientation="vertical" >

    <EditText
        android:id="@+id/editText"
        android:layout_width="wrap_content"
        android:layout_height="50dp"
        android:shadowColor="#00ccff"
        android:shadowDx="1"
        android:shadowDy="1"
        android:shadowRadius="1.5"
        android:hint="Company Name"
        android:textColor="#00ccff"
        android:textSize="30sp" >
    </EditText>

    <Button
        android:id="@+id/submit"
        android:layout_width="fill_parent"
        android:layout_height="wrap_content"
        android:text="Submit">
    </Button>

    <ListView
        android:id="@+id/listView1"
        android:layout_width="fill_parent"
        android:layout_height="wrap_content" >
    </ListView>

</LinearLayout>
```

Listing 13.4 CustomerSearchSctivity Layout File

customerdetails.xml

The *customerdetails* layout file defines the user interface of the `CustomerDetails-Activity` activity. As you can see in Listing 13.5, the code is more complex than the other layout files. It uses a *TableLayout* layout file to organize the other controls on the screen. TEXTVIEW controls are defined for labels and EDITTEXT controls are defined to display customer detail data. Three BUTTON controls are defined that correspond to cancel, update, and delete operations. The resulting screen is shown in Figure 13.17.

Figure 13.17 The CustomerDetailActivity Screen

```
<?xml version="1.0" encoding="utf-8"?>
<LinearLayout xmlns:android="http://schemas.android.com/apk/res/
android"
    android:layout_width="match_parent"
    android:layout_height="match_parent"
    android:orientation="vertical" >

    <TableLayout
        xmlns:android="http://schemas.android.com/apk/res/android"
        android:id="@+id/tableLayout1"
        android:layout_width="fill_parent"
        android:layout_height="fill_parent" >

        <TableRow android:layout_width="fill_parent" >
```

```
    <TextView
        android:id="@+id/textView1"
        android:layout_width="wrap_content"
        android:layout_height="wrap_content"
        android:text="Name" >
    </TextView>

    <EditText
        android:id="@+id/fname"
        android:layout_width="fill_parent"
        android:layout_height="wrap_content"
        android:hint="First Name" >
    </EditText>

    <EditText
        android:id="@+id/lname"
        android:layout_width="fill_parent"
        android:layout_height="wrap_content"
         android:hint="Last Name">
    </EditText>
</TableRow>

<TableRow android:layout_width="fill_parent" >

    <TextView
        android:id="@+id/textView1"
        android:layout_width="wrap_content"
        android:layout_height="wrap_content"
        android:text="Company" >
    </TextView>

    <EditText
        android:id="@+id/companyname"
        android:layout_width="fill_parent"
        android:layout_height="wrap_content"
        android:hint="Company" >
    </EditText>
</TableRow>

<TableRow android:layout_width="fill_parent" >
```

```
        <TextView
            android:id="@+id/textView1"
            android:layout_width="wrap_content"
            android:layout_height="wrap_content"
            android:text="Address" >
        </TextView>

        <EditText
            android:id="@+id/address"
            android:layout_width="fill_parent"
            android:layout_height="wrap_content"
            android:hint="Address" >
        </EditText>
    </TableRow>

    <TableRow android:layout_width="fill_parent" >

        <TextView
            android:id="@+id/textView1"
            android:layout_width="wrap_content"
            android:layout_height="wrap_content"
            android:text="City, State zip" >
        </TextView>

        <EditText
            android:id="@+id/city"
            android:layout_width="fill_parent"
            android:layout_height="wrap_content"
            android:hint="City" >
        </EditText>

        <EditText
            android:id="@+id/state"
            android:layout_width="fill_parent"
            android:layout_height="wrap_content"
            android:hint="State" >
        </EditText>

        <EditText
            android:id="@+id/zip"
            android:layout_width="fill_parent"
            android:layout_height="wrap_content"
             android:hint="Zip" >
```

```
    </EditText>
</TableRow>


<TableRow android:layout_width="fill_parent">

    <TextView
        android:id="@+id/textView1"
        android:layout_width="wrap_content"
        android:layout_height="wrap_content"
        android:text="phone" >
    </TextView>

    <EditText
        android:id="@+id/phone"
        android:layout_width="fill_parent"
        android:layout_height="wrap_content"
         android:hint="Phone" >
    </EditText>
</TableRow>

<TableRow android:layout_width="fill_parent" >

    <Button
        android:id="@+id/button_update"
        android:layout_width="0px"
        android:layout_height="wrap_content"
        android:layout_weight="1"
        android:text="Update" >
    </Button>

    <Button
        android:id="@+id/button_delete"
        android:layout_width="0px"
        android:layout_height="wrap_content"
        android:layout_weight="1"
        android:text="Delete" >
    </Button>

    <Button
        android:id="@+id/button_cancel"
        android:layout_width="0px"
```

```
                android:layout_height="wrap_content"
                android:layout_weight="1"
                android:text="Cancel" >
            </Button>


        </TableRow>
    </TableLayout>

</LinearLayout>
```

Listing 13.5 CustomerDetailActivity Layout File

Listview Row and Header

The layout files in this section define subcomponents of the user interfaces for the `CustomerListActivity` and `CustomerSearchActivity` activities.

customerrow.xml

The code in the *customerrow* layout file, shown in Listing 13.6, defines the structure of a row in the lists shown by the `CustomerListActivity` and the `CustomerSearchActivity`. Each row consists of two TEXTVIEW controls: one to show the company name and one to show the customer name.

```
<?xml version="1.0" encoding="utf-8"?>
<LinearLayout xmlns:android="http://schemas.android.com/apk/res/
android"
    android:layout_width="match_parent"
    android:layout_height="match_parent"
    android:orientation="vertical" >

    <TextView
        android:id="@+id/textView1"
        android:layout_width="wrap_content"
        android:layout_height="40dp"
        android:textColor="#00ccff"
        android:textSize="30sp"
        android:shadowColor="#00ccff"
        android:shadowRadius="1.5"
        android:shadowDx="1"
        android:shadowDy="1"
```

```
        android:text="TextView" >
    </TextView>

    <TextView
        android:id="@+id/textView2"
        android:layout_width="wrap_content"
        android:layout_height="30dp"
         android:textSize="20sp"
        android:text="TextView" >
    </TextView>

</LinearLayout>
```

Listing 13.6 List Row Layout File

customerlistheader.xml

The code for the *customerlistheader* file, shown in Listing 13.7, defines the structure of the header that appears at the top of the lists. It includes a single TextView control.

```
<?xml version="1.0" encoding="utf-8"?>
<LinearLayout xmlns:android="http://schemas.android.com/apk/res/
android"
    android:layout_width="fill_parent"
    android:layout_height="fill_parent"
    android:orientation="horizontal" >

<TextView android:id="@+id/txtHeader"
        android:layout_width="fill_parent"
        android:layout_height="fill_parent"
        android:gravity="center_vertical"
        android:layout_alignParentTop="true"
        android:layout_alignParentBottom="true"
        android:textStyle="bold"
        android:textSize="22dp"
        android:textColor="#FFFFFF"
        android:padding="10dp"
        android:text="Customers"
        android:background="#336699" />
</LinearLayout>
```

Listing 13.7 List Header Layout File

13.6.3 MainActivity

Finally, you are ready to develop the activities. The Project Creation Wizard creates a single activity named `MainActivity`. The `MainActivity` is the first activity shown when the app starts. It has a number of things to accomplish in the sample app. In addition to creating the user interface by instantiating the *activitymain* layout file, `MainActivity` must connect to the Unwired Server, open the local database, and provide the code required to synchronize with the CDB.

Open the `MainActivity` and replace the code with the code shown in Listing 13.8. Leave the line that begins with `Package` at the top of the file.

You need to import the required libraries. If you see an error appear when you enter a line of code, missing required libraries is likely the cause. You can let Eclipse import the correct libraries by right-clicking the background of the editor and selecting SOURCE • ORGANIZE IMPORTS.

You also need to replace the values of the constants to reflect your circumstances. If you have the default installation without a Relay Server, you need to change only the `SERVER`, `USER`, and `ACTIVECODE`.

```
public class MainActivity extends Activity implements Runnable,
OnClickListener {
  private final String SERVER = "smp.example.com";
  private final int PORT = 5001;
  private static String PROTOCOL = "http";
  private static String DOMAIN = "default";
  private static final int REQUEST_DETAIL = 99;
  private static final int REQUEST_LIST = 98;
  private static final int REQUEST_SEARCH = 97;
  private static String USER = "CustomerAndroid";
  private static String ACTIVECODE = "123";
  private static int SYNC_PORT = 2480;
  private static String TAG = "Customers";
  private ProgressDialog pd;

  @Override
  public void onCreate(Bundle savedInstanceState) {
   super.onCreate(savedInstanceState);
   setContentView(R.layout.activitymain);
   Button sync = (Button) findViewById
      (R.id.button_sync);
Button list = (Button) findViewById
```

```
(R.id.button_list);
   Button search = (Button) findViewById
(R.id.button_search);
   Button newCustomer = (Button) findViewById
(R.id.button_create);
   sync.setOnClickListener(this);
   list.setOnClickListener(this);
   search.setOnClickListener(this);
   newCustomer.setOnClickListener(this);

   initApp();
  }


  /**
  * Initialize the app and register/
  * connect to the Unwired Server
  * */
  private void initApp() {
Log.d(TAG, "entering initapp");

//Create the application object
   Application app = App.getInstance(); app.setAppIdentifier("Customers
MBO");
   app.setAppContext( getAppContext());


   // Set synchronization profile
   ConnectionProfile cp = CustomersMBODB
.getSynchronizationProfile();
   cp.setServerName(SERVER);
   cp.setPortNumber(SYNC_PORT);
   cp.setNetworkProtocol(PROTOCOL);
   cp.setAsyncReplay(true);
   cp.setDomainName(DOMAIN);
   cp.save();

   // Set connection properties
   ConnectionProperties connProps =
      app.getConnectionProperties();
   connProps.setServerName(SERVER);
   connProps.setPortNumber(PORT);
   connProps.setLoginCredentials(new
```

```
      LoginCredentials(USER,""));
   connProps.setActivationCode("123");
// Register or connect the app with the Server
   if (app.getRegistrationStatus() !=
  RegistrationStatus.REGISTERED) {
     Log.d(TAG, "registering app");
     app.registerApp();
   } else {
     Log.d(TAG, "connecting app");
     app.startConnection();
   }
}

//Click handlers for buttons
@Override
public void onClick(View v) {
  switch (v.getId()) {
   case R.id.button_sync:
      sync();
      break;
   case R.id.button_list:
      Intent intent1 = new Intent(
MainActivity.this,
      CustomerListActivity.class);
MainActivity.this. startActivityForResult
(intent1, REQUEST_LIST);
      break;
   case R.id.button_search:
      Intent intent2 = new Intent(
MainActivity.this, SearchCustomersActivity.class);
MainActivity.this.
startActivityForResult(intent2,
REQUEST_SEARCH);
      break;
   case R.id.button_create:
      Intent intent = new Intent(
MainActivity.this,
CustomerDetail.class);
      intent.putExtra("sKey", -1);
MainActivity.this. startActivityForResult(intent,
         CustomerDetail.REQUEST_DETAIL);
      break;
   }
```

```
}

/**
 * Start the progress dialog and start the
 * synchronization thread
* */
private void sync() {
pd = ProgressDialog.show(this, "Working..",
   "Synchronizing", true,false);
   Thread thread = new Thread(this);
   thread.start();
}

/**
 * Thread that Synchronizes the specified
 * synchronization group
* */
public void run() {
CustomerSynchronizationParameters
   syncParameter = Customer
      .getSynchronizationParameters();
   syncParameter.delete();
   syncParameter = Customer.
getSynchronizationParameters();
   syncParameter.setState_SP("MN");
   syncParameter.save();
   CustomersMBODB.synchronize();
   handler.sendEmptyMessage(0);
}

/**
 * The handler receives the message from the sync
 * thread and cancels the progress dialog
* */
private Handler handler = new Handler() {
   @Override
   public void handleMessage(Message msg) {
      pd.dismiss();
   }
};

}
```

Listing 13.8 MainActivity Implementation

The following sections describe the portions of the `MainActivity` code that use the Object API.

Registering the App

The code shown in Listing 13.9 creates an instance of the `application` object and associates it with the MBO package you deployed to the Unwired Server. This code must be executed on the main thread, although you may wish to move the registration/connection process into a separate thread. The third line is required for Android. It sets the Android context, which provides information about the app to the Unwired Platform client.

```
Application app = App.getInstance();
app.setAppIdentifier("CustomersMBO");
app.setAppContext(getAppContext());
```

Listing 13.9 Creating an Instance of the Application Object

After the `application` object is created, the connection properties are created and configured with information for the Unwired Server. This code, shown in Listing 13.10, assumes you are not connecting through a Relay Server. If you are using a Relay Server, the port is 80, and you need to configure the farm ID and the URL suffix.

```
ConnectionProperties connProps =
  app.getConnectionProperties();
connProps.setServerName(SERVER);
connProps.setPortNumber(PORT);
connProps.setLoginCredentials(new
  LoginCredentials(USER,""));
connProps.setActivationCode("123");
```

Listing 13.10 Connection Properties

The first time the app runs on the client, it must register itself with the Unwired Server. The code shown in Listing 13.11 registers the app if it has not already been registered. Otherwise, it connects to the Unwired Server.

```
if (app.getRegistrationStatus() !=
  RegistrationStatus.REGISTERED) {
  app.registerApp();
} else {
```

```
  app.startConnection();
}
```

Listing 13.11 Registering the App

Synchronizing the Data

After the app is registered and connected to the Unwired Server, the data can be synchronized. The code in Listing 13.12 retrieves the synchronization profile object from the database and configures the parameters required for synchronization. Port 2480 is used when the network protocol is HTTP. If HTTPS is used, the port is 2481, and if connecting through a Relay Server, the port is 80.

```
ConnectionProfile cp = CustomersMBODB.getSynchronizationProfile();
cp.setServerName(SERVER);
cp.setPortNumber(SYNC_PORT);
cp.setNetworkProtocol(PROTOCOL);
cp.setDomainName("default");
cp.save();
```

Listing 13.12 Connection Profile

Synchronization is handled by the sync function shown in Listing 13.13, which initiates a dialog to show the user that synchronization is in progress and then starts a thread on which the synchronization process runs.

```
/**
 * Start the progress dialog and start the
 * synchronization thread
* */
private void sync() {
pd = ProgressDialog.show(this, "Working..",
   "Synchronizing", true,false);
   Thread thread = new Thread(this);
   thread.start();
}
```

Listing 13.13 Starting a Thread to Synchronize Data

The run function, shown in Listing 13.14, initiates the new thread. Before synchronizing, you must configure the synchronization parameter configured on the CUSTOMER MBO. The first line retrieves a reference to the CUSTOMER MBO `synchronization parameters` object. The next line deletes previously configured values. Synchronization parameter values are cumulative, so each time a value is set, it is added to the previous values rather than replacing them. If you want to replace previous values, you must call the `delete` method.

```
/**
 * Synchronizes the specified synchronization group
* */
public void run() {
CustomerSynchronizationParameters
   syncParameter = Customer
      .getSynchronizationParameters();
   syncParameter.delete();
```

Listing 13.14 Retrieving Synchronization Parameters

After deleting previous values, you must retrieve the `synchronization parameter` object again. The `synchronization parameter` class includes methods to set the values for each of the MBO's synchronization parameters. The code in Listing 13.15 uses the `setState_SP` method to set the `State_SP` synchronization parameter to `MN` and then saves the settings. Of course, we would not normally hard code the parameter value, but it is done here to simplify the code. Synchronization parameters do not take effect unless you call the `save` method, even if a default value is set for the parameter.

```
   syncParameter = Customer.
getSynchronizationParameters();
   syncParameter.setState_SP("MN");
   syncParameter.save();
```

Listing 13.15 Setting Synchronization Parameter

The synchronize method initiates the synchronization process. If you want to synchronize a specific synchronization group, include the name as an argument in the synchronize method. Finally, the call to `sendEmptyMessage` signals the parent thread that the synchronization thread has completed. This causes the `handleMessage` function to execute, which dismisses the progress dialog. The code is shown in Listing 13.16.

```
   CustomersMBODB.synchronize();
   handler.sendEmptyMessage(0);
}

private Handler handler = new Handler() {
   @Override
   public void handleMessage(Message msg) {
```

```
        pd.dismiss();
    }
};
```

Listing 13.16 Synchronizing Data

After you configure the connection and synchronization properties in `MainActivity`, you can perform synchronization by simply calling the `synchronize` method. In this app, synchronization is performed manually when the user taps the SYNC button on the home screen, but you could have the app synchronize automatically when the user modifies the data.

13.6.4 CustomerListActivity

The `CustomerListActivity` displays a list of all the customers in the CUSTOMER MBO. The activity instantiates the user interface, which includes two layout files: *customerlist* and *customerlistheader*. The list data is managed by `CustomerListAdapterActivity`, so `CustomerListActivity` must instantiate that activity and pass it the customer data retrieved from the local database using a dynamic query. The list must be able to respond when the user taps the list by navigating to the `CustomerDetailActivity` and displaying the tapped customer's data.

To create a new activity in Eclipse, right-click the package under the SRC folder and select NEW • CLASS. The name you enter must match exactly the names used in this chapter. The code for the `CustomerListAtivity` is shown in Listing 13.17.

```
public class CustomerListActivity extends Activity {
  private CustomerListAdapter adapter;

  public void onCreate(Bundle savedInstanceState) {
    super.onCreate(savedInstanceState);
    setContentView(R.layout.customerlist);
   viewCustomerList();
  }

  /**
  * Initialize the ListView and attach a click
  * handler
  * */
  private void viewCustomerList() {
    ListView listView =
   (ListView)findViewById(R.id.listView1);
```

```
    View header = (View) getLayoutInflater()
      .inflate(R.layout
      .customer_list_header, null);
    listView.addHeaderView(header);
    adapter = new CustomerListAdapter(this,
      listView, getCustomers());
    listView.setOnItemClickListener(new
    OnItemClickListener()
      {public void onItemClick(AdapterView<?>
      parent, View view, int position,
          long id) {
      Intent intent = new
      Intent(
  CustomerListActivity.this,
          CustomerDetail.class);
        intent.putExtra("sKey",
   .getSK(position-1));
  CustomerListActivity
   .this.startActivityForResult(
Intent, CustomerDetail
  .REQUEST_DETAIL);
      }
      });
  }

  /**
  * Retrieve the data from the Customer MBO
  * using a query object
  * */
  private GenericList<Customer> getCustomers() {
   Query query = new Query();
   SortCriteria sort = new SortCriteria();
   sort.add("company_name", SortOrder.ASCENDING);
   query.setSortCriteria(sort);
   GenericList<Customer> customerIds =
      Customer.findWithQuery(query);
   return customerIds;
  }

  /**
  * Refresh the list contents
  * */
  private void refreshList() {
   getCustomers();
```

```
CustomerListActivity.this.adapter.refreshUI();
  }

  /**
  * When Customer Details returns refresh the list
  * */
  protected void onActivityResult(int requestCode,
   int resultCode, Intent data) {
   if (requestCode ==
      CustomerDetail.REQUEST_DETAIL) {
      if (resultCode == RESULT_OK) {
         refreshList();
      }
   }
  }
}
```

Listing 13.17 CustomerListActivity Implementation

Retrieving Customer Data

The only portion of this code that uses the Object API is the `getCustomers` function shown in Listing 13.18. The code uses a dynamic query to retrieve all the records sorted by company name.

```
/**
  * Retrieve the data from the Customer MBO
  * using a query object
  * */
  private GenericList<Customer> getCustomers() {
      //Create the query object
   Query query = new Query();
      //Create a sort criteria object
   SortCriteria sort = new SortCriteria();
      //Add the criteria
   sort.add("company_name", SortOrder.ASCENDING);
      //Add the sort criteria to the query
   query.setSortCriteria(sort);
   //Execute the query
   GenericList<Customer> customers =
      Customer.findWithQuery(query);
   return customers;
  }
```

Listing 13.18 Retrieving Data from the Local Database

13.6.5 CustomerListAdapter Activity

The `CustomerListAdapter` activity handles the data for the `CustomerListActivty`'s LISTVIEW control. It has no user interface itself but it does instantiate the customer row layout file for each row in the list.

The code for the `CustomerListAdapter` activity is shown in Listing 13.19.

```
public class CustomerListAdapter  extends
  ArrayAdapter<Customer> {
  private Context context;
  private GenericList<Customer> customers;

  public CustomerListAdapter(Activity activity,
   ListView listView, <Customer> customers)
  {
   super(activity,R.layout.customerlist,
      customers);
   this.context=activity;
   this.customers=customers;
   listView.setAdapter(this);
   }

  /**
  * Add the customer records to the adapter
  * */
   public void setCustomers(GenericList<Customer>
   customers)
   {
   this.customers = customers;
   clear();
   for (int i = 0; i < customers.size(); i++)
   {
      this.add(customers.get(i));
   }
   }

  /**
  * Inflate the item layout for each row
  * */
   @Override
   public View getView(int position, View
   convertView, ViewGroup parent) {
```

```
    Customer customer = customers.get(position);
    LayoutInflater inflater = (LayoutInflater)
       context.getSystemService(Context
       .LAYOUT_INFLATER_SERVICE);
    View rowView = inflater.inflate(R.layout
       .customerrow, parent, false);
    TextView textView1 = (TextView) rowView
       .findViewById(R.id.textView1);
    TextView textView2 = (TextView) rowView
       .findViewById(R.id.textView2);
       textView1.setText(customer.getCompany_name());
    textView2.setText(customer.getFname() + " "
       + customer.getLname());

    return rowView;
  }

  /**
  * Retrieve the surrogate key of an item based on
  * its position in the list
  * */
  public long getSK(int position){
     Customer customer = customers.get(position);
     return customer.getSurrogateKey();
  }

  /**
  * Inform the adapter that the data has changed
  * */
  public void refreshUI(){
       CustomerListAdapter.this
.notifyDataSetChanged();
   }

}
```

Listing 13.19 CustomerListAdapter Activity Implementation

Retrieving the Surrogate Key

The only portion of this code that interacts with the Object API is the `getSK` function, shown in Listing 13.20. This function is called by the LISTVIEW control's click handler to retrieve the surrogate key of the selected record. Surrogate keys are created by the Unwired Server and are used to access data instead of the primary keys defined in the EIS. The surrogate key is retrieved here so that the `CustomerDetailActivity` can display the data for the selected customer. The code first retrieves an instance of the customer based on the position of the list that was tapped. It then uses the `getSurrogateKey` function to return the value. There are `getter` and `setter` methods for all of the MBO attributes.

```
/**
  * Retrieve the surrogate key of an item based on
  * it's position in the list
  * */
  public long getSK(int position){
     Customer customer = customers.get(position);
     return customer.getSurrogateKey();
  }
```

Listing 13.20 Retrieving the Surrogate Key

13.6.6 CustomerDetailsActivity

The `CustomerDetailsActivity` displays the data for a selected customer. The user can edit the data, delete the customer, or cancel and return to the list screen. The activity uses the surrogate key passed to it to retrieve the correct customer record, inflates the customer details layout file, and fills the EDITTEXT controls with the customer data. When the user taps the CREATE or UPDATE button, the customer record is updated, and the changes are submitted to the local database. In this app, the changes are not synchronized with the Unwired Server until the user taps the SYNC button on the home screen.

The code for the `CustomerDetailsActivity` is shown in Listing 13.21.

```
public class CustomerDetail extends Activity
  implements OnClickListener {

    public static int REQUEST_DETAIL = 98;
    Customer customer;
    private long surrogateKey;
    private EditText fname, lname, company, address,
   city, state, zip, phone;

    @Override
    public void onCreate(Bundle savedInstanceState) {
```

```
    super.onCreate(savedInstanceState);
    setContentView(R.layout.customerdetails);

    Button update = (Button) findViewById(
R.id.button_update);
    Button cancel = (Button) findViewById(
R.id.button_cancel);
    Button delete = (Button) findViewById(
R.id.button_delete);
    update.setOnClickListener(this);
    cancel.setOnClickListener(this);
    delete.setOnClickListener(this);

    EditText fname = (EditText) findViewById(
R.id.fname);
    EditText lname = (EditText) findViewById(
R.id.lname);
    EditText company = (EditText) findViewById(
R.id.companyname);
    EditText address = (EditText) findViewById(
R.id.address);
    EditText city = (EditText) findViewById(
R.id.city);
    EditText state = (EditText) findViewById(
R.id.state);
    EditText zip = (EditText) findViewById(
R.id.zip);
    EditText phone = (EditText) findViewById(
R.id.phone);

    // Retrieve the surrogate key from the intent
    Bundle bundle = getIntent().getExtras();
    surrogateKey = bundle.getLong("sKey");

    //If surrogate key is > 0 then retrieve the
    //customer
    if (surrogateKey > 0) {
      customer = Customer.find(surrogateKey);
      fname.setText(customer.getFname());
      lname.setText(customer.getLname());
      company.setText(customer.getCompany_name());
      address.setText(customer.getAddress());
      city.setText(customer.getCity());
```

```
      state.setText(customer.getState());
      zip.setText(customer.getZip());
      phone.setText(customer.getPhone());
    }
    }

    /**
    * Click handler for the buttons
    * */
    @Override
    public void onClick(View v) {
    switch (v.getId()) {
      case R.id.button_update:
        EditText fname = (EditText) findViewById(
R.id.fname);
        EditText lname = (EditText) findViewById(
R.id.lname);
        EditText company = (EditText) findViewById(
R.id.companyname);
        EditText address = (EditText) findViewById(
R.id.address);
        EditText city = (EditText) findViewById(
R.id.city);
        EditText state = (EditText) findViewById(
R.id.state);
        EditText zip = (EditText) findViewById(
R.id.zip);
        EditText phone = (EditText) findViewById(
R.id.phone);
        if (surrogateKey > 0) {
       Log.d("Customers","update");
       Customer customer = Customer.find(
  CustomerDetail.this.surrogateKey);
         customer.setFname(fname.getText()
.toString());
         customer.setLname(lname.getText()
.toString());
  customer.setCompany_name(company
.getText().toString());
            customer.setAddress(address.getText()
.toString());
         customer.setState(state.getText()
.toString());
```

```
        customer.setZip(zip.getText()
.toString());
  customer.setPhone(phone.getText()
.toString());
        customer.update();
        customer.submitPending();
      } else {
        Customer customer = new Customer();
customer.setFname(fname.getText()
.toString());
  customer.setLname(lname.getText()
.toString());
  customer.setCompany_name(company
.getText().toString());
customer.setAddress(address.getText()
.toString());
  customer.setState(state.getText()
.toString());
  customer.setZip(zip.getText()
.toString());
  customer.setPhone(phone.getText()
.toString());

        customer.create();
        customer.submitPending();
      }
      break;
   case R.id.button_delete:
      Customer customer = Customer
         .find(CustomerDetail.this
.surrogateKey);
      customer.delete();
      customer.submitPending();
      break;
   }
      CustomerDetail.this.setResult(Activity
.RESULT_OK);
      CustomerDetail.this.finish();
   }
}
```

Listing 13.21 CustomerDetailsActivity Implementation

Retrieving the Customer Record

If the surrogate key is greater than zero, then the CustomerDetail activity was activated when the user selected an existing customer on the LISTVIEW, so the code must retrieve the customer record using the CUSTOMER MBO find method. The get methods for the MBO attributes are used to set the values of the EDITBOX controls.

```
if (surrogateKey > 0) {
     customer = Customer.find(surrogateKey);
     fname.setText(customer.getFname());
```

Updating a Customer

If the user clicks the UPDATE button, the code shown in Listing 13.22 again checks to see if the data is from an existing customer. If it is, the customer object is retrieved, and the CUSTOMER MBO set methods are used to update the attribute values from the values in the EDITBOX controls. After updating the attribute values, the update method is called, which implements the update operation on the local MBO. Finally, the submitPending method is called to create an operation replay record that is uploaded to the server the next time that synchronize is called.

```
if (surrogateKey > 0) {
   Log.d("Customers","update");
   Customer customer = Customer.find(
       CustomerDetail.this.surrogateKey);
   customer.setFname(fname.getText()
     .toString());
...
customer.update();
customer.submitPending();
```

Listing 13.22 Updating a Customer

Creating a Customer

If the surrogate key is less than zero, then a new customer is being created. When the user clicks the UPDATE button, the code shown in Listing 13.23 must first create a new instance of a customer and then set the attribute values. The create method implements the create operation on the MBO, and then submitPending creates the operation replay records uploaded to the server.

```
Customer customer = new Customer();
customer.setFname(fname.getText()
.toString());
```

```
...
customer.create();
customer.submitPending();
```

Listing 13.23 Creating a Customer

Deleting a Customer

Finally, if the user clicks the DELETE button, the code shown in Listing 13.24 finds the customer and then executes the `delete` operation before calling `submitPending`.

```
Customer customer = Customer
        .find(CustomerDetail.this
.surrogateKey);
customer.delete();
customer.submitPending();
```

Listing 13.24 Deleting a Customer

13.6.7 CustomerSearchActivity

The `CustomerSearchActivity` allows the user to enter a company name or a portion of a company name and retrieve a list based on the value entered. Although the interface is different because it includes the EDITTEXT control the user uses to enter a search term, the functionality of this activity is similar to `CustomerListActivity`. It retrieves the customer data and passes it to the `CustomerListAdapterActivity` to manage. It also must respond to taps by navigating to the `CustomerDetailActivity`. The main difference between this activity and `CustomerListActivity` is that it uses an objects query defined in the MBO to retrieve the customer data. The object query includes a parameter used to filter the data retrieved.

The code for the `CustomerSearchActivity` is shown in Listing 13.25.

```
public class SearchCustomersActivity extends Activity
  implements  OnClickListener {

  private CustomerListAdapter adapter;

  public void onCreate(Bundle savedInstanceState) {
    super.onCreate(savedInstanceState);
    setContentView(R.layout.customersearch);

    Button search = (Button) findViewById(
R.id.submit);
```

```
    ListView listView = (ListView) findViewById(
R.id.listView1);
    View header = (View) getLayoutInflater()
.inflate(R.layout.customerlistheader, null);
    listView.addHeaderView(header);

    search.setOnClickListener(this);

 }

 /**
 * Click handler for the buttons
 * */
 public void onClick(View v) {
   switch (v.getId()) {
      case R.id.submit:
      GenericList<Customer> customers =
      getCustomers();
      ListView listView = (ListView) findViewById(
R.id.listView1);
         adapter = new CustomerListAdapter(this,
 listView, customers);
         listView.setOnItemClickListener(new
OnItemClickListener() {
        public void onItemClick(AdapterView<?>
         parent, View view, int position,
long id) {
           Intent intent = new
            Intent(SearchCustomersActivity
.this, CustomerDetail.class);
         intent.putExtra("sKey", adapter
.getSK(position - 1));
SearchCustomersActivity.this
.startActivityForResult(intent,
            CustomerDetail.REQUEST_DETAIL);
         }

         });
      break;
      }
```

```
  }

   /**
   * Retrieve the customers using an object query
   * */
   private GenericList<Customer>
     getCustomers(){
    EditText editText = (EditText) findViewById(
R.id.editText);
    String companyName = editText
.getText().toString() + "%";
    GenericList<Customer> customerList =
Customer.findByCompany(
companyName);
    return customerList;
  }


   /**
   * Refresh the list
   * */
   private void refreshList() {
     getCustomers();
     SearchCustomersActivity.this
.adapter.refreshUI();
  }

   /**
   * Call refresh list when returning from customer
   * detail
   * */
   protected void onActivityResult(int requestCode,
 int resultCode, Intent data) {
     if (requestCode == CustomerDetail
.REQUEST_DETAIL) {
     if (resultCode == RESULT_OK) {
       refreshList();
     }
    }
  }
}
```

Listing 13.25 CustomerSearchActivity Implementation

Retrieving Customer Data

The interesting portion of this code, shown in Listing 13.26, is in the `getCustomers` method, which uses the object query that was created in the CUSTOMER MBO. The code retrieves the value entered in the EDITTEXT box by the user and appends a % wildcard. The SQL uses the `LIKE` operator so the user can enter a partial name. The object query is called `findByCompany`, and the code generator creates a method in the `Customer MBO` class by the same name. The parameters of the object query are passed as arguments of the method.

```
    /**
    * Retrieve the customers using an object query
    * */
    private GenericList<Customer>
      getCustomers(){
     EditText editText = (EditText) findViewById(
R.id.editText);
     String companyName = editText
.getText().toString() + "%";
     GenericList<Customer> customerList =
Customer.findByCompany(
companyName);
     return customerList;
   }
```

Listing 13.26 Retrieving Data Using an Object Query

An alternative to using an object query is to use a dynamic query. The code in Listing 13.27 shows how this works.

```
private GenericList<Customer> getCustomers() {
     EditText editText = (EditText) findViewById(
R.id.editText);
     String companyName = editText
.getText().toString();
        //Create a query object
     Query query = new Query();
        //Create and configure an attributetest
        //object (implements SQL WHERE clause)
     AttributeTest aTest = new AttributeTest();
     aTest.setAttribute("company_name");
```

```
    aTest.setValue(companyName + "%");
    aTest.setOperator(AttributeTest.LIKE);
       //Add the attributetest to query object
    query.setTestCriteria(aTest);
       //Create and configure a sortcriteria object
    SortCriteria sort = new SortCriteria();
    sort.add("company_name", SortOrderType
.ASCENDING);
       //Add the sortcriteria to the query object
    query.setSortCriteria(sort);
    query.from("Customer", "x");
       //Execute the query
    GenericList<Customer> customerList =
        Customer.findWithQuery(query);
       return customerList;

  }
```

Listing 13.27 Filtering Data Using a Dynamic Query

13.7 Running the App

To run the app, create an application connection on the Unwired Server. Log on to the SCC and navigate to APPLICATIONS • APPLICATION CONNECTIONS. Click REGISTER. The properties are shown in Figure 13.18. These properties must match those assigned in the `MainActivity` class. The user (CUSTOMERANDROID) and activation code (123) are assigned in the connection properties, and the application ID must match the application ID assigned to the application instance in `MainActivity`.

Next, create a run configuration in Eclipse. Right-click the project in the Package Explorer and select RUN AS • RUN CONFIGURATIONS. Click the NEW LAUNCH CONFIGURATION icon on the top left. Enter the project name in the PROJECT field and enter a name in the NAME field. On the TARGET tab, enter your preference for the way you want to run the app, emulator, or device. Next, click RUN (Figure 13.19).

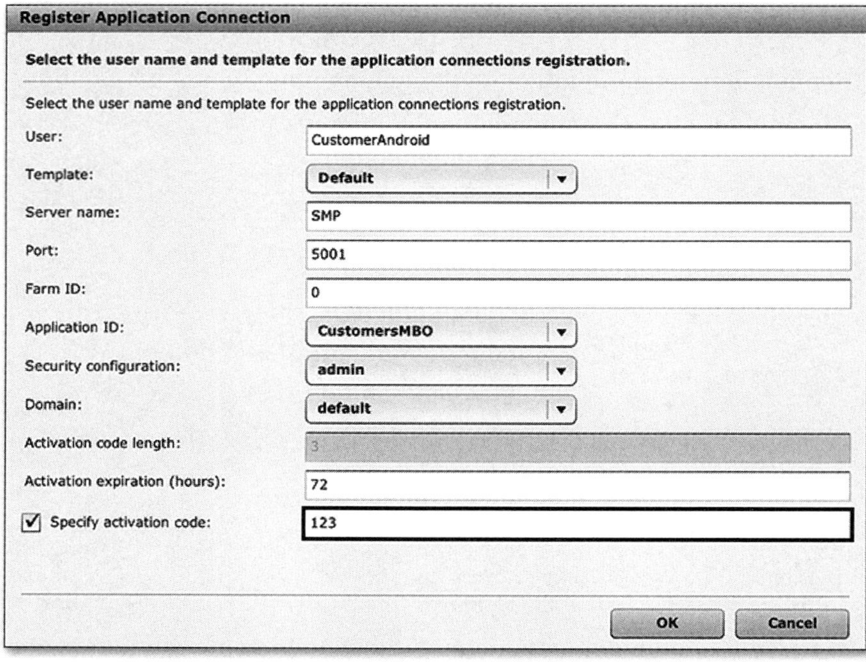

Figure 13.18 Creating an Application Connection on the Unwired Server

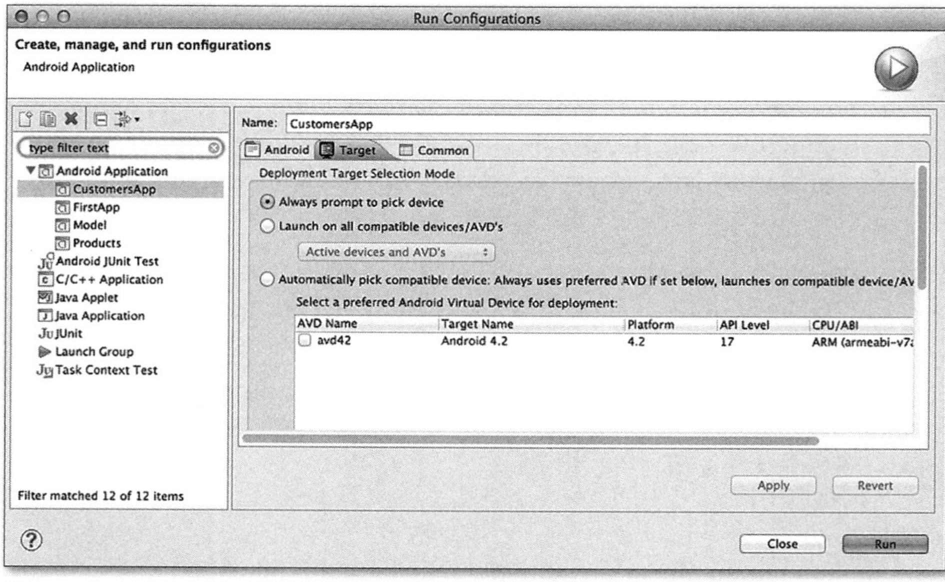

Figure 13.19 Configuring the Run Configuration before Running the App

13.8 Summary

This chapter should provide enough to get you started developing Android Object API apps. There is much more to learn, and a good place to start is the *Developer Guide: Android Object API Apps*, which is available at *http://infocenter.sybase.com/help/index.jsp?topic=/com.sybase.infocenter.dc01726.0221/doc/html/title.html*.

This chapter demonstrates the use of the Object API for iOS apps. The basics of iOS apps are outlined, and a complete example app that illustrates Object API fundamentals is presented.

14 Developing iOS Apps

Although it has faced challenges, iOS remains one of the most dominant platforms on the market. This chapter describes the fundamentals of using the Object API for iOS. As with the previous chapter, the program described in this chapter is simple, but many of the essential elements of the Object API are included.

This chapter demonstrates the fundamentals of developing iOS Object API apps. We start by outlining some basics of iOS development, including a description of the example app we'll be programming. Then, we dive straight into the steps of iOS app development, which are the following:

1. Develop the MBOs and deploy them to the Unwired Server.
2. Generate the Objective-C classes that implement the Object API.
3. Create an iOS project.
4. Import libraries.
5. Import the generated Object API files to the new project.
6. Implement the app.

Background Knowledge

While the code in this chapter produces a working iOS app, the focus is on the techniques for using the Object API. Many of the essential elements of a production app, such as validation, callback handlers, and error trapping, have been left out in order to simplify the code. A familiarity with Objective-C and iOS development is helpful since explanations of the code focus only on the Object API.

14.1 Getting Started

This section gets you started with the basics of iOS app development: the components of an iOS app and how to prepare the development environment. We conclude with a description of the example app we will be coding in this chapter.

14.1.1 iOS App Components

Let's start by describing the components of iOS apps. This description is very brief, so if you are new to iOS development, consider this only a starting point.

Model-View-Controller

The best way to understand the organization of iOS apps is through the *model-view-controller* pattern. In this pattern, each component of an app serves a specific purpose in one of the following categories:

- **Model**
  Model objects contain and manage data and have no knowledge of how the data is used by the user interface. The model objects in an Object API app include the classes generated by the Unwired Workspace that access and manage data in MBOs.
- **View**
  View objects are visible on the user interface. They include individual objects, such as labels and buttons, and other objects that serve to organize other objects. View objects know how to display data but do not know what data they should display.
- **Controller**
  Controller objects provide the app logic. They coordinate the flow of data between the model and view objects. When you create an iOS app from a template, the development environment creates a controller object.

The model-view-controller pattern ensures the separation of data, interface, and processing logic. This produces apps that are easier to understand and maintain and enhances code reusability.

XIB Classes

You can create the user interface in an iOS app programmatically or by using the graphical interface builder. If you choose to use the interface builder, the development environment creates an XIB file for each screen in the app. The XIB file is an XML file that contains the data required to generate the user interface. When the project is built, an NIB file is created for each XIB file; NIB files contain information that describes the visual elements of the user interface as well as any objects that are used to manage the interface. The user interface is generated at runtime using the information in the NIB file.

Figure 14.1 shows an XIB file in the interface builder. For each XIB file representing a view, there is a corresponding controller. The connections between the view and controller occur via *outlets* and *actions*. An outlet is something the view can access in the source code. For example, in Figure 14.1, an outlet called `customerName`, which is a property in the controller class, is shown. The interaction between the XIB and controller is made via a *connection*. In this case, the connection is between the search bar on the interface and the `customerName` property.

Actions are methods in the controller. Connections between the actions and view objects allow the methods to respond to interface events. In Figure 14.1, the list button on the tool bar connects to the list method in the controller.

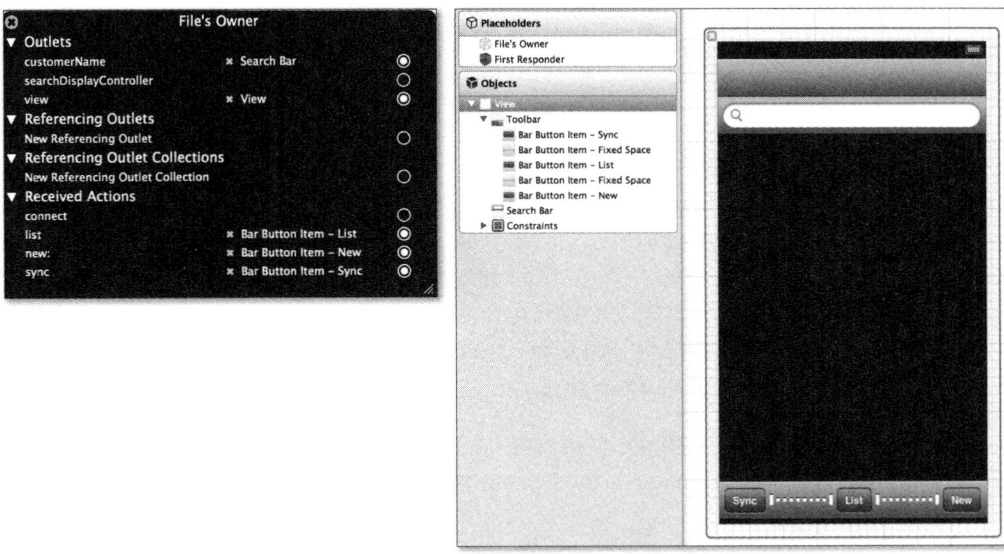

Figure 14.1 XIB File in the Interface Builder

Delegation

Essential to iOS app development is the concept of delegation. Delegation occurs when one object takes responsibility for implementing the methods required by another object. For example, view controllers implement the methods required by view objects to display themselves properly and to interact with the user. One essential delegate in every app is the app delegate, which performs many of the tasks required by the app during startup and shutdown.

14.1.2 Preparing the Development Environment

The development environment for iOS apps is Xcode, which is available from *https://developer.apple.com/xcode/*. The process for installing Xcode and the iOS simulators is described in Chapter 5. Unlike with Eclipse, no further configuration is required to prepare it for developing iOS apps.

14.1.3 Introducing the Example App

The example app developed in this chapter is the same as the one we developed in the previous chapter. It is called CustomerApp, and allows a customer service representative to retrieve, edit, and create customer data while in the field. Once the data is synchronized, the app can be used normally in an offline mode. The simplicity of the app is consistent with the design philosophy that, for mobile apps, simpler is better. Unlike desktop applications, which tend to be complex and multipurpose, users in a mobile environment have come to expect simpler, special-purpose apps. It is a relatively simple matter to integrate the CustomerApp with other device apps, such as contacts, email, and the dialer.

Although the sample app is simple, it is sufficiently complex to demonstrate the fundamental concepts of the iOS Object API. Once the app registers with and connects to the Unwired Server, the user can manually synchronize with the CDB by tapping a button on the first screen. The user can retrieve a complete list of customers or filter the list based on the company name. The user can also edit a customer's data, create a new customer, or delete the customer.

The app consists of a single MBO based on the `Customers` table from the sample database. The MBO includes the object query shown in Figure 14.3.

The app consists of five classes. Screenshots of the three app screens are shown in Figure 14.2. The app opens with the CustomerAppViewController screen. From that screen, the user can tap the SYNC button to synchronize with the Unwired Server. The LIST button opens the ITEMLISTVIEWCONTROLLER screen, showing a list of customers. If the user taps a customer on the list, the DETAILVIEWCONTROLLER screen opens with the customer data, and the user can edit the data or delete the customer. The NEW button opens the DETAILVIEWCONTROLLER screen and allows the user to enter a new customer. Finally, the user can enter a portion of the company name in the search box on the CUSTOMERAPPVIEWCONTROLLER screen to open the ITEMLISTVIEWCONTROLLER screen with a filtered list of customers.

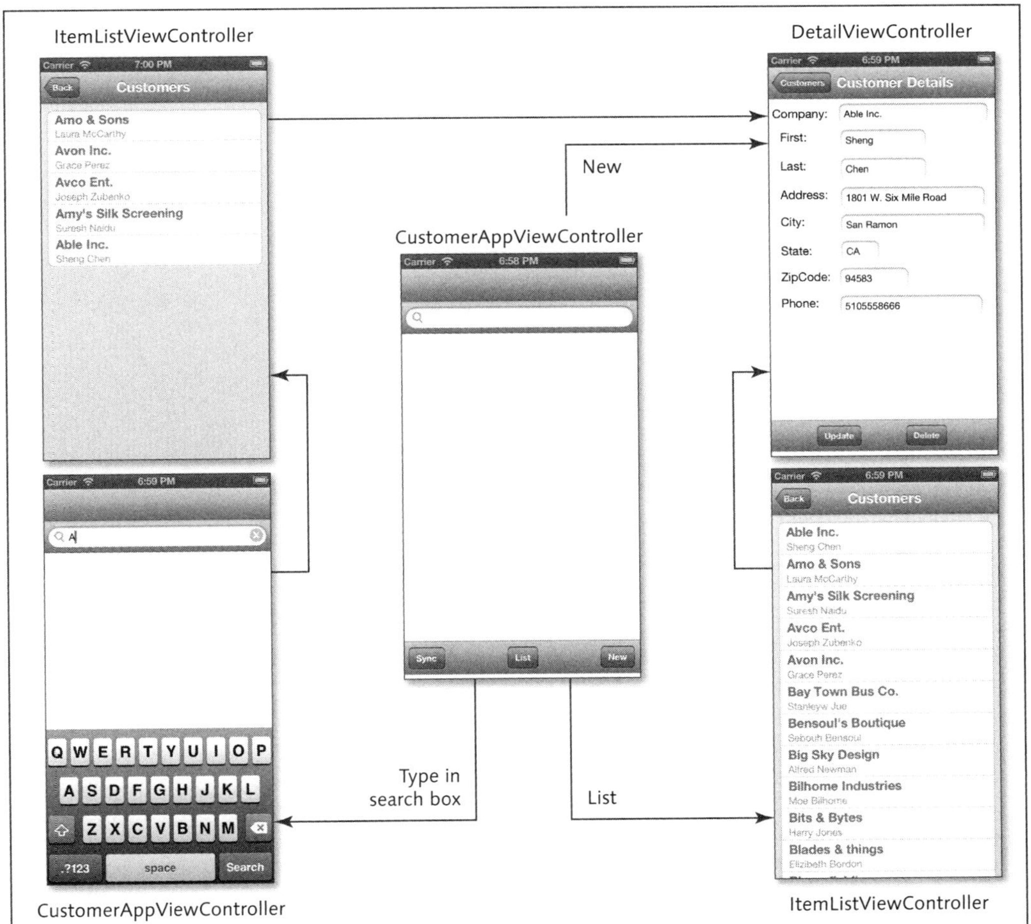

Figure 14.2 Three Screens of the Example App Creating the MBOs

As we explained in Chapter 13, developing MBOs for an Object API app is similar to developing hybrid apps (for more detail on this, see Chapter 7). However, it's important to note that, because they are synchronizing with the CDB, Object API apps must have at least one non-online MBO. Data is not cached in the CDB for online MBOs. Figure 14.3 shows the definition of the MBO.

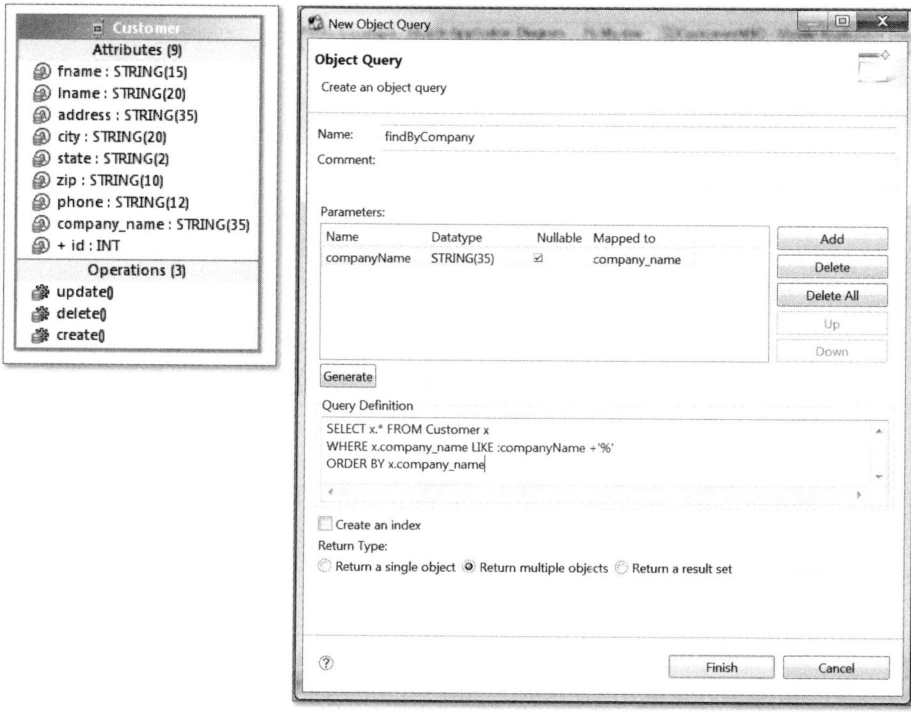

Figure 14.3 Single MBO Based on the Customer Table of the Sample Database

Once the MBOs are created, deploy them to the Unwired Server in the usual way. The app assumes that the MBOs are deployed to the default domain.

14.2 Generating the Object API Classes

By default, the Object API classes are created in a folder called GENERATED CODE in the project workspace. If you intend to create apps for multiple platforms, it is a good idea to create subfolders in the GENERATED CODE folder for each platform.

Simply right-click the GENERATED CODE folder and select NEW • FOLDER to create the folder.

Before starting the Hybrid App Generation Wizard, you must have an active connection to the MBO data sources. To start the wizard, right-click the project name in the WORKSPACE NAVIGATOR or on the background of the Mobile Application Diagram Editor and select GENERATE CODE. The first screen allows you to create a configuration that you can use to recall the settings in the future.

On the next screen, select the MBOs you wish to include. The wizard calculates the page size after you select the MBOs based on the total attribute size. If the total attribute size is larger than the page size setting on the next screen, a warning is displayed.

The last screen of the wizard is shown in Figure 14.4. The following fields can be configured for iOS:

- LANGUAGE
  Select OBJECTIVE-C.
- PLATFORM
  Select IOS.
- UNWIRED SERVER
  Select the Unwired Server connection profile to which you deployed the MBOs.
- DOMAIN
  Select the domain on the Unwired Server to which you deployed the MBOs.
- PAGE SIZE
  The value entered automatically in this field is based on the total size of the MBO attributes. Whatever size is entered here must be larger than the accumulated size of the attributes of any of the MBOs selected and compatible with the database. If you leave the field blank, the page size is set to 4 KB.
- NAME PREFIX
  This field is optional.
- DESTINATION
  You can choose to create the files within the project workspace or an external folder. The default folder is the GENERATED CODE folder in the project workspace. You can also check the box to CLEAN UP DESTINATION BEFORE CODE GENERATION to delete all the files in the destination folder before generating new files.

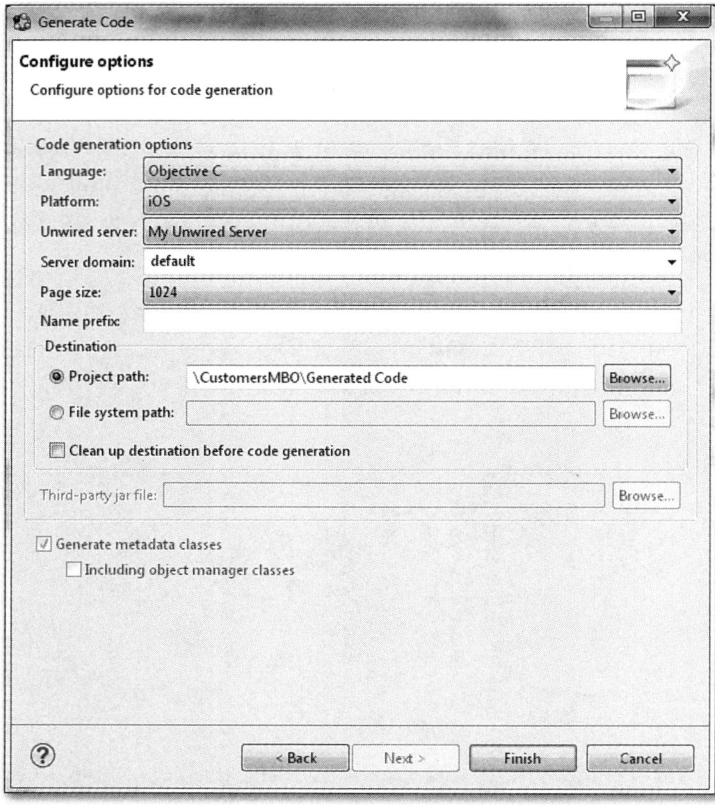

Figure 14.4 Object API Classes Generated by the Generated Code Wizard

Once you click FINISH, the code is generated. Figure 14.5 shows the results in the Workspace Navigator. The specific classes generated depend on the design of the data model. The INCLUDES folder contains the header files, and the SRC folder contains the implementation files. All files are prefixed with the MBO package name.

Among the generated files, the following are referenced in the example app:

- `CustomersMBOCustomersMBODB`
  Implements the methods required to access the device database.
- `CustomersMBOCustomer`
  Implements the attributes, operations, object queries, etc., of the CUSTOMER MBO. A similar class is created for every MBO in the package.

- `CustomersMBOCustomerSynchronizationParameters`
  Implements the synchronization parameter. This is created only because there is a synchornization parameter in this example.
- `CustomersMBOKeyGenerator`
  Implements the generation of surrogate keys used to access MBO instances.

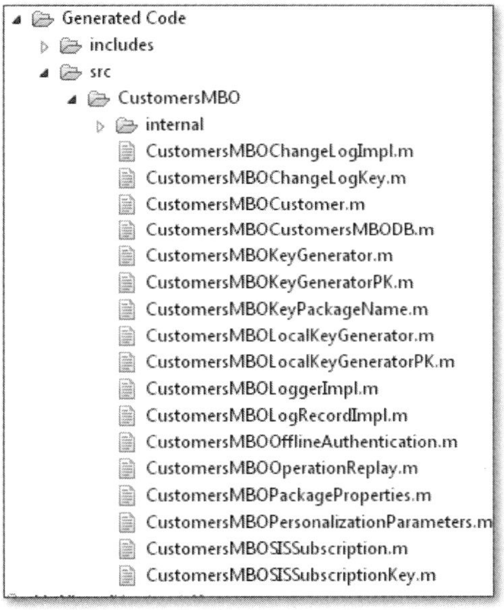

Figure 14.5 Generated Classes

14.3 Creating the Project

In Xcode, select FILE • NEW • PROJECT. Select APPLICATION under the iOS section, and choose a SINGLE VIEW APPLICATION project (see Figure 14.6).

On the next screen, enter "CustomersApp" for the PRODUCT NAME. If you choose to use another name, you have to edit the code later in the chapter to match the name. Enter a company identifier and select UNIVERSAL for DEVICES. Deselect USE AUTOMATIC REFERENCE COUNTING. On the next screen, navigate to the folder in which you wish to create the project and click CREATE.

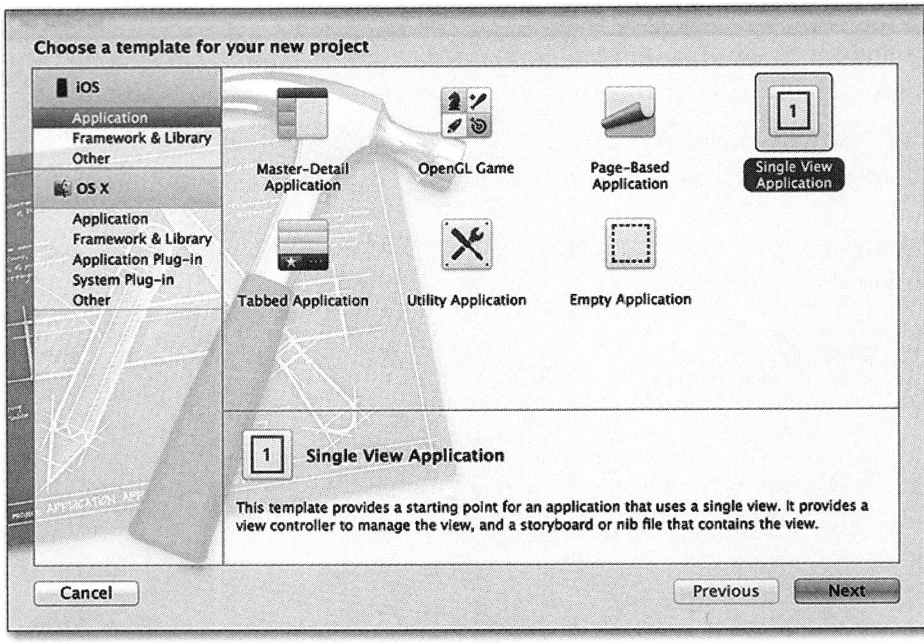

Figure 14.6 Sample App Starting with a Single View App

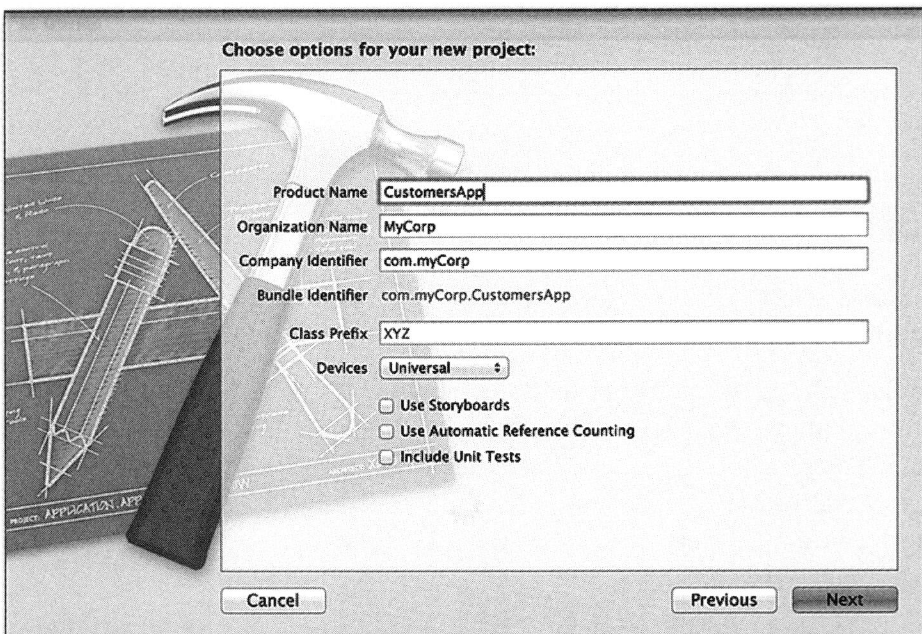

Figure 14.7 The Example App Called CustomersApp

Once you complete the Project Creation Wizard, the project appears as shown in Figure 14.8. The project includes a single view controller with two XIB files and an app delegate. One XIB file is used for the iPhone, and the other is used for the iPad. The example app uses only the iPhone XIB, but you can easily implement the iPad interface by editing the *ViewController_iPad.xib* file.

Figure 14.8 The Initial Project

As part of creating the project, you must also configure it. Select the ARCHITECTURE section of the BUILD SETTINGS. Ensure that the BASE SDK is set to LATEST IOS, the VALID ARCHITECTURES is set to ARMV7, and the SUPPORTED PLATFORMS is set to IOS. Scroll to the DEPLOYMENT section and set the IOS DEPLOYMENT TARGET to the simulator or device version to which you will deploy.

Select CUSTOMERSAPP under PROJECT in the middle pane and then navigate to the APPLE LLVM COMPILER 4.1—LANGUAGE section and make sure OBJECTIVE-C AUTOMATIC REFERENCE COUNTING is set to NO.

14.4 Importing Libraries

Copy the INCLUDES and LIBRARIES folders located in the MOBLESDK213\OBJECTAPI\IOS\RBS folder on your Windows machine to the CUSTOMERSAPP/CUSTOMERSAPP folder on your Mac. Right-click (or control-click) the project name in Xcode in the project navigator and select ADD FILES TO "CUSTOMERSAPP." Navigate to the CUSTOMERSAPP/CUSTOMERSAPP/LIBRARIES/DEBUG-IPHONESIMULATOR folder and select the following files:

- *libclientrt.a*
- *libSUPObj.a*
- *libMO.a*
- *libsupcore.a*
- *libAfariaSLL.a*
- *libDatavault.a*
- *libsupUltralite.a*

Deselect Copy items into destination folder (if needed) and click Add. This assumes you are running the app on a simulator.

Click CustomersApp under Target in the middle pane and then navigate to the Search Paths section and make sure the Library Search Path points to the location of the Libraries folder.

Add a search path to the Header Search Paths that points to the includes folder. Double-click the field and click the plus sign to add a path. Enter the following path:

```
$(SRCROOT)/$(PRODUCT_NAME)/includes
```

Select Recursive so that subfolders are included in the search path. The completed search paths should look similar to those in Figure 14.9.

Select Build Phase and then expand the Link Binary With Libraries section. Click the plus sign and select the following binaries:

- *AddressBook.framework*
- *CoreFoundation.framework*
- *Security.framework*
- *CFNetwork.framework*
- *SystemConfiguration.framework*
- *MobileCoreServices.framework*
- *libicucore.A.dylib*
- *libstdc++.dylib*
- *libz.dylib*

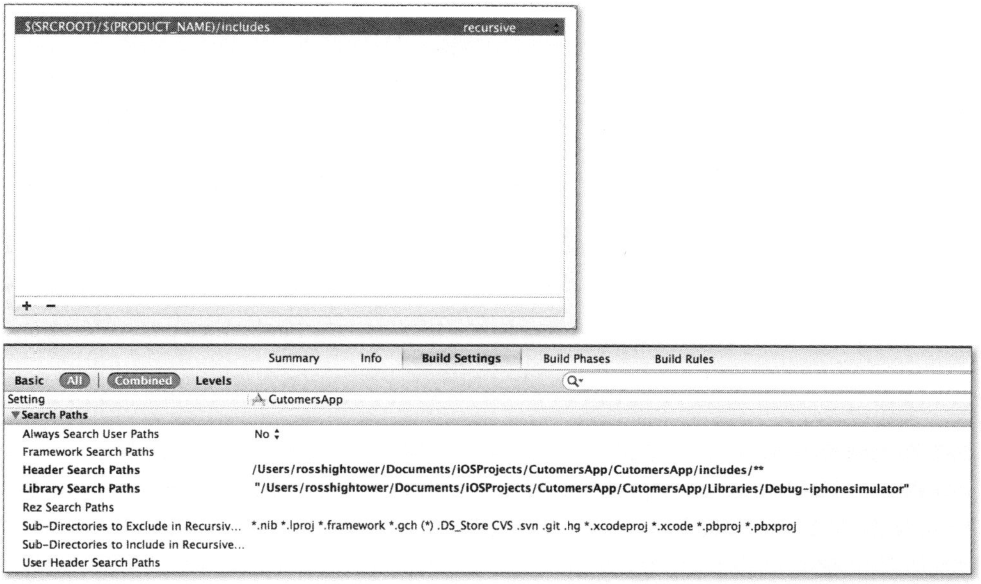

Figure 14.9 Search Paths Pointing to the includes and Libraries Folders

14.5 Importing Generated Code

Copy the GENERATED CODE folder from the project on your Windows machine to the CUSTOMERSAPP/CUSTOMERSAPP folder on your Mac. Right-click (or control-click) the project name in the project navigator and select ADD FILES TO "CUSTOMERSAPP." Select the GENERATED CODE folder, unselect COPY ITEMS INTO DESTINATION FOLDER (IF NEEDED), and click ADD.

The project structure is shown in Figure 14.10. The folders in this view have no effect on how Xcode accesses the files. They are purely for the programmer's organizational preferences.

Select PRODUCT • CLEAN (or `Shift` + `Command` + `K`) and then PRODUCT • BUILD (or `Command` + `B`) to test the configuration. If everything was configured correctly, you should see a message saying that the build succeeded.

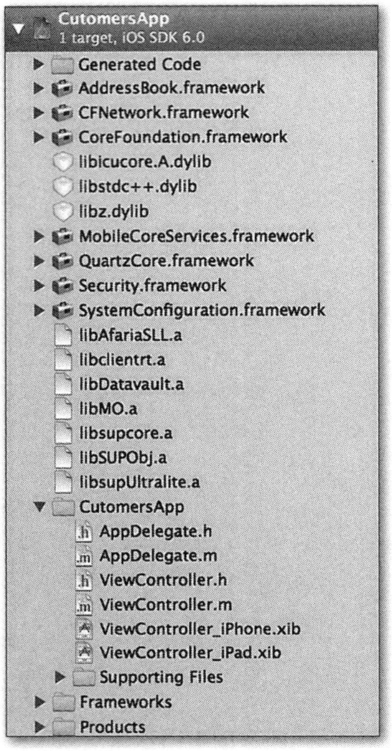

Figure 14.10 Configured Project Structure

14.6 Implementing the App

The following sections include all the code required to implement the app. Two of the screens are built using the interface builder, so you need some basic knowledge of how to use that tool.

Each section begins with the code listing and then includes explanations of those portions of the code that involve the Object API. Each class includes a header file (`*.h`) and an implementation file (`*.m`). The focus is on the minimum functionality required to use the Object API. There are many more options available within the Object API that are described in the documentation.

14.6.1 CustomerAppAppDelegate

`CustomerAppAppDelegate` is the app delegate for the app. Assuming you used the name CustomerApp for your project, the only changes you need to make to the code are to add the import statements and update the `applicationdidFinishLaunchingOptions` method. The following sections show the header (see Listing 14.1) and class (see Listing 14.2) implementations.

CustomerAppAppDelegate.h

```
#import <UIKit/UIKit.h>

@class CustomerAppViewController;

@interface CustomerAppAppDelegate :
      UIResponder <UIApplicationDelegate>

@property (strong, nonatomic) UIWindow *window;

@property (strong, nonatomic)
       CustomerAppViewController *viewController;

@end
```

Listing 14.1 CustomerAppAppDelegate Header File

CustomerAppAppDelegate.m

```
#import "CustomerAppAppDelegate.h"

#import "SUPApplication.h"
#import "SUPApplicationCallback.h"
#import "SUPRegistrationStatus.h"

#import "CustomersMBOCustomersMBODB.h"
#import "SUPEngine.h"
#import "SUPDataVault.h"
#import "SUPConnectionProfile.h"
#import "CustomersMBOLocalKeyGenerator.h"

#import "CustomerAppViewController.h"

@implementation CustomerAppAppDelegate

@synthesize window = _window;
@synthesize viewController = _viewController;
```

```
BOOL wasPreviouslyInBackground = NO;

- (void)dealloc
{
    [_window release];
    [_viewController release];
    [super dealloc];
}

//Runs immediately after the application launches
- (BOOL)application:(UIApplication *)applicaticn
        didFinishLaunchingWithOptions:
         (NSDictionary *)launchOptions
{
    self.window = [[[UIWindow alloc]
          initWithFrame:[[UIScreen mainScreen]
          bounds]] autorelease];

     self.viewController =
         [[[CustomerAppViewController
         alloc] initWithNibName:
          @"CustomerAppViewController_iPhone
         " bundle:nil] autorelease];
    UINavigationController *navController =
         [[UINavigationController alloc]
         initWithRootViewController:
         self.viewController];

    [ConnectionController beginApplicationSetup];

     self.window.backgroundColor = [UIColor
          whiteColor];
    [[self window] setRootViewController:
         navController];
    [self.window makeKeyAndVisible];
    return YES;
}

- (void)applicationWillResignActive:(UIApplication *)application
{

}
```

```
- (void)applicationDidEnterBackground:(UIApplication *)application
{

}

- (void)applicationWillEnterForeground:(UIApplication *)application
{

}

- (void)applicationDidBecomeActive:(UIApplication *)application
{

}

- (void)applicationWillTerminate:(UIApplication *)application
{
    [SUPApplication stopConnection:0];
}

@end
```

Listing 14.2 The CustomerAppAppDelegate Implementation

The part of the code in `CustomerAppAppDelegate` that is relevant to the Object API is the following line:

```
[ConnectionController beginApplicationSetup];
```

This line is included in the `applicationdidFinishLaunchingWithOptions` method, which is invoked immediately after the app launches. The `beginApplicationSetup` method of the `ConnectionController` class contains code that connects the app to the Unwired Server. The result is that the app attempts to connect to the Unwired Server automatically when the user opens it. The code in `ConnectionController` is explained in the following section.

14.6.2 ConnectionController

The `ConnectionController` class handles the initial registration and connection to the Unwired Server. The code below shows the `ConnectionController` header (see Listing 14.3) and class implementation (see Listing 14.4).

ConnectionController.h

```
#import <Foundation/Foundation.h>

@interface ConnectionController : NSObject

+ (void)beginApplicationSetup;

@end
```

Listing 14.3 ConnectionController Header Implementation

ConnectionController.m

```
#import "ConnectionController.h"
#import "SUPApplication.h"
#import "CustomersMBOCustomersMBODB.h"

@implementation ConnectionController

static ConnectionController *appConnectionController;

//This method connects to the Unwired Server
+(void)beginApplicationSetup{
    if(!appConnectionController){
        appConnectionController =
        [[[ConnectionController alloc ] init] retain];
    }

    if([SUPApplication connectionStatus] ==
         SUPConnectionStatus_DISCONNECTED){
       [appConnectionController
        setupApplicationConnection];
    } else {
       NSLog(@"Already connected");
    }

}

//This method configures the connection properties
- (BOOL) setupApplicationConnection{
    SUPApplication *app = [SUPApplication
         getInstance];
    [app setApplicationIdentifier:@"CustomersMBO"];
```

```
    SUPConnectionProperties *properties = [app
         connectionProperties];
    [properties setServerName:@"smp.example.com"];
    [properties setPortNumber:5001];
    [properties setNetworkProtocol:@"http"];

    SUPLoginCredentials *loginCred =
          [SUPLoginCredentials getInstance];
    [loginCred setUsername:@"CustomerIOS"];
    [loginCred setPassword:@""];
    [properties setLoginCredentials:loginCred];
    [properties setActivationCode:@"123"];

    if(![CustomersMBOCustomersMBODB databaseExists]){
        [CustomersMBOCustomersMBODB createDatabase];
    }

    SUPConnectionProfile *connProfile =
          [CustomersMBOCustomersMBODB
          getSynchronizationProfile];
    [connProfile setNetworkProtocol:@"http"];
    [connProfile setPortNumber:2480];
    [connProfile setServerName:@"smp.example.com"];
    [connProfile setDomainName:@"default"];
    [connProfile setUser:@"username"];
    [connProfile setPassword:@"password"];
    [connProfile setAsyncReplay:YES];

    [CustomersMBOCustomersMBODB setApplication:app];

    if([SUPApplication registrationStatus] ==
         SUPRegistrationStatus_REGISTERED){
        [app startConnection:0];
    } else {
        [app registerApplication:0];
    }
    return YES;
}

@end
```

Listing 14.4 ConnectionController Class Implementation

The details of the connection process are included in the `setupApplicationConnection` method. The first line obtains a reference to the SUP `application` object. The second line associates the `application` object with the app deployed to the Unwired Server (see Listing 14.5).

```
SUPApplication *app = [SUPApplication getInstance];
[app setApplicationIdentifier:@"CustomersMBO"];
```

Listing 14.5 Initializing the Application Object

The next step, shown in Listing 14.6, is to configure the connection properties, including the server address, messaging server port, and protocol along with the login credentials. The source of the login credentials and activation code is explained in Section 14.7.

```
SUPConnectionProperties *properties = [app
        connectionProperties];
[properties setServerName:@"smp.example.com"];
[properties setPortNumber:5001];
[properties setNetworkProtocol:@"http"];

SUPLoginCredentials *loginCred =
         [SUPLoginCredentials getInstance];
[loginCred setUsername:@"CustomerIOS"];
[loginCred setPassword:@""];
[properties setLoginCredentials:loginCred];
[properties setActivationCode:@"123"];
```

Listing 14.6 Configuring the Connection Properties

If the database does not already exist, it is created.

```
if(![CustomersMBOCustomersMBODB databaseExists]){
        [CustomersMBOCustomersMBODB createDatabase];
}
```

Listing 14.7 Creating the Database

Next, the parameters required for synchronization are configured in the `synchronization profile` object (see Listing 14.8).

```
    SUPConnectionProfile *connProfile =
         [CustomersMBOCustomersMBODB
         getSynchronizationProfile];
[connProfile setNetworkProtocol:@"http"];
```

```
[connProfile setPortNumber:2480];
[connProfile setServerName:@"smp.example.com"];
[connProfile setDomainName:@"default"];
[connProfile setUser:@"username"];
[connProfile setPassword:@"password"];
[connProfile setAsyncReplay:YES];
```

Listing 14.8 Configuring the Synchronization Parameters

The `application` object created previously is associated with the database.

```
[CustomersMBOCustomersMBODB setApplication:app];
```

Listing 14.9 Associating the Application Object with the Database

Finally, the code in Listing 14.10 checks to see if the app has already been registered with the Unwired Server. If it has, it initiates the connection to the server, and if it has not, it registers the app. The argument passed to `startConnection`, and `registerApplication` is a timeout value. The recommended value is 0, but you should implement a callback handler to monitor the process.

```
if([SUPApplication registrationStatus] ==
     SUPRegistrationStatus_REGISTERED){
    [app startConnection:0];
} else {
    [app registerApplication:0];
}
```

Listing 14.10 Connect to the Unwired Server

Callback Handlers

Although callback handlers have not been included in this example, you will find them indispensable for decoding the connection and synchronization process. For an example of how to use them, see the SUP101 tutorial described in the Unwired Server documentation at *http://infocenter.sybase.com/help/index.jsp?topic=/com.sybase.infocenter.dc01213.0222/doc/html/title.html*.

14.6.3 CustomerAppViewController

The `CustomerAppViewController` controls the initial view displayed by the app. The *CustomerAppViewController_iPhone.xib* file is shown in Xcode's interface builder in

Figure 14.11. The figure also shows the connections between the interface views and the `CustomerAppViewController` elements.

The view contains three buttons and a search bar. The Sync button is used to synchronize with the Unwired Server. The first time the app is run, the user must synchronize, but after the initial synchronization, the user needs to synchronize only if the user wishes to upload or download updates. Once the database is synchronized, the user can tap the List button to see a list of customers or the New button to create a new customer. Or, the user can enter a value in the search bar to search the Customer MBO.

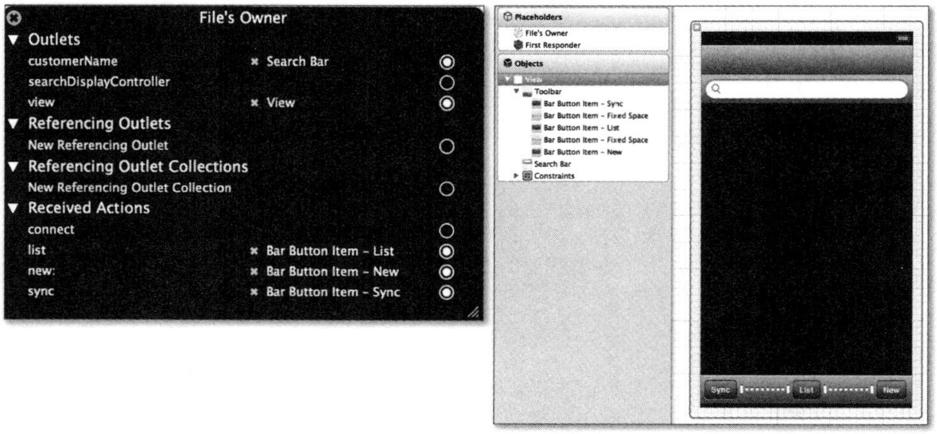

Figure 14.11 CustomerAppViewController_iPhone in Xcode's Interface Builder

The search bar is connected to the property `customerName`. The List button is connected to the list method, which loads the `ItemListViewController` class. This then displays all the customers in the database in a table. The New button connects to the new method, which loads the DetailsViewController screen with empty fields so the user can enter the data for the new customer. The Sync button connects to the sync method, which synchronizes with the Unwired Server. Entering a value in the search field and tapping the Search button will load the ItemListViewController screen with a list of the search results. This action is implemented in code and requires no connection in the interface builder. The sections below show the `CustomerAppViewController` header (see Listing 14.11) and class (see Listing 14.12) implementations.

CustomerAppViewController.h

```
#import <UIKit/UIKit.h>

@interface CustomerAppViewController :
      UIViewController  <UISearchBarDelegate,
      UISearchDisplayDelegate>
{
    IBOutlet UISearchBar *customerName;
}

- (IBAction) sync;
- (IBAction) list;
- (IBAction) new;

@end
```

Listing 14.11 CustomerAppViewController Header Implementation

CustomerAppViewController.m

```
#import "CustomerAppViewController.h"
#import "SUPApplication.h"
#import "SUPRegistrationStatus.h"
#import "CustomersMBOCustomersMBODB.h"
#import "SUPEngine.h"
#import "SUPConnectionProfile.h"
#import "CustomersMBOLocalKeyGenerator.h"
#import "CustomersMBOCustomer.h"
#import "ItemsListViewController.h"


@implementation CustomerAppViewController

//Synchronize the database
- (void) sync{
  [CustomersMBOCustomersMBODB synchronize];
}

//Initialize ItemsListViewController and navigate to the
//that view
- (void) list
{
    ItemsListViewController *itemsListViewController
```

```
        =[[ItemsListViewController alloc] init];
    [[self navigationController] pushViewController:
        itemsListViewController animated:YES];
}

- (void)searchBarSearchButtonClicked:(UISearchBar
        *)searchBar {

    [searchBar resignFirstResponder];
    [self findCustomer];

}

//Initialize ItemsListViewController passing the search
//term entered by the user and navigate to the
//that view
- (void)findCustomer
{

    NSString *aString = [[customerName text]
         stringByAppendingString:@"%"];
    ItemsListViewController *itemsListViewController
         = [[ItemsListViewController alloc]
        initWithName:aString];
    [[self navigationController]
        pushViewController:itemsListViewController
          animated:YES];
}


//Create a new customer object and navigate to the
//DetailViewController
-(void)new
{
    CustomersMBOCustomer *newcustomer =
        [[CustomersMBOCustomer alloc] init];
    DetailViewController *detailViewController =
         [[DetailViewController alloc]
         initWithName:YES];
    [detailViewController setItem:newcustomer];
    [[self navigationController]
         pushViewController:detailViewController
           animated:YES];
```

```
}

- (void)viewDidLoad
{
    [super viewDidLoad];
    customerName.delegate = self;
}

- (void)didReceiveMemoryWarning
{
    [super didReceiveMemoryWarning];
}

- (void)dealloc {
    [customerName release];
    [super dealloc];
}
@end
```

Listing 14.12 The CustomerAppViewContoller Implementation

The `sync` method invokes the `synchronize` method of the `database` class. The synchronization profile was created and configured in the `ConnectionController` class, and as long as there are no changes, you have to create it only once. You could include this statement in the `ConnectionController` class if you want the app to synchronize automatically when it starts. However, in an offline scenario, you should include an option for the user to synchronize manually.

```
- (void) sync{
  [CustomersMBOCustomersMBODB synchronize];
}
```

The new method creates a new customer object and then a `DetailViewController` object, and initializes the item property to the new customer object. It then pushes the `DetailViewController` object onto the navigation controller stack so that the detail screen is visible.

```
-(void)new
{
    CustomersMBOCustomer *newcustomer =
         [[CustomersMBOCustomer alloc] init];
    DetailViewController *detailViewController =
          [[DetailViewController alloc]
```

```
            initWithName:YES];
    [detailViewController setItem:newcustomer];
    [[self navigationController]
            pushViewController:detailViewController
              animated:YES];
}
```

14.6.4 ItemListViewController

The `ItemListViewController` class implements `UITableViewController` and is used to display a list of customers in a table. When the user taps a row in the table, the CUSTOMERDETAILVIEW screen is loaded with the customer data. The interface is implemented programmatically, so there is no XIB file. The sections below show the `ItemListViewController` header (see Listing 14.13) and class (see Listing 14.14) implementation.

ItemListViewController.h

```
import <Foundation/Foundation.h>

@interface ItemsListViewController :
       UITableViewController
{
    NSMutableArray *searchData;
    NSString *searchTerm;
}
@end
```

Listing 14.13 The ItemListViewController header implementation.

ItemListViewController.m

```
#import "ItemsListViewController.h"
#import "DetailViewController.h"
#import "CustomersMBOCustomer.h"
#import "CustomersMBOCustomersMBODB.h"
#import "SUPSortCriteria.h"
#import "SUPAttributeSort.h"
#import "SUPQuery.h"

@implementation ItemsListViewController
```

```
BOOL searching;

//Initialize the object
-(id)init
{
    self = [super initWithStyle:
         UITableViewStyleGrouped];
    searching=NO;
    if(self){
        UINavigationItem *n = [self navigationItem];
        [n setTitle:@"Customers"];
    }
     [self loadData];
    return self;
}

//Initialize the object with a search term and
// boolean variable that indicates whether
//we are showing the entire list or are searching
-(id)initWithName:(NSString *)s
{
    searchTerm = s;
    return [self init];
}

//Load the data from the database into the TableView
//datasource
- (void)loadData
{
    SUPObjectList *customers;
    searchData = [[NSMutableArray alloc] init];

    if ([searchTerm length] == 0)
    {
        SUPQuery *query = [SUPQuery getInstance];
        [query select:@"*"];
        [query  from:@"Customer":@"c"];
        SUPSortCriteria *sort = [SUPSortCriteria
             getInstance];
        [sort add:[SUPAttributeSort ascending:
           @"company_name"]];
        query.sortCriteria = sort;
```

```
        customers = [CustomersMBOCustomer
            findWithQuery:query];
    } else {
        customers = [CustomersMBOCustomer
            findByCompany:searchTerm];

    }
    for (CustomersMBOCustomer * oneRec in customers)
    {
        [searchData addObject:oneRec];
    }
}

//Runs once the view loads
- (void) viewDidLoad
{
    [super viewDidLoad];
}

//Return number of rows in the table
- (NSInteger)tableView:(UITableView *)tableView
          numberOfRowsInSection:(NSInteger)section
{
    return [searchData count];
}

//Create cells for the table
- (UITableViewCell *)tableView:(UITableView *)
            tableView cellForRowAtIndexPath:
                (NSIndexPath *)indexPath
{
    UITableViewCell *cell = [tableView
            dequeueReusableCellWithIdentifier:
                @"UITableViewCell"];
    if(!cell){
        cell = [[UITableViewCell alloc]
            initWithStyle:
            UITableViewCellStyleSubtitle
                reuseIdentifier:
            @"UITableViewCell"];
    }

    CustomersMBOCustomer *c =
```

```
        [searchData objectAtIndex:
        [indexPath row]];
    NSString *string1 = @"";
    NSString *string2 = @"";
    string1 = [string1 stringByAppendingString:
   [NSString stringWithFormat:
     @"%@ ", c.fname]];
    string1 = [string1 stringByAppendingString:
        [NSString stringWithFormat:
          @"%@ ", c.lname]];
    string2 = [string2 stringByAppendingString:
        [NSString stringWithFormat:
          @"%@ ", c.company_name]];
    [[cell textLabel] setText:string2];
    [[cell detailTextLabel] setText:string1];
    [[cell textLabel] setTextColor:
        [UIColor colorWithRed:0.905 green:0.0
          blue:0.552 alpha:1.0]];
    return cell;
}

//If a user selects a row, initialize DetailViewController
//Get the selected customer and navigate to the details view
- (void) tableView:(UITableView *)tableView
        didSelectRowAtIndexPath:
        (NSIndexPath *)indexPath
{
    CustomersMBOCustomer *selectedItem =
        [searchData objectAtIndex:
          [indexPath row]];
    DetailViewController *detailViewController =
        [[DetailViewController
          alloc] initWithName:NO:selectedItem];
    [[self navigationController]
         pushViewController:detailViewController
           animated:YES];
}

-(void)viewWillAppear:(BOOL)animated
{
    [super viewWillAppear:animated];
    [[self tableView] reloadData];
```

```
}

@end
```

Listing 14.14 The ItemListViewController Implementation

The app loads the data from the local database in the `loadData` method. The first part of the code creates an `SUPObjectList` object called `customers`, which holds the returned customer objects. Next, an `NSMutableArray` called `searchData` is created to be the data source for the table used to display the data.

```
//Load the data from the database into the TableView
//datasource
- (void)loadData
{
    SUPObjectList *customers;
    searchData = [[NSMutableArray alloc] init];
```

If the length of the `searchTerm` property is zero, the LIST button is tapped, so the code retrieves all the data.

```
        if ([searchTerm length] == 0)
```

This code shows an example of using a query object to create a dynamic query. First, get a reference to a query object:

```
SUPQuery *query = [SUPQuery getInstance];
```

Next, configure the query object to implement the `SQL SELECT * FROM Customer ORDER BY company_name`.

```
[query select:@"*"];
[query  from:@"Customer":@"c"];
SUPSortCriteria *sort = [SUPSortCriteria
            getInstance];
 [sort add:[SUPAttributeSort ascending:
           @"company_name"]];
query.sortCriteria = sort;
```

Finally, execute the query and store the results in the `customers` object list:

```
 customers = [CustomersMBOCustomer
           findWithQuery:query];
```

An alternative to a dynamic query is to invoke an object query from the MBO, such as `findAll`.

```
customers = [CustomersMBOCustomer findAll];
```

If the user taps the SEARCH button, the `findByCompany` object query is invoked using the search string entered by the user:

```
} else {
  customers = [CustomersMBOCustomer
           findByCompany:searchTerm];
}
```

Finally, the objects in the customer list are added to the `searchData` array:

```
for (CustomersMBOCustomer * oneRec in customers)
    {
        [searchData addObject:oneRec];
    }
 }
```

If the user taps a row on the table, the customer object is retrieved from the `searchData` array and passed to the `CustomerDetailView` object, which is pushed into the navigation controller stack:

```
//If a user selects a row, initialize DetailViewController
//Get the selected customer and navigate to the details view
- (void) tableView:(UITableView *)tableView
        didSelectRowAtIndexPath:
        (NSIndexPath *)indexPath
{
    CustomersMBOCustomer *selectedItem =
        [searchData objectAtIndex:
          [indexPath row]];
    DetailViewController *detailViewController =
        [[DetailViewController
          alloc] initWithName:NO:selectedItem];
    [[self navigationController]
         pushViewController:detailViewController
           animated:YES];
}
```

14.6.5 DetailViewController

The DETAILVIEWCONTROLLER screen shows the details of a customer and allows the user to update or delete the customer. Figure 14.12 shows the *DetailViewController.xib* in Xcode's interface builder. The screen includes eight `UITTextFields` and a toolbar with two buttons.

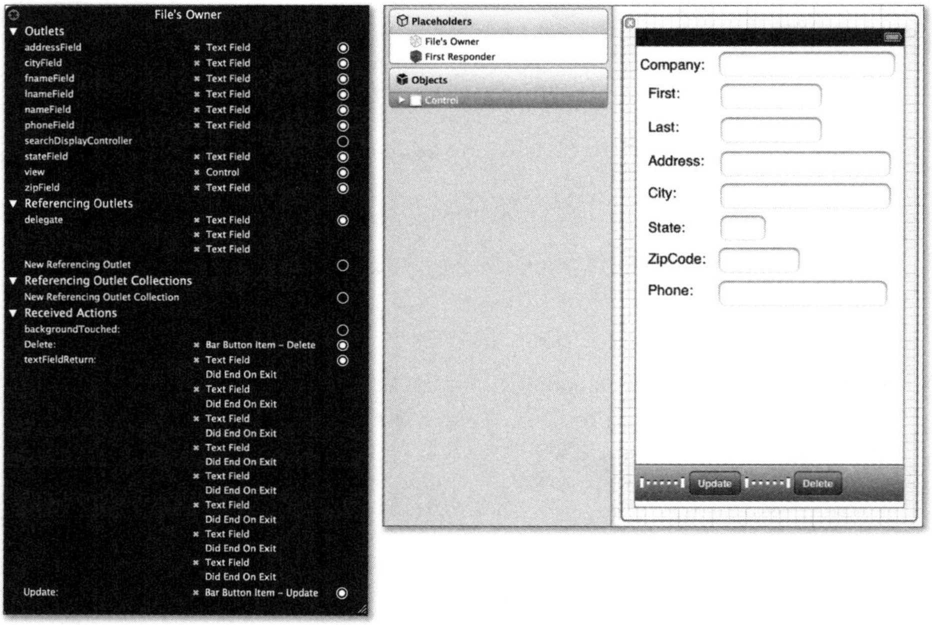

Figure 14.12 The CustomerDetailsView in the Interface Builder

The DETAIL screen can open in two ways. If the user taps a row on a list of customers, the DETAIL view is opened with the selected customer's details showing. If the user clicks the NEW button on the HOME screen, the DETAIL screen is opened with empty fields. The user can edit the data and then tap the UPDATE button to update the database using either the `update` or `create` operation. If the screen shows an existing customer, the user can delete the customer using the `delete` operation. The sections below show the `DetailViewController` header (see Listing 14.15) and class (see Listing 14.16) implementations.

DetailViewController.h

```
import <UIKit/UIKit.h>

@class CustomersMBOCustomer;

@interface DetailViewController : UIViewController
{

    IBOutlet UITextField *fnameField;
    IBOutlet UITextField *nameField;
    IBOutlet UITextField *lnameField;
    IBOutlet UITextField *addressField;
    IBOutlet UITextField *cityField;
    IBOutlet UITextField *stateField;
    IBOutlet UITextField *zipField;
    IBOutlet UITextField *phoneField;
}

@property (nonatomic, strong) CustomersMBOCustomer *item;
@property (nonatomic) BOOL newCustomer;

- (IBAction)Update:(id)sender;
- (IBAction)Delete:(id)sender;
- (IBAction)textFieldReturn:(id)sender;
- (IBAction)backgroundTouched:(id)sender;
@end
```

Listing 14.15 The DetailViewController header implementation.

DetailViewController.m

```
#import "DetailViewController.h"
#import "CustomersMBOCustomer.h"

@implementation DetailViewController

@synthesize item;
@synthesize newCustomer;

//Initialize the object
-(id)init
{
    self = [super init];
```

```
    if(self){
        UINavigationItem *n = [self navigationItem];
        [n setTitle:@"Customer Details"];
    }
    return self;
}

//Initialize DetailViewController with newCustomer set to yes
- (id)initWithName:(BOOL)newCust:
      (CustomersMBOCustomer *)custItem
{
    newCustomer = newCust;
    item = custItem
    return [self init];
}

//Initialize TextViews with values from customer
//object
- (void) viewWillAppear:(BOOL)animated
{
    [super viewWillAppear:YES];

    [nameField setText:[item company_name]];
    [fnameField setText:[item fname]];
    [lnameField setText:[item lname]];
    [addressField setText:[item address]];
    [cityField setText:[item city]];
    [stateField setText:[item state]];
    [zipField setText:[item zip]];
    [phoneField setText:[item phone]];
}

- (void)viewDidLoad
{
    [super viewDidLoad];
    [[self view] setBackgroundColor:[UIColor
          groupTableViewBackgroundColor]];
}

-(void) viewWillDisappear:(BOOL)animated
{
    [super viewWillDisappear:animated];
```

```
    [[self view] endEditing:YES];
    [self UpdateItemFromUI:item];
}

//Load values from TextViews into the customer object
- (void)UpdateItemFromUI:
      (CustomersMBOCustomer *)thisItem
{
    [thisItem setCompany_name:[nameField text]];
    [thisItem setFname:[fnameField text]];
    [thisItem setLname:[lnameField text]];
    [thisItem setAddress:[addressField text]];
    [thisItem setCity:[cityField text]];
    [thisItem setState:[stateField text]];
    [thisItem setZip:[zipField text]];
    [thisItem setPhone:[phoneField text]];
}

//If  this is a new customer, create a new customer
//object update customer object and then submit
//pending to register changes in the MBO
- (void)Update:(id)sender
{
    if (newCustomer) {
        CustomersMBOCustomer *newItem =
         [[CustomersMBOCustomer alloc] init];
        [self UpdateItemFromUI:newItem];
        [newItem create];
    } else {
        [self UpdateItemFromUI:item];
        [item update];
    }
    [item submitPending];
}


//Delete the object
- (void)Delete:(id)sender
{
    [item delete];
    [item submitPending];
}
```

```
//Close the keyboard when user clicks return
-(IBAction)textFieldReturn:(id)sender
{
    [sender resignFirstResponder];
}

- (void)dealloc {
    [nameField release];
    [fnameField release];
    [lnameField release];
    [addressField release];
    [cityField release];
    [zipField release];
    [phoneField release];
    [super dealloc];
}
@end
```

Listing 14.16 The DetailViewController Implementation

The `initWithName` method shown in Listing 14.17 is called to initialize the DETAILVIEWCONTROLLER screen. It takes two arguments. The first is a Boolean argument, which is used to specify whether the customer object is a new customer or an existing customer. The second argument is a `customer` object.

```
//Initialize DetailViewController with newCustcmer set to yes
- (id)initWithName:(BOOL)newCust:
      (CustomersMBOCustomer *)custItem
{
    newCustomer = newCust;
    item = custItem
    return [self init];
}
```

Listing 14.17 Creating a Customer

Immediately before the view appears, the `viewWillAppear` method fills the text fields on the screen using the property values from the item object (see Listing 14.18).

```
//Initialize the textview fields
- (void) viewWillAppear:(BOOL)animated
{
    [super viewWillAppear:YES];
    [nameField setText:[item company_name]];
```

```
    [fnameField setText:[item fname]];
    [lnameField setText:[item lname]];
    [addressField setText:[item address]];
    [cityField setText:[item city]];
    [stateField setText:[item state]];
    [zipField setText:[item zip]];
    [phoneField setText:[item phone]];
}
```

Listing 14.18 Initializing the Text Fields

If the user taps the UPDATE button, the code first updates the item object from the user interface. It then checks the `newCustomer` property to determine whether this is a new customer. If it is, then the `create` operation is invoked. If it is an existing customer, the update `operation` is invoked. Finally, the `submit pending` method creates the replay records, which is uploaded to the Unwired Server during the next synchronization. The code is shown in Listing 14.19.

```
//Load values from TextViews into the customer object
- (void)UpdateItemFromUI:
      (CustomersMBOCustomer *)thisItem
{
    [thisItem setCompany_name:[nameField text]];
    [thisItem setFname:[fnameField text]];
    [thisItem setLname:[lnameField text]];
    [thisItem setAddress:[addressField text]];
    [thisItem setCity:[cityField text]];
    [thisItem setState:[stateField text]];
    [thisItem setZip:[zipField text]];
    [thisItem setPhone:[phoneField text]];
}

//Update button handler
- (void)Update:(id)sender
{
    [self UpdateItemFromUI:item];
    if (newCustomer) {
        [item create];
    } else {
        [item update];
    }
    [item submitPending];
}
```

Listing 14.19 Updating the Database

If the user taps the DELETE button, the `delete` operation is invoked, and the changes are submitted:

```
//Delete the object
- (void)Delete:(id)sender
{
    [item delete];
    [item submitPending];
}
```

14.7 Running the App

Before you can run the app, you must create an application connection on the Unwired Server to which the app can connect. Log on to the SCC and navigate to APPLICATIONS • APPLICATION CONNECTIONS. Click REGISTER. The properties are shown in Figure 14.13. These properties must match those assigned in the `Connection-Controller` class. The user (CUSTOMERIOS) and activation code (123) are assigned in the connection properties, and the application ID must match the application ID assigned to the application instance.

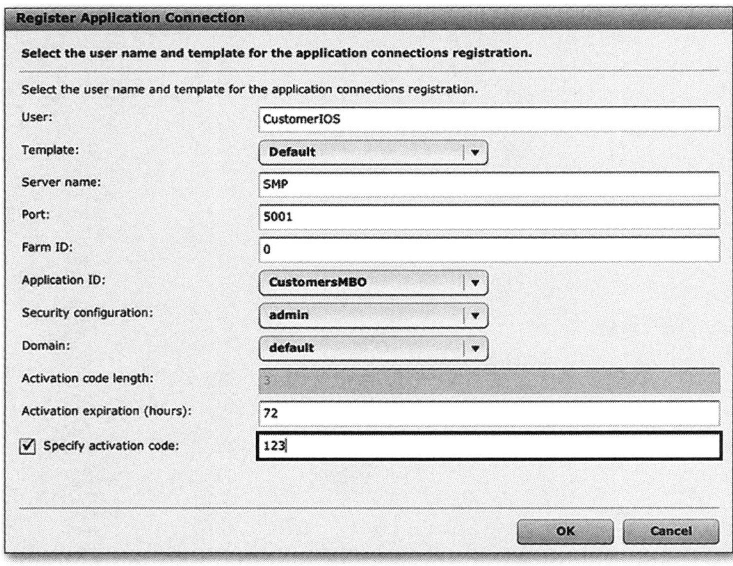

Figure 14.13 Create an Application Connection on the Unwired Server

To run the app, specify the simulator you want to use in the dropdown list at the top left in Xcode, and then click the RUN button.

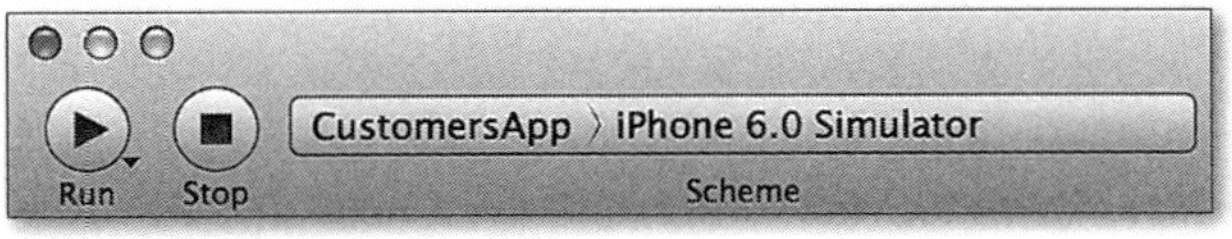

Figure 14.14 Specifying the Simulator You Want to Use

Keep in mind that the code in this chapter implements only the iPhone interface. If you want to use the iPad simulator, you have to implement the interface in *CustomerAppViewController_iPad.xib*.

14.8 Summary

This chapter should provide enough to get you started developing iOS Object API apps. There is much more to learn. A good place to start is the *Developer Guide: iOS Object API Applications*, which is available at *http://infocenter.sybase.com/help/index.jsp?topic=/com.sybase.infocenter.dc01217.0221/doc/html/title.html*.

PART V
Advanced Topics

The Mobile Analytics Kit (MAKit) allows app developers to build apps that visualize data retrieved in the container using HTML5 and JavaScript. End users can graphically view data and modify chart values to examine what-if scenarios.

15 Using the Mobile Analytics Kit

Up to this point, we have developed hybrid apps that retrieve data and display that data to the end user as one or more tables. However, when the amount of data is large, it is often easier to understand the data when it is displayed graphically. In order to build an app that does this, you can use the Mobile Analytics Kit (MAKit), which is a client-side graphic tool for visualizing and analyzing data through mobile apps on mobile devices (primarily tablets).

As a developer, you can use MAKit to design hybrid apps that allow users to graphically view data and modify chart values to examine what-if scenarios. MAKit provides a set of easy-to-use APIs and metadata that allow app developers to build analytical apps that visualize the data retrieved in the Hybrid Web Container.

Two versions of MAKit are available: one for HTML5 and a second one for iOS. The HTML5 version is used for developing hybrid apps, while the iOS version can be used only when you are developing a native iOS app. This chapter focuses on the more commonly used HTML5 version. MAKit is delivered as a JavaScript library that can be used by any HTML5 web app. This library is contained in the SAP Mobile SDK distribution.

The architecture of MAKit, as shown in Figure 15.1, contains the following modules:

- **UI/Presentation**
  Includes the UI with its views.
- **Analytics**
  Processes on-device analytics.
- **Query**
  Provides a standard way to feed data into MAKit.

The APIs offer flexibility and customization options. MAKit is provided as a set of files including JavaScript, CSS, and graphics.

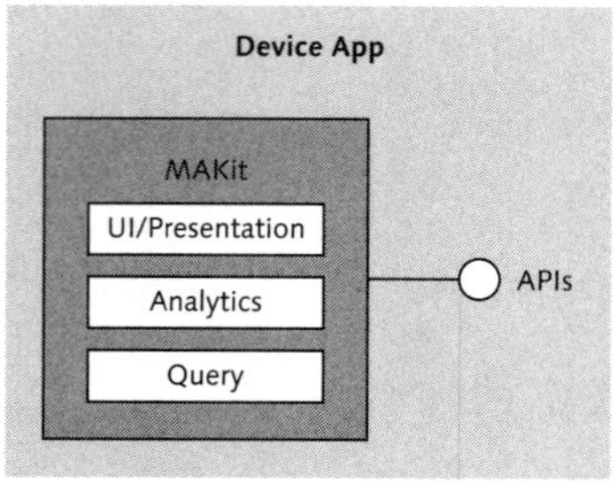

Figure 15.1 MAKit Architecture

In this chapter, we walk you through the process of building an app using MAKit. In Section 15.1 and Section 15.2, we introduce you to the goal of the app we're designing and describe the basic processes of creating the app (most of this is a review of what you've already learned earlier in the book). In Section 15.3, we explain how to copy the files you need in order to generate and manipulate graphs. Finally, in Section 15.4 and Section 15.5, we get into the specifics of how to work with MAKit to create two types of charts: a simple column chart and a zooming chart.

MAChartView Control

The MAKit component of the Sybase Mobile SDK is only one option for including mobile analytics charts in your hybrid app. With this approach you define a chart outside of the Hybrid App Designer by using the Mobile Analytics Kit (MAKit) component of the Sybase Mobile SDK, and then add that chart to the HTMLVIEW control. A second approach, which has less functionality, is to use the MACHARTVIEW control that is presented in Chapter 8.

15.1 Introducing the Example App

This example involves developing an app that will be used to display and analyze sales data. Each data observation represents the sales revenue of a particular item for a period of time. While the app could simply retrieve and display the data as a table, the number of data items is fairly large. The user is not as interested in looking at a single data point as in reviewing and analyzing the data using graphical techniques.

The MAKit library provides us with reusable code that can generate graphs and provide for some analytic techniques for manipulating those graphs.

The specific data we want to graph and analyze is sales revenue data for a series of products over a period of time. These products are all computer related:

- Laptops
- Desktops
- Tablets

We also have data that is available with three levels of aggregation:

- By year, as shown in Figure 15.2
- By quarter, as shown in Figure 15.3
- By month, as shown in Figure 15.4

The first requirement is the development of a simple graph that displays the data by year. The second requirement is to display a zooming chart, which allows us to view this data by year, then by quarter, and then by month.

RevenueByY

No.	product [NVARCHAR(20)]	year [DOUBLE]	revenue [DOUBLE]
1	laptop	2005.0	2000.0
2	desktop	2005.0	1500.0
3	tablet	2005.0	1200.0
4	desktop	2006.0	1700.0
5	laptop	2006.0	2300.0
6	tablet	2006.0	1500.0
7	laptop	2007.0	2500.0
8	desktop	2007.0	1800.0
9	tablet	2007.0	1500.0
10	laptop	2008.0	2800.0
11	desktop	2008.0	2000.0

Figure 15.2 Product Revenue by Year

RevenueByQ

No.	product [NVARCHAR(20)]	year [DOUBLE]	quarter [DOUBLE]	revenue [DOUBLE]
1	desktop	2005.0	1.0	400.0
2	laptop	2005.0	1.0	300.0
3	tablet	2005.0	1.0	200.0
4	desktop	2005.0	2.0	300.0
5	laptop	2005.0	2.0	500.0
6	tablet	2005.0	2.0	400.0
7	desktop	2005.0	3.0	300.0
8	laptop	2005.0	3.0	500.0
9	tablet	2005.0	3.0	300.0
10	desktop	2005.0	4.0	500.0
11	laptop	2005.0	4.0	700.0

Figure 15.3 Product Revenue by Quarter within Year

RevenueByM

No.	product [NVARCHAR(20)]	year [DOUBLE]	quarter [DOUBLE]	month [DOUBLE]	revenue [DOUBLE]
1	desktop	2005.0	1.0	1.0	100.0
2	laptop	2005.0	1.0	1.0	200.0
3	tablet	2005.0	1.0	1.0	100.0
4	desktop	2005.0	1.0	2.0	100.0
5	laptop	2005.0	1.0	2.0	200.0
6	tablet	2005.0	1.0	2.0	100.0
7	desktop	2005.0	1.0	3.0	200.0
8	laptop	2005.0	1.0	3.0	200.0
9	tablet	2005.0	1.0	3.0	200.0
10	desktop	2005.0	2.0	4.0	200.0
11	laptop	2005.0	2.0	4.0	300.0

Figure 15.4 Product Revenue by Month within Quarter within Year

15.2 Creating the Hybrid App

While the focus of this chapter is the use of MAKit, it is necessary to perform some steps to create the basic app. These steps are similar to those used to develop the database app presented in Section 9.1 of Chapter 9, so we do not discuss them in detail. Briefly, they are:

1. Create a project. In this step, create a project with a name such as "MobileAnalytics."
2. Create an MBO for each of these three tables using the simple drag-and-drop method. When creating these MBOs, deselect the three operations. You do not need to create any additional object queries.

3. Deploy the project. Once all three MBOs are created, as shown in Figure 15.5, right-click the background and deploy this project. This causes the project to be deployed to the Unwired Server and generates the project code within the project folder.

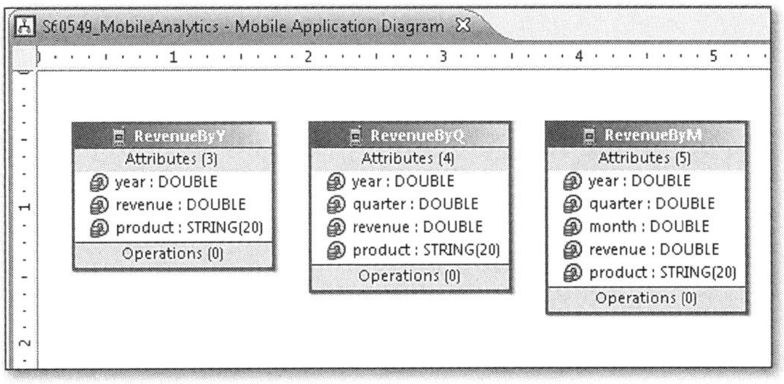

Figure 15.5 MBOs for the Three Database Tables

Now, again using the approach presented in Section 9.1 of Chapter 9, we start with the creation of a client-initiated app by dragging and dropping the REVENUEBYY MBO onto the flow design, as shown in Figure 15.6.

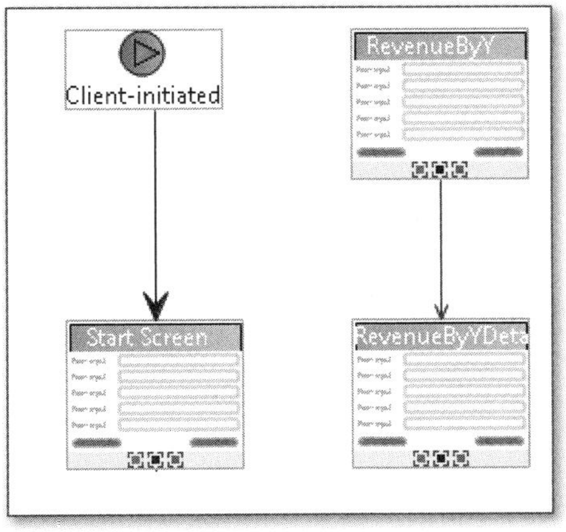

Figure 15.6 Initial Screens after Dragging and Dropping the RevenueByY MBO

Select the screen properties for the START screen and add an HTMLVIEW control, giving it a key of STARTKEY and entering the following for the default value:

```
<h1>Mobile Analytics App - S65049</h1><br><br>This App demonstrates some
```

```
of the features available in mobile analytics provided by the Sybase
Mobile Analytics Kit (MAKit).<br><br>
```

These HTMLVIEW control properties should appear, as shown in Figure 15.7.

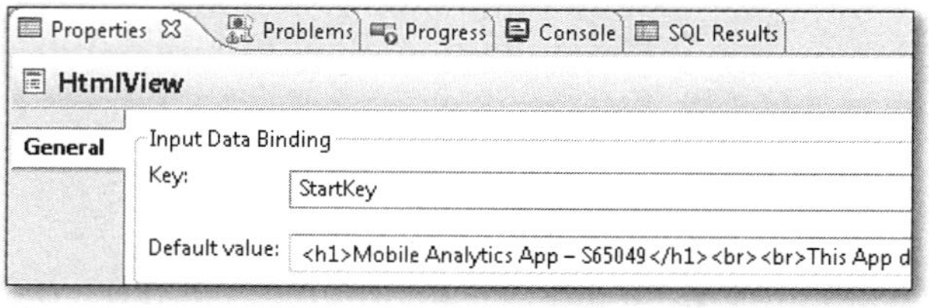

Figure 15.7 Properties for the HTMLView on the Start Screen

Then, add a button below the HTML control. After selecting that button, under properties, give it a name of REVENUE and a key of REVENUE, and then select CENTER for LABEL POSITION, as shown in Figure 15.8. On the bottom half of the properties, change KEY to GETDATAFIRST, change TYPE to ONLINE REQUEST, select REVENUEBYY for the MBO, select FINDALL for OBJECT QUERY, and enter "RevenueByY" for the SUCCESS screen.

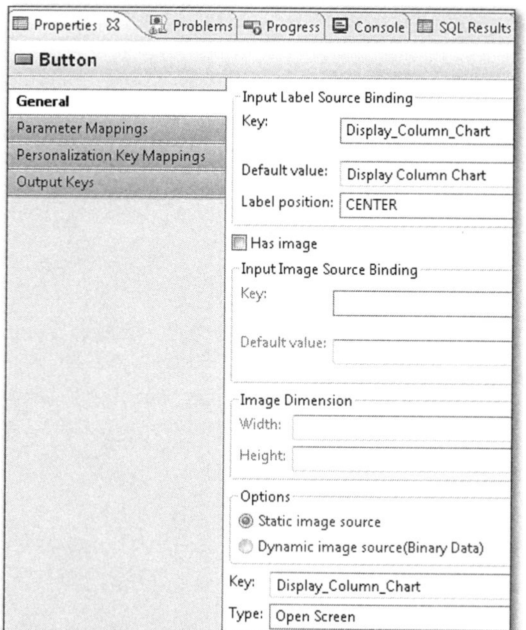

Figure 15.8 Properties for the Button Control on the Start Screen

This completes the initial hybrid app. SAVE all the files, right-click the background, and select GENERATE HYBRID APP.

Make sure that UPDATE GENERATED CODE is selected, but do not select DEPLOY TO UNWIRED SERVER. Click FINISH. At this point, we are only going to generate and save the code in the local project folder. We are not ready to deploy the hybrid app to the server or run the hybrid app on the user's device yet. The Hybrid App Generation Wizard should appear as shown in Figure 15.9.

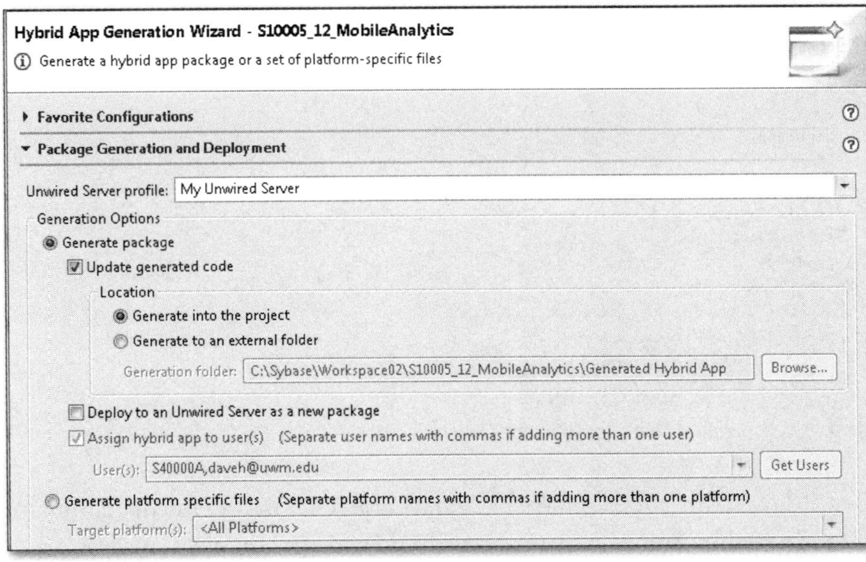

Figure 15.9 Generating and Saving Code to the Project File

15.3 Copying MAKit Files

MAKit is a library that contains a number of JavaScript, CSS, and image files that are required to generate and manipulate the graphs. In this section, we add those files to the hybrid app we have just developed. This step needs to be done for each hybrid app you develop that accesses this MAKit library.

In the previous step, the code for this project was generated and saved in the project folder. The files we are going to copy are all in the GENERATED HYBRID APP folder, as shown in Figure 15.10. The sub-folders shown contain various files that include HTML, CSS, images, JavaScript, XML, etc. It is very important to understand that most of these files are regenerated each time the hybrid app is deployed. Any changes that are made to these files are erased the next time the Hybrid App Generation Wizard is executed. There are a few exceptions. Additional files can be saved in these folders and are not replaced. There are very few unique file names (an example is *custom.js*) where changes made to a file are not lost. For this example, we start by adding some additional files and then limit our changes to adding new files and modifying those few files where changes will not be lost.

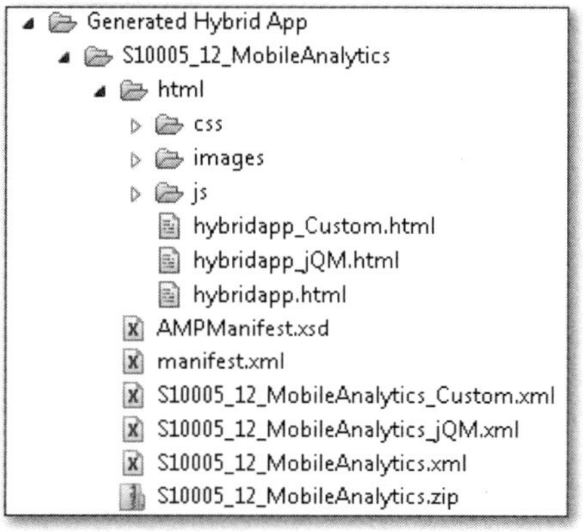

Figure 15.10 Code Generated for the Current Project

A number of JavaScript libraries have been developed to assist in the development of websites. These libraries consist of pre-written JavaScript, which can be incorporated into individual apps. Some examples include: jQuery, MooTools, Prototype, and Dojo Toolkit. To add these libraries to a hybrid app, copy the files containing the code to the appropriate folders and then place calls within the existing code to these functions. In our case, as already explained, we are going to use the Mobile Analytics Kit (MAKit) HTML5.

The MAKit library was automatically copied onto the local workstation when the Sybase Mobile SDK install took place. You can locate the library files by opening Windows Explorer to the following location.

C:\Sybase\UnwiredPlatform\MobileSDK\MAKit\HTML5

The folders in this directory are shown in Figure 15.11.

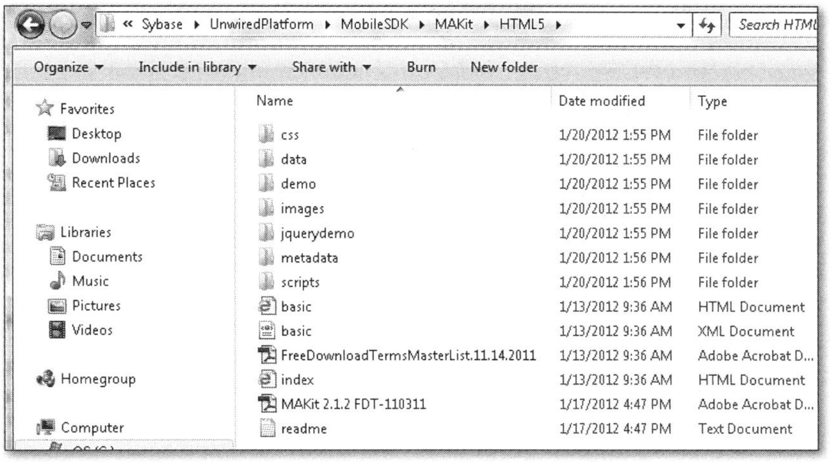

Figure 15.11 Directory with Files Required to Use MAKit HTML5

The following is a series of steps required to copy files from the above directory to your current project. It should be noted again that this process needs to be performed for each project that will utilize the MAKit HTML5. The reason we are using MAKit and copying these files is that we don't want to write all this code ourselves. The concept of using these libraries is called *code reuse*—somebody develops the code, and then others reuse it to improve their apps and reduce development time.

15.3.1 Copying CSS Files

Cascading Style Sheets (CSS) is used to format the way HTML is displayed on the end user's device. The screens generated by MAKit require the following two style sheets. Copy the following two style sheet files from the /CSS subdirectory to the /CSS subdirectory, as shown in Figure 15.12.

- *Default.css*
- *Digitpicker.css*

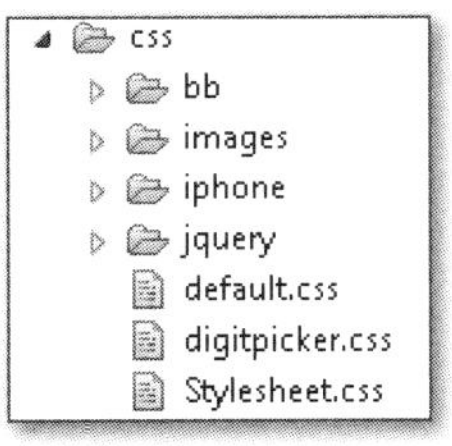

Figure 15.12 CSS Files Copied

15.3.2 Copying Images Associated with CSS

Images are used in conjunction with the CSS to create graphics on the end user's device. Create a new subfolder called IMAGES under /css. Copy all the image files shown below from /CSS/IMAGES to /CSS/IMAGES, as shown in Figure 15.13.

- *chartToolbar.png*
- *da_icon.png*
- *digitpicker-bg-right.png*
- *digitpicker-button.png*
- *e-mail_icon.png*
- *fullscreen_icon.png*
- *rangeselector_icon.png*
- *table_icon.png*
- *whatif_icon.png*

Figure 15.13 Image Files Copied

15.3.3 jQuery Style Sheets

jQuery is a JavaScript library used by SUP. This library was part of our original app, and we are simply adding some additional CSS to be used by jQuery. Copy the following three jQuery-related style sheets shown below from /CSS/JQUERY to /CSS/JQUERY as shown in Figure 15.14.

- *jquery.mobile-1.0.min.css*
- *jquery.ui.datepicker.css*
- *jquery.ui.datepicker.mobile.css*

Figure 15.14 jQuery Style Sheets Files Copied

15.3.4 General Images

The following are some additional images used by the MAKit. Add a new subdirectory called IMAGES to HTML. Copy all the following images from /IMAGES to /IMAGES, as shown in Figure 15.15.

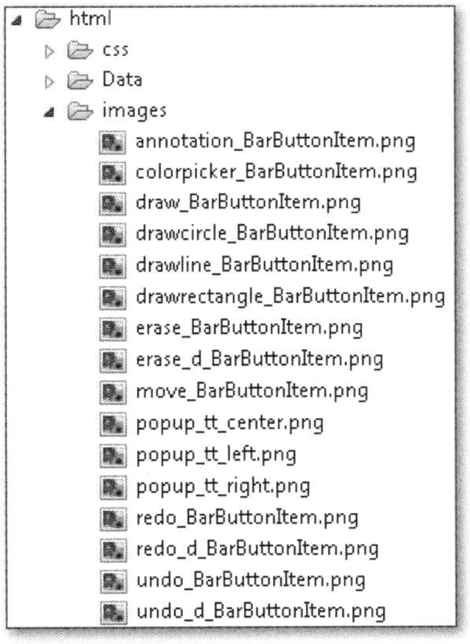

Figure 15.15 Image Files Copied

15.3.5 Copying JavaScript Files

The next files to be copied are the actual JavaScript code. This includes the main MAKit library, three additional JavaScript files, and the LOCINF directory. Copy the following JavaScript files from /SCRIPTS to /JS, as shown in Figure 15.16:

- *Sybase.MAKit-min.js*
- *jquery.ui.datepicker.js*
- *jquery.ui.datepicker.mobile.js*
- *jquery.glob.js*

Figure 15.16 JavaScript Files Copied

Copy the LOCINFO folder from /SCRIPTS to /JS, as shown in Figure 15.17.

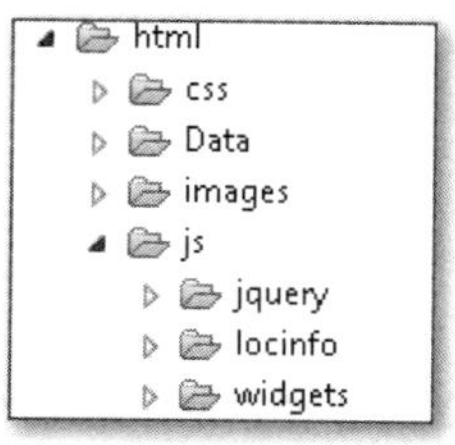

Figure 15.17 Locinfo Folder Copied

15.4 Creating a Simple Column Chart

The first chart we will develop is a simple column chart, which uses the data from the REVENUEBYY MBO. We need to make changes to our current app to connect it to the MAKit code (Section 15.4.1). This includes creating a screen where the column chart is displayed, providing a method for accessing the data, and passing that data to the MAKit code. We then need to add code that describes the chart we are designing and retrieves the data (Section 15.4.2). Finally, we will test the app (Section 15.4.3).

15.4.1 Modifying the Current App

There are a number of changes we need to make to our current app in order to connect it to the MAKit code.

Creating ColumnChart Screen

We first need to create a new screen where the value for both the name and key properties are "ColumnChart," as shown in Figure 15.18. It is important to note in this section that properties can have any value; however, we do need to remember them because most of them will later be referenced in the code. Using the example values we have provided will save time because we have used these same values in the example code that follows.

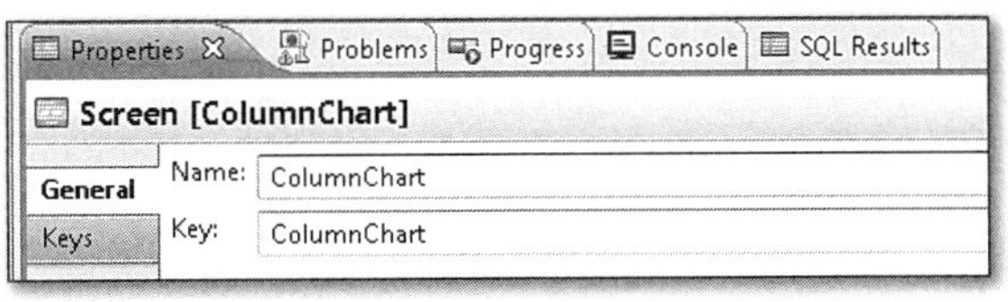

Figure 15.18 Add ColumnChart Screen

The next step is to add an HTMLVIEW control onto the COLUMNCHART screen. It is recommended that you enter "ChartKey" as the value for this control's key.

The default value for this control should be the following HTML, as shown in Figure 15.19:

```
<div id="contents">
<div id="columnChart_1" class="chart"></div>
</div>
```

The `div` tags simply create a container—in this case, an empty container. When the MAKit creates the interactive ColumnChart screen, it places that chart into this container. We then display the ColumnChart screen, which includes the chart. Again, the values `contents`, `column-Chart_1`, and `chart` are used in the code.

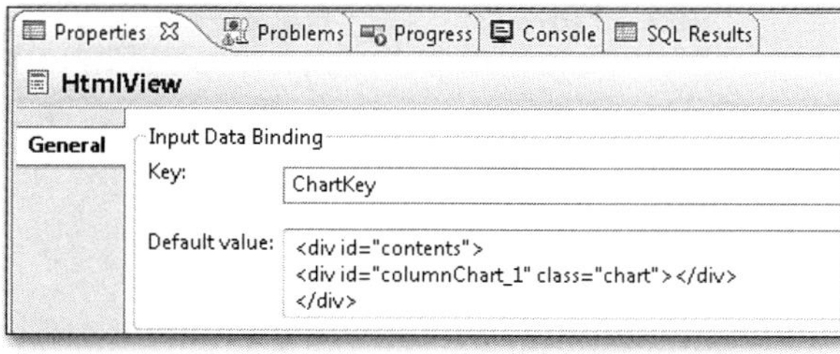

Figure 15.19 Add HTMLView to ColumnChart Screen

Modifying RevenueByY Screen

We next need to modify the RevenueByY screen to remove the data display (ListView) and add a button that is used to display the ColumnChart screen. Select the RevenueByY screen, delete the ListView control, and add a button to the screen with the properties, as shown in Figure 15.20 and Figure 15.21.

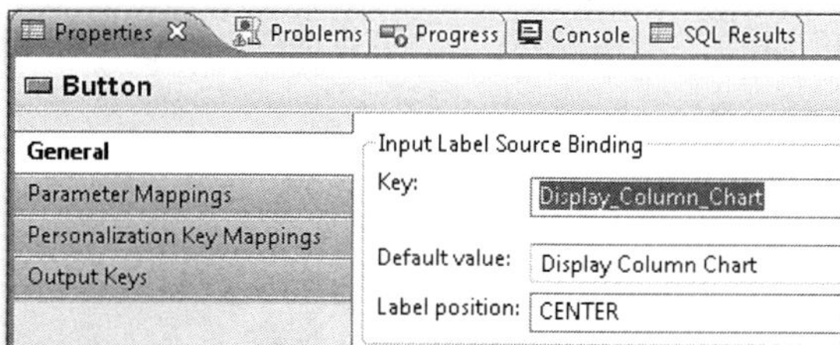

Figure 15.20 Add Button on the RevenueByY Screen

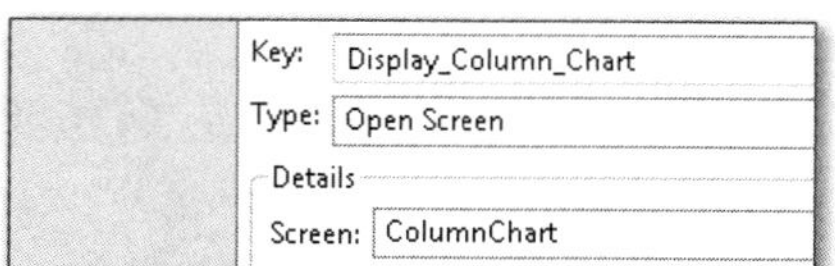

Figure 15.21 Add Button on the RevenueByY Screen

The revised flow design appears as shown in Figure 15.22.

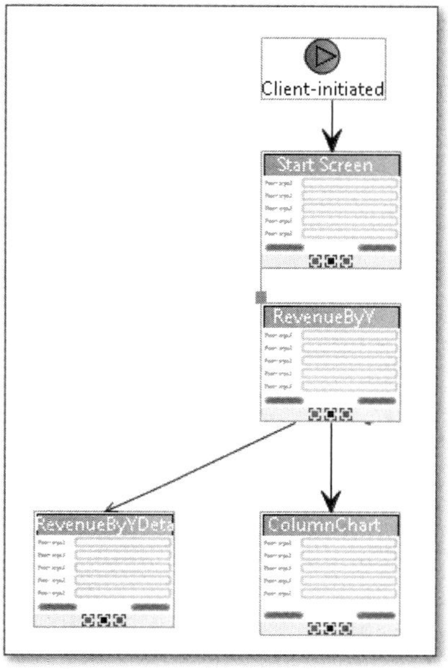

Figure 15.22 Modified Flow Design

15.4.2 Developing Code

We are now ready to develop the code required to create the column chart. Next, we discuss each of the files that need to be created and/or modified.

MyLayout.js

The first file we are going to create actually describes the layout for the column chart. This provided file is written as a JavaScript function that is called by the MAKit HTLM5 code. The file is created as follows:

1. Right-click the JS directory and select NEW • OTHER.
2. Select JAVASCRIPT • JAVASCRIPT SOURCE FILE.
3. Click NEXT.
4. Enter "MyLayout.js" for the file name.
5. Click FINISH.

Add the JavaScript shown in Listing 15.1. Note that this provided standard layout file, as shown in Listing 15.1, can be modified in the future as you gain more understanding of the framework.

```
//used for defining the layout of normal chart on the screen
arrangeCharts = function(cols, rows, contentDivElem, refresh)
{
var width = window.innerWidth;
var height = window.innerHeight;
var contents = contentDivElem;
if (contents == "undefined")
contents = "contents";
var elem = document.getElementById(contents);
elem.style.width = width + "px";
elem.style.height = (height - 40) + "px";
var len = elem.children.length;
var childElem;
for (var i = 0; i < len; i++)
{
childElem = elem.children[i];
childElem.style.width = (95 / cols) + "%";
childElem.style.height = (95 / rows) + "%";
if (refresh != undefined && refresh)
{
if (childElem.chart != undefined)
childElem.chart.refresh();
}
}
};
//used for defining the layout of drilldown chart
arrangeDrillDownCharts = function(cols, rows, contentDivElem, refresh)
{
var offset = 0;
var toolbars = document.getElementsByClassName("toolbar");
if (toolbars != undefined && toolbars.length > 0)
offset = toolbars[0].offsetHeight;
var width = window.innerWidth;
var height = window.innerHeight;
var contents = contentDivElem;
if (contents == "undefined")
contents = "contents"
var elem = document.getElementById(contents);
elem.style.width = width + "px";
elem.style.height = (height - offset - 40 - 40) + "px";
childElem = elem.children[0];
childElem.style.width = (95/cols) + "%";
childElem.style.height = (95/rows) + "%";
```

```
if (refresh != undefined && refresh)
{
if (childElem.drilldowngroup != undefined)
childElem.drilldowngroup.refresh();
}
};
```

Listing 15.1 Defining the Layout of a Normal Chart on the Screen

> **Note**
>
> On line 8, the term `contents` is used, and it matches the first `div` on our HTML control.

Save and then close the file.

MyQuery.js

Add a second file called *MyQuery.js* to the /JS directory. Leave this file empty. We will later insert into this file code that is the query used to retrieve the required data. Save and close the file.

Data and bi_chart_bar.xml

Right-click HTML and add a new folder called DATA. This folder is used to store metadata files that define the various charts.

Right-click DATA, select CREATE A NEW XML FILE (CREATE XML FILE FROM AN XML TEMPLATE) called *bi_chart_bar.xml* and add the code shown in Listing 15.2.

```
<Chart Name="gr_1" ChartType="column" Query="productRevenueByYear"
Title="Revenue By Year">
<Category Column="year"/>
<Series Column="product"/>
<Values>
<Value Expression="revenue" Format="currency"/>
</Values>
</Chart>
```

Listing 15.2 Bar Chart Description

Save and close the file. These last three new files should appear as shown in Figure 15.23.

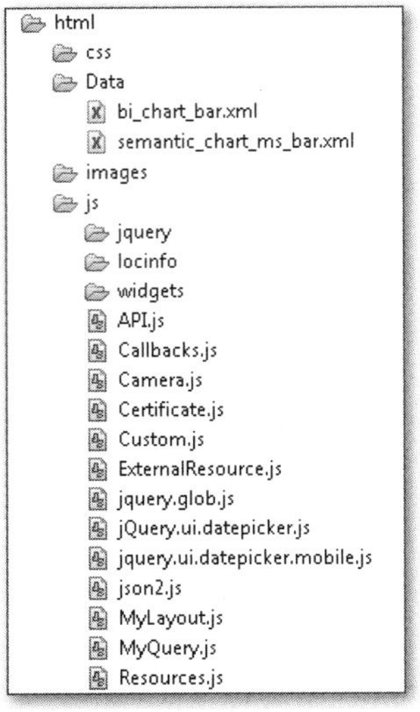

Figure 15.23 The MyLayout.js, MyQuery.js, and bi_chart_bar.xml Files

Custom.js

The code generated for the hybrid app includes a file called *Custom.js*. This file is not updated once it is generated, and we enter code into it to customize the app. This file can be seen in the directory shown in Figure 15.23. The following are segments of code to be added to this file.

The code shown in Listing 15.3 is used for creating the MAKit chart. Locate the `customAfterShowScreen` function in *Custom.js*, as shown in Listing 15.3, and then insert the code into the function so that it appears as shown in Listing 15.4.

```
//Use this method to add custom code after a screen is shown.
function customAfterShowScreen(screenToShow, screenToHide) {
}
```

Listing 15.3 customAfterShowScreen before Changes

```
//Use this method to add custom code after a screen is shown.
function customAfterShowScreen(screenToShow, screenToHide) {
if (screenToShow === "ColumnChart") {
arrangeCharts(1, 1, "contents");//only one chart on the screen
showFullBiChart("columnChart_1", "data/bi_chart_bar.xml");
}
}
```

Listing 15.4 customAfterShowScreen after Changes

> **Note**
>
> `ColumnChart` is the key for the screen, `contents` is the ID defined as the placeholder, `columnChart_1` is the ID for the chart, and *data/bi_chart_bar.xml* is the metadata file. These item names must match those that were used when the screen was created.

Then, add the new function shown in Listing 15.5 after the code entered in the previous step.

```
function showFullBiChart(htmlElementID, metadataFile) {
var elem = document.getElementById(htmlElementID);
var chart = new $MA.Chart(elem.id, true);
chart.load(metadataFile);
}
```

Listing 15.5 Showing the Chart

These changes to the *Custom.js* FILE are now complete for this chart. Save all and close this file. These changes are not replaced the next time you deploy a project. The rule for *custom.js* is "do not replace if it exists."

MyQuery.js

The following code is used to get the required data from the database via the Unwired Server. The MAKit is a query-based solution; that is, the app developer must implement a query that provides data, and the hybrid app developer must add the code that implements the query. A single function exists, called `executeMAQuery`, which is the bridge to return a data table to MAKit. For our example, we are going to place this code in the *MyQuery.js* file. Since this is a file we created, it is not affected by deployment.

Open the *MyQuery.js file* and enter the code shown in Listing 15.6.

```
function executeMAQuery(chartObj, name, args){
var dataTable;
if (name === "productRevenueByYear") {
dataTable = getProductRevenueByYearData();
chartObj.setDataTable(dataTable);
}
}
```

Listing 15.6 Code for Query

Note

productRevenueByYear is the query name in the metadata for the chart. The getProductRevenueByYearData function gets the data for this chart.

The code for the getProductRevenueByYearData function, as shown in Listing 15.7, should also be entered into the *MyQuery.js* file.

```
function getProductRevenueByYearData()
{
//Define the table columns for the chart
var dataTable = new $MA.DataTable();
dataTable.addColumn("product", "string");
dataTable.addColumn("year", "number");
dataTable.addColumn("revenue", "number");
//Get data from SUP MBO -- RevenueByY
var message = getCurrentMessageValueCollection();
var itemList = message.getData("RevenueByY");
var items = itemList.getValue();
var noOfItems = items.length;
if (noOfItems > 0) {
var i = 0;
while (i < noOfItems) {
var theItems = items[i];
//Get column "year" from the MBO
var yearObj=theItems.getData("RevenueByY_year_attribKey");
var year = yearObj.getValue();
//Get column "product" from the MBO
var prodObj=theItems.getData("RevenueByY_product_attribKey");
var prod = prodObj.getValue();
//Get column "revenue" from the MBO
var amountObj=theItems.getData("RevenueByY_revenue_attribKey");
var amount = amountObj.getValue();
//Set the a row of data to the table
dataTable.addRow([prod, year, amount] );
i = i + 1;
}
}
else {
alert("No data is available.");
}
return dataTable;
}
```

Listing 15.7 Function Used to Retrieve Data

Save all and close the file.

15.4.3 Testing the App

Return to the flow design, right-click, and select Generate Hybrid App. This time, select Update Generated Code, Deploy to Unwired Server, and Assign Hybrid App to User.

When we run the hybrid app on an iPhone, as shown in Figure 15.24, we see that the screen constrains part of the chart. While smart phones can be used for the MAKit HTML5, the app works better with a tablet device.

Figure 15.24 Column Chart as Viewed on an iPhone 5

When displayed on an iPad, the same hybrid app appears as shown in Figure 15.25. The color chart shows the sales for three products (desktop, laptop, and tablet) for six years. Select a given year by touching the bars for that year. The actual sales values for each product display in the bar at the top of the chart.

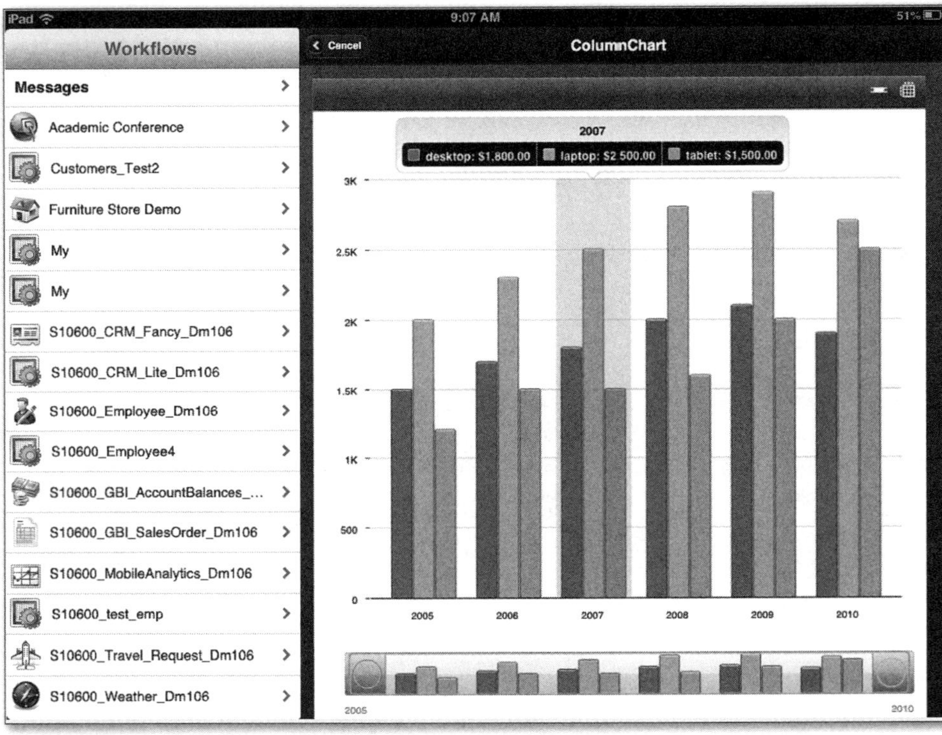

Figure 15.25 Column Chart Displayed on an iPad

You can turn off one or more data series by clicking that button on the top bar. As shown in Figure 15.26, the laptop series has been removed from the chart.

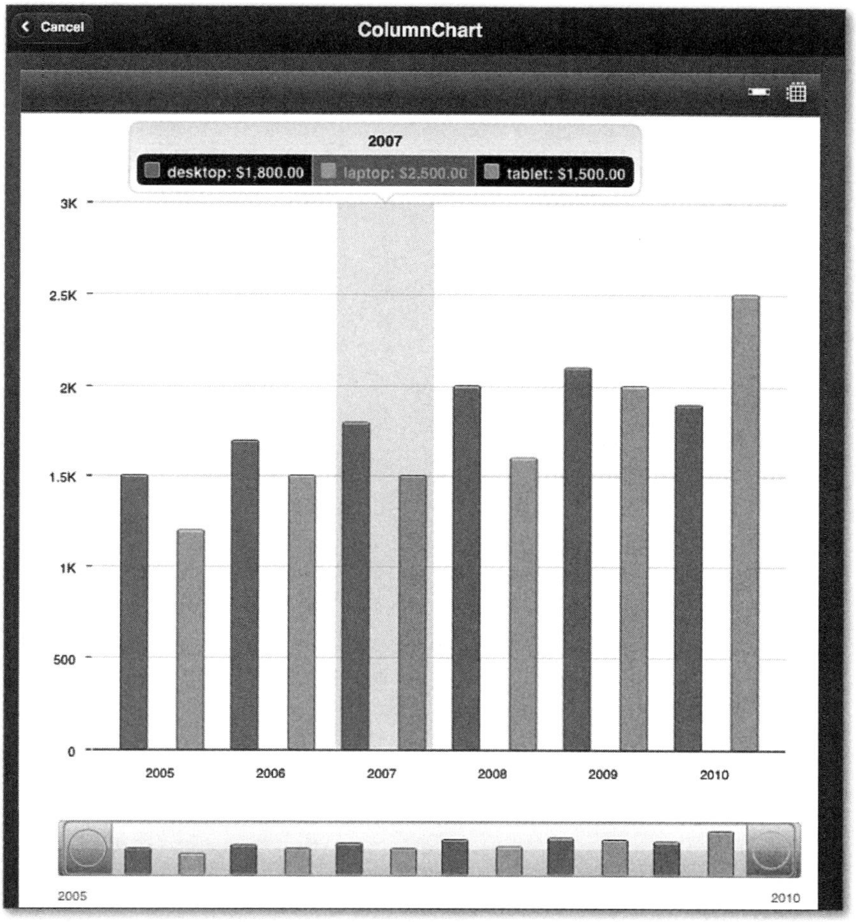

Figure 15.26 Column Chart with Laptop Data Turned Off

The icon at the top of the chart can be used to toggle between displaying the data and the chart, as shown in Figure 15.27.

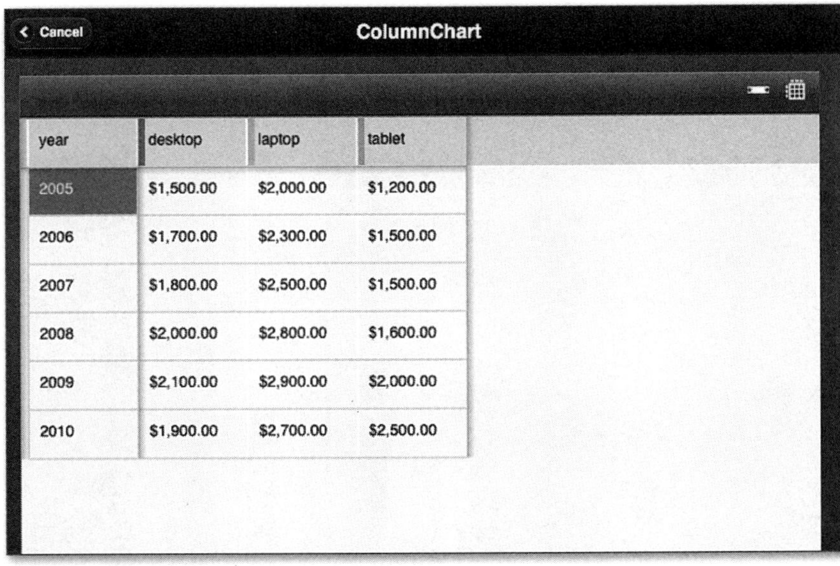

year	desktop	laptop	tablet
2005	$1,500.00	$2,000.00	$1,200.00
2006	$1,700.00	$2,300.00	$1,500.00
2007	$1,800.00	$2,500.00	$1,500.00
2008	$2,000.00	$2,800.00	$1,600.00
2009	$2,100.00	$2,900.00	$2,000.00
2010	$1,900.00	$2,700.00	$2,500.00

Figure 15.27 Data for Column Chart Displayed

15.5 Creating a Zooming Chart

The second type of chart we are going to develop is referred to as a *zooming chart* or sometimes a *drill down chart*. This type of chart allows you to see the big picture and then drill into the data to see more details. The zooming chart for our example provides the ability to view the sales data as a bar chart by looking at sales by month, by quarter or by year. We first need to modify the flow design by adding the two additional MBOs and creating a new screen that will be used to display this new chart. This is followed by code changes that involve creating a *MasterData* XML file and new *MyQuery* file that describes the data to be used to generate this new chart. Finally, a series of changes to the JavaScript is required to actually retrieve the data from the database.

15.5.1 MBOs

We will be using all three of the MBOs we have created: RevenueByM, RevenueByQ, and RevenueByY. While our app contains the first MBO, we need to drag and drop the other MBOS—RevenueByM and RevenueByQ—onto the flow design. While we place them on the screen in order to have a place for storing the data, they are not actually connected in the flow design.

15.5.2 Flow Design Changes

We need to add a new screen for the new chart and a button to display that chart. We describe both processes next.

ZoomingChart Screen

The change we need to make to the flow design is to add another screen called ZoomingChart, which is the screen we use to display this new chart. On this new screen, we place an HtmlView control with the properties shown in Figure 15.28.

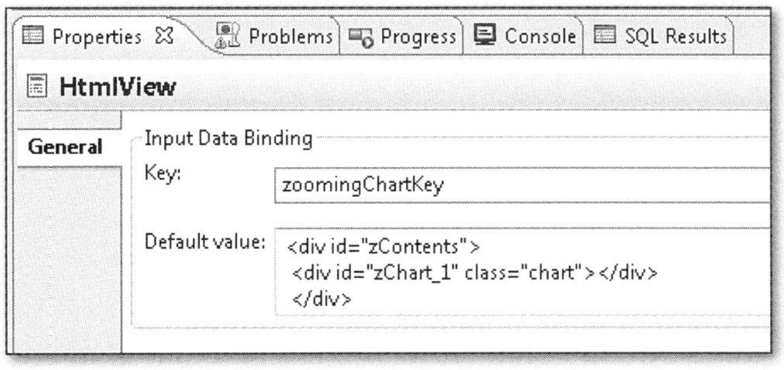

Figure 15.28 HTMLView on the New ZoomingChart Screen

RevenueByY Screen

We finally add a button on the RevenueByY screen to connect to this ZoomingChart screen. The properties are shown in Figure 15.29.

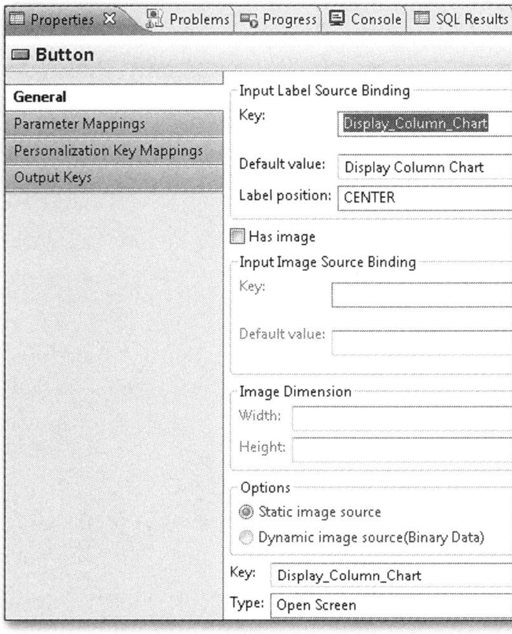

Figure 15.29 Properties for Button to Connect to Zooming Chart Screen

The final flow design now appears as shown in Figure 15.30.

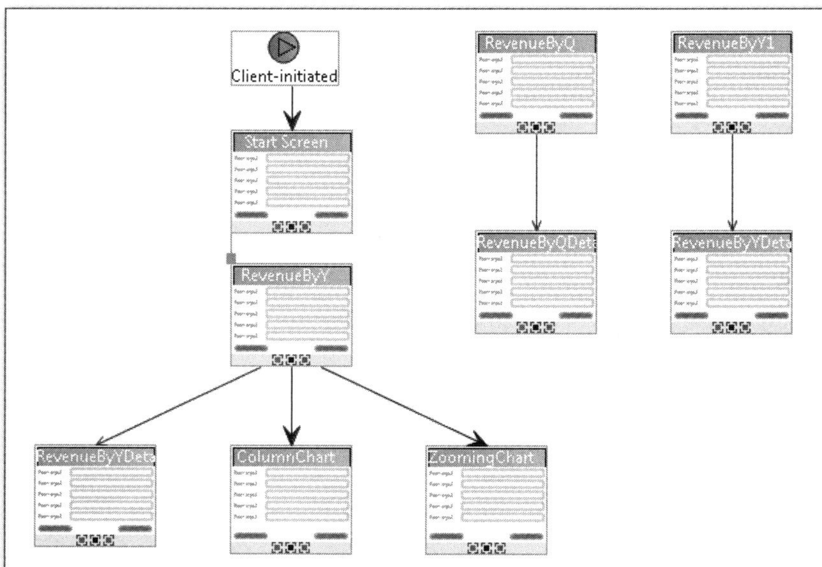

Figure 15.30 Final App with Both Charts

15.5.3 Developing Code

We need to make the following code changes to add this new chart.

Create Metadata File

Create the required metadata file by right-clicking the DATA folder, and create an XML file named *semantic_chart_ms_bar.xml* by adding the code shown in Listing 15.8. Save all and close the file.

```
<Chart Name="chart_1" ChartType="column" Query="productRevenueByYear"
Title="Revenue">
<Category Column="year"/>
<Series Column="product"/>
<Values>
<Value Expression="revenue" Format="currency"/>
</Values>
<ZoomingLevels>
<ZoomingLevel Query="productRevenueByYear" CategoryColumn="year"
Title="Year"/>
<ZoomingLevel Query="productRevenueByQuarter" CategoryColumn="quarter"
Title="Quarter"/>
<ZoomingLevel Query="productRevenueByMonth" CategoryColumn="month"
Title="Month"/>
</ZoomingLevels>
</Chart>
```

Listing 15.8 Metadata for Zooming Chart

Changes to Custom.js

Create the semantic chart. Add the code shown in Listing 15.9 to the existing `customAfterShowScreen` function (before the final `}`) in the *Custom.js* file:

```
if (screenToShow === "ZoomingChart") {
arrangeCharts(1, 1, "zContents"); //only one chart on the screen
showFullZoomingChart("zChart_1", "data/semantic_chart_ms_bar.xml");
}
```

Listing 15.9 Zooming Chart Definition

Add the function shown in Listing 15.10 after the `showFullBiChart` function we added previously.

```
function showFullZoomingChart (htmlElementID, metadataFile) {
var elem = document.getElementById(htmlElementID);
var chart = new $MA.Chart(elem.id, true);
chart.load(metadataFile);
}
```

Listing 15.10 showFullZooingChart

Save all and close.

Modify MyQuery.js

Open the *MyQuery.js* file and replace the existing `executeMAQuery` with the code shown in Listing 15.11. This new file supports both the original graph and the new graph.

```
var yDataOK; //if the data for Year is ready
var qDataOK = false; //if the data for Quarter is ready
var mDataOK = false; //if the data for Month is ready
var myWorkflowMessage; //the incoming message
var ChartObject; //the chart object
var QueryName; // the query name
var Args //possible argument

function executeMAQuery(chartObj, name, args){
var dataTable;
if (name === "productRevenueByYear") { //for the simple chart sample
dataTable = getProductRevenueByYearData();
chartObj.setDataTable(dataTable);
}
else if (name === "productRevenueByQuarter") {
if (qDataOK == false) {
ChartObject = chartObj;
QueryName = name;
Args = args;
retrieveRevemueByQData(); //send request to get data(1)
}
else {
//get the data and pass it to the chart
var dataTable = getProductRevenueByQuarterData(args) //get data (2)
chartObj.setDataTable(dataTable);
chartObj.refresh();
QueryName = null;
```

```
Args = null;
}
}
else if (name === "productRevenueByMonth") {
if (mDataOK == false) {
ChartObject = chartObj;
QueryName = name;
Args = args;
retrieveRevemueByMData(); //send request to get data(3)
}
else {
var dataTable = getProductRevenueByMonthData(args) //get data (4)
chartObj.setDataTable(dataTable);
chartObj.refresh();
QueryName = null;
Args = null;
}
}
else {
}
}
```

Listing 15.11 Retrieving Data and Generating the Graph

Modify MyQuery.js

This file contains the queries used to retrieve the required data. We need to add the queries for the `ByQuarter` and `ByMonth` data. Add the functions shown in Listing 15.12 to the *MyQuery.js* file.

```
function getProductRevenueByQuarterData(args)
{
var dataTable = new $MA.DataTable();
dataTable.addColumn("product", "string");
dataTable.addColumn("year", "number");
dataTable.addColumn("quarter", "number");
dataTable.addColumn("revenue", "number");
//Get data from SUP MBO -- RevenueByQ
var message = getCurrentMessageValueCollection();
var itemList = message.getData("RevenueByQ");
var items = itemList.getValue();
var noOfItems = items.length;
if (noOfItems > 0) {
var i = 0;
```

```
while (i < noOfItems) {
var theItems = items[i];
var yearObj=theItems.getData("RevenueByQ_year_attribKey");
var year = yearObj.getValue();
var prodObj=theItems.getData("RevenueByQ_product_attribKey");
var prod = prodObj.getValue();
var qObj=theItems.getData("RevenueByQ_quarter_attribKey");
var quarter = qObj.getValue();
var amountObj=theItems.getData("RevenueByQ_revenue_attribKey");
var amount = amountObj.getValue();
dataTable.addRow([prod, year, quarter, amount] );
i = i + 1;
}
}
else {
alert("No data is available.");
}
return dataTable;
}
function getProductRevenueByMonthData(args)
{
var dataTable = new $MA.DataTable();
dataTable.addColumn("product", "string");
dataTable.addColumn("year", "number");
dataTable.addColumn("quarter", "number");
dataTable.addColumn("month", "number");
dataTable.addColumn("revenue", "number");
//Get data from SUP MBO -- RevenueByM
var message = getCurrentMessageValueCollection();
var itemList = message.getData("RevenueByM");
var items = itemList.getValue();
var noOfItems = items.length;
if (noOfItems > 0) {
var i = 0;
while (i < noOfItems) {
var theItems = items[i];
var yearObj=theItems.getData("RevenueByM_year_attribKey");
var year = yearObj.getValue();
var prodObj=theItems.getData("RevenueByM_product_attribKey");
var prod = prodObj.getValue();
var qObj=theItems.getData("RevenueByM_quarter_attribKey");
var quarter = qObj.getValue();
```

```
var mObj=theItems.getData("RevenueByM_month_attribKey");
var month = mObj.getValue();
var amountObj=theItems.getData("RevenueByM_revenue_attribKey");
var amount = amountObj.getValue();
dataTable.addRow([prod, year, quarter, month, amount] );
i = i + 1;
}
}
else {
alert("No data is available.");
}
return dataTable;
}
```

Listing 15.12 By-Quarter and by-Month Queries

The original app reads the yearly data, but if the monthly and quarterly data have not been read, then you request the data by adding the functions shown in Listing 15.13 to the *MyQuery.js* file.

```
//Send request to get Quarter data
function retrieveRevemueByQData() {
var rmiKeys = [];
var rmiKeyTypes = [];
var rmiInputOnlyKeys = [];
var rmiInputOnlyKeyTypes = [];
var workflowMessageToSend = getMessageValueCollectionForOnlineRequest('
ZoomingChart', 'GetQData', rmiKeys, rmiKeyTypes);
var inputOnlyWorkflowMessageToSend = getMessageValueCollection
ForOnlineRequest('ZoomingChart', 'GetQData', rmiInputOnlyKeys,
rmiInputOnlyKeyTypes);
doOnlineRequest('ZoomingChart', 'GetQData', 60, 0, '',
null, workflowMessageToSend, inputOnlyWorkflowMessageToSend.
serializeToString());
}
//Send request to get Month data
function retrieveRevemueByMData() {
var rmiKeys = [];
var rmiKeyTypes = [];
var rmiInputOnlyKeys = [];
var rmiInputOnlyKeyTypes = [];
var workflowMessageToSend = getMessageValueCollectionForOnlineRequest('
```

```
ZoomingChart', 'GetMData', rmiKeys, rmiKeyTypes);
var inputOnlyWorkflowMessageToSend = getMessageValueCollection
ForOnlineRequest('ZoomingChart', 'GetMData', rmiInputOnlyKeys,
rmiInputOnlyKeyTypes); doOnlineRequest('ZoomingChart', 'GetMData', 60,
0, '', null, workflowMessageToSend, inputOnlyWorkflowMessageToSend.
serializeToString());
}
```

Listing 15.13 Functions to Load Quarterly and Monthly Data

Async Check

The data communication is asynchronous—when the data is ready (i.e., read in), the system calls `AfterQueryDataReceived`. Add the code shown in Listing 15.14 to the `AfterQueryDataReceived` function in *Custom.js*.

```
function AfterQueryDataReceived(){
myWorkflowMessage = incomingWorkflowMessage;
try
{
var valueCols = incomingWorkflowMessage.getValues();
//check data for year
var yItemList = valueCols.getData("RevenueByY");
if (yItemList != undefined) {
yDataOK = true;
}
else {
yDataOK = false;
}
//check data for quarter
var qItemList = valueCols.getData("RevenueByQ");
if (qItemList != undefined) {
qDataOK = true;
}
else {
qDataOK = false;
}
//check data for month
var mItemList = valueCols.getData("RevenueByM");
if (mItemList != undefined) {
mDataOK = true;
}
```

```
else {
mDataOK = false;
}
//send the specific data back to MAKIT Chart
if (QueryName === "productRevenueByYear") {
var dataTable = getProductRevenueByYearData();
ChartObject.setDataTable(dataTable);
dataTable.refresh();
}
else if (QueryName === "productRevenueByQuarter") {
var dataTable = getProductRevenueByQuarterData(Args)
ChartObject.setDataTable(dataTable);
dataTable.refresh();
}
else if (QueryName === "productRevenueByMonth") {
var dataTable = getProductRevenueByMonthData(Args);
ChartObject.setDataTable(dataTable);
dataTable.refresh();
}
else {
}
QueryName = null;
Args = null;
}
catch (err)
{
QueryName = null;
Args = null;
}
}
```

Listing 15.14 Async Check

Save all and close.

15.5.4 Testing the App

Click Generate Mobile Work Flow Package and run the revised app. The chart showing data by year is essentially the same as Figure 15.25, the chart showing the data by quarter is shown in Figure 15.31, the chart showing the data by month is shown in Figure 15.32, and the actual data is shown in Figure 15.33.

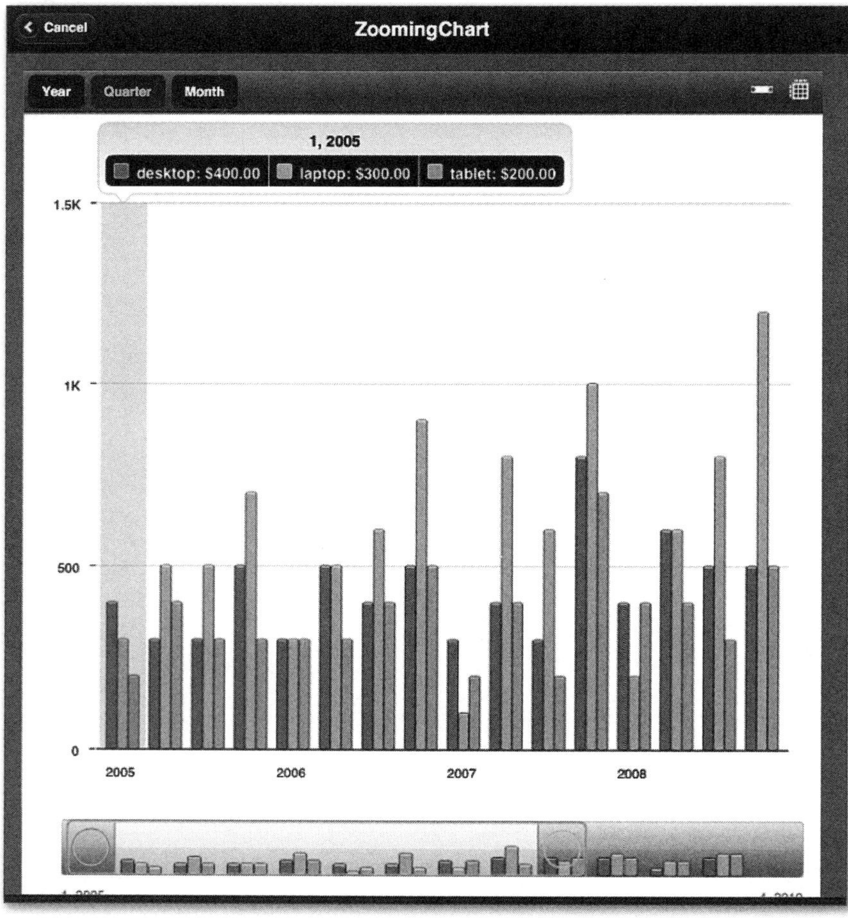

Figure 15.31 Zooming Chart with Data by Quarter

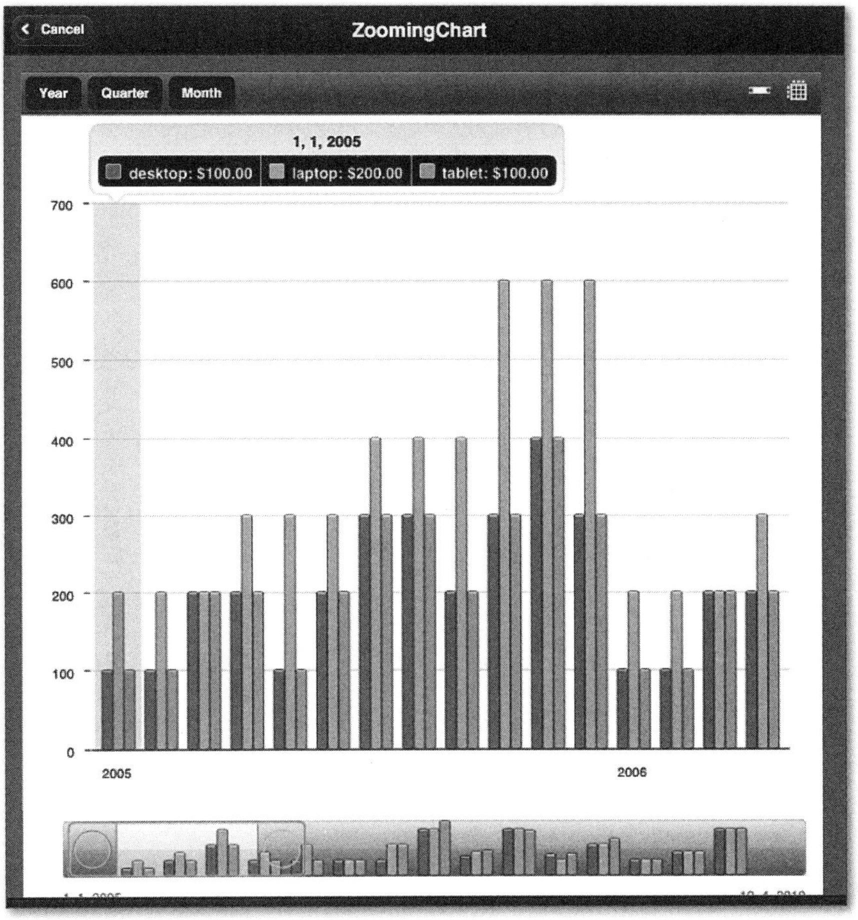

Figure 15.32 Zooming Chart with Data by Month

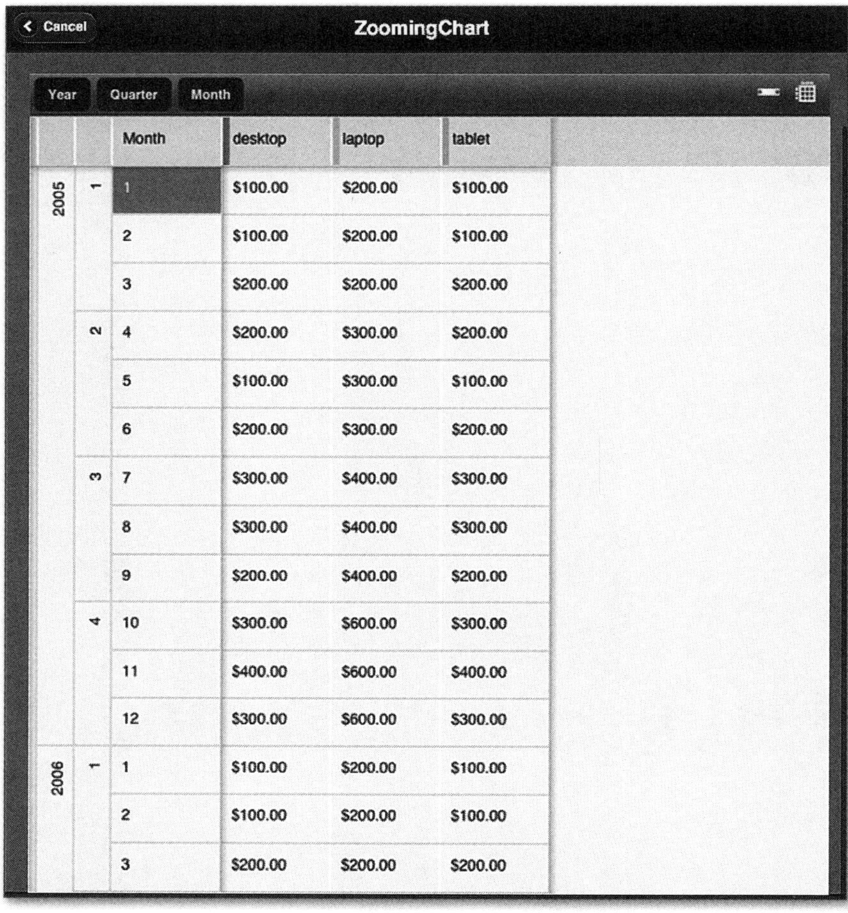

		Month	desktop	laptop	tablet
2005	1	1	$100.00	$200.00	$100.00
		2	$100.00	$200.00	$100.00
		3	$200.00	$200.00	$200.00
	2	4	$200.00	$300.00	$200.00
		5	$100.00	$300.00	$100.00
		6	$200.00	$300.00	$200.00
	3	7	$300.00	$400.00	$300.00
		8	$300.00	$400.00	$300.00
		9	$200.00	$400.00	$200.00
	4	10	$300.00	$600.00	$300.00
		11	$400.00	$600.00	$400.00
		12	$300.00	$600.00	$300.00
2006	1	1	$100.00	$200.00	$100.00
		2	$100.00	$200.00	$100.00
		3	$200.00	$200.00	$200.00

Figure 15.33 Zooming Chart Showing Data View

15.6 Summary

This chapter introduced MAKit HTML5, an example JavaScript library that can be added to a hybrid app. This library provides the ability to add graphics and analytics as screens within a hybrid app. The process involves developing an app that has data, copying the MAKit files into the code developed for this app, and then using the APIs to create the graphics.

In addition to the charting types discussed in this chapter, the MAKit HTML5 code supports a number of different charting types, as shown in Figure 15.34, as well as analytic tools for asking what-if questions. For more information about MAKit charts and metadata, we recommend the *Mobile Analytics Kit (MAKit) Developer Guides: HTML5*, which provided the foundation for the code in this chapter.

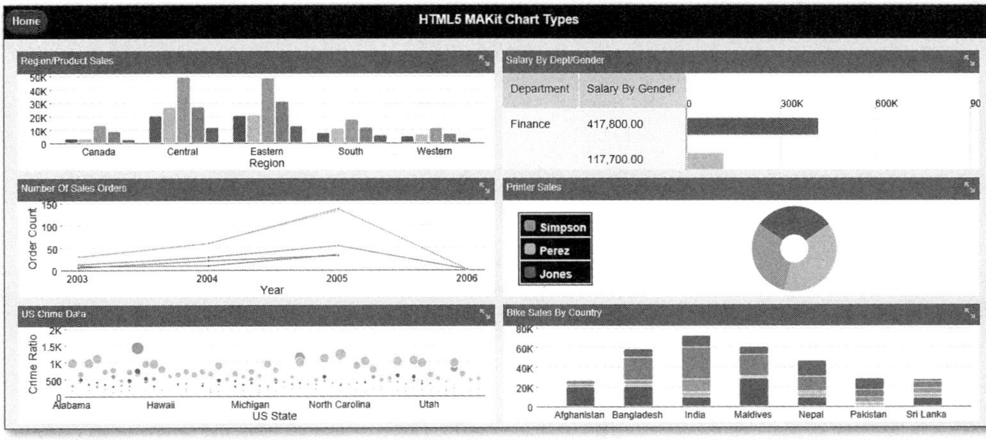

Figure 15.34 Types of Charts

While many developers might wish to access SAP data or use the management and monitoring capabilities of the Unwired Server, not all wish to develop apps using the Unwired Workspace. This chapter describes ways to integrate third-party platforms with SAP Mobile Platform.

16 SAP Support for Open Platforms

The Unwired Workspace is a robust development environment that includes all the tools required to develop cross-platform mobile apps that leverage the features of the Unwired Server. However, SAP Mobile Platform is not a closed environment. SAP has aggressively pursued a policy to open the platform, allowing third-party development platforms to access SAP via the Unwired Server. Developers who wish to leverage the administrative, security, and other features of the Unwired Server but are not interested in MBOs can use the REST Services API introduced in SAP Mobile Platform 2.2. This API allows any HTML-based app to communicate with the Unwired Server and to use proxy connections on the Unwired Server to REST-based data.

Also, SAP has made partnerships that have resulted in tight integration between popular mobile development frameworks and SAP Mobile Platform. This chapter describes ways in which SAP Mobile Platform can be integrated with third-party development environments. Section 16.1 explains the Hybrid Web Container architecture and how third-party frameworks can be integrated into this architecture. Section 16.2 describes three popular third-party development platform partnerships that have a tighter integration with SAP Mobile Platform thanks to SAP partnerships. Finally, Section 16.3 describes SAP Mobile Platform's REST Services API that allows apps to access the Unwired Server using RESTful web services.

16.1 Introduction to Third-Party Frameworks

As we've discussed throughout the book, a hybrid app is a web-based app that runs within the container on a mobile device. Hybrid apps must include the interface and data handling logic, which are built from HTML, CSS, and JavaScript; the hybrid aspect of the app refers to the fact that the app must communicate with the container, which provides some native functionality. In correspondence with this dual nature, the JavaScript API generated for hybrid apps by the Unwired Workspace consists of two parts:

- **Container**
  These APIs are independent of the interface or of whatever framework you choose to build the interface. They are required for building a hybrid app.
- **AppFramework**
  These APIs provide navigation between screens, bind data to interface elements, and provide access to server data. They might be replaced if you choose to use a different framework to build your interface.

Figure 16.1 shows the architecture of the Hybrid Web Container. The JavaScript API container is used to access the container services, while the AppFramework JavaScript API is used to implement navigation, messaging, and data handling in the hybrid app. Apps developed using the Hybrid App Designer (formerly called the Mobile Workflow Forms Editor) rely on the AppFramework.

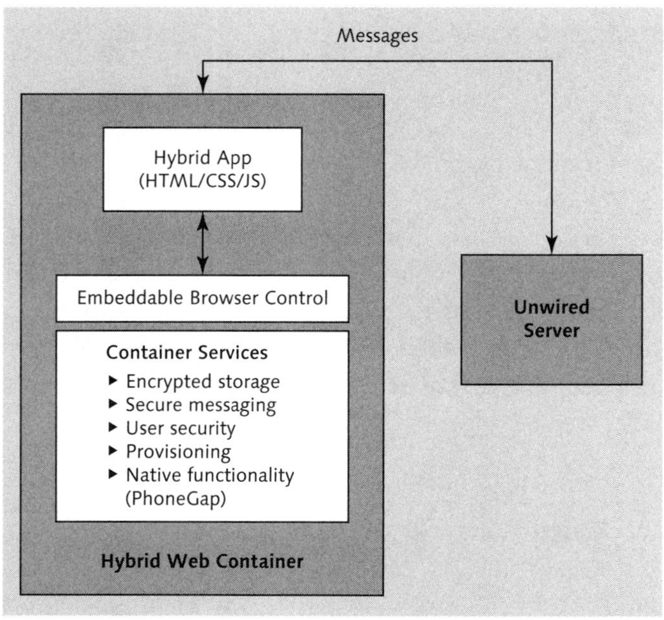

Figure 16.1 JavaScript API

Accessing the JavaScript APIs

The JavaScript APIs are located at *<SUP_HOME>\ UnwiredPlatform\MobileSDK<version>\ HybridApp\API* beginning with version 2.2. Detailed documentation is located at *http://infocenter.sybase.com/help/topic/com.sybase.infocenter.dc01853.0220/doc/html/vhu1349901991724.html* including code samples.

Third-party frameworks fall into two categories. Those in the first category add functionality to the existing JavaScript API but do not replace the AppFramework. Adobe PhoneGap falls into this category. The PhoneGap libraries provide access to native device features and can be used with the AppFramework. As shown in Figure 16.2, the PhoneGap libraries coexist with the container APIs. JavaScript libraries that are compatible with the existing JavaScript API can be added to a project by including it in the HTML/JS folder of a project.

Frameworks in the second category replace the AppFramework API with their own interface and data management tools. This allows developers to leverage the security and data handling features of MBOs and the Unwired Server but develop the app with their preferred framework. Beginning with version 2.2, SAP provides guidance and examples for how to use third-party frameworks.

Documentation for Developing with Third-Party Frameworks

The documentation for using third-party frameworks can be found at *http://infocenter.sybase.com/help/index.jsp?topic=/com.sybase.infocenter.dc01853.0220/doc/html/vhu1343080061068.html.*

The general procedure for developing using third-party frameworks is similar to developing for native apps. The MBOs are created and deployed to the Unwired Server. The Unwired Workspace includes a wizard that generates the required JavaScript APIs, which include the Container APIs and the message handling portion of the AppFramework APIs. Guidance on how to use the APIs can be found in the documentation.

16.2 SAP Partnerships

While it is possible to develop apps using virtually any web-based framework, SAP has developed partnerships with three companies that facilitate the process and provide a tighter integration.

16.2.1 Adobe PhoneGap

Adobe PhoneGap is an open-source framework that allows developers to use JavaScript, HTML, and CSS to access a device's native features. It was originally developed by Nitobi but was purchased by Adobe Systems, who contributed it to the Apache Software Foundation to start the Apache Cordova project. Adobe lists it among its products as Adobe PhoneGap or Adobe PhoneGap Build, and in the Sybase documentation, the libraries are referred to using the name PhoneGap.

The PhoneGap libraries are included with the Unwired Workspace and can be found in the <PROJECT NAME>\Generated Hybrid App\<APPLICATION NAME>\HTML\JS\<PLATFORM> folder in the Workspace Navigator. The libraries are automatically loaded by the `loadPhoneGap()` function in the *API.js* file. No configuration is required to use the PhoneGap API. You simply call PhoneGap functions from an app's custom JavaScript files. Listing 16.1 is an example from the documentation of using a PhoneGap function to create a contact in a device's contacts database.

```
var contact = navigator.contacts.create();
contact.nickname = "Plumber";
var name = new ContactName();
name.givenName = "Jane";
name.familyName = "Doe";
contact.name = name;
contact.save(onSaveSuccess,onSaveError);
```

Listing 16.1 Creating a Contact in a Device's Contacts Database

While SAP Mobile Platform 2.1.3 includes the PhoneGap libraries for only Android and iOS, version 2.2 includes libraries for all four supported platforms. The version of PhoneGap included with SAP Mobile Platform 2.2 is 1.4.1, but it is possible to upgrade to newer versions. The documentation includes instructions on how to upgrade.

The supported features are shown in Table 16.1. You can also use PhoneGap plug-ins to extend this feature set.

Feature	iOS 3GS and newer	Android	BlackBerry 5.x-6.0+	Windows Mobile
Accelerometer	Yes	Yes	Yes	No
Camera	Yes	Yes	Yes	Yes
Capture	Yes	Yes	Yes	No
Compass	Yes	Yes	No	Yes
Connection	Yes	Yes	Yes	No
Contacts	Yes	Yes	Yes	Yes
Device Info	Yes	Yes	Yes	No
Device events	Yes	Yes	No	No
File operations	Yes	Yes	Yes	Yes
Geolocation	Yes	Yes	Yes	Yes
Media	Yes	Yes	No	Yes
Notification	Yes	Yes	Yes	Yes
Database	Yes	Yes	Yes	No

Table 16.1 The PhoneGap APIs Provide Access to Native Device Features

16.2.2 Sencha Touch

Sencha Touch by Sencha is a framework that produces cross-platform apps based on HTML5, CSS3, and JavaScript that reproduce a native look and feel. Not only do Sencha apps look like native apps, but they also respond to gestures (i.e., pinch and swipe) like native apps. Figure 16.2 shows a screen styled with jQuery Mobile using the AppFramework and the same screen using Sencha Touch.

Figure 16.2 Sencha Apps that Replace the jQuery Mobile Styling Used by the AppFramework.

Unlike the AppFramework, Sencha adheres to a model-view-controller architecture, and apps are created primarily in JavaScript. Sencha apps consist of JavaScript files and a small HTML page as a loader. As a result, developing apps in Sencha Touch is a different experience than with many other frameworks, including the AppFramework. Figure 16.3 shows an example of Sencha code used to create a screen. You do not edit HTML or manipulate HTML elements directly, as you do with jQuery. The result is that Sencha development is a more structured experience than that of a framework such as jQuery Mobile.

```
Ext.define("NotesApp.view.NotesListContainer", {
  extend: "Ext.Container",
  config: {
    items: [{
      xtype: "toolbar",
      docked: "top",
      title: "My Notes",
      items: [{
        xtype: "spacer"
      }, {
        xtype: "button",
        text: "New",
        ui: "action",
        id:"new-note-btn"
      }]
    }]
  }
});
```

Figure 16.3 HTML in Sencha Apps Generated from JavaScript

Sencha apps represent data structures as models (see Figure 16.4). When using Sencha Touch with SAP Mobile Platform, MBOs and OData data are represented as models. Models are linked to proxies, which manage server connections and data integration. The result of the partnership with SAP is proxies for accessing MBO data on the Unwired Server and OData data on SAP NetWeaver Gateway.

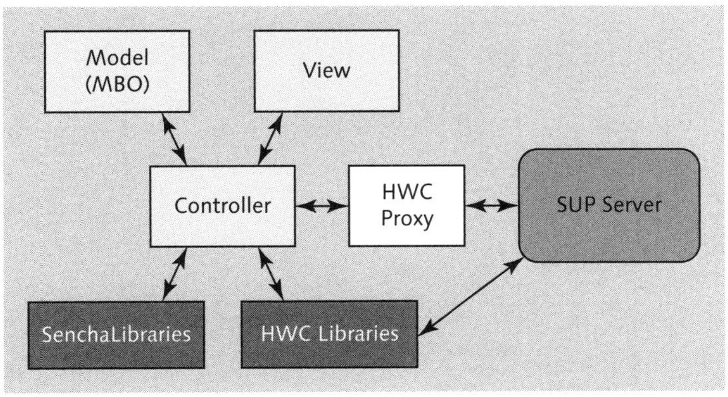

Figure 16.4 Sencha Apps Adhering to a Model-View-Controller Architecture

There is no integration between the Unwired Workspace and the Sencha development environment, Sencha Designer. The developer must manually integrate the Sencha app with the hybrid app in the Unwired Workspace before it is deployed to the Unwired Server. The Sencha app replaces the AppFramework with its own styling and data-handling logic.

To develop an app using Sencha Touch and the Unwired Platform, create MBOs and a hybrid app in the Unwired Workspace. When you generate the hybrid app code, you use the OPTIMIZE FOR PERFORMANCE option to eliminate the jQuery Mobile code. Next, you develop the Sencha app in the Sencha Designer, creating models that reflect the structure of the MBOs. Finally, the Sencha app files are copied into the project in the Unwired Workspace. Figure 16.5 shows the result using a simple app. Notice the *model*, *view*, *controller*, and *proxy* files.

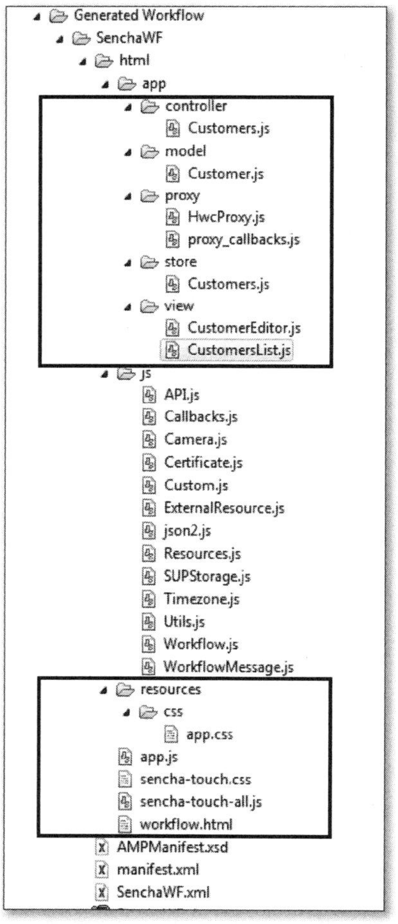

Figure 16.5 Sencha App Files Copied into the Hybrid App Project in the Unwired Workspace

16.2.3 Appcelerator Titanium

Appcelerator Titanium is a framework for developing native apps for Android and iOS using web technologies. The Appcelerator Titanium architecture is shown in Figure 16.6. App development uses the Titanium Studio, which is an Eclipse-based IDE. Similar to Sencha, apps are developed entirely in JavaScript. The JavaScript code is interpreted on the device and translated by the Native Binding Layer to native function calls. The result is that apps are essentially native apps. However, because they are written in JavaScript and translated on the device, the apps are cross platform. Appcelerator Titanium currently supports Android, iOS, and web apps.

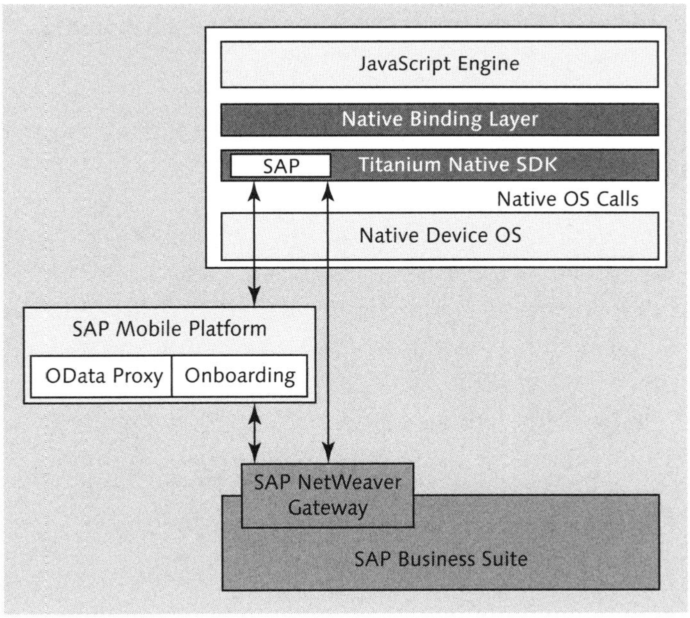

Figure 16.6 Appcelerator Titanium Architecture

Unlike Sencha Touch, Appcelerator Titanium has no native support for accessing MBO data. The result of the partnership with SAP is the SAP OData module, which allows an app to consume services on SAP NetWeaver Gateway. Apps can connect in one of three ways:

- **Directly to SAP NetWeaver Gateway**
  The app uses the URL of the service on SAP NetWeaver Gateway to access the data. The Unwired Server is not involved in the connection.

- **Using the Unwired Server as a proxy**
  The app uses SAP Mobile Platform's REST Services API to connect to the Unwired Server and access SAP NetWeaver Gateway service via a proxy connection on the Unwired Server. In this scenario, the app does not require an application connection on the Unwired Server.
- **Registering with the Unwired Server**
  The app registers with an application connection on the Unwired Server and uses a proxy connection on the Unwired Server to access services on SAP NetWeaver Gateway, as in the previous scenario. The advantage of this scenario is that it leverages the security and administrative features of the Unwired Server.

None of the scenarios require the Unwired Workspace. All the development is performed in the Titanium Studio.

16.3 REST Services Apps

One of the new features provided by SAP Mobile Platform 2.2 is the REST Services API. This API permits any HTTP-based app to access the Unwired Server's REST services. You can build your app using any platform and are not required to use Sybase client libraries. This means that any third-party platform provider, not just SAP partners, can offer tools that integrate with SAP Mobile Platform. An early example is SAP HTML5 Designer by Neptune Software.

REST-based apps cannot use the Unwired Server's data caching and synchronization features, but they can take advantage of app management and monitoring features, including the following:

- Single sign-on
- App registration
- Authentication
- Native push notification
- Configuration

Although you are not required to use libraries from the Unwired Workspace in your apps, there are some libraries that you may use. For example, the libraries used to parse OData and manage the data vault may be used.

REST-based apps are configured in the SCC by creating an app and app template. The app then uses REST services to retrieve and set app information and to register an application connection. Apps can also register with the Unwired Server to receive push notifications.

Apps access data sources through Unwired Server proxy connections (see Figure 16.7). The proxy connection points to a service endpoint and supports read (`HTTP GET`), create (`HTTP POST`), update (`HTTP PUT` or `HTTP MERGE`), and delete (`HTTP DELETE`) operations.

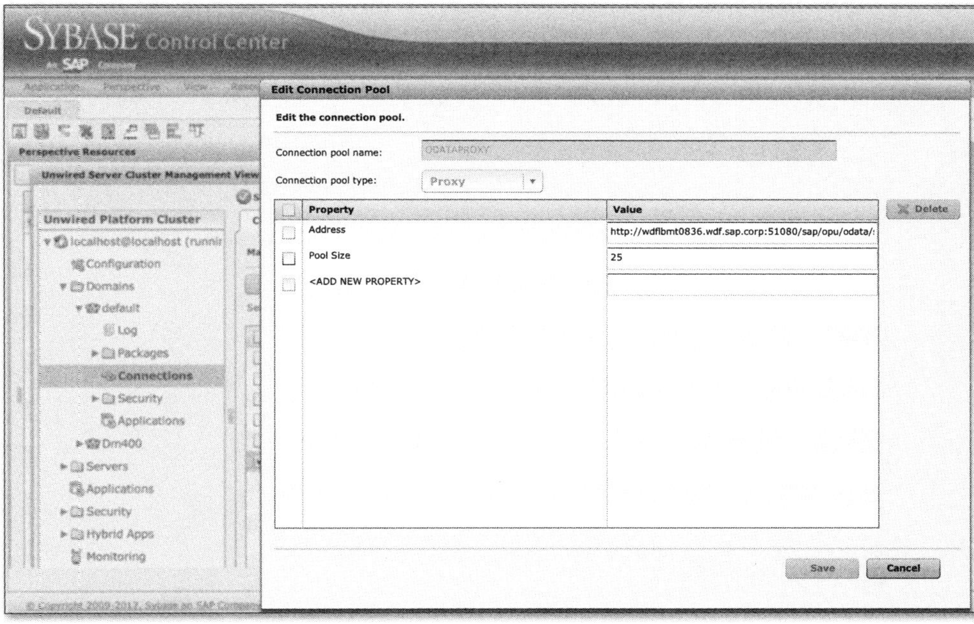

Figure 16.7 Creating a Proxy Connection in the SCC

16.4 Summary

This chapter described ways to integrate third-party platforms with SAP Mobile Platform. By opening the platform, SAP has opened the possibilities for any mobile app developer who wants to take advantage of the data handling and security features of the Unwired Server.

This final chapter brings together many of the concepts described in the book by demonstrating the process of building a hybrid app. The CRM Lite app is used by a customer service rep in the field.

17 CRM Lite Case Study

This chapter describes a hybrid app that includes many of the features described in the book. The chapter begins with a description of the app requirements (Section 17.1) and then describes the data model required by the app (Section 17.2). In the final section, the app is implemented as a standard hybrid app, and then customizations are added.

17.1 The Scenario

The app in this scenario is used by a customer service representative (CSR) for a specialty grocery distributor. The distributor prides itself on customer service and wants its CSRs to be able to provide product information, create sales orders, and respond quickly to customer queries while in the field. When a customer has a query or an issue with a delivery or order, the customer calls a central customer service center. The app should be able to receive and process notifications when customers call the center.

CSRs should be able to retrieve product information, including current inventory levels. If they have questions concerning a product's inventory, the app should allow them to send emails to the inventory manager. The app should also be able to retrieve outstanding product purchase orders, along with quantities and expected delivery dates.

CSRs service customers within a sales region. The app must allow the CSR to personalize it by specifying a sales region and their employee ID. The app should allow CSRs to retrieve records of customers in their sales region, initiate communication with customers, and locate them on a map. CSRs should also be able to retrieve and process sales orders, as well as create new sales orders.

17.2 Defining MBOs

The app is based in part on the sample database that is installed with Sybase Unwired Server and in part of the Northwind sample database delivered with Microsoft SQL Server. The app uses nine MBOs. The MBO flow design is shown in Figure 17.1.

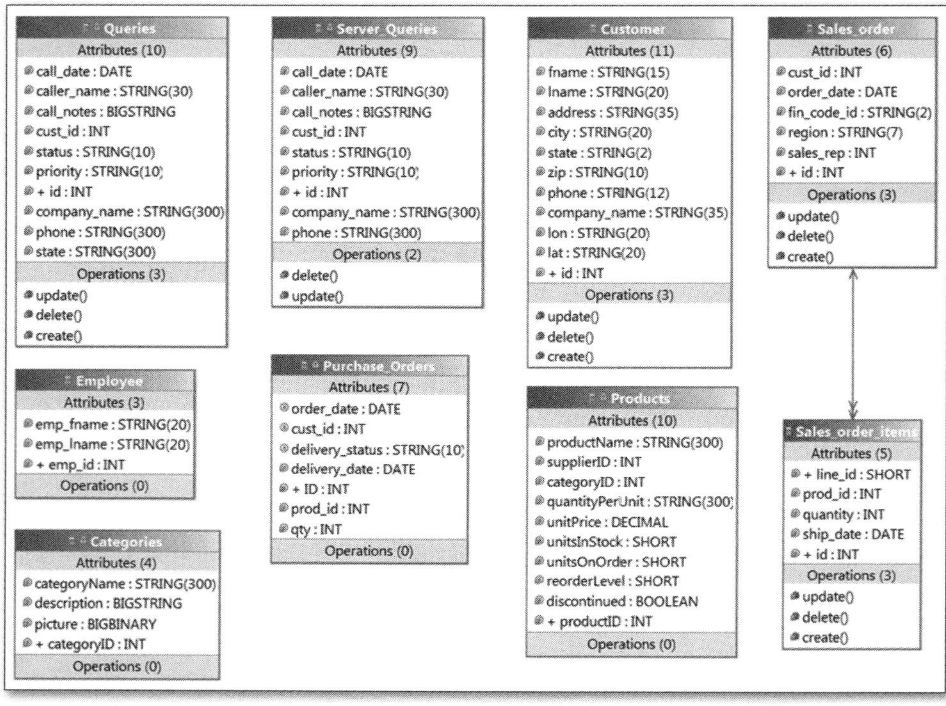

Figure 17.1 The App's Nine MBOs

The app requires the three cache groups shown in Figure 17.2. Most of the MBOs are included in the default cache group, which has an ON DEMAND policy. The SERVER QUERIES MBO is used to send server notifications to the app, so it must be included in the online cache group with an ONLINE policy. Finally, the CATEGORIES and EMPLOYEE MBOs include data that changes rarely, so these are included in a cache group with a SCHEDULED policy with a 24-hour update interval.

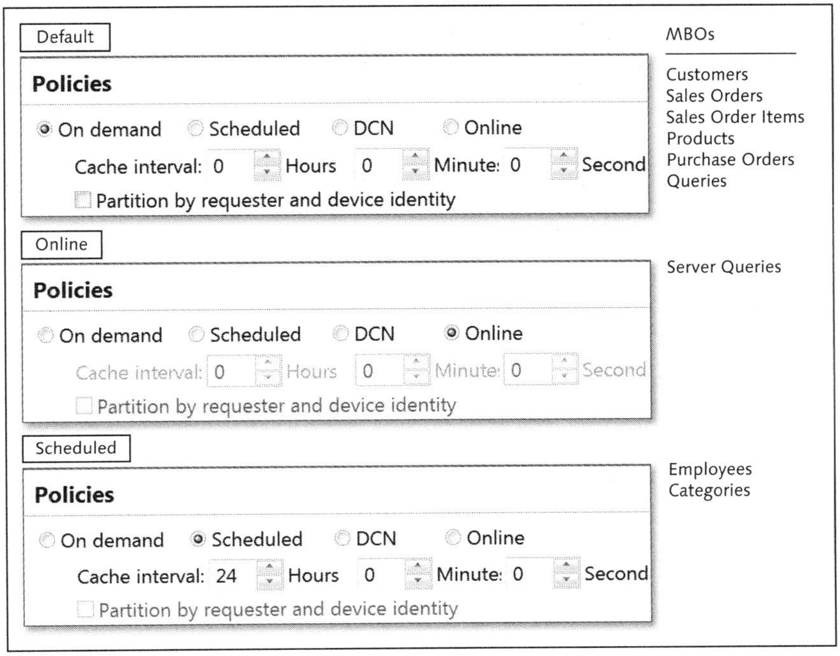

Figure 17.2 Three Required Cache Groups

Next, we describe the MBOs included in the app.

17.2.1 Queries and Server Queries

The two MBOs shown in Figure 17.3 are based on the `call_track` and `customer` tables in the sample database. The SERVER QUERIES MBO is used for server notifications, and the QUERIES MBO is used to retrieve queries within the client-initiated app. The two MBOs are based on the `call_track` table in the Sybase sample database.

There are two versions of the MBO based on the `call_track` table because the SERVER QUERIES MBO is included in an online cache group and can only have the `findByParameter` object query. As a result, this MBO could not be used to retrieve queries in the client-initiated portion of the app. The definition of the SERVER QUERIES MBO is shown in Listing 17.1.

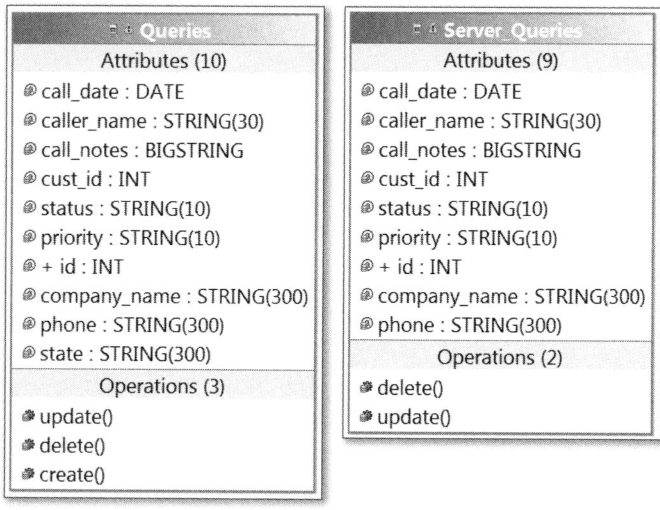

Figure 17.3 Queries and Server Queries MBOs

```
SELECT  x.*, y.company_name,y.phone
FROM sampledb.dba.call_track x INNER JOIN sampledb.dba.customer y
ON x.cust_id = y.id
WHERE x.id = :queryID
```

Listing 17.1 Server Queries MBO Definition

The `join` is necessary because the CSR will need customer contact information to respond to queries. Including it in the QUERIES MBO avoids a trip to the server to download customer information.

The definition of the QUERIES MBO is shown in Listing 17.2. The `WHERE` clause creates a load argument that filters queries on the CSR's sales territory.

```
SELECT  x.*, y.company_name,y.phone,y.state
FROM sampledb.dba.call_track x INNER JOIN sampledb.dba.customer y
ON x.cust_id = y.id
WHERE y.state = :state
```

Listing 17.2 Queries MBO Definition

The `state` load argument is mapped to a personalization key called `statePK`, shown in Figure 17.4.

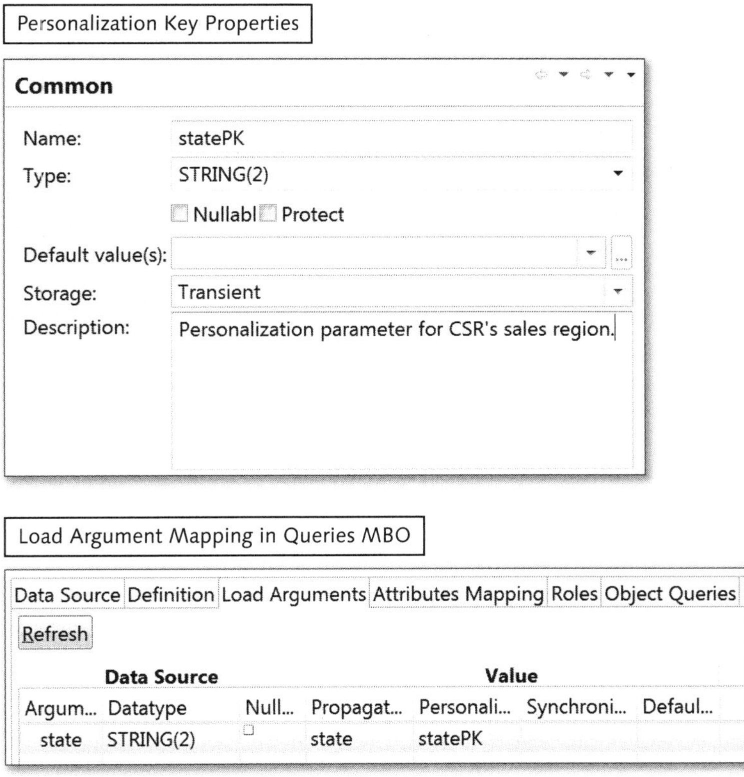

Figure 17.4 StatePK Personalization Key Mapping to the State Load Argument

The QUERIES MBO includes two object queries, both of which we discuss next.

findOpenQueries

The `findOpenQueries` object query retrieves open queries ordered by call date:

```
SELECT x.* FROM Queries x
WHERE x.status <> 'Closed'
ORDER BY x.call_date
```

findCustomerQueries

The `findCustomerQueries` object query is used to retrieve a specific customer's queries:

```
SELECT x.* FROM Queries x
WHERE x.cust_id = :cust_id
ORDER BY x.call_date
```

17.2.2 Customer MBO

Similar to the QUERIES MBO, the CUSTOMER MBO, shown in Figure 17.5, includes a load argument that filters customer records on `state`. This MBO is based on the `customer` table in the sample database.

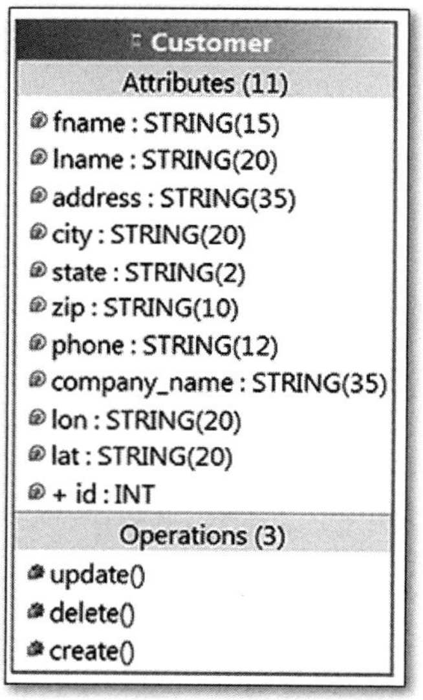

Figure 17.5 Customer MBO

The definition of the CUSTOMER MBO is shown in Listing 17.3.

```
SELECT  id,  fname,  lname,  address,  city,  state,  zip,
phone, company_name, lat, lon FROM sampledb.dba.customer
WHERE state = :state
```

Listing 17.3 Customer MBO Definition

The `state` load argument is mapped to the same `statePK` personalization key used for the QUERIES MBO.

The CUSTOMER MBO includes two additional object queries, which we discuss next.

findByCompanyName

The `findByCompanyName` object query is used to search customers on the `company_name` field. The `LIKE` operator and the `%` wildcard allow the user to search on the leading characters of the company name field:

```
SELECT x.* FROM Customer x
WHERE x.company_name LIKE :company_name +'%'
ORDER BY x.company_name
```

findAllSortedByCompanyName

The `findAllSortedByCompanyName` object query retrieves all customer records sorted by company name. This includes only those records within the CSR's sales region because the records are filtered by the load argument in the definition of the MBO:

```
SELECT x.* FROM Customer x
ORDER BY x.company_name
```

17.2.3 Products and Categories MBOs

The PRODUCTS and CATEGORIES MBOs contain product information and have standard definitions with no load arguments. The products are classified by the categories contained within the CATEGORIES MBO. These MBOs are based on the products and categories tables of the Northwind sample.

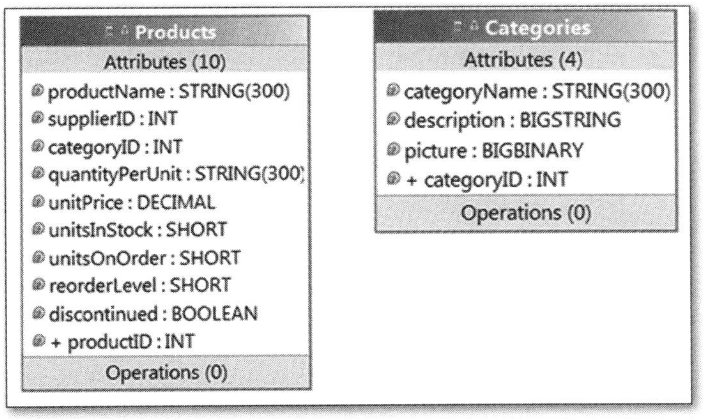

Figure 17.6 Products and Categories MBOs

The only requirement to use MBOs from different data sources is to specify each data source in the SERVER CONNECTION MAPPING dialog when you deploy the MBOs to the Unwired Server. Figure 17.7 shows the dialog for the CRM Lite app.

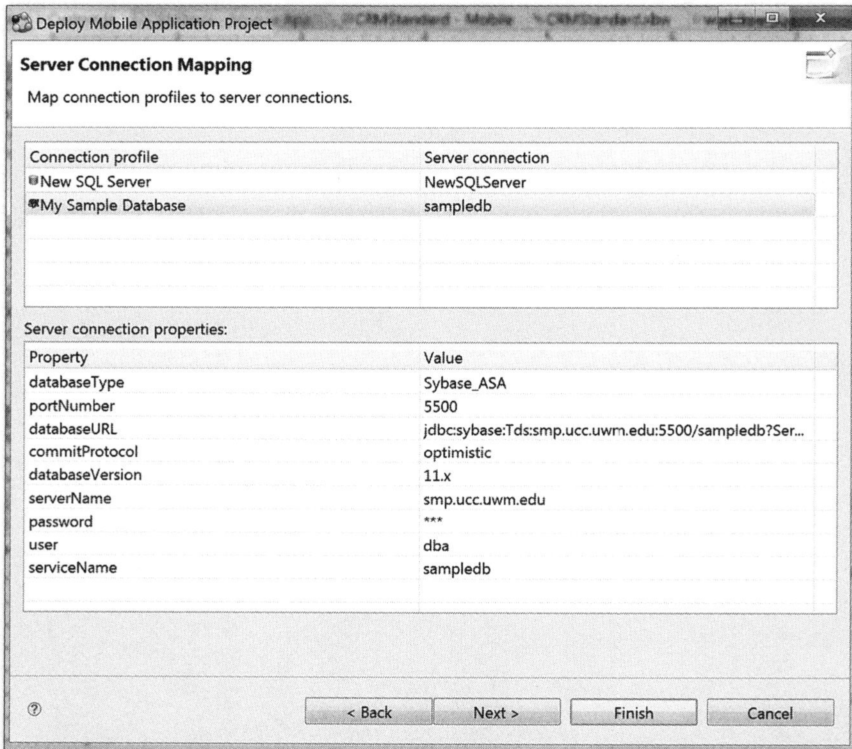

Figure 17.7 Server Connection Mapping with Multiple Data Sources

The PRODUCTS and CATEGORIES MBOs have one object query each, which we discuss next.

findByCategory

The `findByCategory` object query retrieves products within a given category:

```
SELECT x.* FROM Products x
WHERE x.categoryID = :CategoryID
ORDER BY x.productName
```

findAllSortedByName

The `findAllSortedByName` object query retrieves all categories sorted by category name:

```
SELECT x.* FROM Categories x
ORDER BY x.categoryName
```

17.2.4 Employee MBO

The EMPLOYEE MBO, shown in Figure 17.8, is used only to allow the CSR to set the employee ID as a personalization setting. Only the name and ID columns are included in the definition, and no operations are created. This MBO is based on the `employee` table in the sample database.

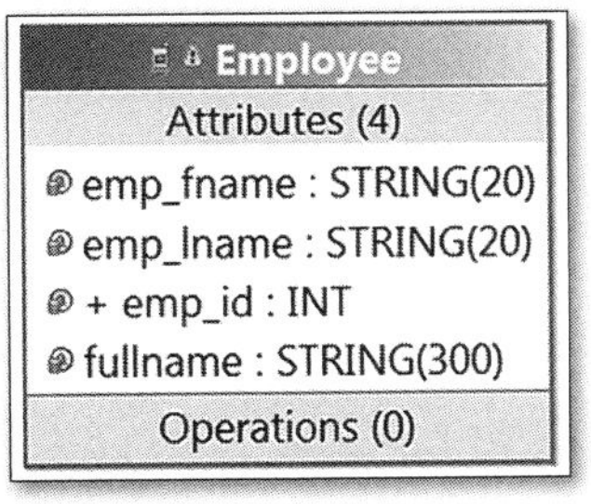

Figure 17.8 Employee MBO

The definition of the EMPLOYEE MBO adds an attribute called `fullname`:

```
SELECT   emp_id,   emp_fname,  emp_lname,
 (trim(emp_fname) + ' ' + emp_lname) AS fullname
 FROM sampledb.dba.employee
```

The EMPLOYEE MBO includes one object query, which we discuss next.

findByLnameSorted

The `findByLnameSorted` object query retrieves all employee records ordered by last name:

```
SELECT x.* FROM Employee x
ORDER BY x.emp_lname
```

17.2.5 Sales Orders and Sales Order Items MBOs

The definition of the Sales Orders and Sales Order Items MBOs is straightforward, but they are the only two MBOs joined by a relationship. These MBOs are based on the `sales_orders` and `sales_order_items` tables in the sample database.

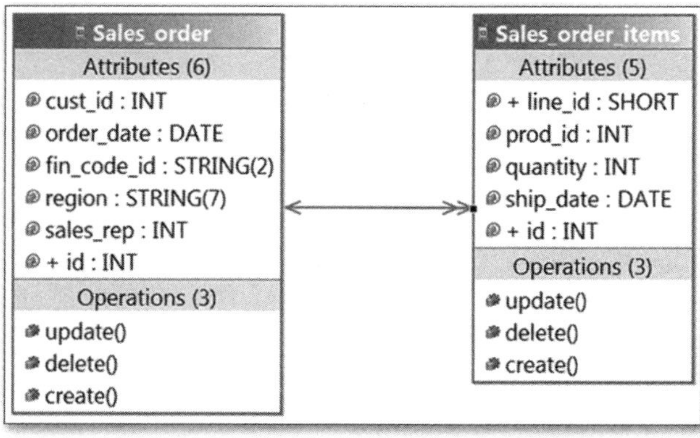

Figure 17.9 Sales Order and Sales Order Items MBOs

Although there are other opportunities to use relationships—for example, between the Customer and Sales Order MBOs—the use of relationships should be considered carefully. There is a limit on the amount of data that can be downloaded to a device at one time, and data retrieved with relationships can exceeded this limit. This is especially true if more than two MBOs are related. For example, if the Customer, Sales Order, and Sales Order Items MBOs were related, every sales order and sales order item for every customer would be retrieved at one time.

Even if the maximum data limit is not exceeded, it still may not be a good idea to use relationships excessively because retrieving large amounts of data results in poor performance. The advantages of having a relationship must be weighed against the effect on performance. The lack of a relationship, in effect, results in a lazy load situation. If the app only occasionally loads child rows, then it does not make sense to load them every time the parent rows are loaded. By contrast, if the child rows are frequently accessed when the parent rows are accessed, then it may make sense to incur to a single, initial long load time.

Another consideration is the advantages of cascading updates permitted by a composite relationship. If cascading updates or deletes are required, the app can become unnecessarily complex without the relationships. How relationships are used depends on specific usage patterns for a particular app.

The SALES ORDER MBO has two additional object queries, which we discuss next.

findByCustomer

The `findByCustomer` object query retrieves the sales orders for a specific customer ordered by `order_date`:

```
SELECT x.* FROM Sales_order x
WHERE x.cust_id = :cust_id
ORDER BY order_date
```

findByEmployee

The `findByEmployee` object query retrieves the sales orders that the CSR created:

```
SELECT x.* FROM Sales_order x
WHERE x.sales_rep = :emp_id
ORDER BY order_date
```

17.2.6 Purchase Orders MBO

The PURCHASE ORDERS MBO, shown in Figure 17.10, allows the CSR to retrieve pending purchase orders so the CSR can determine availability dates for products.

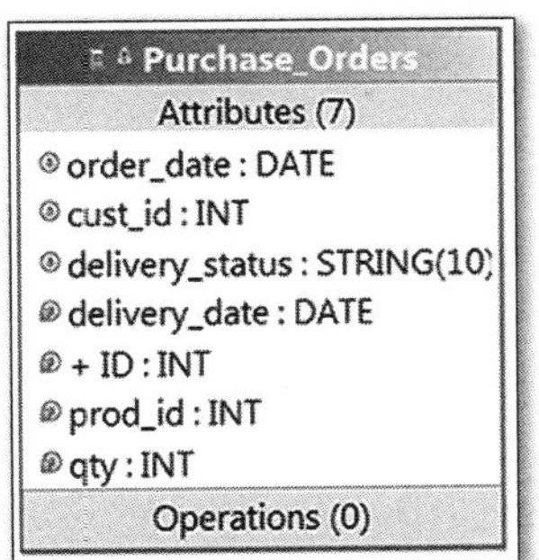

Figure 17.10 Purchase Orders MBO

The definition, shown in Listing 17.4, filters the data on delivery status and joins the purchase orders and purchase order items tables. The tables used to create this MBO were created in the SUP sample database.

```
SELECT  x.delivery_date, x.ID, y.prod_id,y.qty
FROM sampledb.dba.Purchase_orders  x INNER JOIN sampledb.dba.PO_Items y
ON x.ID = y.po_id
WHERE x.delivery_status <> 'Delivered'
```

Listing 17.4 Purchase Orders MBO Definition

The PURCHASE ORDERS MBO includes two additional object queries, which we discuss next.

findAllSortedByProductID

The `findAllSortedByProductID` retrieves all records sorted by product ID:

```
SELECT x.* FROM Purchase_orders x
ORDER BY x.prod_id
```

findByProductSortedByData

The `findByProductSortedByDate` object query is used to find purchase orders by product ID:

```
SELECT x.* FROM Purchase_Orders x
WHERE x.prod_id = :prod_id
ORDER BY x.delivery_date
```

17.3 Developing the Hybrid App

Much, but not all, of the required functionality can be provided using a standard hybrid app without customization. Typically, you should develop the app as a standard app first before beginning to customize it. This allows the Unwired Workspace to generate much of the app and to manage many of the details, such as creating screens and mapping keys to controls. In this section, we describe the standard app and then its customizations.

17.3.1 Designing the Standard App

The flow design of the standard app is shown in Figure 17.11. Even with the simple requirements of the CRM Lite app, the app is quite large. Most of the app can be created by dragging the MBOs onto the flow design workspace and then editing the names of labels, menu items, etc. This section describes only some of the more interesting features.

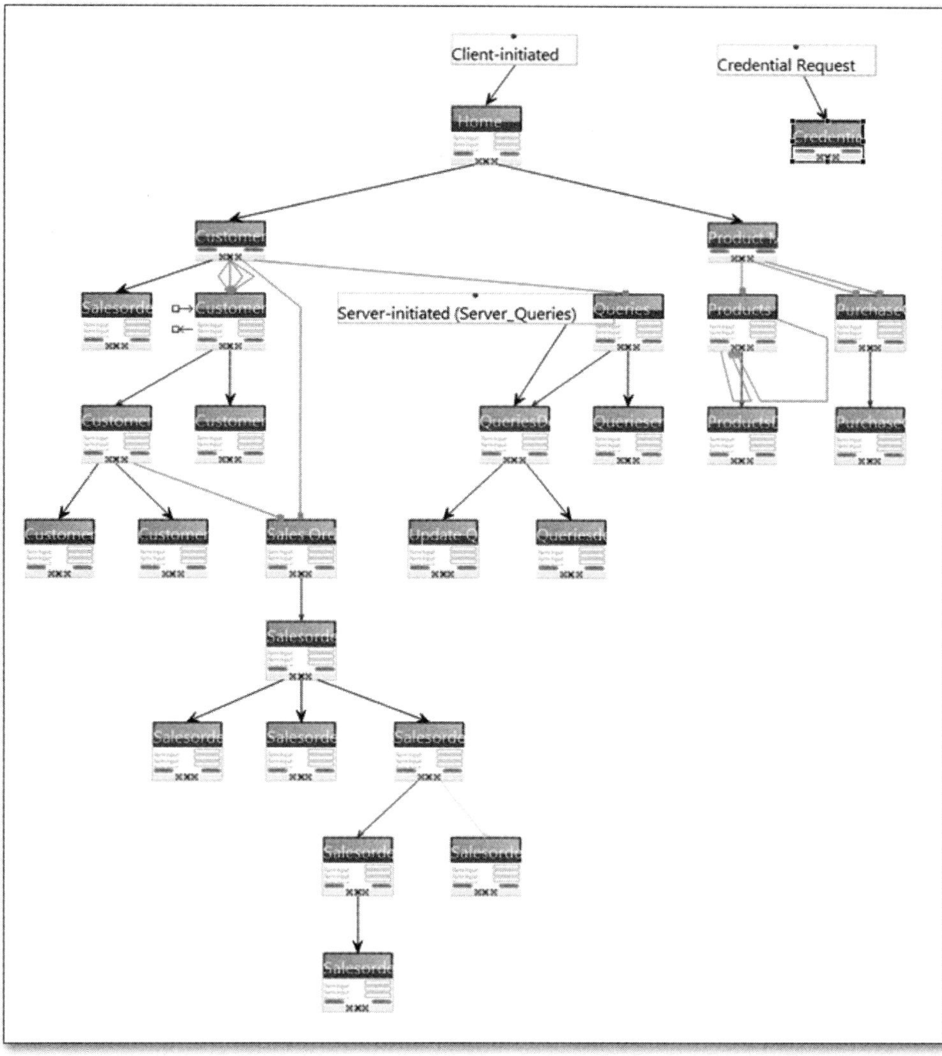

Figure 17.11 Flow Design of the Standard CRM Lite Hybrid App

Header Image

Figure 17.12 shows the app's HOME screen. The screen includes an HTMLVIEW control that is used to insert the image at the top of the screen.

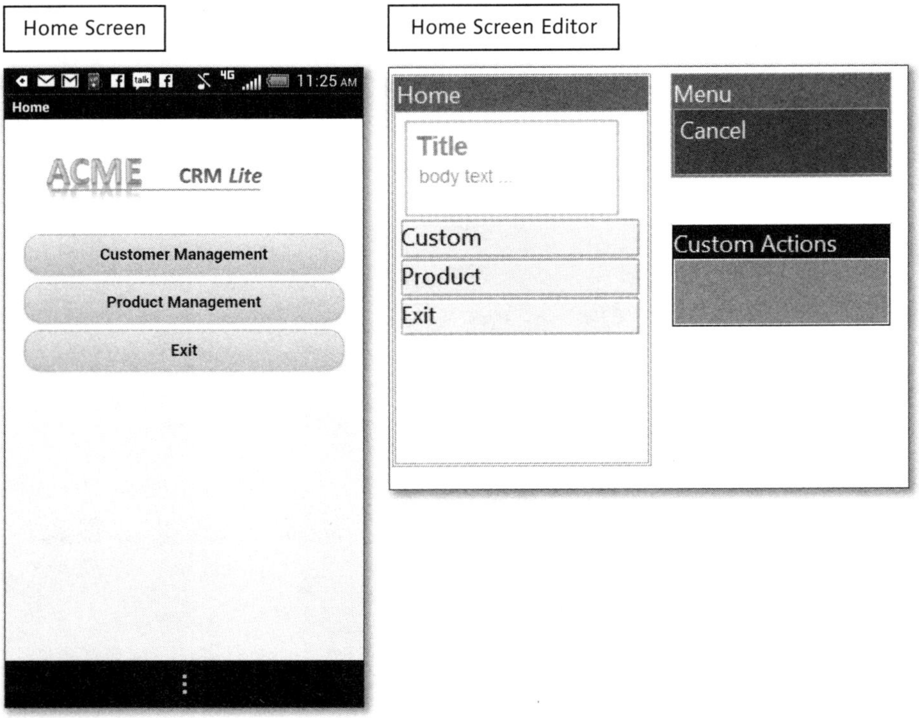

Figure 17.12 Standard App Home Screen

The image file is copied to a folder called IMAGES created in the app's HTML folder. The HTML for the header is as follows:

```
<img src="images\phoneHeater.png" style="width:80%;">
```

The style ensures that the image scales to different device screens. The same HTMLVIEW control is repeated on every screen.

Credentials

The standard app includes a CREDENTIAL REQUEST starting point. It causes the app to request login credentials when valid credentials are not cached on the device. This includes the first time the app runs and whenever the credentials are changed in the SCC. The CREDENTIALS screen configuration is shown in Figure 17.13. Since the configuration of the CREDENTIAL REQUEST starting point was described in Chapter 11, it is not described in detail here.

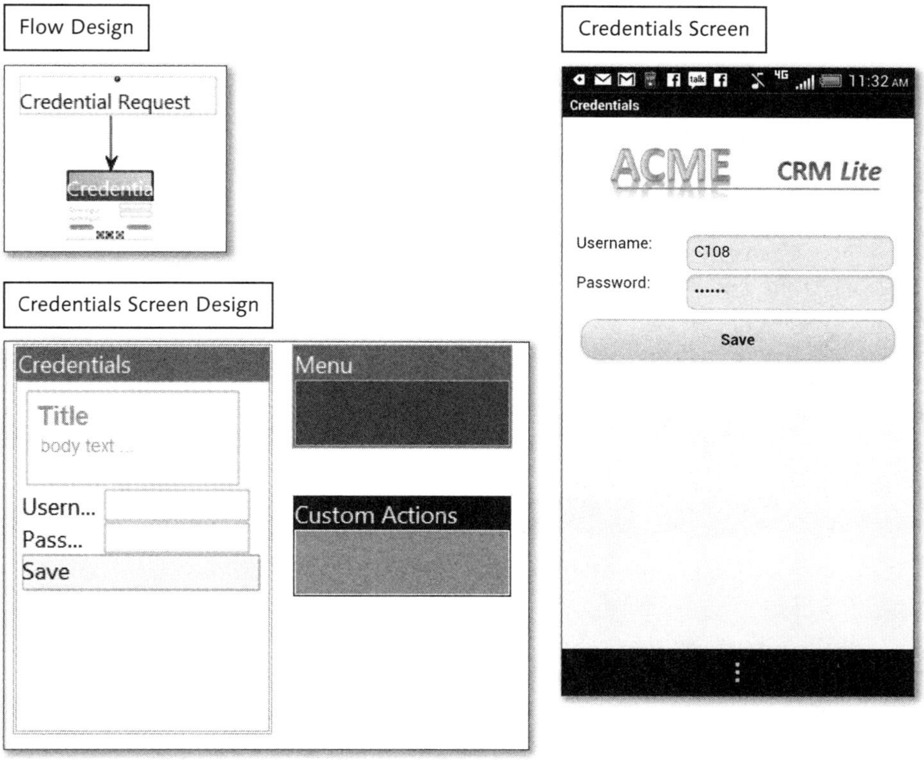

Figure 17.13 Credential Screen

Server Notifications

The server-initiated starting point fulfills the requirement that the user receive notifications of customer queries. The server-initiated starting point was described in Chapter 9. Figure 17.14 shows server notifications in the container app's MESSAGES screen and QUERY DETAIL screen.

When a message is received on the device, the user can tap it to retrieve the query record. If the user can respond to it quickly, the user can tap the UPDATE button to open the UPDATE QUERY screen, as shown in Figure 17.15. The QUERY UPDATE and QUERY CREATE screens include CHOICE controls mapped to the `Queries_status_attribKey` and `Queries_priority_attribKey` attributes. The options are added as a static list.

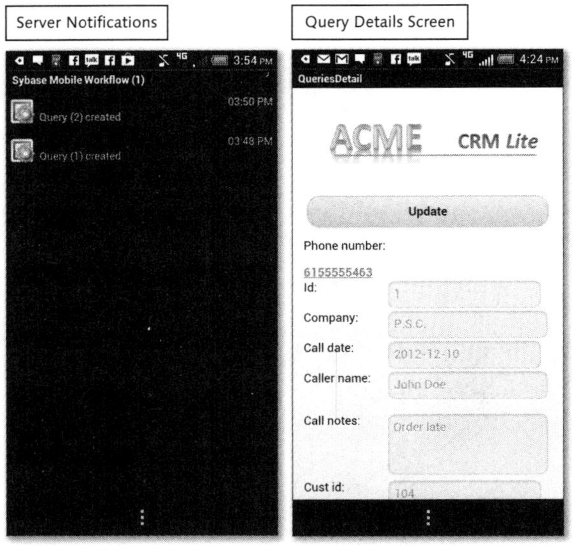

Figure 17.14 Server Notifications

Figure 17.15 Choice Controls on the Query Update Screen

Personalization Parameters

One of the app's requirements is to allow the user to filter the data they retrieve based on the sales region and employee ID. This is accomplished using the two EDITBOX controls on the CUSTOMER MANAGEMENT screen shown in Figure 17.16.

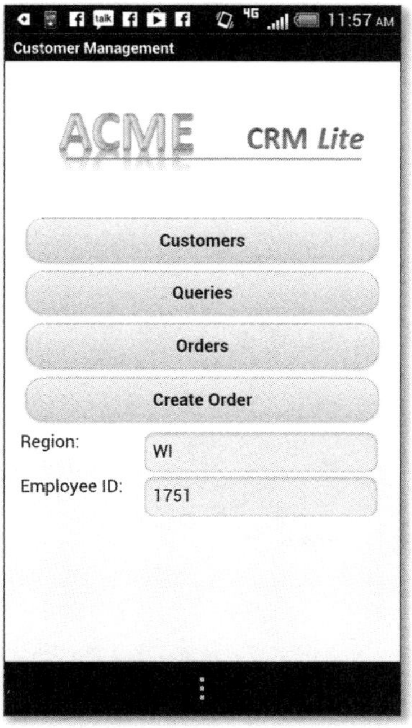

Figure 17.16 Accomplishing Personalization Using EditBox Controls

The CUSTOMERS button retrieves customer records using the `findAllSortedByCompanyName` object query. The `statePK` personalization key is mapped to the region EDITBOX control, filtering the data retrieved from the EIS. The resulting list is shown in Figure 17.17. The CUSTOMER DETAIL screen includes a MENUITEM control that invokes the SALES ORDER MBO's `findByCustomer` object query to retrieve a list of the customer's sales orders.

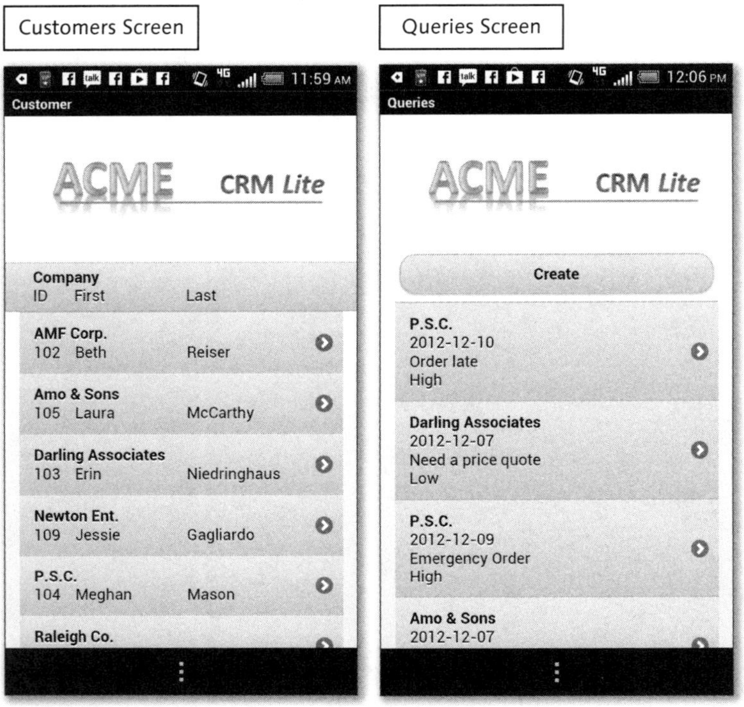

Figure 17.17 The Customers and Queries Lists

The QUERIES button on the CUSTOMER MANAGEMENT screen uses the `findOpenQueries` object query of the QUERIES MBO to retrieve a list of open customer queries (Figure 17.17). As with the CUSTOMERS button, the region EDITBOX is mapped to the `statePK` personalization key to restrict the queries to those linked to the CSR's customers.

Products List

From the PRODUCT MANAGEMENT screen, the user can open the PRODUCTS screen, as shown in Figure 17.18. The PRODUCTS button on the PRODUCT MANAGEMENT screen invokes the `findCategories` object query of the CATEGORIES MBO. The results are used to fill the CATEGORIES CHOICE control on the PRODUCTS screen. Once the user selects a category, the user can tap the PRODUCTS button, which uses the `findByCategories` object query of the PRODUCTS MBO to fill the LISTVIEW control.

The LISTVIEW control shows the PRODUCT NAME and PRICE fields on the first line and the INSTOCK, ONORDER, and reorder fields on the second line.

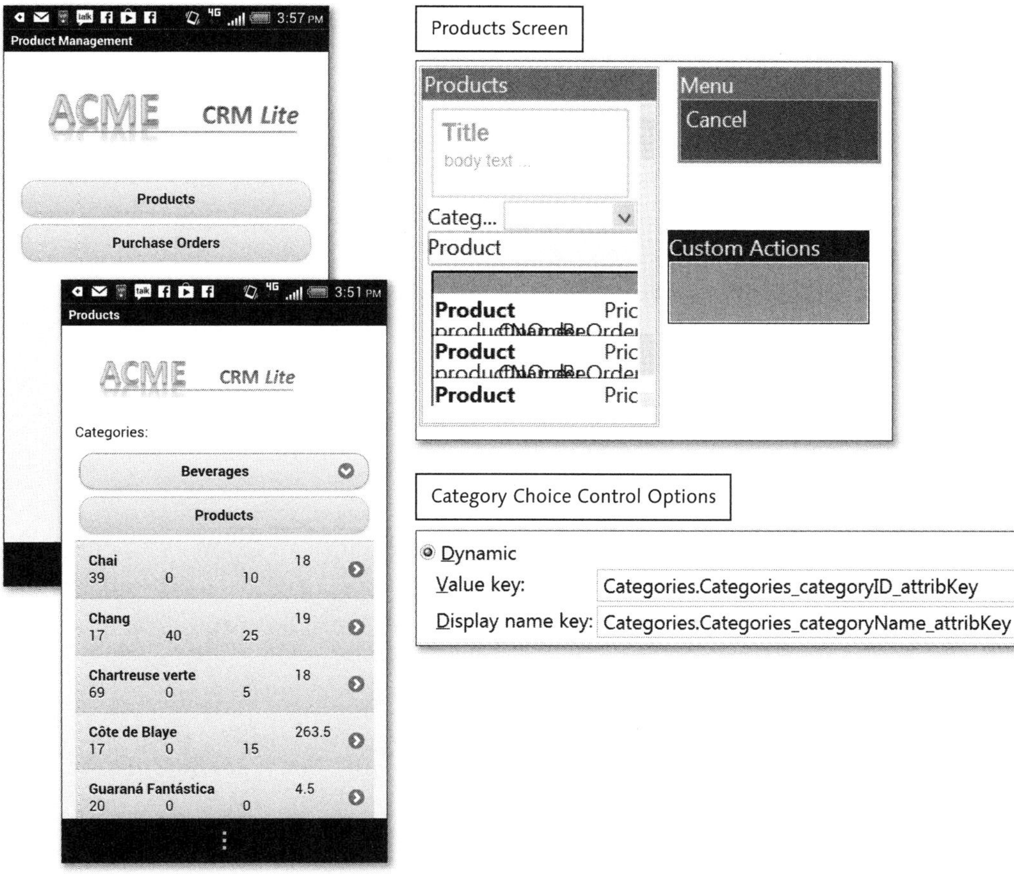

Figure 17.18 Products List Screen

Limitations of the Standard App

There are some requirements that are difficult or impossible to achieve in the standard hybrid app:

- The requirement that the user should be allowed to personalize the app by saving their sales region and employee ID cannot be accomplished with a standard hybrid app. The EDITBOX controls on the CUSTOMER MANAGEMENT screen are not a satisfying solution because they require the user to enter the values every time the app is used.

- Standard apps are restricted to the jQuery look and feel, and the developer has limited control over customizations, such as the header image in the standard app.
- One limitation of the LISTVIEW control is the behavior of the header when the LISTVIEW is not the first control on the screen. If you place other controls above the LISTVIEW control, the SHOW HEADER option is locked. Without a header, the list appears as the products list, as shown in Figure 17.18. Since there is also no way to show labels to identify fields within the list, the list's formatting leaves a lot to be desired. If you select SHOW HEADER before adding other controls, the header appears as it does in the customer list in Figure 17.17. The list in the standard app would be considered unacceptable in a professional app.
- The map requirements are not possible without customization.

17.3.2 Customizing the App

All the customizations described here require custom JavaScript, CSS, or both. The MBOs and the structure of the customized app are the same as the standard app. The only change is the addition of screens used to display maps.

Interface Enhancements

Because the interface is created using HTML5 and CSS, it is possible to create an entirely different look and feel from the jQuery Mobile styling of the standard app. This may be desirable, for example, if a firm wants to mimic its organization's web standards. It is also possible to build on the existing styling to customize the interface in small ways, like those demonstrated here. Figure 17.19 shows the HOME screen of the custom app.

Buttons

Two types of HTML buttons are used in the custom app. Figure 17.20 shows examples of each.

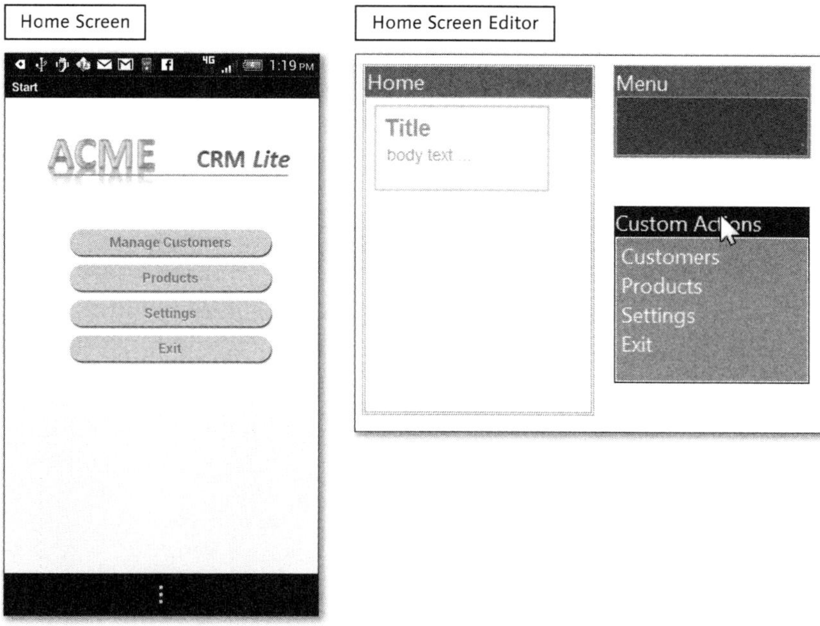

Figure 17.19 Home Screen of the Customized Hybrid App

Figure 17.20 Custom Buttons in the Custom Hybrid App

The first type is a vertical button group, and the second type is a horizontal button group similar to the example described in Chapter 10. As with the example in Chapter 10, the HTML that implements both types of buttons is added to an HTMLVIEW control on each screen. The vertical buttons use code similar to that shown in Listing 17.5.

```
<ul class="menu-css">
    <li><a href="javascript:menuItemCallbackHomeCustomers();">
<center>Manage Customers</center></a></li>
    <li><a href="javascript:menuItemCallbackHomeProducts();">
<center>Manage Products</center></a></li>
    <li><a href="javascript:menuItemCallbackHomeSettings();">
<center>Settings</center></a></li>
    <li><a href="javascript:menuItemCallbackHomeExit();">
<center>Exit</center></a></li>
</ul>
```

Listing 17.5 HTML for the Vertical Button Group

While the `<center>` elements are not the preferred method for centering elements, the normal CSS methods do not work. Unlike the Chapter 10 example, the JavaScript executed in the `<a>` elements is the functions linked to custom actions rather than menu items. This eliminates the need to hide the menu items on iOS devices.

The CSS for the buttons is added to a file called *CustomStyles.css* that is created in the app's css folder. The styling adds a drop shadow and animation effect to provide feedback when the buttons are tapped.

```
.menu-css {
  width: 250px;
  display: block;
  margin-left:auto;
  margin-right:auto;
}

.menu-css li {
  display: list-item;
  list-style: none;
  height: 40px;
}
```

```
.menu-css li a {
  /* Border Radius */
  -webkit-border-radius: 15px;
  -moz-border-radius: 15px;
  border-radius: 15px;
  width: 200px;

  /* Border Shadow */
  -webkit-box-shadow: 1px 2px 2px rgba(0,0,0,0.6);
  -moz-box-shadow: 1px 2px 2px rgba(0,0,0,0.6);
  box-shadow: 1px 2px 2px rgba(0,0,0,0.6);

  /* Animation (Webkit, Gecko & Mozilla) */
  -webkit-transition-duration: 0.20s;
  -webkit-transition-timing-function: ease-out;
  -moz-transition-duration: 0.20s;
  -moz-transition-timing-function: ease-out;
  color: #000;
  background: #fadb4e;
  display: inline-block;
  padding: 5px 15px;
  outline: none;
  text-decoration: none;
}

.menu-css li a:active {
  background: rgba(0,0,0,0.1);
  -webkit-box-shadow: 1px 1px 1px rgba(0,0,0,0.4);
  -moz-box-shadow: 1px 1px 1px rgba(0,0,0,0.4);
  box-shadow: 1px 1px 1px rgba(0,0,0,0.4);
}
```

Listing 17.6 Styles for Vertical Button Groups

An example of the horizontal button groups is shown in Figure 17.21. The difference between this example and the one in Chapter 10 is that, for this app, accommodations are made for handling screens of different sizes.

Figure 17.21 Custom Buttons on the Customer Detail Screen

As with the vertical button group, the HTML for the horizontal button groups, shown in Listing 17.7, is added to HTMLVIEW controls.

```
<div class="inline-menu-css" data-role="controlgroup"
data-type="horizontal">
  <a href="javascript:menuItemCallbackCustomer_DetailUpdate();"
 data-role="button" ><center>Update</center></a>
  <a href="javascript:menuItemCallbackCustomer_DetailDelete();"
data-role="button"><center>Delete</center></a>
  <a href="javascript:menuItemCallbackCustomer_DetailOrders();"
   data-role="button"><center>Orders</center></a>
<a href="javascript:navigateForward('Home');"
data-role="button"><center>Home</center></a>
</div>
```

Listing 17.7 Horizontal Button Group

The styling for the button depends on the size of the screen, so the code in Listing 17.8 is added to the customAfterWorkflowLoad function of the *Custom.js* file.

```
//Retrieve the screen size
var wt = parseInt($(window).width());
//If the screen width is above a certain value then
//use the large style.  If it is less use the small style
//and add the data-mini=true to the inline button groups
if(wt > TABLETWIDTH){
  $(".inline-menu-css").addClass("inline-menu-css-large");
} else {
  $(".inline-menu-css a").attr("data-mini","true");
  $(".inline-menu-css a").addClass("inline-menu-css-small");
}
```

Listing 17.8 Adjusting the Button Styles for the Screen Size

The first line retrieves the width of the window using jQuery. The width is compared with a value determined through experimentation with the population of devices used. In this case, the value was 400. If the screen width is less than this value, then the code adds the `data-mini="true"` attribute and the `inline-menu-css-small` class to all of the horizontal button groups. The class defines which style is applied. The styles shown in Listing 17.9 are added to the *CustomStyles.css* file.

```
.inline-menu-css-small a {
      display: block;
      margin-left:auto;
   margin-right:auto;
   width: 70px;
   font-size: xx-small;
}

.inline-menu-css-large a {
   display: block;
   margin-left:auto;
   margin-right:auto;
   width: 120px;
   font-size: small;
}
```

Listing 17.9 Button Styles

This could be much more complex; however, by choosing only two styles—one for larger screens and one for smaller screens—the implementation is simple and effective.

Header Image

The header image can be formatted more accurately in the custom app because it can be scaled to the specific screen size of the device. First, a `<div>` added to the HTMLVIEW controls is modified to eliminate the style.

```
<div class="header"></div>
```

Next, the code shown in Listing 17.10 is added to the `customAfterWorkflowLoad` function in the *Custom.js* file. This code inserts the header image and scales it to the device's screen width.

```
//Retrieve the width of the window in pixels
var wt =  parseInt($(window).width());
//Code containing the if statement shown above
var headerImage = "phoneHeader.png";
var footerImage = "footer.png";
//Caclulate a width for the header image
var width = 338 * (wt/360);
//Insert the header image
var headerHTML = "<img src='images\\" + headerImage
+ "' style='width:" + width + "px;'>";
   $(".header").html(headerHTML);
//Insert the footer image into the home screen footer
var footerHTML = "<img src='images\\" + footerImage
+ "'>";
$("#bottomOfStartForm").html(footerHTML);
```

Listing 17.10 Adding a Header Image

The code calculates the width of the header image by multiplying the image width in pixels by a ratio of the current device's screen width to the screen width of the device for which the image was originally sized. Next, the header HTML is inserted into the header `<div>`.

Personalized Settings

In the standard app, the CSR filters the customer records and sales orders using EDITBOX controls on the CUSTOMER MANAGEMENT screen. The customized app allows the CSR to configure a personalization setting that is cached on the device so the user does not have to enter it every time.

Saving the Settings

The first step to allowing app personalization is to add an activate starting point to the app. The Settings screen, shown in Figure 17.22, is opened by the activate starting point the first time the app runs on a device.

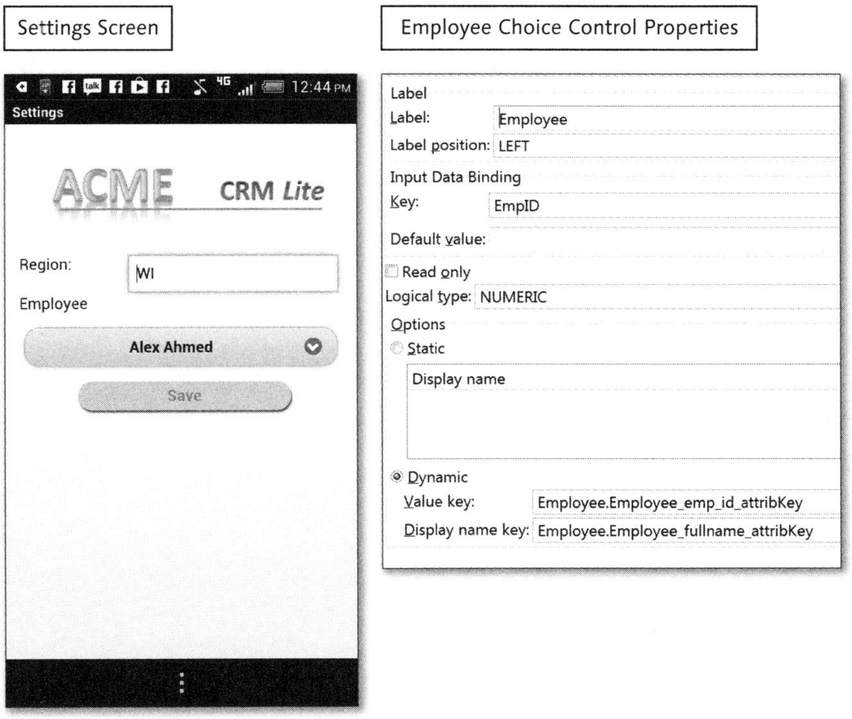

Figure 17.22 Activation of the Settings Screen the First Time the App Runs

The Settings screen includes a choice control that allows the user to save their employee ID. The choice control is filled using a custom action that is invoked when the Settings screen loads. As with the example in Chapter 11, the custom action is invoked by code in the `customBeforeShowScreen` function in the *Custom.js* file.

```
if(screenToShow == "Settings"){
   menuItemCallbackSettingsgetEmployees();
}
```

The properties of the choice control are shown in Figure 17.22. The display key is the `fullname` attribute created in the EMPLOYEE MBO definition.

When the user taps the SAVE button, the settings are saved by the `saveSettings` function that is called by the following code in the `customBeforeMenuItemClick` function in the *Custom.js* file:

```
if(screen == "Settings" && menuItem == "Save"){
   saveSettings();
}
```

The `saveSettings` function, along with many other custom functions, is included in the *CustomJava.js* file that was created in the app's JS folder. The code is shown in Listing 17.11.

```
//Save the user settings
function saveSettings(){
  //Open the settings cache
  settingsStore = initSUPStorage("settings");
  //If the cache was opened successfully
  if (settingsStore){
   //Retrieve a reference to the State EditBox
    var stateBox = $("Input#Region");
    if (stateBox){
      //Save the value
      settingsStore.setItem("region",stateBox.val());
    }
    //Retrieve a reference to the employee id choice
    //control
    var empSelect = $("select#EmpID");
   if(empSelect){
        //Save the value
     settingsStore.setItem("employee",empSelect.val());
   }
    }
}


function initSUPStorage(cache) {
     //implement the getLength prototype method
     SUPStorage.prototype.getLength = function() {
       return store.length();
     }
```

```
    //Instantiate the cache
    return new SUPStorage(cache);
}
```

Listing 17.11 Saving the User's Settings

The SETTINGS screen is shown the first time the user runs the app, immediately after the CREDENTIALS screen. The user can also access the SETTINGS screen using a button on the HOME screen.

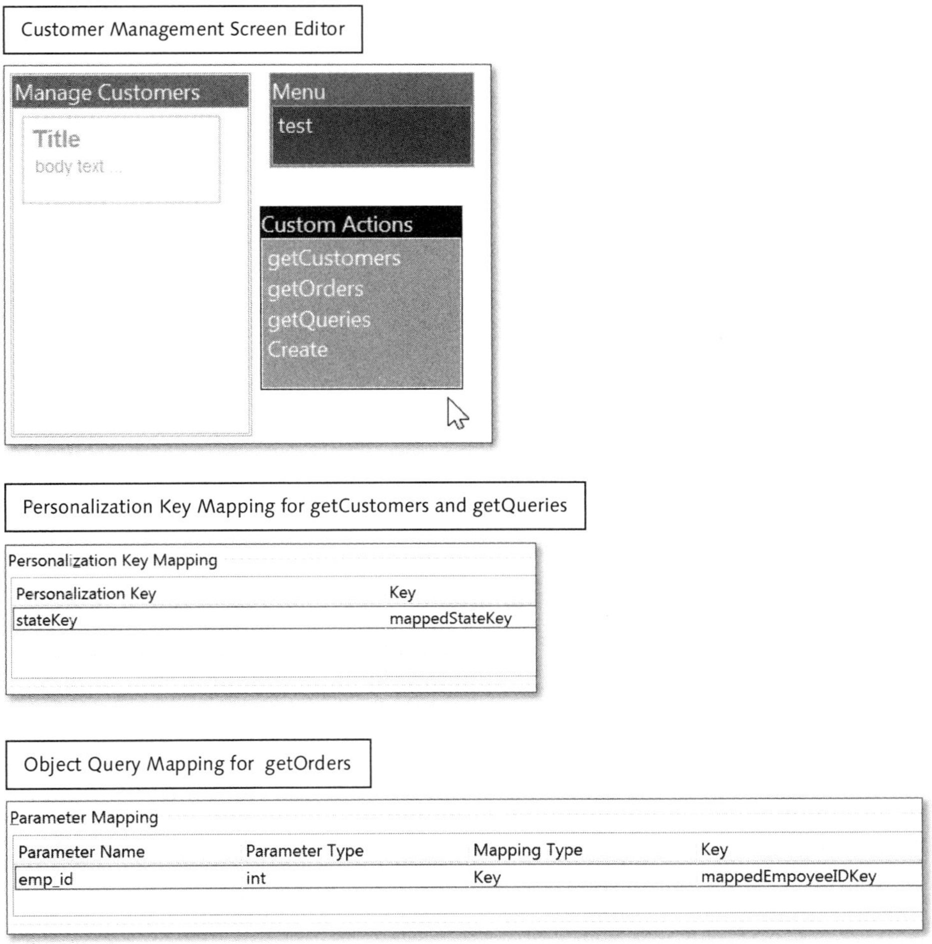

Figure 17.23 Mapping on Customer Management Screen

Applying the Settings

After the settings are saved on the device, you must add some code to apply them. Delete the two EditBox controls on the Customer Management screen. Although custom actions, rather than Button controls, are used in the custom app, the personalization key and parameter mappings remain the same as before:

- The `statePK` personalization key is mapped to the `mappedStateKey` key for the `getCustomer` and `getQueries` custom actions.
- The `mappedEmployeeIDkey` is mapped to the `emp_id` object query parameter for the `getOrders` custom action.

The difference between the custom app and the standard app is that the values of the keys are set by custom JavaScript rather than being handled automatically. The code shown in Listing 17.12 is added to the `customBeforeSubmit` function in the *Custom.js* file.

```
//If the current screen is Manage Customers
if(screenKey == "Manage_Customers"){
  //Retrieve the settings cache
  settingsStore = initSUPStorage("settings");
   if (settingsStore) {
     //If the user tapped customers button
     if (actionName == "getCustomers" || actionName ==
"getQueries"){
     //Retrieve the region setting
  var region = settingsStore.getItem("region");
  //Create a MessageValue object
     var msgVal = new MessageValue();
     //Set the key, value and type
     msgVal.setKey("mappedStateKey");
     msgVal.setValue(region);
     msgVal.setType("TEXT");
     //Retrieve the WorkflowMessage to be sent
     //to the server
     var mvc = workflowMessageToSend.getValues();
     if( mvc ) {
         //Add the MessageValue
      mvc.add( msgVal.getKey(), msgVal );
     }
   }
      ...
   }
```

```
    }
}
```

Listing 17.12 Setting the Value of the statePK Personalization Key

When the workflow message arrives at the server, the server looks for a value for the `mappedStateKey`, since it is mapped to the `StatePK` personalization key. To provide the value, the code creates a `MessageValue` object, sets its key to `mappedStateKey`, and sets its value to the region setting. The code retrieves the region setting from the settings cache. The `MessageValue` object is added to the workflow message that will be sent to the server.

Similar code is used to set the `mappedEmployeeIDKey` key when the user taps the ORDERS button on the CUSTOMER MANAGEMENT screen. Even though the `orders` are filtered by an object query parameter instead of a personalization key, the code, as shown in Listing 17.13, is the same.

```
//Set state personalization key before querying customers
if(screenKey == "Manage_Customers"){
  ...
  if (actionName == "Orders"){
    var emp = settingsStore.getItem("employee");
    var msgVal = new MessageValue();
    msgVal.setKey("mappedEmployeeIDKey");
    msgVal.setValue(emp);
    msgVal.setType("NUMBER");
    var mvc = workflowMessageToSend.getValues();
    if( mvc ) {
   mvc.add( msgVal.getKey(), msgVal );
    }
  }
}
```

Listing 17.13 Setting the Value of the emp_id Object Query Parameter

Custom Lists

The custom app includes a number of custom lists. The list code is added to a file called *lists.js* that is created in the app's JS folder. The lists are inserted in `<div>` elements, which are added to the same HTMLVIEW controls that contain the button groups. The code in Listing 17.14 shows the default value for the HTMLVIEW control on the QUERIES screen. The list is inserted into the `QueryListDiv` `<div>`. The `<div>` immediately above the list `<div>` is a spacer between the button group and the list.

```
<center>
<div class="inline-menu-css" data-role="controlgroup" data-
type="horizontal">
  <a href="javascript:menuItemCallbackQueriesCreate();"
data-role="button" ><center>Create</center></a>
  <a href="javascript:navigateForward('Start');"
data-role="button" ><center>Home</center></a>
</div>
</center>
<div style="height:15px;"> </div>
<div id="QueryListDiv" class="listView"></div>
```

Listing 17.14 The HTMLView Control on the Queries Screen

The JavaScript function that creates the lists, shown in Listing 17.15, is called from the `customAfterNavigateForward` function in the *Custom.js* file.

```
// Create custom lists
if (destScreenKey == "Customers") {
  customerList();
}
if (destScreenKey == "Queries"){
  queryList();
}
if(destScreenKey == "Sales_Orders"){
  ordersList();
}
```

Listing 17.15 Calling the List Creation Functions

The code that generates the lists is similar to the custom list example described in Chapter 10, but there are some differences, which are highlighted in the following sections.

Queries List

The QUERIES and QUERY DETAILS screens are shown in Figure 17.24. The queries list group queries by date and includes an icon linked to the priority attribute. The figure also shows the QUERY DETAIL screen, which includes LINK controls allowing the CSR to call or email the customer from the app.

Figure 17.24 Queries List, which Includes an Informative Icon

The code that generates the queries list is shown in its entirety in Listing 17.16.

```
function queryList(){
    //Retrieve the data
    var message = getCurrentMessageValueCollection();
    var itemList = message.getData("Queries");
    var items = itemList.getValue();
    var numOfItems = items.length;
    var i = 0;
    var htmlOutput = '<ul data-role="listview"
data-autodividers="true" data-theme="e"
id="QueryList" >';
    //The list is grouped by date so pick a date to
    //as an initial comparison
    var callDate = "1901-01-01";
    //Process the array of queries
    while (i < numOfItems) {
     var currItem = items[i];
     var cdate = currItem.getData(
"Queries_call_date_attribKey").getValue();
```

```
     var name = currItem.getData(
"Queries_caller_name_attribKey").getValue();
     var phone = currItem.getData(
"Queries_phone_attribKey").getValue();
     var priority = currItem.getData(
"Queries_priority_attribKey").getValue();
     var notes = currItem.getData(
"Queries_call_notes_attribKey").getValue();
     var comp = currItem.getData(
"Queries_company_name_attribKey").getValue();
     //Extract the date portion of the datatime
     //field
     currcallDate = cdate.substr(0,cdate.indexOf("T"));
     //If the date has changed
     if (currcallDate != callDate) {
     //save the new date
     callDate = currcallDate;
        //Create a list divider
     htmlOutput += '<li data-role="list-divider">'
+ callDate + '</li>';
     }
     //Insert an icon in the list that indicates the
     //query's priority
     if(priority == "High"){
       icon = "images\\Remove.png";
     } else if (priority == "Medium"){
       icon = "images\\Warning.png";
     } else {
       icon = "images\\Help.png";
     }
     //Consrtruct the list element
     htmlOutput += '<li class="listItem"><a class=
"listClickQuery ui-icon-priority"  id ="'
+ currItem.getKey() + '">';
     htmlOutput += '<span style="float:left;width:25%;">
<img src="' + icon + '" style="height:48px;
width:48px;"></span>';
     htmlOutput += '<span ><strong>'+ comp + ', ' + name
+ '</strong> <br>' + notes;
     htmlOutput += '</a></li>';
     i++;
    }
    htmlOutput += '</ul>';
```

```
    //Insert the list into the <div> created in the
    //HTMLView control on the Queries screen
    $("#QueryListDiv").html(htmlOutput);
    //Tell JQuery to reformat the list
    $.mobile.listview.prototype.enhanceWithin(
$.mobile.activePage);
    //Attach a handler for the click event of the list items
   $(".listClickQuery").attr("onclick",
"javascript:listClickHandler($(this), $(this).attr('id'),'Queries');");
}
```

Listing 17.16 Implementing the Queries Custom List

The icon images are copied into the app's HTML/IMAGES folder.

The code snippet shown in Listing 17.17 determines which image to insert based on the query's priority field and inserts the `<img>` element within a `<span>` aligned with the left side of the list item.

```
if(priority == "High"){
   icon = "images\\Remove.png";
} else if (priority == "Medium"){
   icon = "images\\Warning.png";
} else {
   icon = "images\\Help.png";
}
...
htmlOutput += '<span style="float:left;width:25%;">
<img src="' + icon + '" style="height:48px;width:48px;"></span>';
```

Listing 17.17 Selecting an Icon

Another feature that differs from the list in Chapter 10 is how the list item click handler is implemented. The code shown below assigns an `onclick` handler to the `<a>` elements of the list.

```
$(".listClickQuery").attr("onclick",
"javascript:listClickHandler($(this), $(this).attr('id'),'Queries');");
```

The `listClickHandler` function, shown in Listing 17.18, handles `onclick` events for all of the custom lists in the app. You can see the method shown in Chapter 10 in examples from the Internet, but that method can be somewhat finicky at times. This method is more reliable and has the advantage of consolidating the click handlers in one place.

```
function listClickHandler(item,id,list){
   item.addClass("ui-btn-active");
   if (list == "Queries"){
         navigateForward("Queries_Detail",id);
   } else if (list == "Products"){
     navigateForward("Product_Detail",id);
   } else if (list == "Customers"){
      navigateForward("Customer_Detail",id);
   } else if (list == "SalesOrders"){
      navigateForward("Sales_Order_Detail",id);
   }
      If (isBlackBerry()){
        Return;
      }
}
```

Listing 17.18 List Click Event Handler

Customers List

The customer list has one unique feature. This list uses the jQuery `slideToggle` function to create a collapsible list. Although jQuery has a specific collapsible list implementation, the `slideToggle` function produces a smoother animation in hybrid apps. Figure 17.25 shows the CUSTOMERS screen with the collapsible list and the CUSTOMER DETAIL screen.

Figure 17.25 Customer List Using the jQuery slideToggle Function

When the user taps a list divider, it expands, revealing the list items in that section. Tapping it again hides the items. The code in Listing 17.19 shows only the relevant portions of the code required to implement this feature.

```
...
if (currFirstChar != firstChar) {
    // Save the new first character
    firstChar = currFirstChar;
    //If this is not the first time through the loop,
    //close the toggle <div>
    if(i>0){
      htmlOutput += '</div>';
    }
    //and create a list divider
    htmlOutput += '<li class = "group" data-role=
"list-divider">' + firstChar + '</li>';
    //The toggle div contains the elements that will
    //be toggled
    htmlOutput += '<div class="toggle" >';
  }
  //Create a list item
  htmlOutput += '<li  class="listItem"><a class=
"listClickCustomer" data-icon="arrow-l" id ="'
+ currItem.getKey() + '"><h3>' + comp + '</h3>';
  htmlOutput += id + ' ' + lName + ', ' + fName;
  htmlOutput += '</a></li>';
  i++;
}
htmlOutput += '</div></ul>';
//Insert the list into a div created in an HTMLView
//control on the Customers screen
$('#CustomerListDiv').html(htmlOutput);
//Tell JQuery to reformat the list
$.mobile.listview.prototype.enhanceWithin(
$.mobile.activePage);
//Attach a click handler to the list items
$(".listClickCustomer").attr("onclick",
"javascript:listClickHandler($(this),
 $(this).attr('id'),'Customers');");
//Attach a click handler to the list dividers that toggles the
//visibility of the toggle div
$('.group').attr("onclick",'$(this).next(".toggle")
```

```
.slideToggle("fast");');
//Hide the toggle divs
$('.toggle').hide();
```

Listing 17.19 Implementing a Collapsible List

The code encapsulates the `<li>` elements that contain customer records within a `<div>` element that has the toggle class. An `onclick` event handler is assigned to the list divider elements. The event handler selects the toggle `<div>`, which is the element immediately after the clicked `<div>`. It then invokes the `slideToggle` function, which causes the toggle `<div>` to appear or disappear using a slide animation.

```
$('.group').attr("onclick",'$(this).next(".toggle")
.slideToggle("fast");');
```

One issue with this list is that, because the `<li>` elements are contained within a `<div>`, the jQuery Mobile styles are not correctly applied to them. To remedy this, a new style, shown in Listing 17.20, is added to the *CustomStyles.css* file. This style mimics the jQuery `.ui-bar-e` style.

```
.listItem  {
   border: 1px solid   #F7C942;
   background:      #fadb4e;
   color:     #333;
   text-shadow: 0 1px  0 #fff;
   background-image: -webkit-gradient(linear, left top,
 left bottom, from( #fceda7), to( #fadb4e));
}
```

Listing 17.20 Style Applied to the <li> Elements of the Customer List

Product List

The products list, shown in Figure 17.26, contains two unique features. The first is that the categories used for the CHOICE control on the PRODUCTS screen are cached in the same way the postage rates were cached in Section 11.3 of Chapter 11. The second is that the font color is determined based on the inventory situation of the product. If the inventory stock plus the amount on order is less than the reorder point, the font color is changed.

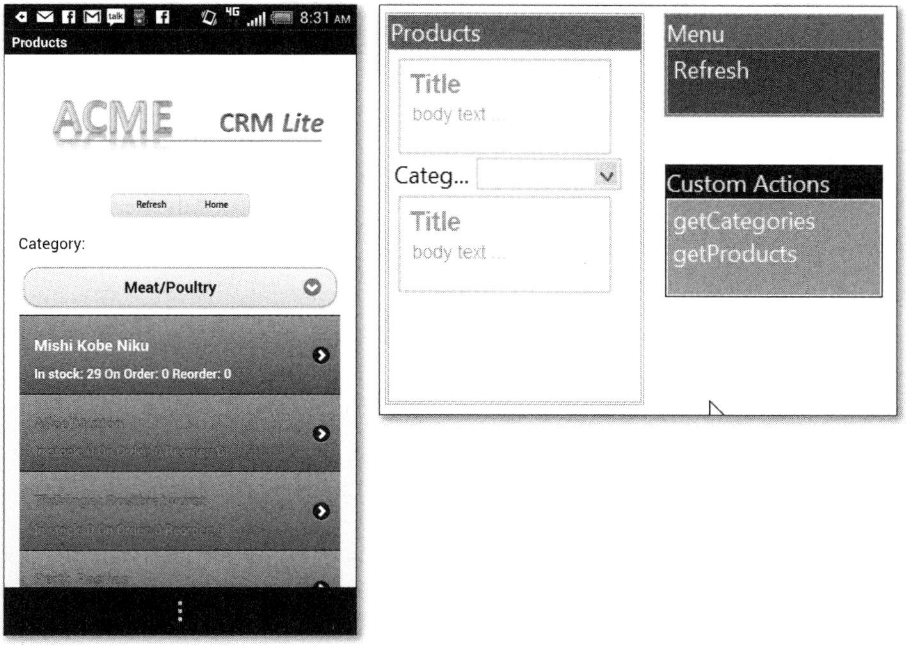

Figure 17.26 Custom Product List Showing the Inventory Situation in the Font Color

The list is inserted into a <div> that is created using the lower HTMLVIEW control, as shown in Figure 17.26.

```
<div style="height:15px;"> </div>
<div id="ProductsList" class="listView"></div>
```

The first <div> provides some space between the CHOICE control and the list. The code in Listing 17.21 shows the relevant portion of the code that generates the products list.

```
var name = currItem.getData
("Products_productName_attribKey").getValue();
var inStock = currItem.getData
("Products_unitsInStock_attribKey").getValue();
var onOrder = currItem.getData
("Products_unitsOnOrder_attribKey").getValue();
var reOrder = currItem.getData
("Products_reorderLevel_attribKey").getValue();
//Calculate the inventory position and set the font color
if((parseInt(inStock) + parseInt(onOrder) -
```

```
parseInt(reOrder)) > 0) {
  invFontClass = "inventory-available";
} else {
  invFontClass = "inventory-unavailable";
}
//Create a list item applying the font color
htmlOutput += '<li><a class="listClickProduct ' +
    invFontClass
+ ' data-icon="arrow-l" id ="' +  currItem.getKey()
+ ">';
```

Listing 17.21 Modifying the Font Color Based on the Inventory Position of the Product

The styles inventory-available and inventory-unavailable are added to the *Custom-Styles.css* file. While it may be simpler to just add the style into the HTML, this method allows you to add more complex styles if you wish.

```
.inventory-available {
   font-color:  #FFF;
}

.inventory-unavailable {
   font-color:  coral;
}
```

Cleaning Up

The last piece of code related to the lists, shown in Listing 17.22, is added to the `customAfterNavigateBackward` function of the *Custom.js* file.

```
if (screenKey == "Queries_Detail" || screenKey=="Product_Detail" ||
screenKey == "Customer_Detail" ||
screenKey == "Sales_Order_Detail"){
   //Remove the ui-btn-active class from the selected
   //<a> element to remove the visual indicator
$("ul li a").removeClass("ui-btn-active");
//refresh the list
   $.mobile.listview.prototype.enhanceWithin(
$.mobile.activePage);
}
```

Listing 17.22 Removing the Active Style from a List Item when the User Navigates back to the List

This executes when the user navigates back to one of app's list screens. It removes the ui-btn-active class from the list's `<a>` elements. The CSS for this class provides the visual indication that the item was selected.

Maps

The CSR can locate a customer or all customers who have open queries on a map (see Figure 17.27). The individual customer map is accessed from the Customer Detail screen, and the queries map is accessed from the Queries screen.

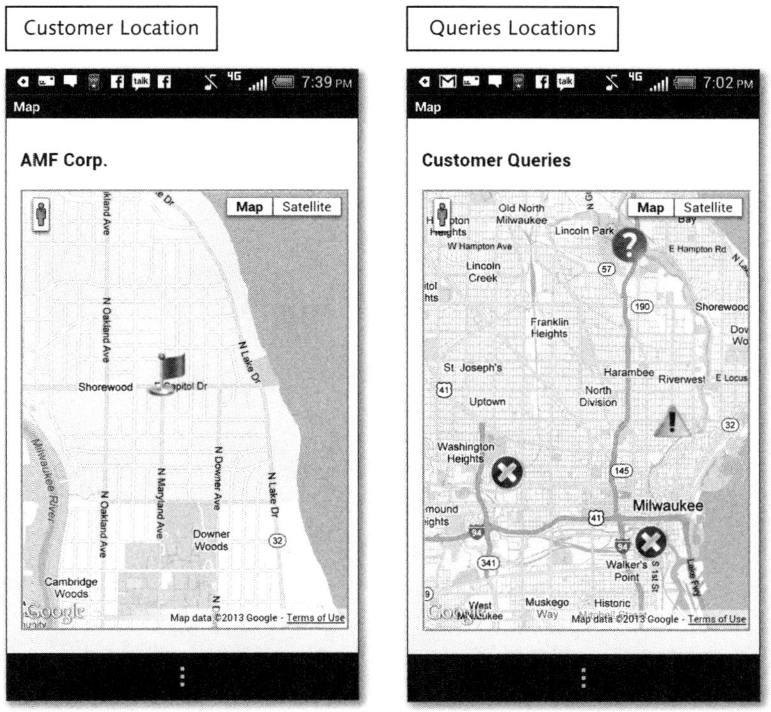

Figure 17.27 The Map Screen Showing Customer Locations

Unlike the example in Chapter 10, this example uses the Google JavaScript Map API directly, rather than the jQuery plug-in. Like the example in Chapter 10, the first step to creating a map is to add a Map screen that contains an HtmlView control that is used to insert a `<div>` with the ID `map_canvas`. When the user taps the Map button on the Queries screen, the code in Listing 17.23, which is the `customAfterNavigateForward` function in the *Custom.js* file, executes. This code extracts the address and location data and constructs and array of the locations to be mapped.

```
if(screenKey == "Queries" && destScreenKey == "Map"){
  var mvc = getCurrentMessageValueCollection();
  var itemList = mvc.getData("Queries");
  var theseItems = itemList.getValue();
  var numOfItems = theseItems.length;
  var i = 0;
  var locations = new Array();
  var address;
  while (i < numOfItems) {
    var currItem = theseItems[i];
    var name = currItem.getData("Queries_caller_name_attribKey")
.getValue();
    var phone = currItem.getData("Queries_phone_attribKey")
.getValue();
    var priority = currItem.getData("Queries_priority_attribKey")
.getValue();
   var notes = currItem.getData("Queries_call_notes_attribKey")
.getValue();
   var comp = currItem.getData("Queries_company_name_attribKey")
.getValue();
   var lat = currItem.getData("Queries_company_lat_attribKey")
.getValue();
   var lon = currItem.getData("Queries_company_lon_attribKey")
.getValue();
   address = comp + "<br>";
   address += notes + "<br>";
   //The marker on the map reflects the priority of the query
   if(priority == "High"){
     icon = "images\\smallRemove.png";
   } else if (priority == "Medium"){
     icon = "images\\smallWarning.png";
   } else {
     icon = "images\\smallHelp.png";
   }
   locations[i]= {lat:lat,lon:lon,add:address,icon:icon}
   i++
 }
 $("#mapCompanyKey").html("<h3>Customer Queries</h3>");
```

```
  drawMap(locations);
}
```

Listing 17.23 Retrieving the Data Required to Draw a Map

The drawMap function, shown in Listing 17.24, is added to the *CustomJava.js* file.

```
function drawMap(loc){

  var map = new google.maps.Map(document.getElementById('map_canvas'),
{
            mapTypeId: google.maps.MapTypeId.ROADMAP,
            center: pyrmont,
            zoom: 15
          });

  var ht = parseInt($(window).height())-100;
  //Use JQuery to set the style of the div that will
  //display the map.
  $('#map_canvas').css({
      'height': ht  + 'px',
      'width': '100%',
      'border-style': 'groove'
  });
  //Cycle through the loc array adding markers to the map
for (var i = 0; i<list.length; i++){
  var cust = new google.maps.LatLng(loc[i].lat,loc[i].lon);
  createMarker(loc[i].add,cust,map);
}
}
```

Listing 17.24 Using the drawMap Function to Draw a Google Map

The drawMap function formats the map_canvas <div> and then loops through the array of locations, adding a marker to the map for each. The createMarker function shown in Listing 17.25 adds the markers to the map.

```
function createMarker(add, cust, map){
var infowindow = new google.maps.InfoWindow();
  var marker = new google.maps.Marker({
        map: map,
        icon : loc[i].icon,
        position: cust
      });
```

```
        google.maps.event.addListener(marker, 'click',
           function() {
             infowindow.setContent(add);
             infowindow.open(map, this);
         });
}
```

Listing 17.25 Adding Markers to the Map

When the user taps the MAP button on the CUSTOMER DETAIL screen, the code shown in Listing 17.26 is executed. The code is similar to the example in Chapter 10.

```
if(screenKey == "Customer_Detail" && destScreenKey == "Map"){
  var form = document.forms[screenKey + "Form"];
  if(form){
    var address = form.Customer_address_attribKey.value;
    address += ", " + form.Customer_city_attribKey.value;
    address += ", " + form.Customer_state_attribKey.value;
    address += ", " + form.Customer_zip_attribKey.value;
    $("#mapCompanyKey").html("<h3>" +
 form.Customer_company_name_attribKey.value + "</h3>");
    var locations = new Array();
    var icon = "images\\redFlag.png";
    locations[0] = {lat:form.Customer_lat_attribKey.value,
lon:form.Customer_lon_attribKey.value,add:address,
icon:icon}
     drawMap(locations);
   } else {
      alert("No address available.");
      return false;
   }
}
```

Listing 17.26 Drawing a Map for One Customer

17.4 Summary

The CRM Lite app example illustrates many features of hybrid apps but barely scratches the surface of what is possible. The standard app can be created in a very short time and, for simple situations, provides an adequate solution. In more complex situations, customization offers an infinite number of possibilities.

The Authors

Dr. Dave Haseman is an IBM professor of IT management and director of the Center for Technology Innovation at the Lubar School of Business at the University of Wisconsin-Milwaukee. He has more than 35 years of experience as a professor, educator, and consultant in the IT field. He has been active in the SAP University Alliance since 2000 and is the founding director of the SAP University Competence Center at UWM, which provides SAP hosting services and help support to more than 130 universities and more than 90,000 students annually in North America. He has developed and taught extensive curricula for the SAP University Alliance in the areas of SAP NetWeaver and mobility. He has published more than 50 academic articles and books and works closely with many SAP customers.

Dr. Ross Hightower is a lecturer and associate director of the Center for Technology Innovation at the Lubar School of Business at the University of Wisconsin-Milwaukee. He received his PhD in Decision Sciences from Georgia State University in 1991 and has 30 years of experience in the IT field. He has been an active member of the SAP University Alliance since 2004 and has been an SAP trainer since 2006. He has published more than 20 academic articles and chapters.

Index

A

Abstract API, 150
Abstraction, 168
Adobe, 27, 151
  PhoneGap, 151, 324, 555
ADT, 424
AJAX, 148, 324
Amazon Web Services (AWS), 72, 91, 107, 108
Analytics, 515
Android, 145, 153, 154, 421
  activity, 422
  application components, 422
  application development, 435
  creating projects, 431
  development tools, 424
  emulator, 118
  example app, 425
  IDE, 151
  importing libraries, 434
  intent, 422
  layout files, 437
  layouts, 422
  MBO, 427
  Object API, 421, 428
  permissions, 422
  retrieving customer data, 457
  retrieving customer record, 464
  running app, 469
  surrogate key, 459
  synchronization, 453
  virtual device, 118
AndroidManifest.xml, 422
Android SDK, 423
  Manager, 118, 423
  tools, 118
API, 144
  JavaScript, 352
  REST Services, 562
  SUPStorage, 369
Appcelerator, 27, 151, 561
AppFramework, 554
Apple, 29
Apple App Store, 29, 134
Application connection, 130
Application Program Interface (see API)
Applications, 24
App store, 144, 153
Asynchronous operation replay, 404
Attributes, 169, 182
AVD, 118

B

Backend data sources, 30
BAPI, 31, 195, 291, 294
BlackBerry, 29, 145, 153, 154
  BlackBerry App World, 29
BOR, 31
Bring your own device (see BYOD)
Business logic, 168
BYOD, 29, 146

C

C#, 53, 145
Cache, 164, 169, 178, 203, 210
  groups, 203
Cache database (see CDB)
Cache policy, 210
Callback handlers, 493
CDB, 59, 60, 206, 387
Close project, 174
Cluster Database, 60
Code generation, 316
Column chart, 535
ConnectionController, 489
Connection profiles, 95
Connectivity, 55

Control properties, 252
Controls
adding, 250
Choice, 360, 361, 370, 373
defining values, 252
CPU/core license, 68
Create, update, delete (CUD), 202
Credential Request Starting Point, 379
Credentials, 379
CRM Lite, 565
customizing app, 584
defining MBOs, 566
developing app, 576
scenario, 565
CSS, 319, 321, 324, 339, 523
CSS3, 147, 150
Custom action, 360, 361, 366
defining values, 260
JavaScript, 362
CustomerAppAppDelegate, 486
CustomerAppViewController, 493
Customer/sales order database example
flow design, 278
review app, 285
screen design, 280
Customization, 315, 316, 347
Custom.js, 315, 323, 337, 362, 367, 532
Custom lists, 595
CustomStyles.css, 586, 604

D

Database systems, 168
Data change notifications, 388
Data Orchestration Engine, 59
DB02, 101
Delegation, 476
Deploy, 184
DetailViewController, 504
Development, 67
Development and test license, 68
DMZ, 56
app deployment, 56
device registration, 56
DMZ (Cont.)
messaging, 57
runtime data access, 57
Dojo Mobile, 149
Dojo Toolkit, 522
Drag and drop, 177, 241
Dynamic credentials, 379

E

Eclipse, 91, 145, 172, 423, 424
EditBox, 368, 379
Employee database example
flow design, 270
MBO, 269
screen design, 271
Encrypted storage, 156
Enterprise Development Server, 68
Enterprise Explorer, 52, 95, 175
Enterprise Information System (EIS), 155, 168
Enterprise Server, 68

F

findAll, 217, 270, 277
findAllSortedByCompanyName, 571
findAllSortedByLname, 277
findAllSortedByName, 573
findAllSortedByProductID, 576
findByCategory, 572
findByCompanyName, 571
findByCustomer, 575
findByEmployee, 575
findByLname, 277
findByPrimaryKey, 270, 277
FindByPrimaryKey, 218
findByProductSortedByData, 576
findCustomerQueries, 569
findOpenQueries, 569
Flow design
case study, 576
Flow Design Editor, 52, 228
Full-featured phones, 29

G

General ledger account balances SAP example, 291
flow design, 296
MBO, 292
screen design, 296
Google, 356
Google Play, 28

H

Hello world, 235
HTML, 315, 319, 321, 326
HTML5, 41, 147, 150, 515
HTML5/JS, 49
HTMLView, 271, 326, 578, 603
Hybrid app, 156
Hybrid App Designer, 230, 318, 319, 322, 324
Hybrid App Generation Wizard, 233
Hybrid apps, 36, 41, 49, 118, 152
architecture, 150
creating, 226
customer/sales order example, 276
defining server notification, 304
development, 225, 269
employee database example, 269
functionality, 347
general ledger example, 291
generating, 265
Hybrid Web Container, 156
patterns, 160
sales order notification database example, 301
screen design, 230, 247
screen flow, 228, 236
testing, 235
Hybrid Web Container, 143, 152, 153, 554
Android, 117, 133
Android emulators, 118, 127
Android SDK, 118
architecture, 153
iOS, 134, 139
iOS simulator, 134
HybridWebContainer.apk, 127

I

Installation, 76
Integrated Development Environment (IDE), 35
iOS, 145, 153, 154, 473
application components, 474
application development, 473
creating project, 481
example app, 476
implementing app, 486
importing generated code, 485
importing libraries, 483
MBO, 476, 477
model-view-controller, 474
Object API, 476, 478
running applications, 510
iPad, 29
iPhone, 29
iPod, 29
ItemListViewController, 498

J

Java, 53, 145, 422
JavaDoc, 430
JavaScript (JS), 41, 147, 149, 150, 315, 319, 320, 322, 324, 328, 335, 343, 515
JDBC, 31, 101
Join, 211
JQuery, 315, 320, 324, 325, 327, 328, 334, 335, 336, 339, 522, 560
jQuery Mobile, 149, 318
JSON, 386

K

Keys, 347
Kit, 149

L

Layout, 422
Library, 149

Link control, 356
Linking, 219
Linux, 28
ListView, 274, 329, 349
Load arguments, 202, 203

M

Mac OS, 134
MAKit, 149, 515
architecture, 515
copying files, 521
creating app, 518
CSS files, 523
example app, 516
images, 525
JavaScript files, 525
jQuery style sheets, 524
simple column chart, 527
zooming chart, 538
Manifest.xml, 319
Many-to-many, 171
Maps, 605
MBO, 39, 158, 167, 192
artifacts, 168
attributes, 347
Categories and Employee, 566
creating, 178
Customer, 570
database operations, 213
data flow and performance, 202
defining, 168
development, 170
development tools, 172
Employee, 573
object queries, 216
operations, 171
Products and Categories, 571
Purchase Orders, 575
Queries, 567
relationships, 171, 219
Sales Order Items, 574
Sales Orders, 574
Server Queries, 566, 567
MenuItem, 260, 360
MessageValue, 351, 356, 365, 368
MessageValueCollection, 351, 365, 368
Messaging Database, 60
Metadata, 168
Microsoft, 154
Microsoft Store, 29
Microsoft Windows Mobile, 145
Mobile Analytics Kit (see MAKit)
Mobile app, 147
Mobile Application Diagram Editor, 52, 176
Mobile apps, 173
architecture, 143
backend data sources, 30
development, 34
types, 32
Mobile Business Object (see MBO)
Mobile development
choosing approach, 45
Mobile development perspective, 91, 225
Mobile market, 23
Mobile packages, 233
Mobile platforms, 28
Android, 28
BlackBerry, 29
iOS, 29
Microsoft, 29
Mobile SDK, 152
Mobility patterns, 170
Model-view-controller, 558
Monitoring and Domain Log Database, 60
MooTools, 522
My Sample Database, 98

N

Native apps, 34, 42, 143, 144
accessing MBO data, 393
advantages, 46
architecture, 143
callbacks, 392
connection profile, 391
disadvantages, 46
dynamic queries, 394
local database, 392
manipulating data, 395
messaging channel, 388
registering, 389

Native apps (Cont.)
replication channel, 388
synchronization, 387
synchronization profile, 391

O

Object API, 49, 53, 387
Afaria APIs, 402
Application API, 397
Authentication API, 405
Callback and Listener APIs, 408
Change Log APIs, 408
Connection API, 402
functionalities, 389
generating classes, 478
included APIs, 397
Log Record APIs, 408
Message-Based Synchronization API, 407
Persistence API, 416
Personalization API, 405
Query APIs, 415
Security APIs, 408
Synchronization API, 406
Objective-C, 53, 145, 479
Object queries, 218, 330, 366
OData, 37, 44, 54, 386, 559
SDK, 54, 386
OEM license, 69
Offline apps, 45, 387
One-to-many, 171
One-to-one, 171
Online apps, 45
Online lookup, 160
Open platforms, 553
Open project, 174
Operating system, 143
Operation replay, 170
Oracle, 101
Output keys, 350

P

Palette, 177, 246
Parameters, 169
PC, 23
PCs vs. smart devices, 24
Personal Development Server, 67
Personalization keys, 203, 208
Personalization parameters, 581
Perspective, 172
Plug-in, 150
Primary key, 180
Production, 67
Project, 178
Properties view, 176
Proprietary, 146
Prototype, 522
Provisioning, 156

Q

Queries, 515
Queries list, 596

R

RBS, 387
Relationships, 169, 219
Relay Server, 57
REST, 32, 187
RESTful, 386
REST Services API, 562
RFC, 31

S

Sales order notification database example
flow design, 308
MBO, 301
test application, 309
SAP, 103
SAP Afaria, 37, 61
SAP Community Network (SCN), 108
SAP GUI, 192
SAP Mobile Platform, 36, 91, 385, 421
architecture, 37
data layer, 38
middleware, 38
mobile app development, 39

SAP NetWeaver Gateway, 37, 54, 386, 561
SAP servers, 103
SAP systems, 168
SCC, 50, 62, 82, 130, 351
administrative nodes, 83
new domain, 87
new user, 84
user interface, 83
Screen design, 225
Screen flow, 225
Secure messaging, 156
Security, 55
Sencha Touch, 27, 149, 557, 561
Server notifications, 162
case study, 579
Shared hosts, 66
Simple Object Access Protocol (SOAP), 32
Single server, 63
Smart phone, 23
SQL, 103, 216
SQL Server, 101
Standalone seat license, 68
Start screen, 229, 239
SUPStorage, 369
SUPStorage.js, 369
Sybase 365, 37
Sybase ASA, 101
Sybase Control Center (see SCC)
Sybase Mobile SDK, 39, 49, 91, 92, 172
backward compatibility, 92
requirements, 92
Sybase Unwired Platform, 37, 39, 49, 82, 143, 153
cloud, 107
editions, 67
environments, 67
installation, 76, 82
landscape, 63
license keys, 73
license types, 68
licensing, 66
load-balancing cluster, 64
Microsoft failover cluster, 65, 66
ports, 80
single-server, 63
Syclo, 27
Syclo Agentry, 37
Synchronization, 59, 169, 178, 386, 387
Synchronization parameter, 203, 206, 427
SynchronizationProfile, 403

T

Tablet, 23
Third-party frameworks, 37, 553
Titanium, 151, 561

U

Unwired Platform Runtime, 49, 55
Unwired Server, 96, 155, 168, 351, 385, 386, 388
Unwired Workspace, 51, 91, 93, 225, 385
connecting to SAP server, 103
connecting to web service, 105
new database connection, 101
sample database, 98
User interface, 515
case study, 584
conditional navigation, 341
CSS files, 324
customizing, 315
custom list, 329
generating project code, 316
HTML files, 320
JavaScript files, 322
jQuery, 325
project files, 319
User security, 156

V

VB.Net, 145
View, 422
ViewGroup, 422
Virtualization, 170
Visual Studio, 145

W

Web apps, 35, 45
  architecture, 147
Web server, 148
Web services, 105, 168, 187
Windows Phone, 29
WorkflowClient log, 351
WorkflowMessage, 351, 365, 595
Workspace Navigator, 52, 174, 319, 430
WSDL, 32, 105

X

Xcode, 134, 145, 151, 476, 481
XIB classes, 475
XIB files, 483
XML, 422
XSLT, 190